Dinosaurs by the Decades

Dinosaurs by the Decades

A Chronology of the Dinosaur in Science and Popular Culture

RANDY MOORE

GREENWOOD

AN IMPRINT OF ABC-CLIO, LLC
Santa Barbara, California • Denver, Colorado • Oxford, England

Library of Congress Cataloging-in-Publication Data

Moore, Randy, author.
Dinosaurs by the decades : a chronology of the dinosaur in science and popular culture / Randy Moore.
pages cm
Includes bibliographical references and index.
ISBN 978-0-313-39364-8 (hardcopy : alk. paper) — ISBN 978-0-313-39365-5 (e-book)
1. Dinosaurs in popular culture—Chronology. 2. Dinosaurs—Miscellanea—Chronology. 3. Science in popular culture—Chronology. I. Title.
QE861.95.M66 2014
567.9—dc23 2013046924

ISBN: 978-0-313-39364-8
EISBN: 978-0-313-39365-5

18 17 16 15 14 1 2 3 4 5

This book is also available on the World Wide Web as an eBook.
Visit www.abc-clio.com for details.

Greenwood
An Imprint of ABC-CLIO, LLC

ABC-CLIO, LLC
130 Cremona Drive, P.O. Box 1911
Santa Barbara, California 93116-1911

This book is printed on acid-free paper ∞

Manufactured in the United States of America

For Darrell and Donna, whose friendship through the decades has been a highlight of my life.

Dinosaurs, more than other fossils, are public property, creatures as much of the public imagination as of scientific resurrection.

—John Noble Wilford, *The Riddle of the Dinosaur*, 1985

Contents

Preface, xi

Acknowledgments, xv

Introduction, xvii

Institutional Abbreviations, xxvii

The Chronology, 1

Appendix: The Geologic Timescale, 417

Glossary, 423

Bibliography, 433

Index, 447

Preface

Dinosaurs—the most famous of all animals—have a privileged place in human imagination. They have been gone for 65 million years, yet they are more popular than most animals still alive around us. Indeed, dinosaurs are virtually everywhere in our society—from museums to marketing campaigns, from science to fantasy, and from entertainment to religion. To some people, dinosaurs are frightening monsters; to others, they are marketing tools and symbols of extinction. They are the stars of movies and petrified trophy rooms, hucksters in advertisements and on billboards, and evangelists of fundamentalist religion. Dinosaurs are the topics of countless songs, books, poems, cartoons, and documentaries; the namesakes of mascots, restaurant chains, and vitamins; the inspirations of video games, toy sets, and playing cards; and the greeters at gas stations, miniature golf courses, roadside attractions, amusement parks, and religious museums (Figure P.1). Few aspects of society have not been invaded by dinosaurs.

Figure P.1 We use dinosaurs to lure people to movies, video games, science museums, parks, gas stations, religious museums, restaurants, and countless other roadside attractions. This photo shows the 80-foot-long *Apatosaurus* that helps attract more than 1.5 million visitors (and more than $10 million) to the eclectic Wall Drug Store in Wall, South Dakota. (Courtesy Randy Moore)

Since British anatomist Richard Owen named them in 1842, dinosaurs have become the most famous and enduring icon of most natural history museums. Their first surge of popularity occurred in the late 1800s as part of a competition between two ambitious scientists—Othniel Marsh and Edward Drinker Cope. A second "bone rush" occurred at the turn of the 20th century when museums—led by Henry Osborn's American Museum of Natural History (AMNH)—realized how dinosaurs could lure visitors to their dusty exhibits. When museums began installing restorations of dinosaurs, attendance skyrocketed. When AMNH announced that it had discovered what became the most famous dinosaur of them all—*Tyrannosaurus rex*—natural history museums everywhere began searching for their own dinosaurs to cash in on dinosaurs' popularity and appeal. Andrew Carnegie's Carnegie Museum of Natural History in Pittsburgh even began exporting molds of their giant sauropod *Diplodocus* throughout the world.

When people realized that the public loved dinosaurs and would pay to see them, it did not take long for dinosaurs to break out of dusty museums and begin starring in theme parks, movies, television shows, and marketing campaigns. Oil mogul Harry Sinclair, who helped fund several dinosaur-seeking expeditions to the American West, used "Dino" the "*Brontosaurus*" to sell gasoline and other products. Although most museums updated their displays with newly discovered dinosaurs and new information about the dinosaurs that were part of their earlier exhibits, many dinosaur-based parks became outdated monuments to past ideas. Some of these parks, such as Dinosaur Park in Rapid City, South Dakota, remain popular tourist attractions, and "dinosaur towns," such as Vernal, Utah, and Glen Rose, Texas, continue to depend on dinosaur-related tourism. Other roadside dinosaurs and dinosaur-based theme parks, like the dinosaurs themselves, have become extinct.

Just as early naturalists explained life by evoking deities and the supernatural, so too did the early discoverers of prehistoric animals such as dinosaurs try to reconcile their findings with religion. For example, fossil collector Thomas Jefferson—who filled the East Room of the White House with fossils and published one of the first papers about American paleontology—refused to accept extinction, because the disappearance of a species would undermine God's perfect creation. Similarly, William Buckland—who wrote the first scientific description of a dinosaur—attributed his discoveries to "the Power, Wisdom, and Wisdom of God" while reconciling fossils with Noah's flood, and Georges Cuvier attributed extinctions to God-directed extinctions. Famed dinosaur-hunter Charles H. Sternberg believed that studying fossils was the way to understand God. Richard Owen used dinosaurs to undermine Darwinian evolution, and AMNH's Henry Osborn—to whom modern exhibits of dinosaurs can be traced—used them to support his predetermined, directional "orthogenetic" view of evolution. Today, dinosaurs are icons of young-Earth creationism, whose advocates claim that humans and dinosaurs were created on the sixth day of creation, and therefore lived together. Not surprisingly, dinosaurs have prominent roles in the many young-Earth creationism-based "creation museums" that thrive across the United States. Just as dinosaurs have been used to market gasoline, vitamins, and countless other products, so too have they been used to promote fundamentalists' religious beliefs and social agendas, including their rejection of evolution, geology, paleontology, physics, and other sciences.

Dinosaurs by the Decades traces how we have discovered dinosaurs, imagined dinosaurs, and used dinosaurs throughout our society. Our many depictions of dinosaurs have created and reinforced a variety of ideas—some accurate, some not—about what dinosaurs were and how they lived. Although dinosaurs are one of the best examples of evolutionary success and adaptation, most early images of dinosaurs portrayed them as maladapted, slow, and stupid; this is why the term *dinosaur* is often used to describe ancient, antiquated individuals, ideas, and technology that have not kept up with the times. This is ironic; for example, we consider dinosaurs to be ancient, but they did not originate until 95 percent of Earth's present history had passed. Similarly, we consider dinosaurs to be failures (because they became extinct), but they ruled Earth for more than 150 million years, which is far longer than any other group of land animals. For comparison, mammals—the group to which we belong—have dominated Earth for less than half that time. Hominids (i.e., humans and our fossil ancestors) have been around for less than 5 percent of that time, and modern humans have been around for less than 0.2 percent of that time. Recorded history covers less than 0.005 percent as much time as the Age of Dinosaurs.

The preeminence in fossil hunting that passed from Great Britain to America in the 19th century has, in turn, now passed to China and elsewhere. Along the way, numerous discoveries have changed many of our earlier ideas about dinosaurs, and popular movies such as *Jurassic Park* have updated the public's images of some dinosaurs. Although our views of dinosaurs have changed drastically, they remain prominent parts of our museums, advertisements, schools, movies, and museums because they fascinate us.

I hope that *Dinosaurs by the Decades* will help you appreciate that fascination.

Randy Moore
March 2014

Acknowledgments

I am grateful to many people for their help while I worked on this book. I thank the many librarians, historians, park rangers, scientists, and others who have advised me, guided me to information, sat for interviews, opened their homes and files for me, and gave me permission to use their images in this book. I am grateful to Kent Kirkby, Paul Sereno, Jack Horner, and David Berman for advising and meeting with me, to officials at Dinosaur Valley State Park and Answers in Genesis for their helpful information, and to researchers at the American Museum of Natural History, Dinosaur National Monument, Natural History Museum, Smithsonian, and other museums for their help. I especially thank Carl Baugh and the Creation Evidence Museum of Texas for their advice and for including me in their excavations along the Paluxy River near Glen Rose, Texas. I thank Terry Dale and Kelly Hoodenpyle for their hospitality and help; R.C. McFall for patiently answering my questions; the late Jane King for access to her notes and photographs; and Betty Gosdin and Anthony Blanco of the Somervell County Heritage Center for providing me with information and photographs of Roland Bird's excavations near Glen Rose. I also thank Sue Hendrickson, Trudy Case of Sinclair Oil, Carolyn Belardo, and Eileen Mathias of the Academy of Natural Sciences, and Mindy McNaugher of the Carnegie Museum of Natural History for their assistance. In every instance, I have tried to accurately quote and portray the authors' ideas; I apologize if I have failed to do so anywhere in this book.

I am grateful to Lorraine Gabbert, Mark Decker, Sehoya Cotner, Glen Kuban, Jennifer Sprague, Kara Witt, Stephanie Cox, Pat MacDonell, Marissa Jane, and Molly for their inspiration, companionship, advice, and travels with me to dinosaur-related sites. I also thank Mark Kane (senior production editor), Sasikala Rajesh (project manager), Ellen Rasmussen (media editor), and Sivakumar Vijayaraghavan (copy editor) of ABC-CLIO for their excellent help, guidance, and patience. I am also indebted to the many other people who assisted me, but whom I can thank only collectively and anonymously.

Introduction

Although iconic dinosaurs such as *Tyrannosaurus, Stegosaurus,* and *Apatosaurus* were never seen by humans and have been extinct for more than 60 million years, their images are everywhere; they entertain us, haunt us, lure us to museums and stores, and convince us to buy toys, soap, clothing, lunchboxes, vitamins, underwear, toilet paper, food, and countless other products. Dinosaurs fascinate us; the Internet contains tens of *millions* of pages about dinosaurs, some of which are real (e.g., *Velociraptor, Triceratops, Apatosaurus*), and others imagined (Raptor Red, Dino).

Virtually everyone has an intuitive idea of what a dinosaur is. For some, dinosaurs resemble the aging roadside dinosaurs that they recall from their childhood vacations; for others they are the monsters and lumbering giants in museums and old paintings, or on the logo at Sinclair gas stations, or the updated bird-like stars of *Jurassic Park.* Regardless of whether these images come from science classes, roadside parks, or blockbuster movies, most people visualize dinosaurs as giant—and often threatening—prehistoric reptiles. Whether docile or terrifying, dinosaurs are familiar. But what, exactly, are they?

WHAT IS A DINOSAUR?

Dinosaurs are archosaurs, which are a group of animals that also includes birds, crocodilians (e.g., crocodiles, alligators, caimans, and their extinct relatives), and pterosaurs (i.e., extinct flying reptiles of the Late Triassic, Jurassic, and Cretaceous). Dinosaurs, which originated in the Triassic from a group of small, active, carnivorous archosaurs called "proto-dinosaurs" (e.g., *Asilisaurus, Marasuchus, Lagerpeton*), share several traits that make them a unique natural group, including an upright and erect posture, perforated hip-sockets, and an enlarged longitudinal ridge on their humerus (the bone that connects the elbow to the shoulder). These traits help define dinosaurs and distinguish them from other animals that the public often confuses with dinosaurs, including wooly mammoths, plesiosaurs (aquatic reptiles), mosasaurs, pterosaurs (flying reptiles), synapsids (e.g., the sail-backed *Dimetrodon*), and ichthyosaurs (aquatic reptiles).

When dinosaurs were first named in 1842, they were distinguished from other reptiles by traits such as their giant size and the large bones in their limbs. By the late 1880s, however, paleontologists realized that some dinosaurs were small and had delicate limbs, and began reconsidering their views of dinosaurs. Today, biologists define dinosaurs based on their ancestry, not their anatomy. Dinosaurs are part of a clade (i.e., a group of organisms that share a single ancestor) that includes the common ancestor of *Triceratops* and birds and all their descendants.

This means that if you start with *Triceratops* and finches (the last members of each branch of dinosaurs' evolutionary tree) and then back up to their last common ancestor, all of the organisms in between are dinosaurs. Birds are the descendants of dinosaurs.

There are two main groups of dinosaurs that are distinguished by the structures of their hips:

> Saurischians, or "lizard-hipped" dinosaurs, are characterized by a pubis that slants down and forward, as in lizards. Saurischians included theropods (e.g., *Tyrannosaurus*) and the long-necked sauropods (e.g., *Apatosaurus*), which were the largest land-animals of all time. Birds evolved from small, meat-eating saurischians.
>
> Ornithischians, or "bird-hipped" dinosaurs, are characterized by a pubis that slants down and backward, as in modern birds, thereby helping to support these dinosaurs' large stomachs. Ornithischians were primarily herbivores and included the dome-headed *Pachycephalosaurus,* crested *Parasaurolophus,* armored *Ankylosaurus, Stegosaurus, Hadrosaurus,* and ceratopsians.

As we have learned more about dinosaurs, our ideas have changed; that is why, since 1980, there have been more than 160 evolutionary trees of dinosaurs. Despite these changes, ornithischians and saurischians remain the two main groups of dinosaurs.

WHEN AND WHERE DID DINOSAURS LIVE?

Geologists divide Earth's history into several large periods called eras (see Appendix). The end of one era, and the start of another, is based on our understanding of how Earth was changed by various cataclysmic events. The Mesozoic ("Middle Life") is the era between the two largest mass extinctions in Earth's history. Dinosaurs originated around 245 million years ago, which was soon after the first of these mass extinctions. This extinction, called the Permian-Triassic mass extinction, was Earth's most intense period of mass death; an estimated 75–95 percent of all species disappeared. These extinctions opened niches for the evolution and diversification of new lineages, including those of turtles, mammals, and dinosaurs. Indeed, about 10 million years after the Permian-Triassic mass extinction, the lineage of animals thought to have led to dinosaurs (e.g., *Nyasasaurus*) took hold in Tanzania and Zambia.

Dinosaurs diversified rapidly after originating near the beginning of the Mesozoic, an era also known as the "Age of Dinosaurs." The Mesozoic lasted from 251 to 66 million years ago.[1] (For comparison, the "Age of Mammals," which began after the demise of the non-avian dinosaurs, has lasted only 66 million years.) The 185 million years of the Mesozoic, which represent approximately 4 percent of Earth's history, are divided into three periods: the Triassic Period, Jurassic Period, and Cretaceous Period (Table 1):

Table 1: The Mesozoic Era consists of the Triassic, Jurassic, and Cretaceous periods. Non-avian dinosaurs evolved in the Triassic and abruptly disappeared at the end of the Cretaceous.

Period	*Start* (mya)*	*End (mya)*	*Duration (my)*	*Examples of Dinosaurs*
Triassic (TR)	251	200	51	*Coelophysis, Herrerasaurus, Eoraptor, Plateosaurus*
Jurassic (J)	200	145	55	*Stegosaurus, Diplodocus, Allosaurus, Apatosaurus, Camarasaurus*
Cretaceous (K)	145	66	79	*Velociraptor, Triceratops, Tyrannosaurus, Microraptor, Argentinosaurus, Deinonychus*

*mya = millions (i.e., 10^9) of years ago; my = millions of years

The Triassic Period lasted from 251 to 200 million years ago (51 million years), which equals 1.1 percent of Earth's history, and 28 percent of the Mesozoic. During the Triassic, Earth's climate was mild to hot and relatively dry, and the predominant vegetation included gymnosperms and ferns. At the start of the Triassic, all of Earth's landmasses were united into a supercontinent called Pangaea, across which the different species of dinosaurs were later dispersed. Near the end of the Triassic, Pangaea began to break apart. A mass extinction that included giant crocodiles at the end of the Triassic paved the way for the rapid evolution of dinosaurs in the Jurassic.

The Jurassic Period lasted from 200 to 145 million years ago (55 million years), which equals 1.2 percent of Earth's history, and 30 percent of the Mesozoic. During this time, Earth's climate was warm and moist, with relatively little seasonal variation. Forests of gymnosperms and ferns were abundant. As Pangaea continued to break apart, dinosaurs were scattered across the globe on separate continents, and new types of dinosaurs evolved in each geographic area. Dinosaurs, the largest of which flourished during this time, have been found on every continent.

The Cretaceous Period lasted from 145 to 66 million years ago (79 million years), which equals 1.7 percent of Earth's history, and 43 percent of the Mesozoic. By the middle of the Cretaceous, angiosperms (i.e., flowering plants) such as magnolias, oaks, maples, and the first grasses dotted the landscape. Continents continued to move toward their present positions, and dinosaurs reached their greatest diversity. The Cretaceous ended with vast volcanic activity and a meteoric impact that resulted in a mass extinction popularly referred to as the K-T extinction.[2] The last non-avian dinosaurs standing included tyrannosaurs, ankylosaurs, ceratopsids, and hadrosaurs.

Contrary to popular belief, no species of dinosaurs lived throughout the Age of Dinosaurs. (The same is true for the Cenozoic ["new life"], which spans from 66 million years ago to the present—i.e., no species of mammal has lived throughout the Cenozoic). For example, *Herrerasaurus* appeared and disappeared in the Triassic, whereas stegosaurs appeared in the Middle Jurassic, but were gone by the Middle Cretaceous. Despite their portrayals together in paintings and movies, *Apatosaurus* ("Brontosaurus") and *T. rex* never saw each other. Indeed, *Apatosaurus* (a Jurassic sauropod) had been extinct for more than 80 million years before *Tyrannosaurus* (a Cretaceous theropod) appeared; this period is longer than the time separating the disappearance of *Tyrannosaurus* and today. Dinosaurs were not evenly distributed, and were never all together in time or place.

WHAT BECAME OF DINOSAURS?

All of the remaining non-avian dinosaurs disappeared at the K-T extinction event, which happened 65.5 million years ago. The only remaining dinosaurs are birds, which are often called "avian dinosaurs."

HOW DIVERSE WERE DINOSAURS?

At the time of the K-T extinction, dinosaurs were diverse, but composed only 1–2 percent of the global fauna. (We remember dinosaurs not because they were overly abundant, but instead because many of them were big.) The diversity of dinosaurs rivaled that of modern mammals, and included dome-headed pachycephalosaurs, bird-like dromaeosaurs, horned ceratopsians, gigantic elephant-like and giraffe-like sauropods, and armor-plated stegosaurs. All non-avian dinosaurs were terrestrial; there were no marine dinosaurs. There are more than 1,400 named species of non-avian dinosaurs, and paleontologists estimate that there are another 700–900 genera of dinosaurs that remain undiscovered (Table 2).

Although iconic dinosaurs such as *T. rex, Apatosaurus* (i.e., *Brontosaurus*), *Stegosaurus,* and *Triceratops* disappeared more than 65 million years ago, dinosaurs' descendants—that is, birds—remain with us. Birds and dinosaurs share several traits, including feathers, oblong hard-shelled eggs, hollow bones, three-toed feet, nests and brooding, wishbones, and legs directly beneath their bodies (in contrast, modern reptiles walk with their legs splayed out, with their knees always bent and their feet pointed out rather than forward). Birds are dinosaurs in the same way that bats are mammals.

Although all dinosaurs were terrestrial, all birds are not; indeed, penguins and flightless cormorants are aquatic birds. This is why modern birds are sometimes called "avian dinosaurs," and why extinct, non-flying dinosaurs are called "non-avian dinosaurs." Pteranodons and pterodactyls were flying reptiles, but were not dinosaurs.

Table 2: The number of known species of mammals, birds, reptiles, and insects. For comparison, there are more than 1,400 named species of non-avian dinosaurs, and hundreds more await discovery.

Group	*Number of Known Species*	*Percentage of Species Discovered*
Mammals	5,700	98
Birds	10,500	96
Reptiles	12,000	80
Insects	1,000,000	20

HOW LONG WERE DINOSAURS ALIVE?

Non-avian dinosaurs were on Earth for about 180 million years. For comparison, modern humans have been on Earth for about 200,000 years, which is 0.004 percent of Earth's history, and only 0.1 percent of the duration of dinosaurs.

HOW ARE DINOSAURS NAMED?

The conventions used for naming dinosaurs (and other animals) are determined by the International Code of Zoological Nomenclature (ICZN). Because of questions about the validity of many names, and in part for convenience, most dinosaurs are known by their generic (rather than their specific) name; for example, most people refer to *Apatosaurus,* not *Apatosaurus excelsus,* and to *Triceratops,* not *Triceratops horridus*. The only exception to this is *Tyrannosaurus rex,* which is commonly known as *T. rex*.[3]

The names of several iconic dinosaurs—for example, *Apatosaurus* and *Tyrannosaurus*—end with *-saurus*. This ending usually denotes a reptile but not necessarily a dinosaur. For example, *Lisboasaurus* and *Aggiosaurus* are crocodilians, *Herbstosaurus* is a pterosaur, and *Unicerosaurus* is a fish. In 2013, *Halticosaurus*—identified as a theropod by Friedrich von Huene in 1932—was determined to be more like a crocodile than a dinosaur. Such revisions continue.

There are several types of questionable, rejected, or forgotten names:

A junior synonym describes the same taxon as a previously published name. When two or more genera are given the same name, the first to be published is the senior synonym, and all of the others are junior synonyms. For example, in 1856, Joseph Leidy used a tooth to name *Troodon formosus*. In 1932, almost 80 years after Leidy's discovery, Charles M. Sternberg found part of a skeleton that he named *Stenonychosaurus inequalis*. The subsequent discovery of more fossils showed that *Stenonychosaurus inequalis* was *Troodon formosus.* Because *Troodon formosus* had been named first, that name had priority, and *Stenonychosaurus inequalis* became a junior synonym of *Troodon formosus.* In this book, you will encounter several other examples of junior synonyms (e.g.,

Anatosaurus and *Anatotitan,* which are junior synonyms of *Edmontosaurus*), the most famous of which is *Brontosaurus,* a junior synonym of *Apatosaurus*.

A *nomen nudum* ("naked name") is a name that has appeared in print but has not been formally published to meet criteria set by ICZN. Although these names are invalid for scientific nomenclature, they are sometimes used as vernacular names. A *nomen ex dissertatione* is a type of *nomen dubium* that appeared in a dissertation and is, therefore, not considered "published" by ICZN.

A *nomen oblitum* ("forgotten name") is a name that has not appeared in the scientific literature for more than 50 years since it was proposed. A famous example of a nomen oblitum is *Manospondylus gigas*. In 1892, Edward Drinker Cope collected two partial vertebrae (one of which has since been lost) that he named *Manospondylus gigas* ("giant porous vertebra," in reference to the many pores in the bone). Because Cope's fossils were so fragmentary, *Manospondylus gigas* was seldom cited. When more of the bones were uncovered, paleontologists realized that *Manospondylus gigas* was the same as *Tyrannosaurus rex,* which had been named in 1905. According to the rules of nomenclature, *Manospondylus gigas* should have had priority over *Tyrannosaurus rex,* because *Manospondylus gigas* was named first. However, effective 2000, ICZN changed its rules of nomenclature so that "the prevailing usage"—in this case, *T. rex*—became the official name because the older synonym (in this case, *Manospondylus gigas*) had "not been used as a valid name after 1899" and because *T. rex* "had been used … as its presumed valid name, in at least 25 works, published by at least 10 authors in the immediately preceding 50 years." *Tyrannosaurus rex*, therefore, is the valid and "protected name" (*nomen protectum*), and *Manospondylus gigas* is a *nomen oblitum.*

A *nomen dubium* ("doubtful name") is a name that describes fossils that are incomplete or that lack diagnostic (i.e., unique) traits, and therefore are too incomplete to be distinguished from similar material of two or more otherwise distinct species. A *nomen dubium* is not invalid; it renders doubtful names to that genus, and few paleontologists will assign more material to it (i.e., the name becomes a historical curiosity).

A *nomen rejectum* is a generic name rejected by ICZN in favor of a different name.

A preoccupied name is a name that is officially published, but which is already used to describe another taxon. These names are invalid and must be replaced.

Although these conventions seem straightforward, navigating the world of dinosaurs' names can be confusing. There is a plethora of names, many of which were based on fragmentary evidence and have been abandoned (i.e., *nomen dubium*). Moreover, several species have been given more than one name (e.g., *Brontosaurus* is a junior synonym of *Apatosaurus*). Interestingly, some names of dinosaurs are typographical errors; for example, "Shuosaurus" is a misspelling of *Shunosaurus,* and "Vitaridrinda" is a misspelling of *Vitakridrinda. Aachenosaurus,* which Gerard Smets described in 1888 as a hadrosaur (i.e., duckbilled dinosaur), was actually a piece of petrified wood. (When the error was pointed out by Louis Dollo, Smets withdrew from science.) To retain the historical context and flavor of the stories, I have used most of the original names coined by the discoverers, including some that are now antiquated synonyms (e.g., *Brontosaurus*).

Paleontologists also wonder how many of the different names assigned to dinosaurs refer to the same species. For example, in 1909, Charles Gilmore identified four species of *Camptosaurus* from one quarry. Were these each a different species, or do they represent normal biological variation of one species? In 2010, Jack Horner and John Scannella argued that the iconic *Triceratops* was a juvenile version of *Torosaurus,*[4] and that *Nedoceratops* was an intermediate between the young adult *Triceratops* and the old adult *Torosaurus* stage (i.e., *Nedoceratops* should be called *Triceratops*). And in 2013, researchers at the University of Pennsylvania announced that three species of *Psittacosaurus* were actually the same species. Several different "species" that paleontologists now study may represent different stages of the same life cycle.

HOW TO USE THIS BOOK

Although the history of dinosaurs in science, religion, and society is a fascinating story, *Dinosaurs by the Decades* is more than a historical narrative. Indeed, this book has a variety of possible uses:

To understand the basics about dinosaurs. In *Dinosaurs by the Decades,* you will learn what dinosaurs are, when and where they lived, how and when they disappeared, and how we imagined and learned about them. A glossary will help you with unfamiliar terms, and an appendix describing geologic history will help you appreciate when dinosaurs roamed the Earth.

To understand the scientific and social contexts of dinosaurs during a particular period. Dinosaurs by the Decades will help you appreciate the historical antecedents, outcomes, and impacts of a variety of dinosaur-related events. You will best appreciate the interrelationships of these events by examining the entries spanning several decades before and after any particular event in which you are interested. For example, some of the early discoveries of dinosaurs were hardly noted, whereas others—for example, the discoveries of *Brontosaurus* and *T. rex*—were major stories that were marketed by museums, corporations, and others. Some dinosaurs became famous, whereas others remain virtually unknown. In *Dinosaurs by the Decades,* you will meet them all.

To supplement the many other books about dinosaurs. Although there are many excellent books about dinosaurs, most of these books focus on science or personalities (e.g., the "Bone Wars" between Edward Drinker Cope and Othniel Marsh) and seldom integrate other events. I hope that *Dinosaurs by the Decades* will provide information and historical context that help readers understand the scientific and societal contexts of dinosaurs.

To understand the history of dinosaurs. Dinosaurs by the Decades includes all of the iconic discoveries and uses of dinosaurs, including their appearances in newspapers, comics, museums, corporate logos, books, movies, world's fairs, and religions. You will also learn about lesser-known events that became harbingers of future trends, including the first dinosaurs displayed to the public, the first appearances of dinosaurs in movies and books, the first woman to be harassed by a dinosaur in a movie, and the first book to feature dinosaurs in a "lost world." Taken together, these events comprise an interesting story in which scientific claims come and go, entrenched ideas are overturned, movie-based portrayals of dinosaurs become inseparable from scientific discoveries, and preachers and their well-funded museums use dinosaurs to reject science while promoting their religious claims and social agendas. Along the way, you will get glimpses of the individuals who created the stories, including many who were brilliant heroes, and others who were eccentric oddballs and shameless frauds. Many of these people and their discoveries have been influenced by religious ideas and historical circumstances (e.g., the "second extinction" of dinosaurs during World War II, when several museums in Europe were destroyed by Allied bombs). I urge readers to resist judging this history with today's criteria, for doing so will blur the importance of yesterday's people and events to today's understanding of dinosaurs.

To locate ideas, people, themes, and discoveries about dinosaurs. The chronological format of this book, as well as an extensive index and icons that trace recurring themes, will help you quickly find information about dinosaurs and related topics:

Fossils and geology

Art and movies

Dinosaurs, birds, and feathers

Books

Coined words

Dinosaurs and religion

Extinction

Firsts

Frauds and fakes

Museums

Tracks

I hope that these features will make *Dinosaurs by the Decades* a useful tool for not only learning more about dinosaurs, but also for discovering lesser-known information related to a topic you are looking for.

Finally, I have tried to tell a story that goes beyond the often-discussed stories of "fossil hunters," "then and now" comparisons, and "Bone Wars." I have not limited discussions to only the people and ideas that have endured; indeed, you will also learn about interesting people, forgotten places, and ideas that failed, but were nevertheless important.

I hope you enjoy the story.

Randy Moore
March 2014

Institutional Abbreviations

AAAS	American Association for the Advancement of Science, Washington, DC, USA
AiG	Answers in Genesis, Petersburg, KY, USA
AMNH	American Museum of Natural History, New York, NY, USA
ANS	Academy of Natural Sciences, Philadelphia, PA, USA
APS	American Philosophical Society, Philadelphia, PA, USA
BAAS	British Association for the Advancement of Science, London, England
BHI	Black Hills Institute of Geologic Research, Hill City, SD, USA
BMNH	Natural History Museum, London, England[5]
BSP	Bayerische Staatssammlung für Paläontologie und Geologie, Munich, Germany
BYU	Brigham Young University Museum of Paleontology, Provo, UT, USA
CMNH	Carnegie Museum of Natural History, Pittsburgh, PA, USA
DPP	Dinosaur Provincial Park, Brooks, Alberta, Canada
DVSP	Dinosaur Valley State Park, Glen Rose, TX, USA
GSC	Geological Survey of Canada, Ottawa, Ontario, Canada
ICR	Institute for Creation Research, Dallas, TX, USA
ICZN	International Commission on Zoological Nomenclature, London, England
IVPP	Institute for Vertebrate Paleontology and Paleoanthropology, Beijing, China
MACN	Museo Argentino de Ciencias Naturales "Bernardino Rivadavia," Buenos Aires, Argentina
MOR	Museum of the Rockies, Bozeman, MT, USA
NMNH	National Museum of Natural History, Washington, DC, USA
NSF	National Science Foundation, Arlington, VA, USA
ROM	Royal Ontario Museum, Toronto, Canada
RTM	Royal Tyrrell Museum of Paleontology, Drumheller, Alberta, Canada
SVP	Society of Vertebrate Paleontology, Bethesda, MD, USA
UCMP	University of California, Museum of Paleontology, Berkeley, CA, USA
UFHNH	Utah Field House of Natural History State Park Museum, Vernal, UT, USA

USGS	United States Geological Survey, Reston, VA, USA
WPA	Works Progress Administration, Washington, DC, USA (closed in 1943)
YPM	Yale Peabody Museum of Natural History, New Haven, CT, USA
ZDM	Zigong Dinosaur Museum, Zigong, China

NOTES

1. Although it is difficult to appreciate the vast length of these periods. this may help: If Earth's history were depicted as a year-long calendar, with Earth forming on January 1, dinosaurs would not appear until December 13, and would go extinct on December 26. (Modern humans would appear just before midnight on New Year's Eve.) Thus, although we consider dinosaurs to be ancient, they did not originate until 95 percent of Earth's present history had passed. Similarly, although we consider dinosaurs to be "failures" because they became extinct, they ruled the planet for more than 150 million years. For comparison, all of recorded human history (i.e., since Egypt's early pharaohs) spans only 5,000 years, which is only 1/32,000 the time that dinosaurs ruled Earth.
2. Tertiary (*T*) was replaced by Paleogene (*Pg*) to designate the time spanning from 65.5 to 23 million years ago, but the abbreviation for Tertiary (i.e., *T*) is still used when discussing extinctions at the end of the Cretaceous (see Appendix). Thus, although K-Pg is the official designation of the boundary, the "K-T boundary" remains common when discussing the fate of dinosaurs.
3. Headlines describing discoveries of new dinosaurs are common, and some paleontologists are concerned about the rapid rate at which dinosaurs are being named. There is pressure to find a new species rather than another example of a previously named species, for these "new" species are more lucrative for collectors, more interesting to the public, more profitable for museums, and more helpful for obtaining funding for research.
4. *Torosaurus* had a larger frill with two holes (*fenestrae,* which is Latin for *windows*).
5. Although this museum is often referred to as the British Museum of Natural History, the museum separated from the British Museum in 1963, at which time it became the British Museum (Natural History). Today, the museum's official name—set by the Museum and Galleries Act 1992—is the Natural History Museum.

The Chronology

1027 Persian naturalist Ibn Sina's (980–1037) *The Book of Healing* describes fossils and petrification.

1506 Italian Renaissance polymath Leonardo da Vinci (1452–1519) begins a manuscript that becomes known as the *Leicester Codex* (named for Lord Leicester, who bought the codex after it was discovered in the late 1690s; it is now owned by Bill Gates). Leonardo interpreted marine fossils on mountaintops as evidence for the vertical movement of landmasses, and may have identified the principle of superposition—"stratified stones of the mountains are all layers of clay, deposited one above the other by various floods"—before it was developed by Nicolas Steno in 1669. Leonardo rejected the two dominant explanations for the location of such fossils, which were (1) the action of the Noachian Flood, and (2) the spontaneous creation of such fossils *in situ*. He reasoned that fossils should be jumbled if they had been deposited by a flood; instead, they are often found in groups, resembling colonies of similar living organisms. Furthermore, dead, heavy-shelled organisms should not rise in the water column but should sink to the bottom of flood waters. Even living organisms, especially those with low mobility, would not have had enough time to migrate hundreds of miles inland from the original ocean during the Flood. Leonardo invoked the idea of a changing planet that experienced mountain building and erosion, ideas that presaged the uniformitarianism of James Hutton and Charles Lyell. Leonardo's interpretation of fossils was cast aside during the Reformation, when a literal interpretation of the Bible, including the Flood, prevailed.

1519 The first vertebrate fossils are seen by Europeans in the Western Hemisphere. These fossils, which were mastodon bones collected by Indians in Tlaxcala, Mexico, were shown to Cortez's army.

1589 French potter Bernard Palissy (1510–1589), one of the first people to argue for the organic origin of fossils, dies of "starvation and maltreatment" in a Bastille prison.

1596 In the third edition of his *Thesaurus Geographicus*, Dutch mapmaker Abraham Ortelius (1527–1598) suggests that the Americas had been "torn away from Europe and Africa . . . by earthquakes and floods." This suggestion is the oldest known statement that anticipated continental drift.

1620 In *Novum Organum*, Sir Francis Bacon (1561–1626) suggests that "it's more than a curiosity" that continents seem to fit together, adding that "the very configuration of the world itself in its greater parts presents Conformable Instances which are not to be neglected." Although Bacon did not compare the opposite shores of Africa and South America (i.e., the opposite coasts of the Atlantic) or explicitly claim that continents had been joined, his observation anticipated continental drift.

1650 Irish prelate James Ussher (1581–1656) completes his influential *Annals of the Old Testament, Deduced from the First Origins of the World. Annals* placed divine creation on the evening of Saturday, October 22, 4004 BCE. Ussher's calculations, which were based on several assumptions (e.g., that the presence of a ripe fruit in the Garden of Eden meant that creation occurred in the autumn), accepted the vast ages of biblical patriarchs (e.g., Adam living to be 930, Seth 912, and Methuselah 969). Ussher's claim, with its corollary that humans lived with dinosaurs, became a foundation of young-Earth creationism.

1665 In *Mundus Subterraneus*, German scholar Athanasius Kircher (1601–1680) attributes giant fossil bones to extinct races of giant humans.

1665 English polymath Robert Hooke (1635–1703) proposes an organic origin for fossils such as ammonites, because "nature does nothing in vain."

1666 While examining a white shark, Danish physician Nicolas Steno (1638–1686) notes that the shark's teeth resemble tongue stones (*glossopetrae*), which were assumed to be the tongues of serpents turned to stone by St. Paul. Steno claimed that glossopetrae were teeth that had once belonged to sharks—that is, that the fossils were remnants of once-living organisms.

1667 The Ashmolean Museum at Oxford, England, opens. This was the world's first natural history museum. Vertebrate paleontology now depends largely on collections housed in universities and museums.

1668 In the moralistic *The Break-Up of the Large and Small Worlds, as Being Demonstrated That America Was Connected before the Flood with the Other Parts of the World*, Rev. François Placet suggests that America originated "either by the conjugation of many floating islands or by the destruction of the island of Atlantis which after sinking down into the deep could have caused the uncovering of a new Earth." Placet claimed that the Atlantic Ocean was created by a biblical flood that destroyed Atlantis (which became an ocean basin) and lifted America. Placet, who believed that "local catastrophes" resulted from original sin (but had negligible impacts when compared to the biblical flood), was among the first to suggest that, based on their shapes, continents were once joined. For the next 250 years, a biblical flood was often invoked to explain the shapes and positions of continents, the distributions of plants and animals (including dinosaurs), and the age of Earth.

1669 In his concise *De solido intra Solidum naturaliter Contento Dissertationis Prodromus*, Nicolas Steno establishes stratigraphy with his principle of superposition, which states that within undisturbed geologic strata, the lower sediments are older than higher sediments. Steno later applied this principle to determine the relative sequence of geologic events in the Tuscany region of Italy; this was the first application of stratigraphic principles to the geologic record. Six years after publishing *De solido*, Steno was ordained as a Catholic priest, after which he gave up science and died penniless. On October 23, 1988, Steno was beatified by Pope John Paul II.

1670 Italian painter Agostino Scilla's (1639–1700) influential *Vain Speculation Undeceived by Sense* depicts fossils as having an organic origin. However, the following year, English physician and naturalist Martin Lister (1639–1712) claimed that fossils were "never any part of an animal."

1677 Oxford University professor of "chymistry" Robert Plot (1640–1696), the first "keeper" (i.e., curator) of Oxford's Ashmolean Museum, discovers in a local quarry the distal end of "a real *Bone*, now petrified" femur from a dinosaur, but suggests that it resembled "exactly the figure of the lowermost part of the Thigh-Bone of a Man, or at least some other Animal." Plot, who became known as Britain's "Genial Father of Country Natural Histories," later suggested that the bone might have belonged to a giant man or woman. Plot described the bone in his *The Natural History of Oxfordshire* (1677), the first illustrated book to mention fossils in England. Plot, whose book popularized regional natural history, noted that the fossil's dimensions far exceeded "the ordinary course of Nature," that "such Stones" formed in rocks, and that the fossils resembled bony parts of humans and other animals, all of which he attributed to the grandeur of God's design. The "enigmatic thighbone" discovered by Plot probably belonged to *Megalosaurus*, a Middle Jurassic theropod that lived in southern England (and elsewhere). *Megalosaurus* later became the first dinosaur bone to be given a scientific description. Plot's fossil, which weighed nearly 20 pounds, has since been lost, and *Megalosaurus* remains poorly understood.

1694 English physicist Isaac Newton (1642–1727) suggests a mechanistic substitute for divine intervention when he wonders whether meteor impacts created geologic formations such as the Caspian Sea. Newton later withdrew his suggestion under pressure from the church. In 1980, Walter and Luis Alvarez blamed a meteor impact for the extinction of dinosaurs.

1695 In *Essay towards a Natural History of the Earth*, English naturalist John Woodward (1665–1728)—a protégé of Robert Plot and professor of "physick" at London's Gresham College—claims that fossils of vertebrates (including dinosaurs) were deposited by a biblical flood, and that the settling of materials in flood waters produced horizontally stratified rocks and fossils. Many young-Earth creationists continue to make this claim. Woodward later helped create the first chair in geology at Cambridge University, and his more than 9,000 fossils, rocks, minerals and other specimens became the nucleus of what is now the Sedgwick Museum.

1699 Oxford professor Edward Lhwyd (1660–1709), the second curator of the Ashmolean Museum, unknowingly illustrates a tooth of a carnivorous dinosaur.

1705 In the posthumous second edition of *The Natural History of Oxfordshire* (which included additions by his stepson John Burman and his successor Edward Lhwyd), Robert Plot argues that fossilized invertebrates in rocks are stones "formed into an Animal Mould" by "some extraordinary plastic virtues latent in the Earth." These "stones" were believed to denote secret places. The first edition of *The Natural History of Oxfordshire* was published in 1677.

1715 English scientist Edmond Halley (1656–1742) proposes that measurements of oceanic salinity could be used to determine Earth's age and show evidence "of the Sacred Writ [that] Mankind has dwelt about 6,000 years." Although Halley never used this technique, others—most notably Irish physicist John Joly (1857–1933)—later used Halley's idea to estimate Earth to be 80–90 million years old.

1719 Yorkshire clergyman William Stukeley (1687–1765) describes a fossil now known to be *Plesiosaurus dolichodeirus*, an ichthyosaur ("fish lizard"). This was the first description of an articulated specimen of a fossil reptile and the first paper devoted entirely to vertebrate paleontology. Stukeley had been told about the fossil (BMNH R.1330) by Robert Darwin (c. 1682–1754), Charles's great-grandfather.

1728 In the posthumously published *A Catalogue of the Extraneous English Fossils*, John Woodward of London's Gresham College describes his fossils. The first specimen listed in the book (Specimen A1) was a broken limb of a dinosaur. This fossil, which is now housed in the Sedgwick Museum at the University of Cambridge, was the first dinosaur bone to be discovered that can still be seen in a museum.

1735 Swedish biologist Carl Linnaeus (1707–1778; known after 1761 as Carl von Linné when granted nobility by the king of Sweden) begins publishing *Species Plantarum*, which in 1753 established the Latin binomial system for naming plants and is considered the beginning of contemporary biological nomenclature. That same year, Linnaeus's *Systema Naturae* (published in Latin, as was customary in Linnaeus's era) outlined his classification of life and established the foundation of systematic biology. *Systema Naturae* revolutionized classification by using common characteristics to categorize life. Linnaeus—the son of a clergyman—used a system of kingdoms, classes, orders, and genera of increasing specificity, each of which was a distinct "archetype" that reflected God's omnipotence. Although Linnaeus placed every known plant and animal into its own unique and static place in the divinely created order of nature, his system was later adapted to evolutionary interpretations. Linnaeus—whose motto was *Deus creavit, Linnaeus disposuit* ("God created, Linnaeus arranged")—had by the 10th edition of *Systema Naturae* (1758) classified 7,700 plant species and 4,400 animal species. Modern taxonomy traces its ancestry directly to this document.

1739 The first recorded fossil-vertebrates are gathered in the United States by Canadian Charles III Le Moyne (1687–1755), second Baron de Longueuil, along the Ohio River near what is now Big Bone Lick State Park in Kentucky. Near the end of the 18th century, these bones were identified as being from an extinct elephant (*Mammut americanum*). In 1807, William Clark's three-week expedition to Big Bone Lick State Park was America's first organized vertebrate paleontology expedition.

1741 Charles Willson Peale (1741–1827) is born in Chester, Maryland. Peale, who became a famous artist, later popularized the modern natural history museum in America. His museum in Philadelphia was the first to display a mounted fossil skeleton of a vertebrate in the United States, and was among the first museums to use Linnaean taxonomy in its exhibits (other museums presented organisms as mysteries and curiosities of the natural world).

1742 French scientist Pierre de Maupertuis (1698–1759) argues that comets' collisions with Earth caused extinctions by altering Earth's environment.

1744 Swiss biologist Albrecht von Haller (1708–1777) coins the term *evolution* (from the Latin *evolvere*, meaning *to unroll*) to describe how embryos originate from preformed structures (*homunculi*) in eggs and sperm. Charles Darwin first used the word *evolution* in *The Descent of Man, and Selection in Relation to Sex* (1871).

1751 Carl Linnaeus proposes that new species within genera can be produced by hybridization, and that this hybridization is guided by God. As was common in his era and since, Linnaeus accepted that God created life in the Garden of Eden and believed that classifying organisms would reveal the pattern of creation. Linnaeus accepted the biblical account of the Flood but did not believe that it could have moved organisms far inland and covered them in sediments in the time available. As he noted, "He who attributes all this to the Flood, which suddenly came and as suddenly passed, is verily a stranger to science and himself blind, seeing only through the eyes of others, as far as he sees anything at all." Most of Linnaeus's collection (on which biological taxonomy was founded) was sold in 1783 to James Edward Smith (1759–1828), and is now housed at the Linnean Society in London.

1753 The museum that was named the British Museum (Natural History; BMNH) in 1963 is established (Figure 1). Today, that museum is the Natural History Museum.

Figure 1 London's famed Natural History Museum, established in 1753, was the world's first national, public, and secular museum. The museum's more than 70 million specimens include items collected by Charles Darwin, Mary Anning, Gideon Mantell, Alfred Wallace, Richard Owen, and other famed scientists. The museum made science accessible to the public (admission is free) and, in doing so, transformed the public's expectations of natural history museums. This sketch of the museum was drawn in 1876 by English painter John William Waterhouse (1849–1917). Today, the Natural History Museum is one of the most renowned museums in the world. (V&A Images, London/Art Resource, NY)

1753 Georges-Louis Buffon's (1707–1788) *L'Histoire Naturelle* suggests that similarities among vertebrates result from common ancestry. Famed American biologist George Gaylord Simpson (1902–1984) later described Buffon as "the pope of 18th-century zoologists."

1755 In Oxfordshire, England, natural history entrepreneur Joshua Platt (1669–1773) finds three vertebrae "of enormous size" of dinosaurs, as well as an incomplete femur weighing 220 pounds. Platt sent them to Quaker merchant and botanist Peter Collinson, who did nothing with them, and today their whereabouts are unknown. Three years later, Platt described a thighbone (probably of *Megalosaurus*, as were the vertebrae), but that bone also has been lost.

1756 In a memoir comparing the geology of North America and Switzerland, Jean Êtienne Guettard (1715–1786) publishes the first illustration of an American vertebrate fossil. The illustration depicted a molar of an American mastodon (*Mammut americanum*). Guettard's memoir also included a crude mineralogical map of North America that noted several fossil-sites near Big Bone Lick, Kentucky.

1758 The 10th edition of Linnaeus's *Systema Naturae* establishes modern nomenclature for organisms.

1759 London's Natural History Museum (BMNH; est. 1753) opens to the public (Figure 1). Except for during parts of World Wars I and II, the museum has been open ever since. BMNH was the first national, public, and secular museum in the world, and today its more than 70 million items span every scientific discipline. It is famed for its skeletons of dinosaurs, especially the 84-foot-long *Diplodocus carnegii* (a gift from American steel magnate—and social Darwinist—Andrew Carnegie to King Edward VII, who was then a trustee of BMNH) that dominates the museum's main hall.

1762 Louis Jean Marie Daubenton (1716–1799) first treats American fossils in a scientific way.

1763 The fossil discovered by Robert Plot in 1676 is illustrated in Richard Brookes's *The Natural History of Waters, Earths, Stones, Fossils, and Minerals*. One end of the fossil resembled a pair of human testicles, promoting Brookes to name it *Scrotum humanum*. Plot's fossil was later determined to be part of the dinosaur *Megalosaurus*. In 1993, two geologists claimed in *Modern Geology* that, by the strictest rules of biological nomenclature, *Scrotum humanum* was the first validly published name applied to a dinosaur. However, most paleontologists rejected this claim because of the informality of Brookes's description. *Scrotum humanum* is now a *nomen oblitum* (i.e., forgotten name) because it has not been used by an author in more than 50 years. Plot's specimen has been lost.

1768 French philosopher Jean-Baptiste Robinet (1735–1820) describes Robert Plot's fossil as a "stony scrotum," adding that such objects are how nature creates human parts. Robinet, who argued that fossils grow inside the Earth, believed that Plot's fossil had the vestiges of a urethra and the musculature of testicles.

1769 The American Philosophical Society (APS) begins publishing its *Transactions*, which later became a major outlet for vertebrate paleontology. Well into the 19th century, personal letters (rather than "papers") were a primary way of dispersing information. Other reports appeared in *Proceedings,* which began publication in 1838. Many of the early experts on paleontology were physicians.

1770 APS asks merchants, sailors, soldiers, and others for "new & curious plants, animals and fossils" that will be displayed in their "cabinet" of curiosities. This became the first museum in America.

1770 As naturalists and others continue to gather fossils of ancient organisms, most people continue to accept the Great Chain of Being and its implication that extinction is impossible. For example, Anglican cleric John Wesley (1703–1791) noted that "Death . . . is never permitted to destroy the most inconsiderable species," and Thomas Jefferson agreed that "such is the economy of nature, that no instance can be produced of her having permitted any one race of her animals to become extinct; of her having formed any link in her great work, so weak as to be broken." Less than 40 years later, Georges Cuvier (1769–1832) documented that extinction occurs.

1770 Dutch miners in Maastricht discover a four-foot-long skull, teeth, and jaws and trigger a debate about the creature's origin. After Napoleon's army captured Maastricht, the fossils were sent to Georges Cuvier in Paris. The sea-going lizard, which Georges Cuvier later named *Mosasaurus,* made headlines throughout Europe. Cuvier, who knew that there is no living lizard with a jaw as large as that of *Mosasaurus*, later used this discovery to argue for extinction. Today, the *Mosasaurus* skull is displayed in the National Museum of Natural History in Paris's famed Jardin des Plantes.

1770 Libertarian poet and physician Erasmus Darwin (1731–1802)—Charles's eccentric grandfather who fathered 14 children—promotes the idea of common descent by printing the allegorical motto *E conchis omnia* ("Everything from shells") on his carriage. Soon thereafter, following a public outcry, he had the motto removed. Erasmus, who died before Charles Darwin was born, helped establish the family fortune that later sustained Charles in a life of privileged comfort.

1772 Sir Christopher Sykes (1749–1801) publishes a catalog that includes Joshua Platt's collection of bones. A thighbone in Platt's catalog cost four shillings.

1778 Swiss geologist Jean-André de Luc (1727–1817) coins the term *geology*. De Luc argued that observable processes such as erosion, deposition, and volcanic eruptions could be used as "natural chronometers" to construct Earth's history. Scientists such as John Phillips, Thomas Huxley, Alfred Wallace, and John Joly later used these techniques to estimate Earth's age.

1784 "Dr. Watson" gives a bone to Cambridge's Woodwardian Museum. This fossil, which is now housed in the Sedgwick Museum, is part of a shoulder blade of *Megalosaurus*.

1784 Cosimo Alessandro Collini (1727–1806) describes the first known specimen of a pterosaur.

1785 Charles Willson Peale opens a museum in Philadelphia displaying bones from Kentucky's Big Bone Lick. This museum, directed by a board of directors chaired by Thomas Jefferson, in 1794 moved to the headquarters of the APS. Peale's mount of a mastodon at the museum in 1806 was the first fossil skeleton mounted in America (and the second in the world).

1785 Scottish geologist James Hutton (1726–1797) reports his ideas about Earth's history to the newly formed Royal Society of Edinburgh in a thirty-page dissertation titled *Concerning the System of the Earth, Its Duration, and Stability*. Hutton claimed that rocks now at Earth's surface were formed from pressurized sediments under the ocean, raised to the surface by pressure and heat, and eroded to the sea, where the process began anew. Hutton's ideas came to be known as *uniformitarianism:* "The past history of our globe must be explained by what can be seen to be happening now. No powers are to be employed that are not natural to the globe, no action to be admitted except those of which we know the principle." Hutton later expanded his ideas into *Theory of the Earth, with Proofs and Illustrations*. Hutton believed that all creatures except humans are ancient, and that all of Earth's changes have been purposeful (i.e., the world has been constantly replenished for humans' use). Hutton, who claimed that geological strata showed no evidence of historical direction, was vilified as an "infidel" by conservative critics. Nevertheless, his ideas became a foundation of modern geology and, according to some scientists, delayed the acceptance of catastrophic events such as meteoric impacts for the extinction of non-avian dinosaurs.

1785 Thomas Jefferson's *Notes on the State of Virginia* discusses discoveries of "mammoths" in Kentucky to refute Buffon's "Theory of American Degeneracy." Jefferson believed that mammoths—"the largest of all terrestrial beings"—quashed Buffon's claim. Some people have claimed that American's preoccupation with the size of dinosaurs is a vestige of concerns produced by Buffon's claim about American degeneracy.

1786 A mastodon tooth at Yale College is the first known vertebrate fossil in an American college. In 1865, famed dinosaur-hunter Othniel Marsh became Yale's first vertebrate paleontologist.

1787 While working in Gloucester County, New Jersey, physician Caspar Wistar (1761–1818) and merchant Timothy Matlack (1730–1829) find what they tell the APS is a "giant's bone." This fossil, which may have been the first dinosaur fossil to be discovered in the United States, was thought to be lost but may have been found in Philadelphia's Academy of Natural Sciences. Wistar, the first American who was a professional vertebrate paleontologist, wrote the first American textbook of anatomy.

1788 James Hutton visits Siccar Point on the rocky Berwickshire coast near Scotland's border with England. The Siccar Point unconformity—a point where two sets of rocks formed at different times come into contact, and described by Hutton as "a junction washed bare by the sea"—shows basal, vertical sediments that are overlain by horizontal layers. Hutton knew that sediments are generally laid down horizontally in water, and concluded that the lower layers must therefore have been tilted and raised by pressure from below, such as during an earthquake. The tilted sediments were subsequently submerged and covered by additional sediments, after which seismic activity again raised the sediments. Hutton published his findings in the inaugural volume of *Transactions of the Royal Society of Edinburgh*. These observations helped Hutton appreciate Earth's vast age.

1788 Museo Nacional de Ciencias Naturales in Madrid premiers a display of *Megatherium*, the first skeletal mount in the world.

1792 Scottish science-writer Robert Kerr (1755–1813) applies a Linnaean name to *Elephas americanus*, a mastodon. This was the first application of a Linnaean name to an American fossil.

1794 Scottish geologist James Hutton's *An Investigation into the Principles of Knowledge* describes Earth as a "beautiful machine" designed for an intended effect. Buried in the 2,138 pages was a chapter about variety in nature that anticipated Charles Darwin's theory of natural selection. However, Hutton provided no data to support his suggestion, and his idea about life's diversity received little attention.

1795 The skeleton of *Megatherium* found near Buenos Aires, Argentina, is re-mounted in Madrid's Royal Cabinet by Jean-Baptiste Bru. The skeleton had first been mounted seven years earlier.

1795 German scholar, explorer, and savant Alexander von Humboldt (1769–1859) discovers fossils in a distinctive layer of limestone in the Jura Mountains in Switzerland. Four years later (i.e., in 1799), von Humboldt named the period corresponding to these fossils the Jurassic period. This period between the Triassic and Cretaceous was later identified as the "middle period" for the dinosaurs. Many of the most famous dinosaurs lived during the Jurassic, including *Apatosaurus* (*Brontosaurus*), *Stegosaurus*, *Brachiosaurus*, and *Diplodocus*. Dinosaurs thrived across the Earth during the Jurassic.

1795 James Hutton publishes his most famous and influential work, the two-volume, 1,100-page *Theory of Earth*, in which he claims that ongoing, endless cycles of erosion, sedimentation, and uplift are constantly, yet imperceptibly, changing Earth, thereby challenging the claim that Earth is only 6,000 years old. Hutton argued that it is unnecessary to invoke extraordinary processes to explain geology. With *Theory of the Earth*, Hutton opened geology to scientific observation, removed it from the influence of Bible-based chronologies (*Theory of Earth* does not mention Genesis), and gave geology its most transforming idea—namely, that Earth is ancient. Although Hutton's writings were full of deistic metaphysics and teleology ("The glove of this earth is evidently made for man"), his conclusions meant that vast periods of time were available for evolution to occur. Hutton's geology was characterized by perpetual change ("rest exists not anywhere") instead of stasis, and invoked no great catastrophes (e.g., a worldwide flood) to explain Earth's geology. Instead, Hutton claimed, "In examining things present we have data from which to reason with regard to what has been . . . The ruins of an older world are visible in the present structure of our planet." Hutton believed that Earth's age was "indefinite" and beyond comprehension, as suggested in the poetic and now-famous sentence that concluded *Theory of Earth*, "We find no vestige of a beginning,—no prospect of an end." Hutton's ideas, which became a foundation of modern geology, were later developed by Charles Lyell and became the basis of uniformitarianism, a foundation of modern geology.

1796 In a paper describing the extinct mammal *Megatherium,* French anatomist Georges Cuvier (1769–1832; Figure 2)—the founder of comparative anatomy—confirms the reality of extinction. To many people, Cuvier's documentation of extinctions, including mass extinctions, was heretical because it "broke" the Great Chain of Being, in which each species had its own ideal and unchanging structure, function, and position in nature. Cuvier knew that the history of life is recorded in layers of rocks containing fossils, that life on Earth had changed over time, and

Figure 2 French zoologist and comparative anatomist Georges Cuvier (1769–1832) founded vertebrate paleontology as a scientific discipline and was also one of the first people to suggest that prehistoric Earth was dominated by reptiles. Cuvier established extinction as fact and, in 1800, described "crocodilian" bones collected near Normandy, France, that were later determined to be *Megalosaurus,* a Jurassic theropod. Cuvier died at the age of 62 during an epidemic of cholera; his is one of the 72 names inscribed on the Eiffel Tower. (Library of Congress)

that life on ancient Earth was different from life today. To reconcile his findings with his religious beliefs, Cuvier claimed that the fossilized organisms in different layers of rocks had died in a series of God-directed catastrophes—that is, there had been "a world previous to ours, destroyed by some kind of catastrophe" and that "life on Earth has been disturbed by terrible events." Each of these catastrophes (Cuvier preferred the less theologically loaded term *revolutions*, which were analogous to the traumatic political changes in France) eliminated some species, and each catastrophe was followed by separate creations or immigrations of new species. Cuvier, who established the discipline of paleontology, realized that an alliance between stratigraphy and paleontology could help explain Earth's history. However, he rejected evolution, claiming that it was "contrary to moral law, to the Bible, and to the progress of natural science itself." No human fossils were known to Cuvier, who believed that humans came into existence after the biblical Flood. Cuvier's catastrophism was popular for several decades.

1796 Surveyor and canal-excavation engineer William Smith (1769–1839) observes that within layers of sedimentary rock, "each stratum contained organized fossils peculiar to itself, and might, in cases otherwise doubtful, be recognised and discriminated from others like it, but in a different part of the series, by examination of them." This is the first statement of the principle of faunal succession, a scientific theory stating that fossils in undisturbed strata of sedimentary rocks succeed each other vertically in a reliable, specific order over vast areas of land. This principle enabled Smith to use fossils from one part of England to predict and identify fossils in particular strata of other parts of England. Smith did not publish his idea until 1816 in *Strata Identified by Organized Fossils*.

1796 Thomas Jefferson, who in five years will become the third president of the United States, hires William Clark (later of Lewis and Clark fame) to search for remains of extinct mastodons in Kentucky. When Congress authorized the Lewis and Clark expedition of 1803, Jefferson helped finance the expedition and asked the explorers to send bones of large animals back to Washington. At one point, the unfinished East Room of the White House was filled with bones of ancient animals, and was called the "Mastodon Room." (George Washington was the first U.S. president to be interested in vertebrate paleontology; Washington kept in his house a mastodon "grinder which was found on the Ohio" River.) In 1797, while en route to his inauguration as vice president, Jefferson delivered a wagonload of fossils to the APS. The APS, founded in 1743 by Benjamin Franklin (who also had an interest in vertebrate fossils), was the nation's first scientific organization.

1797 A jaw and several teeth of what would later be named *Megalosaurus* are obtained by the Oxford Anatomy School at Christ Church College.

1797 Science patron Thomas Jefferson becomes president of the APS. During his 17 years as president of the APS, Jefferson promoted vertebrate paleontology.

In 1797, Jefferson read a paper at a meeting of the APS titled "A Memoir on the Discovery of Certain Bones of a Quadruped of the Clawed Kind in the Western Parts of Virginia"; that paper is among the first technical presentations on vertebrate paleontology in America. Jefferson's presentation was based primarily on three claws—each more than twice as large as those of African lions—found in a cave in western Virginia. Jefferson was succeeded as president of the APS by Caspar Wistar, who a decade earlier had found a "giant bone" that may have been from a dinosaur.

1799 Caspar Wistar publishes the first technical description of professional quality about American vertebrate fossils (*Megalonyx*). This was also the first such publication in America. Wistar's only other publication about vertebrate paleontology, which appeared in 1818, described more bones "found in the morass called the Big Bone Lick and presented to the [American Philosophical Society] by Mr. [Thomas] Jefferson." Despite Wistar's meager number of publications regarding vertebrate paleontology, famed paleontologist George Gaylord Simpson argued in 1942 that "in the long roster of American vertebrate paleontologists, the name of Caspar Wistar should stand first in line."

1799 Thomas Jefferson publishes a paper in *Transactions of the American Philosophical Society* describing "Megalonyx," a North American fossil ground-sloth similar to the extant South American sloth. Jefferson inaccurately concluded that the bones were those of a cat three-times larger than a lion. Jefferson's "Megalonyx" has no taxonomic validity; he used the word as a vernacular equivalent of "great claw" rather than as a Linnaean name. (*Megalonyx jeffersonii* was formally named in Jefferson's honor by Richard Harlan [1796–1843] in 1825.) Jefferson's famous paper (his only publication to identify a fossil) was one of the first American paleontology publications. Despite Jefferson's enthusiasm for fossil animals, he rejected extinction: "For if one link in nature's chain might be lost, another and another might be lost, till this whole system of things should vanish by piece-meal."

1799 William Smith circulates a paper describing how to trace "the strata by organized fossils embedded therein." That same year, Smith and others named more than 20 strata near Bath, England. At the junction of two strata they had named Pennant Street and Millstone, they noticed a distinctive change in the fossils: "Beneath this bed no fossil, shells, or animal remains are found; above it no vegetable impressions." Today, the lower strata with "vegetable impressions" are the Carboniferous, and the layers above the Permian and Triassic. The boundary between the Permian and Triassic marks the most catastrophic mass extinction in the fossil record.

1799 Workers discover a massive femur and other bones in the Hudson River Valley; this discovery triggers "Mammoth Fever" in the United States.

1800 French anatomist Georges Cuvier (Figure 2) first describes "crocodilian" bones from near Normandy, France. These bones were later determined to be *Megalosaurus*, a Jurassic theropod.

1801 Artist and showman Charles Willson Peale, the inventor of the modern natural history museum in America, excavates three mastodons near Newburgh, New York. Peale's display of one of these mastodons in Philadelphia stimulated the public's interest in fossils and was the first fossil skeleton to be mounted in America (and one of the earliest in the world). This began the close association of public museums and vertebrate paleontology. Peale, who is best remembered for his paintings of George Washington, founded several museums, many of which were later bought (and split up) by showman P. T. Barnum. In 1955, the U.S. Postal Service commemorated Peale's museums with a 3¢ stamp.

1801 Alexander von Humboldt argues "what we call the Atlantic Ocean is nothing else than a valley scooped out by the sea." Although von Humboldt noted the matching shapes and geological features of coastlines of opposing continents, he did not suggest how continents could have moved. Von Humboldt, whom Charles Darwin described as "the greatest traveling scientist who ever lived . . . I worship him . . . ," was one of the most prominent people in the world.

1801 William Smith, who two years earlier had published the first geological map (a circular map of the area around Bath), sketches the first geological map of an entire country. This 8.5-foot-by-6-foot map (scaled at 5 miles per inch), which was based on fossils and the mineral composition of rocks, became known as "the Map That Changed the World." When George Bellas Greenough began selling a cheaper map (that many claimed was stolen from Smith), Smith's income dropped and he was forced to sell his fossils to BMNH. In 1819, Smith was sent to debtor's prison for 11 weeks. However, by 1831, Smith had rehabilitated his reputation and was awarded a lifetime pension by King William IV and the Wollaston Medal of the Geological Society of London.

1802 James Hutton's friend John Playfair (1748–1819)—a famed mathematician and ordained minister of the Church of Scotland—publishes *Illustrations of the Huttonian Theory of the Earth*, a book that brought Hutton's principle of uniformitarianism to a wide audience. Playfair noted, "The theory of Dr. Hutton stands here precisely on the same footing with the system of Copernicus." Playfair's book, which urged full separation of religion and science, accompanied Charles Darwin aboard the *Beagle*. Ironically, *Illustrations* contained no illustrations.

1802 While plowing a field, Pliny Moody—a teenager living near South Hadley, Massachusetts—discovers a row of five huge tracks resembling those of three-toed birds. Religious authorities attributed the tracks to a giant, biblical bird called "Noah's Raven." (Noah had released a raven after the flood to determine if land had dried.) The tracks were later purchased by Elihu Dwight, and in 1839 were acquired by Edward Hitchcock of Amherst College, who had begun studying fossilized tracks from the area. The tracks were not recognized as dinosaurian until after 1860.

1803 American artist Rembrandt Peale's (1778–1860) 92-page *Historical Disquisition* references Georges Cuvier's work while dramatizing catastrophic extinctions for the public. Peale, who later exhibited a mastodon skeleton in his Baltimore museum, was acclaimed—like his father, Charles Willson Peale—for his portraits of historical figures such as George Washington, Alexander von Humboldt, and Thomas Jefferson. Peale's famed "Porthole" painting of Washington, titled *Patriae Pater* (Latin for *Father of Our Country*), was purchased by Congress in 1832 for $2,000 and now hangs in the Old Senate Chamber. Another of Peale's portraits of Washington hangs in the Oval Office.

1803 U.S. president Thomas Jefferson appoints Meriwether Lewis (1774–1809) and William Clark (1770–1838) to explore the uncharted American West and search for "the remains and accounts of any [minerals and animals] which may be deemed rare or extinct." Lewis and Clark expected to find mountains of salt, unicorns, giant beavers, and living mastodons. They found none of these, but did find fossils that challenged ideas about Earth's history. That same year, Jefferson and James Monroe negotiated the purchase of more than 820,000 square miles of France's Territory of Louisiana for $15 million (approximately 3¢ per acre). This so-called "Louisiana Purchase" encompassed all or part of 14 U.S. states and two Canadian provinces, including several areas (e.g., Montana, Wyoming) that later yielded numerous iconic dinosaurs collected by Barnum Brown, Othniel Marsh, Edward Drinker Cope, and other famous dinosaur-hunters. The Louisiana Purchase, which removed the ability of France and Spain to block the port of New Orleans, was commemorated on the Louisiana state quarter in 2002, a 10¢ U.S. stamp in 1904, and a 37¢ U.S. stamp in 2003.

1804 Georges Cuvier (Figure 2), a devout Christian who considered Genesis as history, ponders Earth's age when he claims that fossils found near Paris are "thousands of centuries" old. In a separate paper, Cuvier again confirmed extinction by describing fossilized animals unlike anything still living. Cuvier believed that animals' similarities resulted from common functions, not common ancestry, and that an animal's function was determined by its form (his so-called "correlation of parts"). Cuvier, who claimed that nature had a "repugnance" for changes in design, was convinced that every species needs all of its parts just the way they are, and that any change would cause death.

1804 In the first volume of a three-volume work titled *Organic Remains of a Former World*, James Parkinson (1755–1824)—the discoverer of Parkinson's disease and a founder of the Geological Society of London—catalogs all of the known fossil reptiles as the remains of Noah's Flood. He later referred to fossils as the remains of a pre-human world. Parkinson, who treated the days of creation as vast periods of time, united Genesis with Cuvier's catastrophism. The last volume of *Organic Remains of a Former World* appeared in 1811.

1804 Lewis and Clark begin their expedition to the American West. In what is now Montana, Clark later named the fossil-rich Judith River after a young woman he later married. In 1854, the first remains of dinosaurs in the Western Hemisphere were found in central Montana; Montana is also where the first dinosaur eggshells in the Western Hemisphere, the first nest of baby dinosaurs in the world, and the first dinosaur embryos in the world were discovered. One-fourth of Montana's land exposes Mesozoic rock.

1806 Charles Willson Peale's (1741–1827) museum in Philadelphia, Pennsylvania, displays a mounted mastodon. Peale's exhibit was the first mounted skeleton of a vertebrate in the United States; as Rembrandt Peale (Charles's son) noted, this was "the first of American animals in the first of American Museums." The legs of this creature are visible in Peale's 1822 painting *The Artist in His Museum*. Peale's famous museum was later sold to P. T. Barnum and turned into a penny arcade. The museum's historic skeleton was destroyed by fire.

1806 William Clark finds an exposed rib in the Hell Creek Formation along the Yellowstone River nearing Billings, Montana. Clark believed the bone was a rib from a big fish, but some paleontologists later claimed that this was the first documented discovery of a dinosaur fossil in North America. The next year, Clark and 10 workers excavated 300 fossil bones, which they shipped to Thomas Jefferson (and which he exhibited in the White House the following year). Although many people later noted Jefferson's contributions to paleontology, Jefferson was not a vertebrate paleontologist, never a collected a vertebrate fossil, never gave a technical name to a vertebrate fossil, and made only meager contributions to science. However, Jefferson's interest in fossils helped popularize vertebrate paleontology in the United States.

1807 After two Yale geologists claim that two meteors had recently hit Earth, President Thomas Jefferson claims that he "would rather believe that those two Yankee professors would lie than to believe that stones fell from heaven." In 1980, Luis and Walter Alvarez claimed that a meteor had triggered the Cretaceous-Tertiary (K-T)[1] mass extinction that eliminated the non-avian dinosaurs.

1807 The Geological Society of London—the first scientific society for geologists—is established to promote discoveries, acquaint geologists with each other, and stimulate "their zeal." The society originally was a dinner club that met at an upscale tavern to isolate itself from lower levels of society, but in 1825 it began admitting any man interested in geology. The society included many geologists who discovered dinosaurs, including William Buckland (1811), William Conybeare (1811), and Gideon Mantell (1820). In 1859, the society awarded Charles Darwin (who had joined in 1836) its highest honor.

1807–1809 While doing a geological survey of the British Isles, William Smith discovers fossil bones of what would later be named *Iguanodon* from Cuckfield in Sussex. This is one of the earliest records of a dinosaur discovery.

1809 A vertebra of a sauropod is found near Dorchester-on-Thames, Oxfordshire, and given to the Woodwardian Museum at Cambridge. This was the first discovery of a sauropod bone.

1809 Charles Robert Darwin (1809–1882) is born into a life of privilege in Shrewsbury, Shropshire, England. Darwin shared his birthdate with Abraham Lincoln, both of whom later appeared on the currency of their countries. In 1859, Darwin's *On the Origin of Species* famously proposed evolution by natural selection, which philosopher Daniel Dennett (b. 1942) later called "the single best idea anyone ever had."

1809 Jean-Baptiste Lamarck (1744–1829) publishes his most famous book, *Philosophie Zoologique*, in which he shocks Cuvier and much of Europe by proposing life's "tendency to progression" and "tendency to perfection." Lamarck claimed that the fixity of life is an illusion, that life is instead in a constant state of advancement and improvement—"there is no species which is absolutely constant"—driven by spontaneous generation, and that this advancement is too slow to be perceived except with the fossil record. Lamarck argued for the inheritance of acquired traits and claimed that evolution is driven by organisms' needs as they strive to fulfill their way of life. According to Lamarck, species are little more than arbitrary points along a continuum, and environmental changes alter the needs of organisms living in those environments. In turn, these altered needs changed the organisms' behaviors, and these altered behaviors then led to the greater or lesser use of structures (this change in structural utility occurred through the involvement of a "nervous fluid"). The more an organism used a part of its body, the more that part developed (similarly, disuse of a part resulted in its decay). Lamarck was his era's most renowned advocate of evolution, and his godless model for evolution was the first testable hypothesis to explain how a species could change over time. Lamarck was exceedingly confident about his idea, referring to it as a "permanent truth, which can only be doubted by those who have never observed or followed the operations of nature." However, Lamarck's conclusions were not accompanied by a persuasive mechanism, which prompted many of the leading scientists of Lamarck's time to reject his ideas. Nevertheless, Lamarck's proposals were popular with the public (they reinforced the self-congratulatory Victorian emphasis on progress produced by diligence and hard work), so much so that Charles Darwin alluded to them in later editions of his *On the Origin of Species*.

Soon after the publication of *Origin*, there was a revival of Lamarckism, but the claims associated with this resurgence were based on flawed experiments and fraudulent data (e.g., the "midwife toad" fraud in 1919). Although Darwin privately described *Philosophie Zoologique* as "veritable rubbish," he later acknowledged that Lamarck was a "justly celebrated naturalist" who "was the first man whose conclusions on the subject excited much attention." Lamarck's idea was later often depicted in textbooks as giraffes stretching to reach leaves atop trees, with the trait of lengthened necks then being passed to their offspring. However, *Philosophie Zoologique* included relatively little about giraffes. Lamarck's ideas were often invoked by paleontologists to describe the evolution of dinosaurs.

1811 Twelve-year-old Mary Anning (1799–1847) unearths the first complete *Ichthyosaurus* from the limestone (Mesozoic) cliffs near her home in the southern England coastal town of Lyme Regis. The self-taught Anning became one of the most famous fossil hunters in history, finding an almost perfect skeleton of *Plesiosaurus* (which became the basis for one of the "reconstructions" of the Loch Ness Monster) and the first *Pterodactylus* (later renamed *Dimorphodon marconyx* by Richard Owen). Although Anning published only one paper ("Note on the Supposed Frontal Spine in the Genus Hybodus," which appeared in 1839 in the *Magazine of Natural History*), she became a legendary figure in paleontology. Anning's many discoveries, some of which are displayed today at BMNH, helped scientists decipher the geological record, and convinced people of the reality and extent of past extinction events. Despite their fame and importance, none of Anning's early discoveries were named after her.

1812 Georges Cuvier's *Researches on Fossil Bones* begins modern vertebrate paleontology. That same year, Cuvier discovered that some strata in the Paris Basin contain fossils of marine mammals, others contain fossils of freshwater animals, and others bear no fossils at all. Cuvier concluded that these ancient forms of life had appeared and disappeared suddenly due to the effects of catastrophes. After each catastrophe, there were new creations and immigrations of organisms that had avoided the catastrophes because they lived elsewhere. In *Discourse on the Revolutionary Upheavals on the Surface of the Earth and on the Changes Which They Have Produced in the Animal Kingdom*, Cuvier noted, "Man, to whom a mere instant has been granted on earth, would have the glory of reconstructing the history of thousands of centuries which predated his existence, and thousands of beings which have never been his contemporaries . . . Would it not be glorious for geologists to burst the limits of time?" Cuvier concluded, "Lands once laid dry have been reinundated several times . . . Life in those times was often disturbed by these frightful events. Numberless living things were victims of such catastrophes: some inhabitants of the dry land, were engulfed in deluges; others, living in the heart of the seas, were left stranded when the ocean floor was suddenly raised up again; and whole races were destroyed forever, leaving only a few relics which the naturalist can scarcely recognize." Cuvier also correctly identified the first known fossil remains of a pterodactyl, and claimed that pterosaurs were flying reptiles. In 1812, Cuvier published *Inquiry into Fossil Remains*—the first major work of paleontology—in which he described extinct species and classified them with Linnaeus's system, but claimed that they were the products of earlier, separate creations triggered by catastrophes instead of ancestors of living organisms. Cuvier's views became known as *catastrophism*. Cuvier was not the first person to discover or describe vertebrate fossils, but he was the first person to reasonably explain them.

1812 Gideon Algernon Mantell (1790–1852), whose work began the scientific study of dinosaurs, publishes his first geological essay, "On the Extraneous Fossils Found in the Neighborhood of Lewes." (Lewes is a town near the southern coast of England where Mantell lived.)

1812 The Academy of Natural Sciences (ANS), the oldest natural science institution in the Western Hemisphere, is founded in Philadelphia, Pennsylvania, "for the encouragement and cultivation of the science, and the advancement of useful learning." The academy opened its doors to the public in 1828, and has been at its present location since 1876. ANS was responsible for several important discoveries of dinosaurs, and in 1868 was the first place in the world to display a fully articulated dinosaur. In 2011, ANS of Philadelphia linked with Drexel University to become the Academy of Natural Sciences of Drexel University.

1814 Mary Anning's fossils of *Ichthyosaurus*, which inspired the early interest in creatures from "former worlds," are displayed at BMNH. Anning's fossils are in Gallery 30.

1814 The second of Charles Willson Peale's skeletons of mastodon is mounted in Rembrandt Peale's Baltimore Museum.

1815 Flamboyant cleric and professor William Buckland (1784–1856; Figure 3), the first Reader in Mineralogy at Oxford University (for an annual salary of £100), discovers an unusual pelvis, hind limbs, vertebrae, jaw, and teeth in a quarry at Stonefield. Buckland did not describe the discovery, but kept the specimens. In 1818, Cuvier visited Buckland at Oxford, and three years later, Irish naturalist Joseph Pentland (while working in Cuvier's lab) asked Buckland in a letter, "Will you send your Stonefield reptile or will you publish it yourself?"

1815 Inspired by discoveries by Mary Anning, English obstetrician and paleontologist Gideon Mantell publishes his first paper about fossils of the Lewes area. Mantell later discovered that the forelimbs of *Iguanodon* were shorter than its hind limbs, and therefore that the dinosaurs were not built like mammals, as claimed by Richard Owen.

Figure 3 In 1824, British geologist, cleric, and paleontologist William Buckland (1784–1856) produced the first scientific description of a dinosaur, which he named *Megalosaurus*. In this lithograph, Buckland is holding a large bone, with other fossils and skulls on the bench beside him. Buckland had eclectic interests, and, in 1845, became dean of Westminster. (SSPL/Getty Images)

1815 William "Strata" Smith's *A Delineation of the Strata of England and Wales with part of Scotland* establishes modern stratigraphy, the scientific discipline of describing and interpreting rock successions and their characteristic fossils. Although Smith had little interest in determining the ages of rocks he examined (he accepted the biblical creation story), his work transformed the work of Steno and Hutton into practical geology while providing key insights to modern evolutionary studies.

1816 William Smith's *Strata Identified by Organized Fossils* announces his famous principle of faunal succession—that is, that there are fossils "peculiar to each stratum." Smith, who claimed that his method of tracing strata by the fossils they contained was a "science not difficult to learn," noted "each layer of these fossil organized bodies must be considered as a separate creation, or is an undiscovered part of an older creation."

1817 The word *paleozoologie* is coined by French anatomist Henri Marie Ducrotay de Blainville (1777–1850), a student of Georges Cuvier and editor of the French scientific journal *Journal de Physique*. De Blainville used the word *paleontologie* (later Anglicized to *paleontology*) to describe the study of ancient animals by using fossils and to refer to Cuvier's reconstructions of extinct animals from fossil bones.

1818 Bone fragments from the Connecticut Valley are considered to be human, and later (1855) to be reptilian.

1818 Georges Cuvier's *Essay on the Theory of the Earth* uses the fossil record to confirm extinctions and popularize catastrophism. When Cuvier's book was translated into English by Edinburgh University's Robert Jameson (1774–1854), Cuvier's idea was presented as if the most recent catastrophe was the biblical flood. Later, William Buckland praised Cuvier's "inestimable Essay" and hoped to show that geological sediments were produced by the flood. In 1818, Cuvier examined *Megalosaurus* and concluded that it was a mammal. Cuvier later admitted that he was wrong. *Megalosaurus* was a dinosaur.

1818 Yale chemist Benjamin Silliman (1779–1864), the first person to distill petroleum, establishes the *American Journal of Science*. Although this journal was not the first scientific journal in the United States (that was Samuel Mitchill's *The Medical Repository*, founded in New York in 1797), *American Journal of Science* was the most influential journal for early vertebrate-paleontological discoveries in North America, and its first issue included instructions for collecting fossils. Silliman, who published several studies of dinosaurs, was the first person to scientifically describe an American meteorite, and is commemorated by a statue in front of Yale's Sterling Chemistry Laboratory. Yale's Silliman College is named for him.

1818 Tracks and bones of the prosauropod *Anchisaurus* are found by Solomon Ellsworth Jr. in Late Triassic sandstone while blasting a well in the Connecticut Valley. Ellsworth gave the fossils to Yale College. Some of these tracks are preserved in Yale's Peabody Museum of Natural History (YPM, as Specimen 2125); these are the earliest-discovered dinosaur specimens still preserved in a museum in North America (casts of some of these bones are also displayed at Dinosaur State Park in Rocky Hill, Connecticut). In 1820, Nathan Smith (1762–1829, who organized the Yale University Medical School) described these bones in *The American Journal of Science*, noting "they might be human bones," but did not consider the specimens as "sufficiently distinct to form the basis of a certain conclusion." In 1857, Professor Jeffries Wyman (1814–1874) recognized the bones as reptilian, and in 1915, the bones were recognized as dinosaurian and named *Anchisaurus colurus* by Richard Lull (1793–1864). Rev. Edward Hitchcock—the third president of Amherst College—described tracks from the same area that were later determined to have been made by a dinosaur.

1820 Eccentric Anglican clergyman William Buckland (1784–1856; Figure 3) returns from Europe convinced that similar successions of geologic strata in continental Europe and in Britain contain similar fauna. That same year, Buckland's *Vindiciae Geologicae; or the Connexion of Geology with Religion Explained* used gap creationism to reconcile geology with Noah's Flood and biblical creation. Buckland claimed that the word *beginning* in Genesis referred to an unspecified period between the formation of the Earth and the creation of today's organisms, during which there were many extinctions and successive creations. Buckland's so-called *diluvium theory*, which was based on Cuvier's "successive catastrophes theory," was embraced by theologians searching for evidence of a worldwide flood. In 1824, Buckland published the first scientific description of a dinosaur.

1820 Fragments of feathers in Upper Jurassic limestone in Germany are mentioned in a book, and seven years later, a bird's head is reported from the same sediments. These specimens were lost before being formally described.

1821 Gideon Mantell shows some of his fossils to the Geological Society of London, but his display arouses little interest. Most of Mantell's colleagues believed that a curious tooth (later shown to be from *Iguanodon*) found by Mantell's wife came from a mammal or fish.

1821 William Buckland (Figure 3) finds large bones on the Isle of Wight, which are reported as being from whales. However, Buckland later described the bones as being from the bulky herbivore *Iguanodon* ("iguana tooth"). The following year, Buckland claimed that he discovered an ancient human fossil belonging to a witch (he later named the skeleton "Red Lady of Paviland").

1821 William Conybeare (1787–1857) and Henry Thomas de la Beche (1796–1856) publish the first complete anatomical description of an ichthyosaur (a fossil marine reptile). This paper, which mentioned Buckland's "Huge Lizard," was followed three years later by the first reconstruction of an ichthyosaur. At about this time, Conybeare (a geologist who coined the term *plesiosaur*) urged Buckland to name the fossil *Megalosaurus*. The following year, James Parkinson's *Outlines of Oryctology: An Introduction to the Study of Fossil Organic Remains* described *Megalosaurus* (including a sketch of one tooth) as "an animal apparently approaching the Monitor in its mode of dentition, and not yet described . . . The animal must in some instances, have attained a length of 40 feet, and stood eight feet high." Although this was not a scientific description, Parkinson is sometimes credited with naming *Megalosaurus* (the first dinosaur to be named).

1822 Belgian statesman and amateur geologist Jean Baptiste Julien D'Omalius d'Halloy (1783–1875) names the so-called *French Chalk* that is older than the Cenozoic and younger than the Jurassic the *Terrain Crétacé* (Cretaceous System), thus providing the source of the name for the Cretaceous period (*creta* is Latin for *chalk*). The Cretaceous (i.e., between the Jurassic and Tertiary) later became associated with the earliest flowering plants and the last dinosaurs. Famous dinosaurs that lived during the Cretaceous included *Hadrosaurus, Triceratops*, and *Tyrannosaurus*. Non-avian dinosaurs lived during the Cretaceous on all continents, but disappeared at the end of the Cretaceous.

1822 Gideon Mantell (1790–1852) and Mary Ann (Woodhouse) Mantell (1795–1869; his 26-year-old wife of six years) discover several large fossil bones and teeth "distinct from any that had previously come under [our] notice" in a quarry near Whiteman's Green in Cuckfield, England. Renowned anatomists Richard Owen and Georges Cuvier dismissed the finds as belonging to known mammals. While examining collections in London's Hunterian Museum (at the Royal College of Surgeons), Mantell met naturalist and assistant curator Samuel Stutchbury (1798–1859), who showed Mantell a recently prepared skeleton of an iguana. Mantell believed that the teeth he'd found were similar to those of the iguana—but 20-times larger—and, in a paper read before the Royal Society, Mantell named his creature *Iguanosaurus*. Soon thereafter, however, Mantell received a letter from geologist William Conybeare criticizing that name: "Your discovery of the analogy between the Iguana and the fossil teeth is very interesting—but the name you propose *Iguanosaurus* will hardly do" because *Iguanosaurus* ("iguana lizard") was equally applicable to modern iguanas. Conybeare suggested several other names, including *Iguanodon* ("iguana tooth"). In 1825, Mantell's discovery was announced to the Royal Society of London by Davies Gilbert, the vice president of the society (Mantell did not attend), and published in the society's journal (*Philosophical Transactions of the Royal Society of London*). By this time, Cuvier—admitting "I am quite convinced of my mistake"—agreed that Mantell's fossils were from an unknown monster (as Mantell later recalled, Cuvier asked, "Do you not agree that it looks as if we have here a new animal, an herbivorous reptile?").[2] Mantell's find—announced "in the hope that, imperfect as are the materials at present collected, they will be found to possess sufficient interest to excite further and more successful investigation, that may supply the deficiencies which exist in our knowledge of the osteology of this extraordinary animal"—was the second formally recognized dinosaur (although the word *dinosaur* would not be coined by Owen until 1842). In 1822, Mantell's well-illustrated *The Fossils of the South Downs; or Illustrations of the Geology of Sussex*—the first publication dealing primarily with giant fossil reptiles—described several teeth and bone fragments from "unknown animals" collected in Tilgate Forest: "The teeth, vertebrae, bones and other remains of an animal of the lizard tribe, of enormous magnitude, are perhaps the most intriguing fossils that have been discovered in the county of Sussex." Mantell considered *The Fossils of the South Downs*—his first book—to be his best, and spent the rest of his life updating the book. Although Mantell had a publisher, he paid all of the production costs himself; the book sold only 152 copies, and Mantell lost £300 on the book.[3] Mantell—who geologist Robert Bakewell claimed would "ride on the back of his *Iguanodon* into the Temple of Immortality"—became obsessed with dinosaurs, and in 1839, his wife left him and his medical practice failed. Facing mounting debt, Mantell sold his fossils to BMNH. Mantell was involved in a carriage accident in 1841 that damaged his spine; four years later, Mantell began taking opium for the pain caused by the severe lumbar scoliosis produced by his accident. Near the end of his life, Queen Victoria gave Mantell a civil gratuity of £100 per year. When Mantell died in 1852 of an opium overdose, Owen removed part of Mantell's spine, which was displayed at the Hunterian Museum of London's Royal College of Surgeons to show "the severest degree of deformity" (the spine and an accompanying cast were destroyed in 1969 because of lack of space). In 1949, civic leaders in Maidstone (a village near where the Mantells discovered their fossils) incorporated the Mantells' dinosaur into the town's coat of arms.

1823 Physician Richard Harlan (1796–1843), America's first professional vertebrate paleontologist, becomes the first American to apply Linnaean names to fossil vertebrates. Harlan, who was familiar with the evolutionary ideas of Lamarck and Erasmus Darwin, believed that species were immutable. However, he cared little for religious authority, noting, "All theological considerations aside, we would be inclined to take the facts as they are." In 1824, Harlan named the only fossil that survived from the Lewis and Clark expedition, a fish named *Saurocephalus lanciformis*, which had been found in 1804 (Specimen 5516 at the ANS). Harlan's short, but bitter, feud with physician Isaac Hays (1796–1879) about the classification of fossils was a prelude to the more famous "Bone War" between Edward Drinker Cope and Othniel Marsh.

1823 Reverend William Buckland's *Reliquiae Diluvianae, or Observations on the Organic Remains Contained in Caves, Fissures, and Diluvial Gravel Attesting the Action of a Universal Deluge*, argues that a flood had covered the Earth and marks the boundary of today's human world and the pre-human world of the remote past. Buckland—Oxford's first geologist—later noted that it is "impossible to ascribe the formation of these strata to . . . the single year occupied by the Mosaic deluge . . . The strata [document] periods of much greater antiquity." That same year, Buckland notified Cuvier that he would soon publish his studies of the "Stonesfield Monitor": "My Dear Baron, Herewith I send you Proof Plates of the great Animal of Stonesfield, to which I mean to give the name of Megalosaurus & which I shall publish in . . . the Geological Transactions."

1824 In the same year that he became president of the Geological Society, and at the same meeting of the Geological Society of London at which Mary Anning described and displayed some of her famous fossils of plesiosaurs, William Buckland publishes the first scientific description of a dinosaur. In a paper titled "Notice on the *Megalosaurus* or Great Fossil Lizard of Stonesfield," which appeared in *Transactions of the Geological Society of London*, Buckland wrote "Whilst the vertebral column and extremities much resemble those of quadrupeds, the teeth show the creature to have been oviparous, and to have belonged to the order of Saurians or Lizards . . . [This creature] would have equaled in height our largest elephants, and in length fallen but little short of the largest whales[4] . . . I have ventured, in concurrence with my friend and fellow-labourer, the Rev. W. Conybeare, to assign it the name *Megalosaurus*." (In 1826, Cuvier gave the dinosaur a binomial: *Megalosaurus bucklandi*, or "Buckland's Big Lizard.") Buckland based his conclusions primarily on part of a lower jaw (with teeth) and some postcranial bones. (Robert Plot, a chemistry professor at Oxford University, had described a similar bone in 1677, but did not recognize it as belonging to a dinosaur.) Bones of *Megalosaurus* had been found earlier by Gideon Mantell at Cuckfield, but had not been described. Buckland, whose work led to the recognition of dinosaurs as extinct reptiles, acknowledged a scholarly debt to Cuvier by adding "a length exceeding 40 feet and a bulk equal to that of an elephant seven feet high have been assigned by Cuvier to the individual to which [the thigh bone] belonged." Buckland didn't recognize that the serrated teeth of *Megalosaurus* were set in sockets, as are those of crocodiles (in contrast, the teeth of lizards are attached directly to the jawbone). When Buckland later discovered coprolites (fossilized excrement), he suggested that carnivores had existed since the Creation, but later realized that this contradicted a consequence of Original Sin. Buckland rationalized this by claiming that carnivores are an example of God's benevolence because carnivores reduced the suffering of old and sick individuals by killing them. Buckland did not know that his discovery was from a dinosaur, instead believing that it was from a large, extinct reptile. By 1837, Buckland—who tried to reconcile science with biblical accounts of creation and the Noachian Flood—was a science celebrity. Replicas of *Megalosaurus* footprints adorn the lawn of the Oxford University Museum of Natural History.

1824 After earlier describing "a large tooth, which from the worn, smooth, and oblique surface of the crown had evidently belonged to a herbivorous animal," Gideon Mantell asks friend Charles Lyell to take his curious fossils to Paris for examination by famed anatomist Georges Cuvier. The tooth was large and worn, suggesting it came from a herbivore, but at that time no mammals were known from the Cretaceous. And if the tooth was from a reptile, there were no such reptiles known from any geologic age. After examining the fossils, Cuvier tentatively identified the tooth as being from a rhinoceros, but later told Mantell "these teeth are most certainly unknown to me." Mantell also noted, "No known existing reptiles are capable of masticating their food, [so] I could not assign the tooth in question to a saurian." The fossilized teeth collected by Mantell later became the means of identifying *Iguanodon*, a discovery that stirred much public interest. William Buckland considered *Iguanodon* to be "the most gigantic of all quadrupeds that have ever trod upon the surface of our planet."

1824 Richard Owen suggests that *Megalosaurus* is an amphibian. *Megalosaurus* was originally believed to have resembled a giant lizard.

1824 The U.S. Army Corps of Engineers is authorized by Congress to survey roads and canals. This was the first federally-funded geologic and geographic survey of the United States.

1825 French comparative anatomist Georges Cuvier (Figure 2) declares that the Mesozoic marine reptile *Plesiosaurus* was "the most heteroclite . . . and the most monstrous [animal], that had yet been found amid the ruins of the former world." That same year, Cuvier's admission of his mistake about *Iguanodon*, along with the name *Iguanodon*, was first published in a letter titled "Notice on the Iguanodon, a Newly Discovered Fossil Reptile, from the Sandstone of Tilgate Forest, in Sussex." This letter, which was from Gideon Mantell to his friend David Gilbert, was printed in *Philosophical Transactions of the Royal Society*. (The letter, which describes seven teeth, had to be presented to the Royal Society by its treasurer because Mantell was not yet a member.) Mantell's *Iguanodon*—an "extraordinary animal" that was not given a species name—was the second dinosaur (and the first herbivorous dinosaur) to be described. Mantell estimated that *Iguanodon* could have been 200 feet long, but later modified his estimate to 60 feet. *Iguanodon* became one of the most famous dinosaurs, appearing in Arthur Conan Doyle's *The Lost World*, an episode of *Dinosaur Planet*, the television documentary miniseries *Walking with Dinosaurs*, and other books and films.

1825 Gideon Mantell sketches what *Iguanodon* might have looked like (Figure 4). Mantell never published his sketch, but it likely represented the first drawing of a dinosaur. *Iguanodon* was the first plant-eating dinosaur, and the

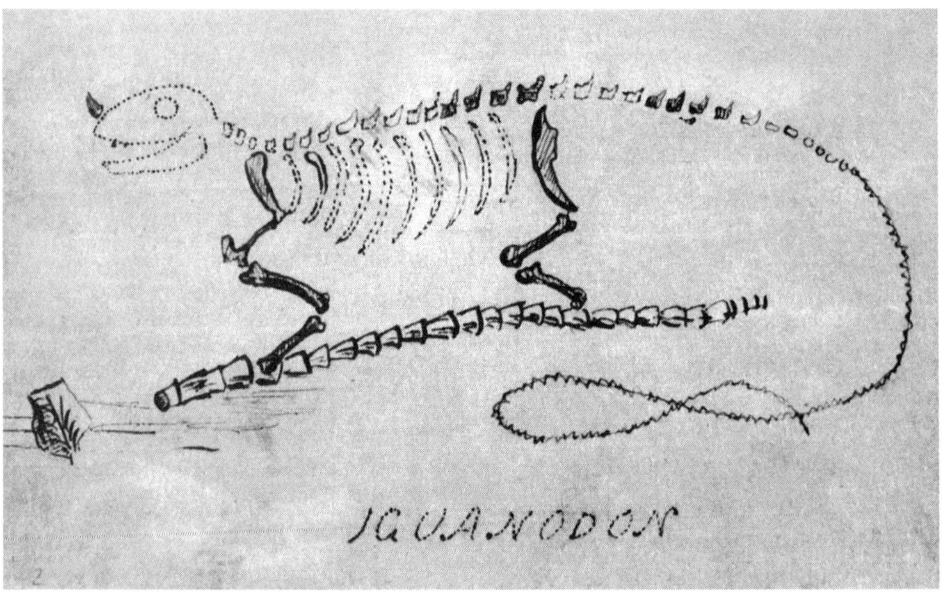

Figure 4 British obstetrician, geologist, and paleontologist Gideon Algernon Mantell's (1790–1852) reconstruction of the structure and life of *Iguanodon* began the scientific study of dinosaurs. Mantell, whose *Iguanodon* was the second dinosaur to be described, discovered four of the first five genera of dinosaurs then known. (Louie Psihoyos/Corbis)

first ornithischian dinosaur, to be discovered. Mantell's discovery of *Iguanodon* secured his election as a fellow of the Royal Society. Mantell, like Buckland, viewed dinosaurs as lizard-like animals; people had no background or experiences to think of them any other way. Buckland, however, did not believe that Mantell found an ancient lizard; instead, Buckland claimed that Mantell's tooth had fallen from a younger stratum and was therefore probably from a mammal or fish. Mantell disliked Buckland, often describing Buckland's behavior as "unmerited" and "unhandsome."

1826 George Bugg's (1769–1851) massive *Scriptural Geology; or, Geological Phenomena Consistent Only with Literal Interpretations of Sacred Scriptures* attacks the antediluvian ideas of Cuvier and Buckland. Bugg argued, "The SCRIPTURAL ACCOUNT of the DELUGE, will alone account for the phenomena of the fossil strata."

1827 Gideon Mantell's 92-page *Illustrations of the Geology of Sussex: The Fossils of Tilgate Forest* confirms that at least four saurians lived in the Weald: ancient crocodiles, *Megalosaurus, Iguanodon*, and *Plesiosaurus*. Mantell's book, which included 22 plates, had a frontispiece showing Mantell, Lyell, and Buckland at Whiteman's Green Quarry. *Illustrations of the Geology of Sussex* was the first book devoted primarily to saurians. Mantell's expensive book, which was subsequently described as "the rarest and most historic dinosaur book in English," attracted little interest; only 50 of the 150 copies were sold.

1828 While working near Normandy, Arcisse de Caumont (1801–1873) finds the first dinosaurs in France (*Megalosaurus*).

1828 William H. Sleeman (1788–1856) discovers sauropod vertebrae near Jabalpur, India. Richard Lydekker subsequently named Sleeman's discovery *Titanosaurus*, but this identification is now questioned. Sleeman's fossils were the first dinosaurs to be discovered in central India.

1829 Gideon Mantell opens his fossil-filled home to the public as a museum, which later became known as the Sussex Scientific Institution and Mantellian Museum. Geologist Robert Bakewell (1768–1843) noted, "It is the remains of large animals evidently formed for walking on land that renders the museum of Mr. Mantell so unique." One side of the museum displayed Mantell's fossils of *Iguanodon* and other bones of reptiles "of enormous magnitude." Mantell, however, wrote "In truth, I am now sick of the cold-blooded creatures I am surrounded by." Mantell later sold his museum to BMNH.

1829 When James Louis Macie Smithson (1765–1829; the nephew of British scientist James Smithson) dies in Genoa, Italy, without an heir, Smithson's will leaves his estate (valued at $508,318) "to the United States of America, to found at Washington, under the name of the Smithsonian Institution, an establishment for the increase and diffusion of knowledge among men." Although Smithson had never visited the United States, the Smithsonian Institution became the world's largest museum and research complex. The U.S. Congress gave Smithsonian Institution control of the National Cabinet of Curiosities; these became part of the National Museum of Natural History (NMNH), one of 19 museums in complex. The Smithsonian Institution, which opened in 1910 under the directorship of Joseph Henry, later housed many important fossils of dinosaurs, several of which were collected by Othniel Charles Marsh. For example, in the Dinosaur Hall, visitors encounter the 21-foot-long *Allosaurus fragilis*, which was collected by Marsh's explorers in 1883. This specimen—the most complete *Allosaurus* now displayed anywhere in the world—was featured on the cover of cartoonist Gary Larson's *The Prehistory of the Far Side* (1989). Today, NMNH is the world's most visited museum of natural history; attendance exceeds seven million people per year.

1829 Noting Mary Anning's observations of "bezoar stones" found in the abdomens of ichthyosaurs, William Buckland coins the term *coprolite* to describe fossilized feces.

1830 English geologist Henry Thomas de la Beche's *A More Ancient Dorset* is the first reconstruction of a paleo-ecosystem. De la Beche's watercolor was based on evidence from Jurassic fossils. De la Beche, who was one of the few scientists who acknowledged Mary Anning's contributions to paleontology, was among the first to visually convey complex paleontological ideas to the public. Some of the creatures in de la Beche's painting are defecating (coprolites were discovered where the skeletal remains were found).

1830 Charles Lyell (1797–1875) publishes the first (of three) volumes of his monumental and ambitiously titled *Principles of Geology: Being an Attempt to Explain the Former Changes of the Earth's Surface by Reference to Causes now in Operation*. In *Principles*, a name chosen to echo Isaac Newton's *Principia* (1687), Lyell used the work of James Hutton and John Playfair as a starting point to describe an ancient Earth whose features had been formed by "the slow agency of existing causes" operating for long periods of time. Although Lyell did not dispute the occurrence of local catastrophes, he popularized and refined Hutton's ideas about uniformitarianism while showing that Earth is something far more than the natural home for humans; rather, humans have been on a ceaselessly changing Earth for only a tiny portion of its history. Although Hutton had dealt only with the physical aspects of geology, Lyell realized that geology would be incomplete if it ignored fossils, which record the changes encountered by life on Earth. Lyell suggested that the periodicity in climate would produce cycles of species through time, so that "huge *Iguanodon* might reappear in the woods, and the ichthyosaurs in the sea." *Principles* was the standard reference for geologists for several decades. Although Lyell encouraged Darwin to publish *On the Origin of Species*, he initially rejected Darwin's claims about life's history, and was the most reluctant of Darwin's confidants to publicly support evolution (early editions of *Principles* argued that organisms had been created perfectly adapted for local conditions). However, Lyell's own discoveries validated Darwin's ideas, prompting Thomas Huxley (1825–1895) to note that Lyell was "doomed to help the cause he hated." By the 10th edition of *Principles* (1867), Lyell endorsed Darwin's theory, noting its "strong presumption in favour of the truth." Lyell linked an incomprehensibly old Earth with natural theology, but not with biblical cosmology: "We discover everywhere the clear proofs of a Creative Intelligence, and of His foresight, wisdom, and power." However, by the end of his life, Lyell—who believed that scriptural geologists were "wholly destitute of geological knowledge"—questioned traditional theism. *Principles of Geology*—the most influential geology book in history—forever raised geology's public image to a science by synthesizing vast amounts of data with elegant and compelling rhetoric.

1831 Anatomist Richard Owen (Figure 5), who will invent the clade Dinosauria in 1842, publishes his first zoological paper in *Proceedings of the Zoological Society of London*. The paper was titled "On the Anatomy of the Ourang-outang."

1831 Gideon Mantell describes the Mesozoic as the "Age of Reptiles."

1831 Othniel Charles Marsh is born in Lockport, New York. Marsh, a founder of vertebrate paleontology in America, published almost 300 papers, reports, and books between 1861 and 1899, describing 225 new genera, 469 species, 64 families, 8 suborders, and 1 subclass. Marsh's "bone war" with Edward Drinker Cope—one of the great sagas of science—stimulated the public's interest in dinosaurs and produced many remarkable discoveries (e.g., at the start of the feud, there were nine named species of dinosaurs in North America; at the end, there were almost 150). Marsh and Cope were both rich, well connected, ambitious scientists, and their bitter dispute damaged both participants while producing often-dubious taxa; many of their new species were described in rushed, poorly-documented papers based on fragments of bones, poor illustrations, and inadequate descriptions of the animals (e.g., Cope named five species of *Laelaps* based on minor differences in teeth from a small area of the Montana's Judith River Formation). Several of these publications produced little more than confusion and misconceptions that linger today. Although Marsh spent few seasons in the field, he named 80 species of dinosaurs (Cope named 64); this total, which exceeded that of any other paleontologist, included iconic genera such as *Brontosaurus, Stegosaurus*, and *Triceratops*. Many of the discoveries of Marsh and Cope were abandoned later, however, because of the poor materials and descriptions produced in the collectors' haste to outcompete each other. For example, although Marsh is credited with

Figure 5 In 1842, Richard Owen (1804–1892) coined the word *dinosaur*. Owen, who rejected Charles Darwin's theory of evolution by natural selection, was an expert on interpreting fragmentary remains, and he perfected his anatomical skills by dissecting dead animals from London's Regent's Park Zoo (the world's oldest scientific zoo). Owen studied many birds (most notably, *Dinornis maximus*, an extinct and flightless moa) but did not recognize that birds are the descendants of dinosaurs. Owen was the first director of London's Natural History Museum; Figure 1). (Owen, Richard. *The Life of Richard Owen,* 1894)

the names *Apatosaurus* and *Brontosaurus*, most of his publications about these dinosaurs are now of purely historical interest.

1831 The BAAS is founded as a rival to the Royal Society (established in 1660).

1832 German vertebrate paleontologist Christian Erich Hermann von Meyer (1801–1869) classifies the carnivorous *Megalosaurus* and the herbivorous *Iguanodon* as "Saurians with Limbs similar to those of the heavy Land Mammals." When von Meyer's paper was translated into English five years later by George F. Richardson (1796–1848), Gideon Mantell noted that *Hylaeosaurus* "probably [also] belongs to this division."

1832 Gideon Mantell's reconstruction of a "horned" *Iguanodon* is the first known scientific attempt to reconstruct a dinosaur. Mantell's reconstruction was not published until 1968.

1832 William Smith is awarded the Wollaston Medal by the Geological Society; this is the first formal recognition of Smith's groundbreaking geological work. Smith was later awarded an annual civic pension of £100 for his retirement. Six years later, Richard Owen—who in 1842 would coin the word *dinosaur*—was also awarded the Wollaston Medal.

1832 Workers at Whiteman's Green Quarry uncover bones that they give to Gideon Mantell for examination. One block—4.5 feet long and 2.5 feet wide—contained ribs, 12 vertebrae, dermal spines, and other bones of *Hylaeosaurus* ("forest lizard"). Although Mantell, who told friends that "I have made a grand discovery", announced *Hylaeosaurus* to the Geological Society, his description of *Hylaeosaurus* was published in his *Geology of the South-East of England* the following year. *Hylaeosaurus*, a large armored reptile with conical spines on its neck and shoulder, was the first ankylosaur to be described, and the first dinosaur named from more than a few bones or teeth (Richard Owen later used Mantell's specimens to conclude that *Hylaeosaurus* was a large terrestrial mammal, similar to a rhinoceros.) Mantell's discovery is now displayed at BMNH (in the limestone in which it was found), and was the most obscure of Richard Owen's three founding members of the clade Dinosauria. Before his death in 1852, Mantell described two other genera of dinosaurs.

1833 George Scharf's (1788–1860) restoration of Britain's Tilgate Forest region (including *Hylaeosaurus* and *Megalosaurus*) is the earliest known attempt to reconstruct the external appearance of dinosaurs and their environment.

1833 Gideon Mantell, near bankruptcy, moves to fashionable Brighton and makes his home a museum. The museum failed, largely because Mantell waived the entrance fee for most visitors. The museum's first curator was George F. Richardson.

1833 London socialist and wine merchant William Devonshire Saull (1784–1855) opens a museum in London that later houses an *Iguanodon* on which "the characters of the Order Dinosauria were mainly founded" by Richard Owen. Saull's *Iguanodon*, which came from the Isle of Wight, had several fused sacral vertebrae, which Owen later recognized as a distinguishing trait of Dinosauria. Today, Saull's *Iguanodon* is housed at the BMNH (Fossil Reptilia No. 37685).

1833 *Penny Magazine* publishes a reconstruction of *Megalosaurus*, the first dinosaur (and first theropod) to be named. This was one of the first reconstructions to be published. Buckland's *Megalosaurus* jaw is displayed at the Oxford University Museum of Natural History.

1833 The star of Swiss draftsman Rodolphe Toepffer's (1799–1846) *Journeys and Adventures of Doctor Festus* is launched into space by a windmill, where he sees giant pits at Earth's poles. Magma at the bottom of these pits provided enough heat at the poles to create grasslands inhabited by prehistoric animals (e.g., mastodons and mammoths). This theme—that is, the discovery of ancient life in "lost worlds" on Earth—later became the basis for countless books and movies about dinosaurs, including Arthur Conan Doyle's *The Lost World* and Michael Crichton's *Jurassic Park*.

1834 Friends of Gideon Mantell purchase for £25 a rock collected "in the lowermost strata" of Shanklin Sandstone near Maidstone (about 40 miles northeast of Brighton) containing bones and a tooth similar to his *Iguanodon*. This "Maidstone *Iguanodon*," which was found in a quarry owned by amateur geologist William Bensted (1802–1873), was described by Richard Owen in an article titled "Report on British Fossil Reptiles" in the *Report of the British Association for 1841*, and more extensively in 1851 (*Fossil Reptilia of the Cretaceous Formations*) and 1855 (*Reptilia of the Wealden and Purbeck Formations*). Mantell included this fossil in his home-based museum, and today a similar specimen is displayed near the exit of BMNH's Gallery 21 opposite a print of Mary and Gideon Mantell.

1834 German geologist Friedrich August von Alberti (1795–1878) names the Triassic after finding three different rock-sequences in Germany that were older than the Jurassic but younger than the Permian. The oldest of these three was sandstone, the intermediate one was mostly limestone, and the youngest one was mostly shale. The Triassic (from the Latin word *triad*, meaning *three*) spanned from 250 to 208 million years ago. Dinosaurs appeared about 228 million years ago in the Late Triassic, at about the same time as the first crocodiles, plesiosaurs (long-necked marine reptiles), mammals, pterosaurs (flying reptiles), and turtles. Examples of Triassic dinosaurs include *Plateosaurus* (a herbivore) and *Coelophysis* (a carnivore).

1834 Gideon Mantell's "A Lecture on Geology and on the Organic Remains of a Former World, Discovered in Sussex" attracts hundreds of attendees. The *Brighton Guardian* noted, "One of the most interesting portions of the lecture was the juxtaposition of the thigh bones of the small Iguana, and of the mighty Iguanodon . . . [Mantell] concluded that there existed formerly in England a reptile 100 feet in length!" Today we know that *Iguanodon* was only about 30-feet long. Mantell collected *Iguanodon* throughout his life; more than 250 of his specimens are now housed at BMNH.

1834 Johann Friedrich Engelhardt discovers Triassic fossils in Germany that in 1837 will be named *Plateosaurus* by Hermann von Meyer. This sauropod was the first dinosaur reported from Germany, the first sauropodomorph, and the first large Triassic dinosaur to be discovered.

1834 While aboard HMS *Beagle*, Charles Darwin notes, "Certainly no fact in the world is so startling as the wide and repeated exterminations of its inhabitants." The exterminations are now called mass extinctions, the most famous of which closed the Cretaceous and claimed the non-avian dinosaurs as its most famous victims (dinosaurs such as *Diplodocus* and *Brachiosaurus* were extinct before the Cretaceous began). In 1980, Luis and Walter Alvarez proposed that the so-called K-T extinction was triggered by an asteroid hitting the Yucatán Peninsula that produced 300-foot-high tsunamis and a Magnitude 13 earthquake that was one-million-times stronger than the strongest earthquake recorded in human history.

1835 James Deane (1801–1858), a physician of Greenfield, Massachusetts, informs Amherst College's Edward Hitchcock that he has found slabs of sandstone from the Connecticut River "distinctly marked with the tracks of a turkey (as I believe) in relief. . . . The tracks, four in number, are perfect, and must have been made when the materials were in a plastic state, and at a period I leave you to tell." Hitchcock obtained and began studying the tracks, suggesting that they were made by flightless birds similar to the extinct New Zealand moa (*Dinornis*) or Madagascar's elephant bird (*Aepyornis*), both of which were described while Hitchcock was studying his tracks. Hitchcock's visions of giant, extinct birds were noted by several writers, including Henry Wadsworth Longfellow (whose father-in-law helped pay for Hitchcock's work), who wrote in his poem "To the Driving Cloud" (1845) about Native Americans who "stalked those birds unknown, that have left us only their footprints." Similarly, Herman Melville's *The Encantadas* (1854) describes himself as being similar to "an antiquary of a geologist, studying the bird-tracks and ciphers upon the exhumed slates trod by incredible creatures whose very ghosts are now defunct." Hitchcock became the first great modern dinosaur-tracker in North America.

1835 One year after receiving an honorary degree from Yale College, Gideon Mantell is awarded the Wollaston Gold Medal, the highest award of the Geological Society of London, "for the discovery of two genera of fossil reptiles, *Iguanodon* and *Hylaeosaurus*." (The only previous recipient of the medal had been stratigrapher William Smith, in 1831.) The following year, William Buckland told people attending a meeting of the BAAS that as they travel across Sussex, the former "Country of the Iguanodon," they would "crush beneath your carriage-wheels the remains of creatures which, had you lived a hundred thousand years ago, might have turned the tables, and crushed you." Mantell's rival, Richard Owen, was awarded the Wollaston Medal in 1838.

1836 English geologist Samuel Stutchbury (1798–1859) and Henry Riley (1797–1848) name the small sauropod *Thecodontosaurus* ("socket-tooth lizard") discovered two years earlier in Bristol, the United Kingdom. This early plant-eater was the fourth-named species of dinosaur, and the first dinosaur to be described from the Triassic. The following year, Hermann von Meyer—considered the founder of vertebrate paleontology in Germany—officially named *Plateosaurus* ("flat lizard"), the fifth-named species of dinosaur.

1836 In his book *How to Observe—Geology*, Sir Henry Thomas de la Beche describes a new way to excavate and transport fragile fossils: "It may even be desirable to go to the expense of preparing plaster of Paris on the spot, and cover the fossil. . . . By this process the exposed part of the skeleton becomes set in a block . . . so that by working carefully beneath it and the fossil in the friable rock, the skeleton is eventually on the surface of plaster of Paris, from which it may eventually be freed." This technique, which became common among British paleontologists, was invented independently in the United States in the 1870s.

1836 Richard Owen (Figure 5) is appointed the prestigious Hunterian Professor of Anatomy at the Royal College of Surgeons at an annual salary of £120. Owen, a vocal opponent of Charles Darwin's theory of evolution by natural selection, later coined the term *dinosaur*.

1836 Thanks to a tip from physician James Deane (with whom he would feud until Deane's death in 1858), Edward Hitchcock (1793–1864)—a professor of theology and geology and president of Amherst College—discovers "remarkable footmarks in stone in the valley of Connecticut River, which have since awakened so much interest among intelligent men." Hitchcock invented a term to describe the tracks and a new scientific discipline: "I include all the varieties of tracks under the term *ornithichnites* . . . signifying stony bird tracks: and if it be convenient to speak of the subject as a distinct branch of knowledge, I should call it Ornithichnology." In his first publication about the tracks, Hitchcock attributed these tracks to seven species of "a biped animal, a bird perhaps" that he regarded as "the giant ruler of the valley." Hitchcock noted that the tracks in the mud were deep "as if an Elephant had passed over it. I could not persuade myself, until the evidence became perfectly irresistible, that I was examining merely the track of a bird." Hitchcock believed his conclusions were accurate ("Four out of five, I presume, would draw this conclusion at once"), but the prints were later shown to have been made by bipedal dinosaurs. Hitchcock linked birds with dinosaurs by noting that the organisms responsible for the tracks walked and ran almost exclusively on their hind feet with long strides on limbs close to the their bodies. Hitchcock concluded that the narrow tracks indicated that they were not made by lizards, bears, or *Iguanodon*-like creatures. Hitchcock also realized that the track-makers were social organisms because they moved in groups. Hitchcock was later the founding chairman of the Association of American Geologists and Naturalists, which later became the American Association for the Advancement of Science (AAAS). During his lifetime, Hitchcock collected more than 20,000 fossil footprints, many of which are displayed at Amherst College in the world's largest fossil footprint museum. Hitchcock noted, "When I speak of species here, I mean species in oryctology, not in ornithology. And I doubt not, that . . . these different tracks were made by birds that were generically different." ("Oryctology" was a common name for "fossil zoology," which is what we call "paleontology" today.) In 1841, Hitchcock acknowledged that "these names, implying only a resemblance, leave the real nature of the tracks open to discussion," and then proposed using the Linnaean system of classification: "I propose the term Ichnolite . . . to be the name of the Class [and] I would divide this Class into Orders . . ." Four years later, Hitchcock noted, "Hitherto names have been given to the footmarks and not to the animals. But since all geologists now admit that these impressions are real tracks, this paper attempts to name the animals that made them, and to classify and describe them." Hitchcock's poem "The Sandstone Bird," published with a pseudonym in *The Knickerbocker* magazine, was a tribute to the dinosaurs' tracks.

1836 William Buckland (Figure 3), who had first described a dinosaur, writes *Geology and Mineralogy Considered with Reference to Natural Theology*, a *Bridgewater Treatise* to promote "the Power, Wisdom and Goodness of God, as manifested in the Creation." Buckland's book discussed various types of creationism (e.g., day-age creationism, gap creationism, and theistic evolution) and noted that Noah's Flood lacked geological evidence. Richard Owen's "Report" announcing "Dinosauria" later quoted Buckland's *Bridgewater Treatise*.

1837 Christian Erich Hermann von Meyer (1801–1869) uses fossils found near Bristol, Germany, to describe and name *Plateosaurus* ("Broadway lizard"), a plateosaurid dinosaur that lived in the Late Triassic in central and northern Europe. *Plateosaurus*, a bipedal herbivore, was the first dinosaur found in Germany and the fifth named dinosaur-genus that is still considered valid.

1837 Edward Hitchcock describes another 14 types of "fossil footsteps," noting that some of the tracks "bear so near a resemblance to the feet of living Saurians, that I have denominated them Sauroidichnites. I have sometimes thought that they might have been made by Pterodactyls." These tracks were later attributed to dinosaurs.

1837 Richard Owen publishes his first paper about paleontology. That paper described a fossil mammal found in Argentina by Charles Darwin during his voyage aboard HMS *Beagle*.

1837 Charles Darwin begins describing his "dangerous" idea in a secret notebook labeled "Transmutation of Species." Page 36 of this notebook contained Darwin's first "tree of life," in which—beneath the words "I think"—he depicted life not as a hierarchical and philosophical ranking of "higher" and "lower" forms (as Aristotle and other naturalists had claimed), but instead as a branching tree showing shared origins. The tree's branches, which were extended by common ancestry, did not necessarily lead anywhere; they just spread. Darwin's tree showed that species evolved, and in some cases, one species gave rise to many species. Although Darwin did not publish his theory for 22 years, his "tree of life" became a metaphor for his view of how species evolve.

1838 A three-man committee (one of whom was Owen's father-in-law) of the BAAS pays Richard Owen £200 to produce a study of the fossil reptiles of Britain. Although Owen had done relatively few studies of reptiles, his report the following year was hailed by William Lucas as work "of the greatest comparative anatomist living." In 1842, in a footnote in the second part of his report, Owen boldly invented the clade Dinosauria.

1838 Charles Darwin reads "for amusement" the sixth edition of Thomas Malthus's *An Essay on the Principles of Population* and notes, "Being well prepared to appreciate the struggle for existence . . . it at once struck me that under these circumstances favourable variations would tend to be preserved and unfavourable ones destroyed. The result of this would be the formation of a new species. Here then, I had at last got a theory by which to work." To Darwin, natural selection was the force that constantly adjusts the traits of future generations.

1838 English Romantic artist John Martin (1789–1854)—famous for his apocalyptic portrayals of biblical stories—produces *The Country of the Iguanodon* (Figure 6), the first of several dramatic paintings of dinosaurs (including *Hylaeosaurus*). Martin's work, which was featured as the frontispiece of Gideon Mantell's *Wonders of Geology* (which went through six editions in 10 years), showed *Megalosaurus* attacking *Iguanodon*. Martin's painting was criticized by scientists as being inaccurate, but was popular with the public. Benjamin Waterhouse Hawkins gave *Iguanodon* a horn on its snout (making the dinosaur look somewhat like a rhinoceros), as Martin did in *The Country of the Iguanodon.*

1838 French paleontologist Jacques-Amand Eudes-Deslongchamps (1794–1867) describes *Poekilopleuron bucklandii* ("mottled rib"), which he reconstructs from fossils found near Caen. Deslongchamps's *P. bucklandii* was destroyed during the Allied liberation of Normandy in 1944.

1838 Lamenting in his journal that "And so passes away the labor of 25 years!!!", Gideon Mantell sells his museum, including all of its fossils, to BMNH for £4,087. The following year, Mary Ann Mantell left her husband Gideon who, by this time, had stopped crediting her for finding the original *Iguanodon* tooth. Soon thereafter, Gideon Mantell moved to Clapham in south London; his former home there is now a dental office. Although Mantell remains most famous for *Iguanodon*, he later published substantial papers about mollusks, turtles, fruits, and birds. The Mantells never divorced.

1839 Richard Owen (Figure 5) publishes *Report on British Fossil Reptiles, Part I* for BAAS. (BAAS also commissioned Owen to prepare similar reports about British mammals and birds, but asked Louis Agassiz [1807–1873] to prepare the report about fish.) Owen's report described 16 plesiosaurs and 10 ichthyosaurs, most of which were new species named by Owen. Two years later, Owen lectured

Figure 6 This adaptation of English Romantic artist John Martin's (1789–1854) *The Country of the Iguanodon* was created in 1837 and appeared the following year as the frontispiece of Gideon Mantell's popular *The Wonders of Geology* (1838). Martin had visited Mantell's home in 1834, after which the latter noted, "Mr. Martin was deeply interested in the remains of the *Iguanodon,* etc. I wish I could induce him to portray the country of the *Iguanodon*. No other pencil but his should attempt such a subject." Martin's 12-inch by 17-inch original watercolor image hangs in the Museum of New Zealand Te Papa Tongarewa, the national museum of New Zealand; this museum also houses an *Iguanodon* tooth collected by Mantell. (Museum of New Zealand Te Papa Tongarewa/Gift of Mrs. Mantell-Harding, 1961/The Bridgeman Art Library)

about reptiles during BAAS's annual meeting in Plymouth. Although many people have assumed that Owen coined the term *Dinosauria* in this lecture, there is no evidence to support this claim. Indeed, in 1841, Owen included all of the dinosaur genera in his discussion of lizards, suggesting that he either had not yet given them a specific name or had not yet separated them as a separate group. Owen coined the term *Dinosauria* as he was preparing *Part II* of his report, which was published in 1842; this monograph marked the first public appearance of the word *dinosaur*. The small print-run of Owen's monograph (25 copies for Owen's personal use and distribution) incorrectly listed the publication date of Owen's 1842 report as 1841. Owen's reports to BAAS prompted William Buckland to bestow on Owen the title "the British Cuvier." Owen, who used the Dinosauria to argue against the popular version of evolution, defined Dinosauria as "an order of extinct reptiles, characterized by cervical and anterior dorsal vertebrae with par- and diapophyses [certain protuberances], articulating with bifurcate ribs; dorsal vertebrae with a neural platform, sacral vertebrae exceeding two in number; body supported on four strong unguiculate [with nails or claws] limbs."

1840 Edward Drinker Cope is born into a wealthy family of Philadelphia-based shipping merchants. By the time he was six years old, Cope was visiting the Philadelphia Academy of Natural Science, and at 18 he authored the first of his more than 1,400 scientific publications. Cope, who became one of the world's greatest collectors of dinosaur fossils, later had a famous feud with Othniel Marsh that became known as "the Bone Wars." Cope's proposals about the gradual enlargement of mammalian species over geologic history became known as "Cope's Law." By the end of his life, Cope had spent most of the fortune he inherited.

1841 British geologist John Phillips (1800–1874), a nephew of William Smith, names the Mesozoic and Kainozoic (later changing the spelling to Cainozoic, and now referred to as Cenozoic) geologic eras. Boundaries of these eras, like those of Sedgwick's Paleozoic, were marked by mass extinctions that produced sharp breaks in the fossil record. Phillips defended catastrophists' claims that fauna often appeared and disappeared from Earth, after which they were replaced by divine creations. After Charles Darwin published *On the Origin of Species* in 1859, Phillips claimed that his study of marine deposits "showed no evidence of evolution having occurred."

1841 Gideon Mantell is hurt in a carriage accident, and begins taking opium to manage the pain. Eleven years later, Mantell died of an opium overdose.

1841 In his 831-page *Final Report on the Geology of Massachusetts*, Edward Hitchcock describes all of the "stony bird tracks" that he's found. Later, Hitchcock found the type-specimen of *Ornithoidichnites sillimani,* which had been used for 60 years as a flagstone in Middletown, Connecticut. Today, that track is "the gem of the collection" at Amherst, Massachusetts.

1841 In what London's *The Times* described as a "very long" lecture about British fossil reptiles to attendees of the annual meeting of BAAS, Richard Owen promotes his anti-Darwinian views by claiming that *Iguanodon* and similar animals "never exhibit any indications that their forms graduated or by any process passed into another species. They appeared to have sprung from one creative act." The *Literary Gazette* reported that Owen saw "no gradation or passage of one form into another, but that [each] were distinct instances of Creative Power, living proofs of a divine will and the work of a divine hand, ever superintending and ruling the existence of our world." The *Athenaeum* reported that Owen had grouped *Megalosaurus, Iguanodon*, and *Hylaeosaurus* as "gigantic forms of terrestrial Saurians," but not in any new order (i.e., Owen had not yet grouped them as Dinosauria). Owen's lecture was reported in newspapers and magazines in England, France, Germany, the United States, and other countries.

1841 Orra L. White (1796–1863), the wife of Edward Hitchcock, sketches some of the first scientific illustrations of dinosaur prints. White was the first woman to be recognized for her work in American paleontology.

1841 Richard Owen (Figure 5) uses fossilized vertebrae, ribs, and arms from the Isle of Wight to describe the first sauropod (*Cetiosaurus*, meaning "whale lizard"), but believes it is a whale-like crocodile. Since then, sauropods have been described from across the world. Sauropods had the smallest brains (relative to body size) of any dinosaur; their brain occupied only about 20 percent of the volume expected of a similarly sized alligator.

1841 Two years after acquiring them, Edward Hitchcock describes the fossilized tracks discovered in 1802 by Pliny Moody. Hitchcock noted the fossilized tracks' resembled those of the American coot (*Fulica americana*), and therefore attributed them to *Ornithoidichnites fulicoides*.

1842 In April, two months after Roderick Murchison told the Geological Society of London that British anatomist Richard Owen "will shortly lay before the world the results of his researches into the extinct Saurians of our island," Richard Owen (1804–1892; later Sir Richard Owen) publishes *Report on British Fossil Reptiles, Part II* for BAAS. The edition included 1,500 copies of the book, each priced at 13s/6d. Owen recognized that *Iguanodon, Megalosaurus*, and *Hylaeosaurus* share several traits that distinguished them from other ancient or living creatures. For example, they all were giant ("far surpassing in size the largest of existing reptiles") terrestrial animals that had pillar-like legs tucked beneath their bodies (unlike the splayed position of reptiles' legs); they also each had five fused vertebrae welded to their pelvic girdle. On p. 103 of the *Report*, Owen began dinosaur classification and coined the term *Dinosauria*: "The combination of such characters . . . all manifested by creatures far surpassing in size the largest of existing reptiles will, it is presumed, be deemed sufficient ground for establishing a distinct tribe or suborder of Saurian Reptiles, for which I would propose the name *Dinosauria*." In a footnote, Owen defined the term to mean "fearfully great, a lizard." Owen used the superlative form of *deinos* (as Homer had in the *Iliad*), which means "fearfully great." Many other people used *deinos* as an adjective to mean "terrible," despite the fact that few dinosaurs were terrible (and none were lizards). Owen grouped dinosaurs according to several traits, including teeth in bony sockets, ribs having two heads, sacra made of five fused vertebrae, a complex shoulder girdle, long and hollow limb bones, and mammal-like feet. Owen endowed his reconstructed dinosaurs with mammalian ("advanced") traits that modern-day reptiles—that is, their purported descendants—lack. This contradicted Lamarckian evolution, for it implied that the "highest" reptiles lived in the Mesozoic, and that only "degraded" lizards are alive today. (These ideas were speculative because there were no complete skeletons available to Owen.) Owen did not include other British fossil reptiles that were later shown to be dinosaurs. For example, Owen identified the sauropod *Cetiosaurus* as a giant marine crocodile, the theropod *Poekilopleuron* as a crocodile, and the prosauropod *Thecodontosaurus* as a lizard. Owen, a bitter opponent of Darwinian evolution, added that dinosaurs had traits "originally impressed upon them at their Creation" that was orchestrated by God via a "divinely plotted path of adaptive changes." Owen's claim that dinosaurs were perfectly fitted for a primitive Earth (the "height of reptilian achievement") fit well with the religious orthodoxy of early Victorian England. In 1888, British anatomist Harry Seeley (1839–1909) used hip structure to divide Dinosauria into two clades: Ornithischia and Saurischia.

1842 In *Geology for Beginners*, artist John Martin—famous for his paintings of biblical subjects—depicts dinosaurs as belching giants. Martin's paintings were inaccurate, but dramatic and popular.

1843 Richard Owen (Figure 5) names *Cetiosaurus* ("whale lizard"). This giant sauropod, which Owen suspected to be aquatic, was not known to be a dinosaur until 1869, when it was declared to be a dinosaur by Thomas Henry Huxley. *Cetiosaurus* was the sixth dinosaur to be named, and the first sauropod to be discovered.[5] Sauropods dominated the Jurassic. Although not all sauropods were giants, all giants were sauropods.

1844 Charles Darwin expands his earlier outline of natural selection into a 231-page essay that rejects the fixity of species. The table of contents of this essay, as well as much of the essay's text, was similar to that of *On the Origin of Species*, which did not appear for another 15 years.

1844 Eight years after describing dinosaur tracks in the Connecticut Valley, Edward Hitchcock writes, "'Not a track remains,' says Dr. Buckland, 'or a single hoof, of all the countless millions of men and beasts whose progress spread desolation over the Earth. But the reptiles that crawled upon the half finished surface of our planet, have left memorials of their passage enduring and indelible.' And we may add, that the proudest monuments of human art will moulder down and disappear; but while there are eyes to behold them, the sandstone of the Connecticut Valley will never cease to remind the observer of the gigantic races that passed over it while yet in an incipient state."

1844 In *Medals of Creation*, Gideon Mantell describes fossils as medallions created by the Creator to mark the success of epochs. *Medals of Creation*, which summarized knowledge about dinosaurs, was one of the first English compilations of paleontology.

1844 Massachusetts surgeon James Deane, who had told Edward Hitchcock of the giant tracks at Turners Falls, Massachusetts, claims to have been the first to appreciate the importance of the fossil footprints described by Hitchcock. Later the same year, an angry Hitchcock responded with "Rejoinder to the 'discovery of fossil footmarks' by J. Deane." Hitchcock named 39 genera and 82 species from the Turner Falls tracks. Deane continued his studies of fossilized tracks, and the feud between Deane and Hitchcock raged for several years.

1845 Gideon Mantell's "Journal," which Mantell had kept since 1818, lists the first of many complaints about Richard Owen. Such comments continued until Mantell's death in 1852.

1845 William Buckland (Figure 3)—the first person to formally describe a dinosaur—is appointed dean of Westminster, thereby ending most of his scientific research. Buckland was an Anglican divine by title, as well as the first officially appointed geologist at Oxford. Charles Darwin acknowledged Buckland "though very good humoured and good-natured, [he] seemed to me vulgar and almost coarse man. He was incited more by a craving for notoriety, which sometimes made him act like a buffoon, than by a love of science."

1845 W. G. Atherstone and A. G. Bain uncover the first traces (a partial skull and several limb bones) of plated, spiky stegosaurs in Early Cretaceous rocks near Grahamstown, South Africa. The bones were sent to Richard Owen, who confused them with those of armored reptiles called pareiasaurs. In 1910, paleontologist Robert Broom (1866–1951) argued that some of the fossils belonged to an ankylosaur. In 1929, Franz Nopcsa classified the fossils as belonging to a stegosaur, and in 1981, these findings were confirmed. The fossils are now named *Paranthodon africanus*.

1846 French geologist Pierre Émile Philippe Matheron (1807–1899) describes the 25-foot-long *Hypselosaurus* ("highest lizard") based on bones discovered in Late Cretaceous deposits in France near the mouth of the Rhône River, along with fragments of fossilized eggshells. In subsequent decades, numerous eggs—some up to a foot long—of *Hypselosaurus* were found in the region. *Hypselosaurus*, which Matheron believed was a giant crocodile, was formally named in 1869.

1846 U.S. president James K. Polk signs a bill chartering the Smithsonian Institution. The first building ("The Castle"), in which the remains of James Smithson are interred, was completed in 1855.

1846 Mary Anning, whose reputation bordered on legendary even during her short lifetime, is made an honorary member of the all-male Geological Society of London. Late in her life, William Buckland helped Anning secure an annual pension of £25, thereby making Anning the first woman to receive government support for scientific research.

1847 Mary Anning dies of breast cancer and is buried at St. Michael the Archangel Church in Lyme Regis. Her obituary was published in the *Quarterly Journal of the Geological Society*, an organization that barred women from its membership until 1904. After Anning's death, The Royal Society donated funds for a stained-glass window honoring Anning's memory in the Parish Church at Lyme Regis. Anning, the subject of Terry Sullivan's 1908 tongue-twister "She sells seashells by the seashore," remains one of the most influential and interesting people in the history of paleontology. As lamented by English novelist John Fowles (1926–2005) in *The French Lieutenant's Woman* (1969), Mary Anning never had a new species named after her until after her death. Today, Anning's discoveries line the walls of Gallery 30 of BMNH.

1847 More than a decade before Charles Darwin announces his theory of evolution by natural selection, Joseph Leidy—who collected and described hundreds of new species (including many dinosaurs) but seldom speculated about them—endorses what later becomes a tenet of Darwin's theory: "The study of the earth's crust teaches us that very many species of plants and animals became extinct at successive periods, while other races originated to occupy their places. This probably was the result, in many cases, of a change in exterior conditions incompatible with the life of certain species and favorable to the primitive production of others. . . ."

1847 Richard Owen and others found the Palaeontographical Society, which later publishes many of Owen's papers about dinosaurs.

1848 Gideon Mantell is given a fossil jaw bought from local miners by Lambert Brickenden. This specimen, which was found near and in the same geological horizon as Mantell's *Iguanodon* teeth, was one of the first fossilized complete jaws of a dinosaur found anywhere in the world. Mantell labeled the jaw as being from *Iguanodon*, but in 2010, American student Andrew McDonald reclassified it as *Kukufeldia tilgatensis*, a name that honors Kukufeld (the Old English name for Cuckfield) and Tilgate Forest, the area surrounding Cuckfield that has yielded several fossils of dinosaurs.

1848 In a paper titled "An Attempt to Discriminate and Describe the Animals That Made the Fossil Footprints of the United States," Edward Hitchcock concludes that the fossilized tracks he had found were made by giant three-toed birds, "some ten or twelve feet high." Hitchcock described his interest in the fossils this way: "I have devoted much time, and labor, and thought, to these mementos of the races that, in the dawn of animal existence in the Connecticut Valley, tenanted the shores of its rivers and estuaries. Whatever doubts we may entertain as to the exact place on the zoological scale which these animals occupied, one feels sure that many of them were peculiar and gigantic: and I have experienced all the excitement of romance, as I have gone back into those immensely remote ages. . . . Strange, indeed, is this menagerie of remote sandstone days; and the privilege of gazing upon it, and of bringing into view one lost form after another, has been an ample recompense for my efforts, though they should be rewarded by no other fruit." These prints, which are today displayed at the Amherst College Museum of Natural History (formerly the Pratt Museum of Natural History), were later shown to be footprints of bipedal dinosaurs. They are the first known discovery of dinosaur remains in North America.

1848 In *Archetype and Homologies of the Vertebrate Skeleton*, Richard Owen (Figure 5) presents an updated argument from design while explaining homologies as variations of vertebrate archetypes. Owen famously depicted the vertebrate archetype—an idealized form of all vertebrates—as a skeletal zeppelin hovering above humans, birds, fishes, and other vertebrates. Owen, who tutored Queen Elizabeth's children, noted the similar organization of human hands, bat wings, and whale flippers. Owen's explorations of homology, along with the establishment of BMNH (Figure 1), became his most lasting legacy to modern biology.

1848 Lord Alfred Tennyson (1809–1892), who was Poet Laureate during much of Queen Victoria's reign, says that the best way to appreciate giant, prehistoric animals is to stand by a railroad track at night as a train passes by. Tennyson's "Nature, red in tooth and claw," from the 56th stanza of *In Memoriam, A.H.H.* (1849), became a famous description of the struggle for existence.

1849 After reading about fossil bones discovered near Fort Laramie, Wyoming, geologist David Dale Owen (1807–1860) sends his colleague John Evans (1812–1861) and a small crew to examine the fossil beds. Owen wrote from the badlands that "embedded in the debris, lie strewn, in the greatest profusion, organic relics of extinct animals." The following year, T. A. Culbertson was sent by the Smithsonian Institution to collect fossils in the area.

1849 APS gives its fossils to ANS in Philadelphia.

1849 Gideon Mantell describes *Hypsilophodon* from fossils found on the Isle of Wight, but identifies it as a juvenile *Iguanodon*. That same year, despite opposition from Richard Owen, Mantell was awarded the Royal Society's Gold Medal. When Thomas Huxley described *Hypsilophodon* in 1869, he became the first person to note that such dinosaurs had a pubic bone pointing backward, like birds.

1849 Gideon Mantell discovers what he calls "the most stupendous humerus of a terrestrial reptile ever discovered; it is 4 and a half feet in length." This fossil, found eight miles from Oxford, became the basis for Mantell's description (in 1850) of *Pelorosaurus* ("monstrous lizard"). *Pelorosaurus* was the first sauropod to be identified as a dinosaur, but it was not the first to be discovered; in 1841,

Richard Owen had discovered *Cetiosaurus*, but he incorrectly identified it as a crocodile-like marine reptile (Mantell identified *Pelorosaurus* as a dinosaur who lived on land). *Pelorosaurus*, which was up to 50 inches long, lived during the early Cretaceous.

1849 The Department of the Interior is created to develop and preserve the natural resources of the United States. Many of the activities of this department will affect the collection of dinosaur fossils.

1850 Charles Hazelius Sternberg (1850–1943) is born near Cooperstown, New York. After studying at Kansas State University with paleontologist Benjamin Mudge (1817–1879), Sternberg began collecting fossils for Edward Drinker Cope, and later with his three sons created a family dynasty in paleontology. Fossils collected by the Sternbergs are displayed in museums throughout the world.

1850 Gideon Mantell names *Pelorosaurus*, the first-named sauropod recognized as being terrestrial.

1850 Thanks to lobbying by Charles Lyell and others, Gideon Mantell is awarded the Royal Society's Royal Medal for his work on dinosaurs. Mantell's award was opposed by Richard Owen, who had received the medal in 1846 for his studies of extinct mollusks.

1851 Edward Hitchcock's *The Religion of Geology and Its Connected Sciences* reconciles science and religion. Hitchcock tried to remain objective; as he said about the many fossilized tracks of dinosaurs that he discovered, "the real question is, not whether these hypotheses accord with our religious views, but whether they are true."

1851 Franz Unger's (1800–1870) *Primitive World* is the first significant book containing drawings of "prehistoric animals."

1851 In a discussion of the teeth of *Iguanodon*, Gideon Mantell refers to "my first discovery of a tooth in a quarry near Cuckfield." These and other statements prompted some people to question Mantell's earlier claim that the first *Iguanodon* teeth had been found by his wife Mary.

1851 In his presidential address to the Geological Society of London, Charles Lyell supports Edward Hitchcock's claim that the fossilized footprints described in Connecticut "should, I think, remove all skepticism in regard to the ornithic nature of most of these [Connecticut Valley] bipeds." Hitchcock's views were presented in virtually all geology books, including Lyell's *Principles of Geology*, Mantell's *Medals of Creation*, Louis Agassiz's *Geological Sketches*,

Buckland's *Bridgewater Treatise,* and Hugh Miller's *Testimony of the Rocks*, as well as in the works of popular writers such as Oliver Wendell Holmes, Herman Melville, and Henry David Thoreau. Today, museums and private collections contain more than 35,000 tracks gathered from 40 sites in Connecticut and Massachusetts.

1851 In Franz Unger's (1800–1870) *The Primitive World in Its Different Period of Formation*, Austrian artist Josef Kuwasseg (1802–1877) illustrates Gideon Mantell's *Iguanodon* as a giant lizard with misplaced horns.

1851 Just a year before his self-induced death, Gideon Mantell laments "the spirit of self-aggrandizement and jealousy [by which Richard Owen] had exerted [his] baneful influence over this department of paleontology." Thomas Huxley had similar feelings about Owen, noting that same year that "it is astonishing with what an intense feeling of hatred Owen is regarded by the majority of his contemporaries, with Mantell as arch-hater. The truth is, [Owen] is the superior of most, and does not conceal that he knows it, and it must be confessed that he does some very ill-natured tricks now and then."

1852 After years of suffering from pain and depression, Gideon Mantell—who had discovered four of the five genera of dinosaurs then known—dies in London of an opium overdose and is buried in West Norwood Cemetery in outer London beside his third child, Hannah Matilda. No one was invited to Mantell's funeral which, in accordance with his wishes, was as "plain as possible." Before Mantell was buried, Mantell's nemesis Richard Owen removed part of Mantell's spine for the Royal College of Surgeons. Mantell's spine (Specimen 4808.1), which was mounted and displayed there until 1969, when it was destroyed.[6] Mantell's death, followed by Buckland's in 1956,[7] left Richard Owen as the only one of the first three namers of dinosaurs. In *Literary Gazette*, Owen's anonymous obituary noted Mantell's "want of exact scientific knowledge . . . [that] compelled him to have recourse to those possessing it." (Owen's harshness likely cost him the presidency of the Geological Society the following year.) At the time of his death, Mantell had discovered four of the five genera of dinosaurs then known. Today, Mantell's former home in Lewes is adorned with a brass plate noting Mantell "discovered the fossil bones of the prehistoric *Iguanodon* in the Sussex Weald." Ironically, Owen's famed identification of the moa (*Dinornis giganteus*), an extinct flightless bird from New Zealand, was aided by shipments of bones from New Zealand by Walter Mantell (1820–1895), the émigré son of Gideon and Mary Mantell. Walter took his father's fossils to New Zealand, where one can see them in Wellington's Museum of New Zealand Te Papa Tongarewa. Mantell's famous tooth of *Iguanodon* (Specimen MNZ GH 004839) is also in this museum, but is not displayed. Mantell's will did not mention his estranged wife, Mary.

1852 Although Gideon Mantell had earlier suggested a relationship between *Pelorosaurus* and dinosaurs, German paleontologist Friedrich August von Quenstedt (1809–1889) formally lists *Pelorosaurus* in Dinosauria.

1852 Paul Gervais (1816–1879) uses a humerus (now lost) to name *Aepisaurus*, the first sauropod from outside England.

1852 The board of directors of the Crystal Palace Company resolves "that a Geological Court be constructed containing a collection of full-sized models of the Animals & Plants of certain geological periods, and that Dr [Gideon] Mantell be requested to superintend the formation of that collection." After being turned down by Mantell, the board of directors hired Richard Owen. Others such as H. T. de la Beche and artist John Martin had "restored to life" dinosaurs on paper, but Owen's dinosaurs were to be first life-size models for the public. The models were built by Benjamin Waterhouse Hawkins (1807–1889) to "illustrate and realize—the revivifying of the ancient world—to call up from the abyss of time and from the depths of the earth, those vast forms of gigantic beasts which the Almighty Creator designed with fitness to inhabit and precede us in possession of this part of the earth called Great Britain." Hawkins's models of *Iguanodon* (with their thumb-spikes), an elephant-like *Megalosaurus* (with a lizard-like head), and *Hylaeosaurus* (which resembled an ankylosaur)[8] depicted Richard Owen's belief that dinosaurs were unique, huge (*Iguanodon* was almost 40 feet long), slow, lumbering, unprecedented elephantine animals that resembled giant mammals more closely than reptiles. (Hawkins's 15 models at Crystal Palace, which came to be known as "the Dinosaur Court," also included non-dinosaurian reptiles, including *Mosasaurus, Pterodactylus*, and *Ichthyosaurus*.) Hawkins arranged the animals in the order in which their fossils had been discovered. Two years after being commissioned to do the work, public displays of Hawkins's models of dinosaurs—some of which were made of more than 30 tons of clay—were displayed on three islands representing the Paleozoic, Mesozoic, and Cenozoic eras. This exhibit forever changed the way people view ancient life on Earth. Today, Hawkins's models (which were classified as "Grade One Buildings" in 2007)—which include no bones or fossils of the animals they depict—dominate several islands in Sydenham, just outside of London.

1853 In the opening paragraph of *Bleak House*, Charles Dickens includes a *Megalosaurus* walking down a London road: "As much mud in the streets as if the waters had but newly retired from the face of the earth and it would have been wonderful to meet a Megalosaurus, forty feet long or so, waddling like an elephantine lizard up Holborn Hill. . . ." This was the first mention of a dinosaur in a published work of fiction.

1853 On New Year's Eve, Richard Owen hosts a dinner for 21 scientists (and directors of the Crystal Palace Company) in the belly of one of Hawkins's

40-foot-long *Iguanodon* sculptures.[9] On the same day, *The Illustrated London News* told readers, "At Sydenham we are not to be contented with either pictures or dry bones." Invitations for the publicity stunt from Hawkins (which helped make the dinosaurs a sensation before they were all completed) were printed on paper shaped like a wing of a pterodactyl and featured banners proclaiming the names of Owen, Cuvier, Buckland and Mantell, "those great names whose high position in the science of palaeontology and geology would form the best guarantee for the severe truthfulness of [Hawkins'] works." Owen sat in the head of the *Iguanodon*, and Mantell's name hung at the rear of *Iguanodon*, opposite that of Owen. In an article titled "Fun in a Fossil," the magazine *Punch* reported, "If it had been an earlier geological period they might perhaps have occupied the *Iguanodon*'s inside without having any dinner there." The party was enlightened by a song in which Hawkins's dinosaur was brought back to life: "A thousand ages underground, His skeleton had lain, But now his body's big and round, And there's life in him again!" The American company Ward's Natural Science Establishment later sold plaster-cast miniatures of Hawkins's Crystal Palace models; these were the first dinosaur models to be marketed to the public. For the next 40 years after the opening of Owen's "Mausoleum to the memory of a ruined world" at the Crystal Palace exhibit, no one tried to recreate dinosaurs.[10]

1854 After examining fossils collected the previous year by J. M. Orpen in South Africa, Richard Owen (Figure 5) describes the mid-sized (12 feet long) Triassic prosauropod, *Massospondylus*. Although this was the first dinosaur to be discovered in the Southern Hemisphere, Owen believed it was a "large, extinct, carnivorous reptile." In 1866, Thomas Huxley identified *Euskelosaurus*, another Triassic prosauropod from South Africa.

1854 American geologist Ferdinand Vandeveer Hayden (1829–1887) finds fossilized teeth near the mouth of the Judith River in what would later be central Montana. Hayden,[11] who gave up work as a physician after serving in the Civil War, gave these fossils to anatomist Joseph Leidy of ANS in Philadelphia, who described them as being from fossil reptiles similar to *Iguanodon* and *Megalosaurus*. Leidy concluded that some of the teeth came from *Deinodon* (which resembled *Megalosaurus*), *Palaeoscincus, Trachodon* ("rough tooth"; "a herbivorous lacertilian reptilian allied to the *Iguanodon*"), and *Troodon* (then spelled *Troödon*). These were the first dinosaur fossils to be formally recognized in North America; of these names, only *Troodon* remains in use.

1854 English dinosaur-hunter Samuel Beckles (1814–1890) describes bird-like tracks that he believes were made by dinosaurs. Eight years later, Beckles suggested the tracks were made by *Iguanodon*.

1854 Joseph Leidy, who in 1846 had identified *Trichinella spiralis* as the creature that causes trichinosis, describes a fossilized jaw and teeth that resemble those of *Megalosaurus* as *Bathygnathus borealis* ("northern deep jaw"). Although Leidy did not claim that *Bathygnathus borealis* was a dinosaur, he did note that it was "one of the bipeds, which made the so called 'bird tracks' of the New Red Sandstone" in Connecticut. The following year, Leidy compared *B. borealis* with dinosaurs, prompting other scientists to consider *B. borealis* to be a dinosaur. *B. borealis* is now known to be a pelycosaur, a reptile from the Permian. Leidy named 375 species, 14 of which were dinosaurs (three of these names remain valid).

1854 The world's first life-size models of dinosaurs, commissioned in 1852 and sculpted at a cost of £13,729 by Benjamin Waterhouse Hawkins, greet Queen Victoria and 40,000 other people attending the opening ceremonies of London's Crystal Palace Exhibition in London's Hyde Park, thereby providing the public with the first opportunities to experience dinosaurs (Figure 7). (The Crystal Palace, which in 1851 hosted the first World's Fair, was the largest glass and iron building

Figure 7 English sculptor and natural historian Benjamin Waterhouse Hawkins (1807–1894) drew this sketch in 1854 to show the planned placement of dinosaurs in the "Dinosaur Court" in Crystal Palace Park. Hawkins arranged the animals in chronological order (with the oldest to newest positioned from right to left) to match the rock layers in which the fossils had been found. Hawkins's models for the reconstructions were never on public view, and none have survived. The sketch shown here originally appeared in 1854 as *Diagram of the Geological Restorations at the Crystal Palace*. Today, Hawkins's reconstructions can be seen just outside of London along Sydenham's "Prehistoric Monster Trail." (Private Collection/The Bridgeman Art Library)

ever constructed.) Until Hawkins created his models, dinosaurs were poorly understood and of little interest to anyone except a few paleontologists. *The Illustrated London News* described Hawkins's models as "gigantic restorations of the Extinct Inhabitants of the Ancient World" that included *Ichthyosaurus, Plesiosaurus, Iguanodon*, and *Megalosaurus*. To produce the scaly, stumpy-legged dinosaurs, Hawkins worked closely with Richard Owen, the first director of BMNH and originator of the word *dinosaur*. Hawkins's models were based on composites of familiar animal-traits—for example, *Iguanodon* resembled a scaly rhinoceros, and *Hylaeosaurus* resembled a dragon. More than one million people visited the exhibit during each of the next 50 years as dinosaurs moved to the forefront of the public's imagination. The 1854 display of dinosaurs at Crystal Palace produced the first massive wave of public interest in dinosaurs. Hawkins's sculptures were classified by governmental authorities as buildings, not sculptures; as Hawkins commented years later, "in the instance of the *Iguanodon*, it is not less than building a house upon four columns, as the quantities of material of which the standing *Iguanodon* is composed, consist of four iron columns 9 feet long by 7 inches in diameter, 600 bricks, 650 5-inch half-rounded tiles, 900 plain tiles, 38 casks of cement, 90 casks of broken stone, making a total of 640 bushels of stone." The reproductions reflected the best scientific knowledge of the time. Owen, who rejected Darwin's ideas about evolution, made the models at Sydenham as mammalian as possible; this made modern reptiles appear less advanced than their fossil predecessors, thereby refuting the simple chain of progressive change.[12] During the Victorian Age, Hawkins was one of the most effective popularizers of natural history; his models are the only major surviving feature of the Crystal Palace. Hawkins, who spent much of the rest of his life lecturing about pterosaurs, dragons, and the shortcomings of evolution, sold lithographs of various animals that he imagined; his opposition to evolution was ironic, for he had worked with Joseph Leidy (who accepted evolution), had helped Charles Darwin illustrate Darwin's account of his voyage aboard HMS *Beagle*, and had illustrated the iconic comparison of human and ape skeletons for Thomas Huxley's book *Man's Place in Nature*. In 2002 and 2007, Hawkins's models of *Iguanodon* were restored and once again drew visitors to the small park in the London suburb of Sydenham. Today, 14 models adorn the park's "Prehistoric Monster Trail."

1855 Edward Hitchcock displays his collection of 8,000 "bird tracks" at Amherst College in Amherst, Massachusetts. The tracks (i.e., the Hitchcock Ichnological Cabinet) were displayed in the lower floor of a two-story museum called the "Appleton Cabinet," so named because it was built with money provided in the will of Samuel Appleton of Boston.

1855 The "Haarlem specimen" (TM 6428, also known as the "Teyler specimen") of *Archaeopteryx* is discovered near Riedenburg, Germany, and described in 1875 by Hermann von Meyer as *Pterodactylus crassipes*. In 1970, John Ostrom reclassified the fossil as *Archaeopteryx*, and today the specimen is in the Teylers

Museum in Haarlem, the Netherlands. (Teylers Museum is the first and oldest museum in the Netherlands.) The specimen, which is the oldest specimen of *Archaeopteryx*, consists mostly of limb bones, some cervical vertebrae, and ribs.

1855 While working for the United States Geological Survey (USGS) in Sioux country of the Nebraska Territory (in what is now central Montana), geologist Ferdinand Vandeveer Hayden finds more teeth of a duckbill dinosaur near the mouth of the Judith River. Local Native Americans, who referred to Hayden as "the man who picks up stones while running," did not bother Hayden because they believed he was a harmless madman. Hayden gave his specimens to Joseph Leidy, an anatomist at the University of Pennsylvania, who—noting that the teeth were different from other teeth ever found in North America, but similar to teeth that British scientists had referred to as dinosaurs—later classified the organisms as dinosaurs. These organisms—later named *Trachodon mirabilis* ("rough tooth"), *Deinodon horridus* ("terror tooth"), and *Troödon formosus* ("wound tooth")—were among the first discoveries of fossilized dinosaurs. Leidy is considered by many to be the founder of American paleontology.

1855 The Albany Museum is established in Grahamstown, South Africa. Today, the museum houses fossils of the prosauropods *Massospondylus* and *Euskelosaurus*, which was first described by Thomas Huxley in 1866.

1856 Joseph Leidy describes *Deinodon horridus* from the Judith River Beds of what would be Montana. *Deinodon* ("terror tooth") was the first dinosaur recognized in North America. That same year, Joseph Leidy used teeth to name *Trachodon*, a dinosaur similar to *Iguanodon*, and *Troödon* ("wounded tooth"). Leidy later suggested that dinosaurs with this dentition (e.g., *Iguanodon, Trachodon*, and *Hadrosarus*) resembled kangaroos, not the heavy quadrupeds proposed earlier by Richard Owen. Today, *Trachodon* is not considered a valid name because the teeth are not sufficient for identifying genera, and the name *Troödon* was changed to *Troodon* (i.e., without the diaeresis) in 1876. Little more material of *Troödon* was discovered until 1902, when Lawrence Lambe reported similar teeth in Alberta, Canada. Lambe named 20 dinosaurs (several of which were discovered by Charles H. Sternberg and his family), 11 (i.e., 55%) of which remain valid.

1856 Ludwig Rütimeyer (1825–1895) describes some remains of a plateosaurid and names it "Dinosaurus." When it was pointed out that this name had already been given to a lizard, Rütimeyer renamed the fossil *Gresslyosaurus. Gresslyosaurus* is now considered a synonym of *Plateosaurus*.

1856 Richard Owen (Figure 5)—a famed anatomist, friend of Queen Victoria, and foe of Charles Darwin—becomes the first Superintendent of the Natural History departments of BMNH at a salary of £800 per year. He immediately began campaigning for a new building (as Owen put it, "a cathedral to nature"), which opened in 1881 (Figure 1). *The (London) Times* declared that "there is not a more distinguished man of science in the country" than Owen. Owen, who had named dinosaurs in 1842, was lauded as the equal of France's Georges Cuvier and Germany's Alexander von Humboldt.

1856 Sixteen human-like bones are discovered by limestone workers in a cave in the Neander Valley (Neanderthal in Old German) near Düsseldorf, Germany. These remains, which were characterized by a low brow, thick jaw, and thick bones, were given by local preacher Johann Fuhlrott to Hermann Schaaffhausen, who in 1858 published a detailed description of the bones; this description noted that the skull resembled that of a large ape. Schaaffhausen's work was translated into English in 1861 by British surgeon George Busk, who in 1864 nominated Charles Darwin for membership in the Royal Society. In 1863, these 41,000-year-old fossils became the type-specimen of *Homo neanderthalensis* (*Neanderthal 1*), now known to be an extinct cousin of *Homo sapiens*. Neanderthals were often depicted as "cavemen" in movies about dinosaurs.

1856 Two teeth from *Astrodon* are the first sauropod remains named in North America.

1856 William Buckland's death is memorialized with a plaque near the cloisters in Westminster Abbey that reads, "He applied the Powers of his mind to the Honour and Glory of God." Buckland, who published the first scientific description of a dinosaur, was a former dean of Westminster.

1857 Henry Fairfield Osborn (1857–1935) is born into a wealthy, well-connected family in Fairfield, Connecticut. Osborn, whose father was a railroad baron, later used his family's connections—especially those with his uncle J. Pierpont Morgan and his brother Frederic's friendship with Theodore Roosevelt—to build AMNH into one of the world's greatest museums. Today, AMNH houses one of the greatest collections of dinosaur fossils in the world.

1858 Amateur naturalist William Parker Foulke (1816–1865) and a crew from Philadelphia's ANS excavate a dinosaur skeleton (49 individual bones and teeth) from a ravine in Haddonfield, New Jersey. Joseph Leidy (1823–1891), who became known as the "father of North American vertebrate paleontology," later described the skeleton and, in doing so, became the first person to describe the anatomy of a dinosaur (which he named *Hadrosaurus foulkii*). This, the first dinosaur named in North America (and the first hadrosaur named) from good skeletal material, began the study of dinosaurs in North America. *H. foulkii*, an ornithischian dinosaur, was found about 10 feet below the surface; when alive, the 73-million-year dinosaur was 25 feet long, 12 feet tall, and weighed 3–4 tons. The skeleton was the most complete yet found in North America; its pelvis and limbs were intact, and there were nine teeth, fragments of jaws, and 28 vertebrae. Leidy noted the specimen's similarities to *Iguanodon* and believed that it "was most probably amphibious." Soon thereafter, Leidy suggested, "It is not improbable that *Hadrosaurus* retained the ordinary prostrate condition." After noting that the hind limbs of *Hadrosaurus* were longer than the forelimbs, Leidy suggested that *Hadrosaurus* and other dinosaurs might have been bipeds. (This suggestion was anticipated by Gideon Mantell for *Iguanodon*.) When *H. foulkii* was displayed publicly a decade later at the Philadelphia Academy of Sciences (with several reconstructed parts that were missing), it was the first such display in the world. Thomas Jefferson tried to procure the bones, but was rebuffed. When biologist Othniel Marsh (1831–1899) later viewed the Haddonfield site with Edward Drinker Cope (1840–1897), Marsh offered the local miners money if they would send bones to him instead of Cope; this triggered the bitter, public, and spectacular "Bone War" between Marsh and Cope, which destroyed the finances of both participants. Charles Willson Peale (1741–1827) excavated a second skeleton at Haddonfield and took it to Philadelphia, where it was later displayed at the APS. From 1868 to 1883, *H. foulkii* was the only articulated dinosaur mounted anywhere in the world. (The next articulated dinosaur to be mounted was *Iguanodon*, exhibited in Brussels by Louis Dollo and preparator Louis De Pauw.) In 1994, the Haddonfield site where *H. foulkii* was discovered was made a national monument. Today, near downtown Haddonfield, you can visit the 30-foot-deep wooded ravine (and the "Hadrosaurus Run" stream that runs through it) where *Hadrosaurus* was found. In town you will find an 8-foot-tall bronze *Hadrosaurus foulkii* sculpted by John Giannotti and erected there in 2003.

1858 Edward Drinker Cope begins working with Joseph Leidy at Philadelphia's ANS.

1858 Edward Hitchcock's large, 220-page monograph *Ichnology of New England: A Report on the Sandstone of the Connecticut Valley, Especially Its Fossil Footmarks*, published by the state of Massachusetts, describes Hitchcock's collection of fossilized tracks, noting the possibility (based on a few tail impressions) that the "giant birds" that left the tracks shared some traits with reptiles. As Hitchcock noted on page 190, "What a wonderful menagerie! Who would believe that such a register lay buried in the strata? . . . At first men supposed that the strange and gigantic races which I had described, were mere creatures of imagination, like the Gorgons and Chimeras of the ancient poets. But now that hundreds of their footprints, as fresh and distinct as if yesterday impressed upon the mud, arrest the attention of the sceptic on the ample slabs of our cabinets, he might as reasonably doubt his own corporeal existence as that of these enormous and peculiar races." As authors of the *Bridgewater Treatises* had tried to do in England, Hitchcock linked geology with theology, noting that the animals that had left the tracks were evidence of "the same infinitely wise and benevolent Being as the Author of all." Hitchcock, who died six years after the book's publication, wrote, "I have done what I could in laying the foundation, and in gathering a storehouse of materials. Let others, with better light to guide them, carry up and complete the structure."

A decade later, Thomas Huxley claimed that the creatures that left the tracks in Connecticut "were either birds or reptiles, or more probably both. . . ." Hitchcock, who in 1830 had become the first director of the Massachusetts Geological Survey, opposed Darwinian evolution and was one of the most influential Congregational theologians in the first half of the 19th century. Hitchcock was also the first person to describe the remains of dinosaurs in North America.

1858 In *La Création et ses Mystères Dévoilés* (*Creation and Its Mysteries Unveiled*), French geographer Antonio Snider-Pellegrini (1802–1885) becomes the first person to unequivocally claim that modern continents were once a single landmass that was separated by several catastrophes, including Noah's Flood. Snider-Pellegrini was also the first person to note that similar fossils and rock formations occur on adjacent continents. Snider-Pellegrini's work, which linked and split continents, anticipated Alfred Wegener's theories about Pangaea by several decades.

1858 James Deane dies, but Edward Hitchcock continues to claim that he (i.e., Hitchcock) was the first to study the tracks discovered in the Connecticut Valley.

1858 Richard Owen publishes a study of the 13-foot-long, spined *Hylaeosaurus*, which was described earlier by Gideon Mantell. Owen subsequently published papers about the plated *Scelidosaurus* (1859), the plated *Omosaurus* (1877), the sauropod *Cetiosaurus* (1875), the sauropod *Bothriospondylus* (1875), and *Megalosaurus* (1883).

1858 Wilhelm Schulz (1805–1877), known in Spain as Guillermo Schulz, discovers the first dinosaurs and marine reptiles in Spain.

1859 Edward Drinker Cope publishes his first scientific paper while working part-time at ANS in Philadelphia. Two years later, Cope entered the University of Pennsylvania and began studying anatomy under Joseph Leidy. During his lifetime, Cope published more than 1,400 papers and described 64 species of dinosaurs (as well as 300 species of fish). However, many of Cope's papers included errors, and many of the scientific names that he assigned were later disqualified or withdrawn.

1859 Fossils of a turkey-sized, bipedal coelurosaur that will later be named *Compsognathus* are discovered in Bavaria. Thomas Huxley, who considered *Compsognathus* to be the closest relative to dinosaurs, but not a dinosaur itself, used *Compsognathus* as his primary evidence to argue in 1868 that birds had evolved from reptiles, noting that "it is impossible to look at the conformation of this strange reptile and to doubt that it hopped or walked, in an erect or semi-erect position, after the manner of a bird, to which its long neck, slight, head, and small anterior limbs must have given it an extraordinary resemblance." For more than a century, *Compsognathus* was celebrated as the smallest dinosaur.

1859 Geologist and Roman Catholic priest Abbé Jean-Jacques Pouech (1814–1892) finds several "very large egg shells" These fragments, found in the French Pyrenees, were the first reported dinosaur eggshells. (Pouech later questioned whether the fragments came from the eggs of "gigantic birds," and the fragments were not recognized as belonging to dinosaurs until the early 20th century.) Since this discovery, dinosaur eggs and eggshells have been found worldwide. The richest sites have been Upper Cretaceous strata, especially in dry, arid regions. In 1982, Kenneth Carpenter (b. 1949) suggested that this distribution of dinosaur eggs and eggshells might be due to taphonomic bias.

1859 In a canyon just south of Moab, Utah, geologist John Strong Newberry (1822–1892) finds bones of what was probably a sauropod in Cañon Pintado. The bones were given to Joseph Leidy, after which they were described by Edward Drinker Cope (an adventurous loner and former student of Leidy's). In 1877, Cope named the species *Dystrophaeus viaemalae* ("wasted one") and claimed that this was the first dinosaur to be discovered in the Morrison Formation in western North America. The next documented fossil from the famed Morrison Formation (named after the town of Morrison, Colorado, which was founded in 1874 by George Morrison) was not collected for another 10 years. Dinosaurs comprise approximately one-third of the 90 types of vertebrates in the Morrison Formation.

1859 John Murray Publishing of London, which had published all of Charles Lyell's books, publishes Charles Darwin's 502-page *On the Origin of Species by Means of Natural Selection, or The Preservation of Favoured Races in the Struggle for Life*. Murray printed 1,250 copies of *On the Origin of Species*, 139 of which were distributed as promotional copies. Booksellers bought all of the 1,111 remaining copies (for 15 shillings—about £35 today—apiece) on the first day they were for sale. Darwin's theory was based on an ancient Earth, and he wrote that anyone not grasping Earth's antiquity "may at once close this book." *Origin* shifted thought from a foundation of untestable awe of special creation to a science-based examination of the natural world based on natural mechanisms and historical patterns. Whereas scientists before Darwin had often invoked purpose to explain biology (e.g., that a particular structure was present because it was pleasing to a deity), Darwin's idea replaced purpose with function and history.

1859 Joseph Leidy names *Astrodon* (later *A. johnstoni*), the first sauropod named in the United States. His specimen, which consisted of a crown of one tooth from the Early Cretaceous, was unearthed in an iron mine at Muirkirk, Maryland, by Baltimore dentist Christopher Johnston. *A. johnstoni* (YPM 798) became Maryland's state dinosaur in 1998.

1859 Richard Owen (Figure 5) describes early Jurassic remains of *Scelidosaurus*, one of the first complete dinosaur skeletons (and the first Jurassic ornithischian dinosaur) to be discovered. The fossils were discovered in Dorset of southern England. Two years later, Owen described more bones of *Scelidosaurus*, and in 1863 he described a nearly complete skeleton of *Scelidosaurus*.

1860 Benjamin Waterhouse Hawkins's restorations of prehistoric animals are featured on a set of six posters. Small models of Hawkins's restorations were also sold in Ward's catalog of scientific equipment and supplies.

1860 During the meeting of the BAAS at Oxford, Thomas Huxley and Anglican Bishop Samuel Wilberforce (1805–1873) clash over the topic of evolution. Huxley—who had implored other scientists to defend Darwin's idea "if we are to maintain our position as the heirs of Bacon and the acquitters of Galileo"—had pledged to the publicity-shy Charles Darwin that "as for your doctrines I am prepared to go to the stake if requisite," and the Oxford meeting provided that opportunity. Although not an actual debate, the legendary exchange between Wilberforce and Huxley was later hailed as a turning point in the acceptance of evolution by the scientific community and represented a serious challenge to the Church's authority.

1860 Edward Hitchcock updates *Elementary Geology*, a bestseller that uses gap creationism to reconcile geology with Genesis. Although earlier editions of *Elementary Geology* had disparaged the evolutionary claims of Lamarck, the 1860 edition questioned Darwin's theory. Hitchcock's book, which eventually went through 30 editions, argued that "scientific truth is religious truth," accepted an ancient Earth, and argued that the Bible and science must be related. Hitchcock collected more than 20,000 fossilized footprints of dinosaurs, and believed that the prints were left by huge, extinct birds. Hitchcock was America's leading advocate of catastrophism-based gap-creationism and a Charter Member of the National Academy of Sciences (NAS).

1860 Nineteen-year-old Edward Drinker Cope publishes 31 papers. While still a friend of Othniel Marsh, Cope named a fish *Ptyonius marshii*.

1860 Richard Owen's *Paleontology* textbook includes moa and the Connecticut Valley footprints in the same illustration to highlight their similarities.

1860 Thanks to financial support from his uncle and wealthy self-made businessman George Peabody, Othniel Charles Marsh—described by his landlady as "very odd" and with whom getting acquainted was "running against a pitchfork"—graduates from Yale College. Two years later, Marsh graduated from Sheffield Scientific School (as would famed paleontologist William Bell Hatcher in 1884) before going to Europe for additional study. Peabody, along with J.P. Morgan, Chicago retailer Marshall Field, and Andrew Carnegie, sponsored much of the early work in paleontology in the United States.

1860 Thanks to lobbying by Joseph Leidy and others, Charles Darwin is elected a member of ANS in Philadelphia. The Academy was the first scientific institution in America to recognize Darwin's contributions to science. When Leidy read *On the Origin of Species*, he wrote to Darwin, "I feel as though I had hitherto groped about in darkness and that all of a sudden a meteor flashed upon the skies."

1861 Edward Hitchcock argues that the discovery of *Archaeopteryx* validates his conclusions about giant tracks in the Connecticut Valley.

1861 German paleontologist Johann Andreas Wagner (1797–1861) reports *Compsognathus* ("dainty jaw") and notes its bird-like features, but denies any actual kinship between birds and reptiles. The 24-inch-long bird-like dinosaur, which had been found in Solnhofen, Germany, eventually got to YPM, where it was rediscovered in 1961 by John Ostrom. *Compsognathus* was the first complete theropod, and one of the first small dinosaurs, to be discovered.

1861 German physician Karl Häberlein discovers fossils of the chicken-sized *Archaeopteryx lithographica* ("ancient wing of the printing stone") in Upper Jurassic limestone just north of Solnhofen, Germany. The fossil, which was known in Germany as *Urvögel* ("first bird"), was named by paleontologist Hermann von Meyer: "I consider *Archaeopteryx lithographica* a suitable name for this animal." (The name *lithographica* came from the fine-grained limestone that surrounded the fossil and that was used in the printing industry.) The 150-million-year-old fossil was an unusual feathered bird-like creature that had dinosaur-like traits; for example, it had long feathers on its arms (as do birds), but its fingers were clawed and separate (like those of carnivorous dinosaurs). Richard Owen, with the help of English naturalist and negotiator George Robert Waterhouse (1810–1888), bought the specimen (today known as BMNH 37001, or the "London specimen"), along with 1,702 other specimens, for £700 for BMNH (for comparison, a typical house at the time sold for £100, and £700 was double the museum's annual acquisitions budget). Owen bought the fossil from Häberlein, who—needing to raise a dowry for his daughter—had acquired the fossil from quarrymen in return for medical services. Owen, who argued that *Archaeopteryx* was not an intermediate between birds and reptiles, described the specimen in 1863 and noted that "the bones of the present specimen . . . indicate [it] once had been a bird." The following year, Darwin told a colleague that the *Archaeopteryx* fossil is "by far the greatest prodigy of recent time. It is a grand case for me, as no group was so isolated as birds." The "London specimen" of *Archaeopteryx*, which is missing its head, is housed in BMNH.

1861 In a letter published in *Neues Jahrbuch für Mineralogie* (*New Yearbook for Mineralogy*), prominent German paleontologist Christian Erich Hermann von Meyer (1801–1869) describes a small (2.4-inch-long and 0.5-inch-wide), isolated impression of "an ancient feather from the lithographic limestone" in Upper Jurassic limestone near Solnhofen in southern Germany. At the time, there were no birds known before the Tertiary. Thus, von Meyer—one of the first to recognize that similarities between animals did not necessarily mean close relationships (i.e., what today is known as convergence)—noted that the feather "need not necessarily be derived from a bird" and could be from a "feathered animal differing essentially from our birds." The next year, von Meyer published another letter claiming "this is a veritable fossilization in the lithographic stone that fully corresponds with a bird's feather." That same year, the feather—the first evidence of birds in the Jurassic—was linked to the raven-sized *Archaeopteryx*, the earliest known bird. The main slab of the feather, which is still the best-preserved feather of *Archaeopteryx*, is housed in the Humboldt Museum für Naturkunde in Berlin, and the counterslab is in the Bavarian Paleontological State Collections in Munich.

1861 In a paper titled "A New Reptile Supposedly Furnished with Bird Feathers," Johann Andreas Wagner—who had earlier named *Compsognathus*—describes *Archaeopteryx*, but names it *Griphosaurus problematicus* ("puzzling mythical lizard"). Wagner, who rejected Darwin's theory of evolution by natural selection, had not seen the fossil; instead, he based his observations on a sketch made from memory by German paleontologist Albert Oppel (1831–1865). Wagner declared the fossil to be a strange reptile unrelated to birds, adding, "I must add a few words to ward off Darwinian misinterpretation of our new Saurian. At first glance of the *Griphosaurus* we might certainly form the notion that we had before us an intermediate creature, engaged in the transformation from the Saurian to the bird. Darwin and his adherents will probably employ the new discovery as an exceedingly welcome occurrence for the justification of their strange views upon the transformation of animals. But in this they will be wrong." In 1862, Wagner's claims were translated and published in English in *Annals and Magazine of Natural History*.

1861 Physician James Harrison finds dinosaur bones near Charmouth in southern England. That same year, Richard Owen described these bones as *Scelidosaurus* ("limb lizard"). This dinosaur, described in *The Armored Dinosaurs* (2001) as "the first complete dinosaur," appeared in Nintendo's *Jurassic Park: Park Builder* video game and in the children's show *Harry and His Bucket Full of Dinosaurs*.

1861 *Punch* magazine features a cartoon showing antediluvian creatures staring at a comet.

1862 Benjamin Waterhouse Hawkins's painting of *Hylaeosaurus* appears on a souvenir poster. Hawkins later included a group of *Hylaeosaurus* in his painting *Jurassic Period in England*, which he completed while at Princeton University in the mid-1870s. *Hylaeosaurus* was neglected by most paleoartists.

1862 In *The Geologist*, S. J. Mackie (1832–1902) describes the newly discovered *Archaeopteryx* as a "sensation," and Henry Woodward labels it as "startling" and a "wonderful discovery."

1863 Abraham Lincoln establishes the NAS to accommodate the government's requests to "investigate, examine, experiment, and report upon any subject of science or art."

1863 At a meeting of the Royal Society, Richard Owen (Figure 5) describes *Archaeopteryx* as the "by-fossil-remains oldest-known feathered Vertebrate." Although University of Munich professor Andreas Wagner had claimed in 1861 that the fossil was of a reptile he called *Griphosaurus*, Owen upheld von Meyer's original name of *Archaeopteryx*.

1863 Edward Drinker Cope meets Othniel Marsh in Berlin, while Cope is touring Europe and Marsh is studying at the University of Berlin. In subsequent years, the relationship between Marsh and Cope became famously hostile. The only record of Cope's and Marsh's meeting in Berlin was made by Marsh, who in 1890—while trying to convince *New York Herald* readers that Cope was mentally and morally unfit—noted, "I had some doubts of [Cope's] sanity." Cope, who was often more speculative than Marsh, soon fired back, claiming that Marsh "was using me . . . to furnish him with brains." Cope understood that he was not popular with everyone; when he described the Miocene mammal *Anisonchus cophater*, he wrote to Henry Osborn, "It's no use looking up the Greek derivation of cophater . . . I have named it in honor of the number of Cope-haters who surround me."

1863 French scientist and writer Louis Figuier's (1819–1894) *The World before the Deluge* (Figure 8) includes sketches by French painter Édouard Riou (1833–1900) of battling *Megalosaurus* and *Iguanodon* (with a nasal horn) that were based on Benjamin Waterhouse Hawkins's models from a decade earlier (Figure 7). Riou later provided 56 illustrations for the seventh edition of Jules Verne's *Journey to the Center of the Earth*; this was the first edition of Verne's famous book that included several new paleontological passages.

1863 In a letter to Charles Darwin, mammalogist Hugh Falconer (1808–1865) notes that "had the Solnhofen quarries been commissioned—by august command—to turn out a strange being à la Darwin—it could not have executed the behest more handsomely—than in the *Archaeopteryx*." Although Darwin did not strongly invoke *Archaeopteryx* in defense of his ideas, he did believe that *Archaeopteryx* exemplified the incompleteness of the fossil record; as Darwin noted in the fourth edition of *On the Origin of Species*, "Hardly any recent discovery shows more forcibly than his how little we as yet know of the former inhabitants of the world."

1863 In an article titled "The Lithograph of the Lithographic Pterosaur," German scientist Christoph Giebel (1820–1881) is the first to claim that the "London specimen" of *Archaeopteryx* is a fake.

Figure 8 French artist Édouard Riou's (1833–1900) sketch titled *Ideal Scene of the Lower Cretaceous Period*—Plate XXI in the English version of Louis Figuier's *The World before the Deluge* (1866)—depicted "*Iguanodon* and *Megalosaurus* [battling] for the mastery in the centre of a forest." In fact, both of these dinosaurs were bipedal, and the horn on the snout of *Iguanodon* was on its thumb. (Bettmann/Corbis)

1863 Thomas Huxley begins teaching students that birds are "so essentially similar to Reptiles in all the most essential features of their organization, that these animals may be said to be merely an extremely modified and aberrant Reptilian type." That same year, Huxley told the Royal Society that birds originated in dinosaurian reptiles. Huxley, who grouped birds and reptiles into a group called Sauropsida, supported his claim by citing more than 30 traits shared by birds and non-avian dinosaurs (e.g., hollow bones, elongated neck bones). Although most modern paleontologists accept Huxley's claim, a group named BAND (Birds Are Not Dinosaurs) rejected Huxley's claim, as do young-Earth creationists. For example, in 2008, young-Earth creationist Ken Ham claimed, "If we look at God's Word to find the truth, we'll see that [birds being related to dinosaurs] could not be possible. Why? Because the birds, the flying creatures, were created on Day 5 [and] . . . the dinosaurs were made on Day 6" of Creation Week.

1863 Upon seeing *Archaeopteryx* while in Germany, Edward Drinker Cope notes, "The collections from Solnhofen are exceedingly interesting and numerous. . . . It is very bird-like, it has a long, slender bill as *Conchiosaurus* [a dinosaur]. . . . It is nearer bird than reptile but is neither."

1863 Writing in *Philosophical Transactions of the Royal Society*, Richard Owen describes *Archaeopteryx* as a primitive bird. Owen dismissed its four teeth as being part of a fish.

1864 Edward Hitchcock, who had been the first person to publish a detailed description of dinosaur fossils from North America, dies believing that he had been studying the tracks of antediluvian birds. That same year, Edward Hitchcock Jr. embraced the discovery of the Jurassic bird *Archaeopteryx* as a confirmation of his father's work.

1864 In his fanciful *A Journey to the Center of the Earth*, French writer Jules Verne (1828–1905) describes the adventures of Professor Otto Lindenbrock (in the common English translation of Verne's book, Lindenbrock was replaced by Professor Von Hardwigg) and his fellow travelers, who reach the center of the Earth by descending the Sneffels Volcano.[13] While descending through geologic epochs, the travelers witness the days of biblical creation, including an interior sea in which an ichthyosaur battles a plesiosaur (several of the Mesozoic reptiles in Verne's book were restored at Crystal Palace). Verne's book, which converted prehistoric discoveries into high drama, explored the dissonance between Genesis and the scientific claims about evolution and an old Earth. Verne's book, published originally in French (Verne did not read or write English), was translated into English in 1872, and was the first novel to include prehistoric animals. The first illustrated edition of *A Journey to the Center of the Earth* appeared in 1867 and was modeled after Guillaume Louis Figuier's popular *The World before the Deluge* (1863). When the movie version of *A Journey to the Center of the Earth* premiered in 1959, the battling marine reptiles were replaced by dinosaurs (played by iguanas with attached horns and sails).

1864 University of Jena embryologist and vertebrate anatomist Karl Gegenbaur (1826–1903) classifies the late Jurassic theropod *Compsognathus longipes* as an intermediate between birds and other reptiles.

1865 Edward Hitchcock posthumously describes the 6-foot-long *Megadactylus* from bones discovered in 1818 in Manchester, Connecticut. In 1882, Othniel Marsh renamed the dinosaur *Amphisaurus*, and then *Anchisaurus* in 1885 when it was discovered that the name *Amphisaurus* had already been used to describe another animal. The site where Hitchcock collected many of his tracks is now the Springfield Armory National History Site. That same year, Edward Hitchcock Jr. published descriptions of more than 20 ornithoid reptiles and more than 30 thick- and thin-toed birds.

1865 Industrialist and amateur archaeologist Sir John Evans (1823–1908) claims that *Archaeopteryx* seems "to link together the two great classes of Birds and Reptiles [and that] its extreme importance as bearing upon the great question of the Origin of Species must be evident to all."

1865 Joseph Leidy unearths what he names *Coelosaurus* (in 1979, *Coelosaurus* was renamed *Ornithomimus*). Leidy also published additional details and several sketches of *Hadrosaurus* in the monograph *Cretaceous Reptiles of the United States*. Leidy, noting that its hind limbs were much larger than its fore limbs, reported, "The great disproportion of the size between the fore and back parts of the skeleton of *Hadrosaurus* lead me to suspect that this giant extinct herbivorous lizard may have been in the habit of browsing, sustaining itself kangaroo-like in an erect position on its back extremities and tail." (Gideon Mantell had noted that the forelimbs of *Iguanodon* were smaller than the hind limbs, but these observations were discounted by Richard Owen.) Leidy's conclusion that *Hadrosaurus* was bipedal differed from Richard Owen's reconstructions of *Iguanodon, Hylaeosaurus*, and *Megalosaurus*, which Owen depicted as quadrupedal.

1865 Othniel Marsh (Figure 9) visits Charles Darwin at Down House, Darwin's home just outside of London. By 1874, Marsh was a committed evolutionist.

1865 Rev. William Fox (1813–1881) discovers a headless spinal column that is designated *Polacanthus* the following year. *Polacanthus* was restored as a quadruped in 1905 by Caleb Barlow (working with Baron Franz Nopcsa), and popularized by Neave Parker's restoration on cards sold by BMNH.

1866 At the commencement of Yale College's Sheffield Scientific School, Othniel Marsh (Figure 9) is appointed professor of paleontology, the first such professorship in the New World. Marsh believed that his appointment, which involved no teaching and was "without salary from existing funds," would place "him on a level with men all of greater age and experience than myself." Marsh named 80 dinosaurs, including *Allosaurus* (1877), *Camptosaurus* (1885), *Ceratops* (1888), *Diplodocus* (1878), *Stegosaurus* (1877), and *Triceratops* (1889).

1866 Charles Henry Hitchcock (1836–1919), a son of Edward Hitchcock, publishes "A Description of a New Reptilian Bird from the Triassic of Massachusetts." Hitchcock helped found the Geological Society of America.

1866 Edward Drinker Cope uses anatomical features of the lower leg and ankle to sort dinosaurs into four orders: Orthopoda (straight feet, including duckbills and armored dinosaurs), Goniopoda (ankle feet, including carnivores), Symphopoda (grown-together feet, including small carnivores such as *Compsognathus*), and Opisthocoela (a name invented by Richard Owen that was unrelated to leg anatomy and including large, swamp-dwelling dinosaurs). Cope's system was confusing; for example, Symphopoda included herbivores and carnivores, and Orthopoda included armored quadrupeds and armorless bipeds. That same year (i.e., 1866), Cope's article titled "Discovery of a Gigantic Dinosaur in the Cretaceous of New Jersey" used scant remains to describe the theropod *Laelaps aquilunguis* ("eagle-clawed terrible leaper"; in Greek mythology, *Laelaps* was a dog that always caught its prey, but was turned to stone in

Figure 9 Othniel Charles Marsh (1831–1899) founded vertebrate paleontology in America. Marsh, who feuded bitterly with Edward Drinker Cope, argued convincingly that birds are evolutionarily linked with dinosaurs. Marsh named more dinosaurs (80) than anyone else in history; these dinosaurs included the now-iconic *Stegosaurus, Diplodocus, Triceratops,* and *Brontosaurus* (*Apatosaurus*). (Library of Congress)

mid-leap.) Cope described his discovery—the first partial skeleton of a predatory dinosaur found in the United States—as "the most formidable type of rapacious terrestrial vertebrata of which we have any knowledge" and "the devourer and destroyer of Leidy's *Hadrosaurus*." The fossils had been collected from the Late Cretaceous by workers at the West Jersey Marl Company about 12 miles south of Haddonfield, New Jersey, and Cope portrayed the animal as actively attacking *Hadrosaurus*. When it was discovered that the name "Laelaps" had already been used to designate a genus of mite, *Laelaps* was renamed (in 1877) *Dryptosaurus* by Cope's rival, Othniel Marsh. Cope viewed dinosaurs as active and energetic; an active, leaping *Dryptosaurus* was later featured in a famous painting by Charles Knight, the most influential of all artists who restored extinct animals (Knight's early images portrayed dinosaurs as active animals). During his lifetime, Knight—the first artist to work with paleontologists to ensure scientifically accurate images—produced more than 1,000 drawings, sculptures, and paintings of ancient animals. Countless future re-creations of ancient animals—in encyclopedias, papers, books, movies, museums, and elsewhere—were based on Knight's work.

1866 Miniature models of Benjamin Waterhouse Hawkins's Crystal Palace dinosaurs are advertised in the Ward's Natural Science Establishment catalog. Ward's (established in 1862 by Henry A. Ward) specialized in sales of fossil casts for museums and collectors.

1866 The first dinosaur-tracks are reported west of the Mississippi.

1866 George Peabody (1795–1869), Othniel Marsh's uncle and the greatest philanthropist of the mid-Victorian era, endows $150,000 (up from $100,000 promised three years earlier) for "the foundation and maintenance of a Museum of Natural History, especially for the departments of Zoology, Geology and Mineralogy, in connection with Yale College." The first trustees were Professors Benjamin Silliman, Sr., James Dwight Dana, Othniel Marsh, and Senator James Dixon. Although the museum (Yale's first museum building) originally focused on minerals collected by geologist Benjamin Silliman, Marsh's discoveries later composed most of the original exhibits, and the museum's Great Hall of Dinosaurs was dominated by *Apatosaurus*, a sauropod discovered by Marsh in 1877 (but named *Brontosaurus*). This was the first mounted skeleton of *Apatosaurus*, and it remains the most famous sauropod, despite the fact that others (e.g., *Brachiosaurus*) were larger and heavier. Peabody had amassed his fortune as an overseas merchant, and later, as a London banker.

1867 Andrew Carnegie leaves Pittsburgh, and never lives there again. However, Carnegie never forgot that Pittsburgh was a source of his wealth, and supported the city generously for the rest of his life. CMNH houses a world-class collection of dinosaurs.

1867 Congress authorizes geological explorations of the American West. These explorations, which included the King Survey, the Hayden Survey, the Powell Survey, and the Wheeler Survey, were the first federally-funded geological explorations. In 1879, the U.S. Congress consolidated these surveys into the USGS.

1867 Famed fossil-hunter Edward Drinker Cope's paper in *Proceedings of the Academy of Natural Sciences of Philadelphia* titled "An account of extinct reptiles that approach birds" suggests that dinosaurs and birds were evolutionarily related ("gave an account of the extinct reptiles which approached the birds in their structure.") That same year, Cope—repeating Robert Chambers's arguments from *Vestiges of the Natural History of Creation* (1844)—claimed that the path of evolution was "conceived by the Creator according to a plan of His own, according to His pleasure." Cope also claimed that "the most bird-like of the tracks of the Connecticut sandstone" had been made by dinosaurs.

1867 Thomas Huxley describes 14 features shared by birds and reptiles (but not by mammals) and suggests a category called Sauropsida to include both groups of animals. Huxley also suggested that dinosaurs, like birds, were hot-blooded. That same year, Huxley—who named three of Great Britain's dinosaurs—described *Acanthopholis*. In 1999, *Acanthopholis* was made a *nomen dubium*.

1868 A near-complete skeleton of *Cetiosaurus* is discovered. The following year, Thomas Huxley argued that *Cetiosaurus* was a dinosaur. Discoverer Richard Owen had earlier believed that *Cetiosaurus* resembled a giant marine crocodile.

1868 After examining *Archaeopteryx* and the small, bipedal theropod *Compsognathus* (both of which were found in the same limestone sediments in Germany), Thomas Huxley proposes in "On the Animals Which Are Most Nearly Intermediate between Birds and Reptiles" that "there can be no doubt that the hind quarters of the *Dinosauria* wonderfully approached those of birds in their general structure, and therefore that these extinct Reptiles were more closely allied to birds than any which now live." Huxley justified his separation of *Compsognathus* from other dinosaurs this way: "Notwithstanding its small size, this reptile must, I think, be placed among, or close to, the Dinosauria; but it is still more bird-like than any of the animals which are ordinarily included in that group. . . . Birds are evolved from dinosaurs and the proof is here in these fossils. These are your missing links. . . . Surely there is nothing very wild or illegitimate in the hypothesis that the phylum of the class Aves has its roots in the dinosaurian reptiles." Huxley's suggestion would not be taken seriously for nearly a century. Huxley's description of *Archaeopteryx* focused on Owen's errors, including Owen's inability to distinguish the animal's left and right feet.

1868 Benjamin Waterhouse Hawkins mounts, at no charge, *Hadrosaurus* at the Academy of Natural Sciences in Philadelphia. (Before this time, the fossils had been exhibited in boxes.) The skeleton lacked a skull, so Hawkins sculpted one using modern iguanas as his model. Hawkins believed that *Hadrosaurus* had a kangaroo-like posture, which was radically different from the lumbering, quadruped posture proposed earlier by Richard Owen. The skeleton's pose was based on Joseph Leidy's observation in 1865 that "the enormous disproportion between the fore and hind parts of the skeleton of *Hadrosaurus* has led me to suspect that this great herbivorous Lizard sustained itself in a semierect position on the huge hinder extremities and tail." The 26-foot-long, 14-foot-tall skeleton on display at the academy increased annual attendance from 30,000 visitors to more than 100,000 visitors in only two years (despite the fact that the museum was usually open to the public only two afternoons per week, and closed in August). The popularity of the *Hadrosaurus* exhibit forced ANS to begin charging admission fees to limit attendance, despite the fact that, in 1868, few people in the United States had heard the word *dinosaur*. The display also produced orders for duplicates (called "plasterotheres") from several other museums (e.g., Smithsonian, Chicago, and Princeton's Nassau Hall, the nation's capital for four months in 1783 when the Congress of the Confederation met there). With this display—the first of a fully articulated dinosaur displayed anywhere in the world for the next 15 years—Hawkins brought dinosaur-mania to America and forever changed the way dinosaurs were displayed to the public.

1868 Cope and Marsh spend a week searching for hadrosaur bones in Leidy's quarry in Haddonfield, New Jersey. (By this time, Cope lived in Haddonfield.) During this visit, Marsh privately offered pit-managers money to send bones they found to him in New Haven. That same year, Cope noted, "If ["*Laelaps*" was] warm-blooded, as Prof. Owen supposes the Dinosauria to have been, he undoubtedly had more expression than his modern reptilian prototypes possess. He no doubt had the usual activity and vivacity which distinguishes the warm-blooded from the cold-blooded vertebrates." That same year, Cope—a Neo-Lamarckian—argued in *On the Origin of Genera* that natural selection acting on random variation is not enough to explain life's diversity. Instead, Cope proposed a "steady progressive development of organization [through] a continual crowding backward of the successive steps of individual development." In his later years, Cope placed a greater emphasis on adaptation with less involvement of a Creator.

1868 In a letter to German embryologist Ernst Haeckel (1834–1919), Thomas Huxley outlines how birds could have evolved from reptiles: "The road from Reptiles to Birds is by way of *Dinosauria* to the *Ratitae* [flightless birds such as emus and ostriches]—the Bird "Phylum" was Struthious, and wings grew out of rudimentary fore limbs." *Archaeopteryx* was relatively insignificant to Huxley; although he believed that *Archaeopteryx* showed that birds could have evolved from reptiles, he conceded that it was "more remote from the boundary-line between birds and reptiles than some living Ratitae are."

1868 In *The Fossil Reptiles of New Jersey*, Edward Drinker Cope (Figure 10) depicts *Laelaps* (the first dinosaur that he discovered) in a kangaroo-like pose. In the same sketch, Cope showed the plesiosaur *Elasmosaurus*, on which he placed the skull atop its tail. This mistake was later noted by Othniel Marsh, who was Cope's rival.

1868 In the same year that he debunked the discovery of "fossil human remains" in Nebraska, Othniel Marsh discovers bones of a small horse (*Equus parvulus*, now *Protohippus*) while excavating in Nebraska. Marsh later teamed with Thomas Huxley to describe the evolution of modern horses from a four-toed ancestor and, in the process, provided strong evidence for how species evolve. Several months after Marsh and Huxley predicted that a more ancient, five-toed animal probably existed, fossils of *Eohippus* ("the dawn horse") were discovered. In 1871, Marsh found the first American pterosaur fossils, and in 1878 he named *Diplodocus* ("double beam"), casts of which were distributed throughout the world by Andrew Carnegie (including the one currently on display at BMNH). Marsh's discoveries made him one of the most famous paleontologists of his era and convinced many people of the validity of Darwin's theory.

Figure 10 American paleontologist Edward Drinker Cope (1840–1897), shown here at age 55, lost almost all of his family's fortune during his legendary and bitter "Bone War" with Othniel Marsh. Cope discovered, described, and named more than 1,000 species of vertebrates, 64 of which were dinosaurs. However, many of Cope's (and Marsh's; Figure 9) descriptions of dinosaurs were hastily prepared, incomplete, and poorly documented, and now have only historical significance. The deeply religious Cope, a neo-Lamarckian, hoped that his discoveries of dinosaurs would vindicate his belief in progress and perfectibility. (Louie Psihoyos/Corbis)

1868 Medicine Bow, Wyoming, is founded as a Union Pacific Station for providing water and other supplies for its locomotives. (In 1902, Medicine Bow was featured in Owen Wister's novel *The Virginian*.) Numerous iconic dinosaurs (e.g., *Diplodocus*) displayed today in YPM, NMNH, and other museums were discovered in nearby Como Bluff and shipped to museums on the Union Pacific Railroad. In 1868, on his first transcontinental trip aboard the Union Pacific Railroad, Othniel Marsh stopped briefly at Como Station, where he examined a sauropod bone found at a nearby ridge. Marsh was not impressed, and the vast numbers of dinosaur fossils at Como Bluff remained undiscovered. Subsequent discoveries at Como Bluff—an east-west ridge located between the towns of Rock River and Medicine Bow, Wyoming—later packed museums and exhibit halls throughout the world with fossils of dinosaurs and other vertebrates. The dinosaurs discovered at Wyoming's Como Bluff—a low ridge near the rails of Union Pacific Railroad—became some of the most familiar ones known to the public, including *Allosaurus, Apatosaurus* (*Brontosaurus*), and *Stegosaurus*.

1868 Othniel Marsh (Figure 9) is appointed Vertebrate Paleontologist of the USGS. During the next 14 years, Marsh spent more than $200,000 of personal funds to collect dinosaurs, including nearly all of the iconic stegosaurs and sauropods from Morrison, Colorado, Cañon City, Colorado, and Como Bluff, Wyoming. Thomas Huxley later described Marsh's collection of fossils this way: "There is no collection of fossil vertebrata in existence which can be compared to it."

1868 Thomas Huxley claims that the curious three-toed tracks reported by Edward Hitchcock from the Connecticut Valley show that "at the commencement of the Mesozoic epoch, bipedal animals existed which had the feet of birds, and walked in the same erect or semi-erect fashion."

1868 With the fame of the Crystal Palace dinosaurs in mind, Andrew H. Green (Comptroller of New York City's Central Park) writes to Benjamin Waterhouse Hawkins: "Recognizing the interest that has long attended your restoration of the forms of extinct animals in Europe, the Commissioners of the Central Park have thought that a similar work . . . reconstituting the phenomena of the ancient epochs of this continent would be of equal scientific value, and of special interest in an education point of view. . . [We ask you] to undertake the resuscitation of a group of animals of the former periods of the American continent. . . . The sympathy and support of the scientific men of this country, and that museums and collections of fossil treasures . . . will be freely opened for such examinations as you may wish to make in the prosecution of this interesting undertaking." Hawkins accepted the invitation, noting his "restorations . . . in the Crystal Palace at Sydenham were the first efforts of the kind ever attempted, and their acknowledged success . . . encourages me to hope that a similar enterprise may meet with equal favor on this side of the Atlantic. . . . The interest in the remains of ancient animal life which Geology has revealed within the last century is worldwide, and almost romantic in its influence upon the imagination." The next year, Hawkins planned to build two plesiosaurs, "gigantic *Hadrosaurus* of the exact dimensions (one twenty-six feet, other thirty-one feet, long), [and] . . . models of Laelap's Aquilunguis." Hawkins's museum was to be at the southwest corner of Central Park opposite 63rd Street. Before trying to create this "Paleozoic Museum," Hawkins set up a studio at what is now AMNH, but the studio was subsequently destroyed by vandals hired by politician William "Boss" Tweed (1823–1878), and the planned museum never opened. Hawkins, who illustrated the fish and reptiles for Charles Darwin's *Zoology of the Voyage of the HMS Beagle*, later told the *New York Times* that Darwin's theory of evolution "could not be believed by any comparative anatomist" and that the "unity of design in nature shows the hand of God himself." Although none of Hawkins's models of dinosaurs survived (they were allegedly buried in the southern part of Central Park), a mold of Hawkins's *Hadrosaurus* survived. One replica was sent to New York, where it was destroyed in 1871; another went to Princeton University in 1874; another went to the Smithsonian Institution, and then to the Field Columbian Museum in 1984; and the final replica went to Philadelphia for the Declaration of Independence Centennial in 1876. Later in 1876, the Philadelphia replica was sent to Edinburgh, Scotland, where it became the first mounted dinosaur in Europe. Today, 15 of Hawkins's 17 paintings of the geological epochs he was to illustrate are displayed at Princeton University.

1869 At a meeting of the Royal Society, Thomas Huxley defends his claim of an evolutionary connection between dinosaurs and birds. Huxley argued that small, birdlike dinosaurs such as *Compsognathus* resembled the true ancestors of what birds might have looked like.

1869 Edward Drinker Cope names the family Hadrosauridae, sometimes called the duck-billed dinosaurs. That same year, in an article in *American Naturalist*, Cope included a scene of life during the Cretaceous; this was one of the earliest scenes from deep time by an American artist.

1869 Geologist Philippe Matheron (1807–1899) finds fragments of eggshells and nearly complete eggs in Upper Cretaceous sediments in southern France. Matheron wondered whether the eggs were from the sauropod *Hypselosaurus* or "a gigantic bird."

1869 Harry Govier Seeley (1839–1909) discovers the type-specimen of *Giganto-saurus megalonyx* (BMNH 32498–99, "giant lizard"). The name "Gigantosaurus" later became embroiled in taxonomic controversies (involving *Barosaurus, Tornieria, Janenschia*, and other genera of dinosaurs) and is considered by many paleontologists to be a *nomen dubium*. During his career, Seeley named 35 dinosaurs, but only four (i.e., 11%) remain valid.

1869 The Union Pacific Railroad and the Central Pacific Railroad meet at Promontory Point, Utah, thus creating the Trans-Continental Railroad. This railroad, one of the greatest technologic feats of the 19th century, later enabled paleontologists working in the American West to ship their fossils to museums in the Midwest (e.g., Chicago's Field Museum) and East (e.g., AMNH, CMNH, YPM). In 1944, the U.S. Postal Service commemorated the 75th anniversary of the Trans-Continental Railroad by issuing a 3¢ stamp.

1869 Theodore Roosevelt, Morris Jesup, J. Pierpont Morgan, and Albert S. Bickmore (1839–1914), a one-time student of Louis Agassiz, found AMNH in the Upper West Side of Manhattan in New York City. Although AMNH stayed out of the Cope-Marsh dispute, it later hired Henry Osborn—a rich, ambitious Princeton-educated scientist—to found its Department of Vertebrate Paleontology. Osborn, who had met Charles Darwin in Thomas Huxley's lab in 1879, often flaunted his and his museum's status. Osborn disliked Marsh, and became a supporter of Cope. Today, AMNH—which has done as much as any institution in the world to help the public appreciate Earth's vast history—consists of 25 interconnected buildings, houses more than 30 million specimens, and hosts four million visitors annually. Its exhibits—which include more dinosaur bones than any museum in the world—cover more than 700,000 square feet of floor space. Its two dinosaur halls contain more than 100 dinosaur specimens, of which more than 80 percent are actual fossils (i.e., not casts). The most popular of these specimens include a mummified *Hadrosaurus*, an *Oviraptor* egg with an embryo, the world's only intact *Velociraptor* skull, a 4-foot-wide vertebra of *Diplodocus*, famous skeletons of *Apatosaurus* and *T. rex*, and a *Coelophysis bauri* specimen containing bones of another dinosaur in its stomach.

1869 Thomas Huxley describes *Hypsilophodon foxii*, the first small ornithopod discovered. *Hypsilophodon foxii* was one of the first dinosaurs used to support the evolutionary link between dinosaurs and birds.

1869 While surveying for the USGS, American geologist Ferdinand Vandeveer Hayden is shown a partial caudal vertebra found near Steamboat Springs, Colorado. The specimen was eventually sent to Joseph Leidy in Philadelphia, who in 1870 named the animal *Poicilopleuron* (then *Antrodemus*). Later studies suggested that the *Antrodemus* bones were probably *Allosaurus*. By this time, Leidy had retreated from vertebrate paleontology because of his relative lack of money and his distaste for competition with colleagues. Today, a statue honoring Leidy (erected in 1907) stands in front of Philadelphia's Academy of Natural Sciences.

1869 Workmen digging a well at William Newell's farm in Cardiff, New York, unearth a 10-foot-long statue claimed to be a "fossilized human being" that becomes known as the Cardiff Giant. The giant, which became a national sensation, was later declared a fraud by Othniel Marsh. Today, the Cardiff Giant is displayed at Farmers' Museum in Cooperstown, New York.

1869 Edward Drinker Cope restores *Laelaps*, but none of his three sketches will be published during his Cope's lifetime. In 1877, Benjamin Waterhouse Hawkins painted *Laelaps* attacking a herd of *Hadrosaurus* in Cretaceous New Jersey, and two decades later (i.e., in 1897), Charles Knight produced his famous painting of two fighting *Laelaps* for AMNH. Although Cope had once been a talented artist, he produced relatively few restorations of dinosaurs (his kangaroo-like *Laelaps* was an exception).

1870 As Joseph Leidy nears the end of his distinguished career as America's leading vertebrate paleontologist, he writes to a British colleague, "Formerly, every fossil one found in the State came to me, for nobody else cared to study such things, but Professors Marsh and Cope, with long purses, offer money for what used to come to me for nothing, and in that respect I cannot compete with them." That same year, Leidy noted, "Prof. Cope has described the skeleton [of *Elasmosaurus*] in a reverse position and in that view has represented it in a restored condition. . . ." (Othniel Marsh sarcastically claimed that Cope should have named *Elasmosaurus* "Streptosaurus" ["twisted reptile"]). Henry Osborn, Cope's primary biographer, later wrote that "the decade of 1870 to 1879. . . was the golden period in Cope's life; the period of his greatest achievement, of his incomparable western expeditions, of the acquirement of the greatest part of his fossil collections, of the foundation of his principal scientific discoveries."

1870 Edward Drinker Cope shows the affinities between dinosaurs and the fossilized footprints from the Connecticut Valley collected by Edward Hitchcock.

1870 Othniel Marsh (Figure 9) begins an expedition with 11 students from Yale to Wyoming and Utah; this was the first expedition to the American West specifically to collect fossils. Thanks to support provided by 43 troops and letters from U.S. Army commander George Sherman, which opened the way for help from almost every fort along the way (e.g., Fort McPherson in Nebraska), Marsh's expedition was successful (e.g., during this expedition, Marsh uncovered the first North American pterosaur). For a while, the expedition was accompanied by Buffalo Bill Cody. That same year, Marsh began shipping vertebrate fossils to Yale University. During the next 22 years, Marsh sent more than 3,000 shipments, and spent more than $200,000 of personal funds for his work (as well as $150,000 from USGS).

1870 Surveyor Ferdinand Hayden describes Othniel Marsh as "more ambitious than Cope ever was; he is raging ambitious."

1870 Thomas Huxley (Figure 11) argues in "Further Evidence of the Affinity between the Dinosaurian Reptiles and Birds" that birds descended directly from dinosaurs. Huxley noted that if the leg bones of baby chickens were enlarged and fossilized, "there would be nothing in their characters to prevent us from referring them to the Dinosauria." Huxley rejected Edward Drinker Cope's claim that ankle bones are the critical feature for classifying dinosaurs, and instead used a 12-item list of traits (including teeth, hipbones, legs, and jaws) to reinvent Dinosauria as three different families: Scelidosauridae (armored dinosaurs such as *Hylaeosaurus*), Iguanodontidae (herbivores such as *Iguanodon* and *Hadrosaurus*), and Megalosauridae (carnivorous dinosaurs such as *Dryptosaurus* and *Megalosaurus*). When problems arose (e.g., *Compsognathus* fit into all three groups), Huxley added a fourth order he named Ornithoscelida ("bird leg"), which included Dinosauria and *Compsognathus*. Huxley also argued that dinosaurs, like birds, were warm-blooded and had bird-like anatomical traits (e.g., wings). That same year (i.e., 1870), Huxley told the Geological Society that it is "very doubtful"

Figure 11 Thomas Henry Huxley (1825–1895), Charles Darwin's most vocal protagonist, was a towering figure in biology. Soon after its discovery, Huxley claimed *Archaeopteryx* as a transition between birds and reptiles. Starting in 1868, Huxley compared *Archaeopteryx* with prehistoric reptiles and concluded it most closely resembled dinosaurs such as *Compsognathus* and *Hypsilophodon*. Huxley was one of the first people to propose an evolutionary relationship between birds and dinosaurs. (Huxley, Thomas Henry and Leonard Huxley. *Life and Letters of Thomas Henry Huxley*, 1903)

that dinosaurs were "the actual linear types by which the transition from lizard to bird was affected." Huxley believed that dinosaurs' real ancestors were "hidden from view in the older [geologic] formations." Near this time, Huxley's studies of paleontology began to decline as he became increasingly involved with lecturing, writing, and politics.

1871 Although most paleontologists no longer consider dinosaurs to be oversized lizards, Benjamin Waterhouse Hawkins's painting "Jurassic Life of Europe" for Princeton's Museum of Natural History depicts them as such. Three years later, Hawkins returned to England to find his giant models of dinosaurs neglected and "sadly depreciated."

1871 Charles Darwin's *The Descent of Man, and Selection in Relation to Sex* uses biogeography and other evidence to suggest that humans originated in Africa.

Henry Osborn, who later promoted orthogenesis and eugenics, responded to Darwin's suggestion by proposing that "Dawn Man" originated in Asia. Osborn sent Roy Chapman Andrews to Asia to find fossils of human ancestors, but Andrews instead found dinosaurs there that catapulted him to fame.

1871 During a second drip down the Green River, famed explorer John Wesley Powell (1834–1902) notices "reptilian remains" in the Uinta Mountains near Vernal, Utah. In 1893, AMNH's Olaf A. Peterson (1865–1933) found dinosaur bones there in the Morrison Formation. This area later became famous as Dinosaur National Monument (DNM), and the Uintah Basin as "Utah's Dinosaurland." The Morrison Formation is famous for its collection of long-necked dinosaurs, including the iconic *Apatosaurus, Diplodocus*, and *Camarasaurus*.

1871 During his first trip to the American West, Edward Drinker Cope writes to his five-year-old daughter Julia, "I have been traveling about in this country looking for fossil bones of huge animals such as I bring home in my bag to Haddonfield [New Jersey] sometimes." That same year, Cope began prospecting in Kansas. For these expeditions, Cope hired American lawyer and geologist Benjamin Franklin Mudge (1817–1879; one of Marsh's guides) as an assistant. Mudge, who led the first geological survey of Kansas in 1864, helped train Charles H. Sternberg and became a strong opponent of evolution.

1871 In *The Geology of Oxford and the Valley of the Thames*, John Phillips (1800–1874)—a vocal critic of Darwin's theory of evolution by natural selection—describes a skeleton of *Cetiosaurus*. Phillips's description, which provided the first accurate image of a sauropod, compared *Cetiosaurus* to crocodiles, lizards, and other dinosaurs. Phillips also noted that *Megalosaurus* was "essentially reptilian; yet not a ground crawler like the alligator, but moving with free steps chiefly, if not solely, on the hind limbs, and claiming a curious analogy, if not some degree of affinity, with the ostrich."

1871 John Whitaker Hulke (1830–1895) begins dinosaur paleoneurology with a description of the braincase of *Iguanodon*.

1871 Near Jabalpur, India, a "Mr. Medlicott" finds vertebrae and a 46-inch-long portion of a dinosaur femur. This discovery, from which British paleontologist Richard Lydekker (1849–1915) named *Titanosaurus indicus* in 1877, was the first major discovery of a dinosaur in the southern hemisphere. Lydekker's subsequent sketch of *Stegosaurus* was an important part of the plot in Arthur Conan Doyle's *The Lost World* (1912).

1871 While in Kansas on his second western expedition with 10 Yale students, Othniel Marsh finds a headless, flightless, "large fossil bird, at least five feet in height" that he names *Hesperornis regalis* ("ruling western bird"). Two years later, he found another skeleton of *Hesperornis*, this one with a skull. When Marsh realized that the skull included teeth, he announced that his discovery and the toothed bird *Ichthyornis* (discovered by Benjamin Mudge) "does much to break down the old distinction between Birds and Reptiles, which the *Archaeopteryx* has so materially diminished." *Hesperornis* later appeared in several television shows, including *Primeval* and *Dinosaur Train.*

1872 Benjamin F. Mudge, a chemist living in Kansas, sends Othniel Marsh a collection of fossilized bones, which includes hollow bones of a bird and the jaws of a saurian. Although Marsh initially described the specimens as belonging to different animals (*Ichthyornis dispar* for the bird, *Colonosaurus mudgei* for the saurian), he later realized that the bones were from one animal—a bird with teeth. Another Kansas geologist, Samuel Williston (1851–1918)—who subsequently worked for Marsh—later noted that Professor Mudge "had been sending his vertebrate fossils previously to Professor Cope for determination." After learning Marsh (a boyhood acquaintance of Mudge) was interested in these fossils, Williston sent the fossils to Marsh. Had Cope received the box, he would have been the first to announce the discovery of "birds with teeth."

1872 Charles Darwin notes Edward Drinker Cope's neo-Lamarckism in the sixth (and final) edition of *On the Origin of Species* and recognizes the "most unexpected manner" in which ostriches, *Compsognathus*, and *Archaeopteryx* form a transition between birds and reptiles. That same year, Darwin wrote in a letter to another neo-Lamarckian (Alpheus Hyatt) that "it has quite annoyed me that I do not clearly understand yours and Professor Cope's views, and the fault lies in some slight degree, I think, with Professor Cope, who does not write very clearly." The sixth edition of *On the Origin of Species*, which was one-third longer than the first edition, contained an additional chapter dealing with objections that have been raised about Darwin's theory. During Darwin's lifetime, John Murray Publishers sold more than 25,000 copies of the English version of *Origin*, and more than 56,000 copies by 1899.

1872 Edward Drinker Cope publishes 56 papers and begins gathering evidence of Othniel Marsh's scientific errors. Cope later accused Marsh of fraud, mismanagement, and scientific incompetency, thereby making the "bone wars" a public spectacle. That same year, *American Naturalist* relegated the ongoing accusations by Cope and Marsh to appendices published "at the expense of the author." In one issue, Marsh bought nine pages, which he used to question Cope's ethical standards, noting "for this kind of sharp practice in science, Prof. Cope is almost as well known as he is for the number and magnitude of his blunders."

1872 English writer William Davenport Adams's (1828–1891) *Life in the Primeval World* includes dramatic (but inaccurate) restorations of *Megalosaurus, Iguanodon*, and *Hylaeosaurus*. Adams was influenced by French writer Victor Meunier's (1817–1903) *Les Animaux d'Autrefois* (1869).

1873 Barnum Brown (1873–1963) is born in Carbondale, Kansas, and named after circus showman P. T. Barnum, whose traveling circus had recently visited his state. Brown, who was later known as "the Father of the Dinosaurs" and "Mr. Bones," became the greatest dinosaur collector of all time. He discovered *T. rex*, and his study of the growth patterns of the small, horned dinosaur *Protoceratops* was the first to consider the importance of ontogeny in dinosaurs. Brown shares his birthdate (February 12) with Abraham Lincoln and Charles Darwin.

1873 Henry Woodward (1832–1921), a founder-editor of *Geological Magazine* (established in 1864) and an assistant at BMNH who had examined *Archaeopteryx* soon after its arrival there, notes, "There is perhaps no single point along the whole line of defences raised by the opponents of the theory of Evolution which has been more warmly contested than the . . . relationship of birds to reptiles."

1873 Othniel Marsh writes to Edward Drinker Cope that "the information I received . . . made me very angry, and had it come at the time I was so mad with you for getting away [with a collector who had worked for Marsh] I should have 'gone for you,' not with pistols or fists, but in print. I came very near publishing this with some other transgressions including a certificate from Mr. Kinne but my better judgment prevailed. I was never so angry in my life." That same year, at a party in New York for English scientist John Tyndall, Cope met Marsh. Cope later remarked that Marsh "stuck to me like a leech and I hope became fully satisfied that I was not a thief." Other scientists learned of Marsh's feud with Edward Drinker Cope when, in an article titled "Note on the Dates of Some of Prof. Cope's recent Papers, by O. C Marsh," published in the *American Journal of Science*, Marsh claimed that Cope antedated papers describing newly discovered vertebrates. When Marsh convinced the APS to censure Cope for printing papers before they were read, Cope was "disgusted at the meanness of the proceedings." In the summer of 1873, Cope wrote in *American Naturalist* "most of Prof. Marsh's criticisms are misrepresentations, his systematic innovations are untenable, and his statements as to the dates of my papers are either criminally ambiguous or untrue." When the Cope–Marsh feud erupted in newspapers in 1890, it was much more vicious.

1874 Alexander Agassiz (1835–1910), the son of famed biologist (and Darwin opponent) Louis Agassiz, writes to Thomas Huxley that Cope has earned "the contempt of all the scientific men of the country."

1874 Charles Robert Knight (1874–1963) is born in Brooklyn, New York, and later studies at the New York's Metropolitan Museum of Art. Although his first job as an artist was to draw plants and animals for stained-glass windows sold to churches, he later became the most influential of all artists who re-created ancient life, especially dinosaurs.

1874 Despite having 12 field crews in western fossil-fields, Othniel Marsh returns to Yale to supervise construction of YPM. In future years, Marsh's administrative responsibilities increasingly curtailed his field work.

1874 Former English clergyman Arthur Lakes (1844–1917) sends a tooth (YPM-VP 4192) found near Golden, Colorado, to Othniel Marsh, who places the specimen in Yale's collection. Marsh never published a description of the tooth, but its size and the beds in which it was found suggested that it may have been the first specimen of *Tyrannosaurus* ever collected. Lakes's original quarry on the west side of Dinosaur Ridge on Alameda Parkway (overlooking Morrison, Colorado) is marked with a historical landmark.

1874 George Mercer Dawson (1849–1901), son of famed geologist Sir William Dawson of McGill University, reports the first Cretaceous dinosaur bones in western Canada. Dawson found the fossils in the Wood Mountain area of southern Saskatchewan.

1874 Just three years before the discovery of *Stegosaurus*, H. G. Seeley describes a partial skull of *Craterosaurus pottonensis*, but Seeley's discovery was not declared a stegosaur until the 1980s. In 1874, another stegosaur (*Dacentrurus*) was discovered, but Richard Owen could not determine if the armored plates belonged to a dinosaur or were the head plates of a giant fish. Since these discoveries, stegosaurs have been found in Middle Jurassic to Early Cretaceous sediments of North America, Europe, Africa, and Asia. About 10 percent of *Stegosaurus* tail-spikes have healed breaks at their tips.

1874 Michigan geologist Alexander Winchell's *Sketches of Creation* includes three scenes by Édouard Riou from *The World before the Deluge* (1863). The cover of Winchell's book featured a gold-embossed image of Benjamin Waterhouse Hawkins's *Hadrosaurus*.

1874 Othniel Marsh (Figure 9) goes west to search for fossils near the Black Hills, where he tells Oglala Lakota (Sioux) Chief Red Cloud (1822–1909) that if Red Cloud would help him obtain fossils, Marsh would present Red Cloud's concerns about governmental fraud and dishonesty to federal officials in Washington, DC. Red Cloud agreed and Marsh later presented Red Cloud's concerns to administrators serving President U.S. Grant, and then to Grant himself (and several newspapers). Later, the Secretary of the Interior and others in the Bureau of Indian Affairs resigned or were fired because of corruption and their poor treatment of Native Americans. Red Cloud later recalled that Marsh "came here and I asked him to tell the Great Father something. He promised to do so, and I thought he would do like all white men, and forget me when he went away. But he did not . . . he is the best white man I ever saw."

1874 Princeton University awards Benjamin Waterhouse Hawkins an honorary doctorate of science.

1874 U.S. president Ulysses S. Grant lays the cornerstone for the first building of AMNH along the 77th Street in New York City. That building opened two years later.

1875 Alfred Cope dies at age 69 and his 35-year-old son, Edward Drinker Cope, inherits $250,000. Cope will spend almost all of this money in a ruthless and unseemly competition with Othniel Marsh to discover and name fossils.

1875 Friedrich von Huene (1875–1969) is born in Tübingen, about 60 miles west of Solnhofen, German. During his 61-year professional career, von Huene discovered *Proceratosaurus*, *Antarctosaurus*, and a herd of more than 35 *Plateosaurus*, and named 71 dinosaurs (of which 18 remain valid); this was more than anyone in history except Othniel Marsh (who named 80).

1875 Hermann von Meyer describes a specimen of *Pterodactylus crassipes*, which had been found in 1855 near Riedenburg, Germany. In 1970, this specimen was reassigned by Yale University paleontologist John Ostrom as *Archaeopteryx*, and is now the "Haarlem specimen" of that genus.

1875 Richard Owen (Figure 5) describes the first remains of a stegosaurid, an incomplete skeleton of *Omosaurus* (in 1902, *Omosaurus* was renamed *Dacentrurus*, meaning "pointed tail"). The bones had been discovered the previous year by Welsh naturalist William Davies (1814–1891), who believed they were from a "huge dragon." Two years later, nearly complete skeletons of *Stegosaurus* were found in Quarry 13 at Wyoming's Como Bluff. *Stegosaurus* later became one of the most popular dinosaurs, appearing on postage stamps and greeting cards, and in toy sets, television shows, and creation museums.

1876 After his application for a fossil-hunting job with Benjamin Mudge is rejected, Charles Hazelius Sternberg asks Edward Drinker Cope for a job. Sternberg described what happened next: "Fortunately, the Professor responded promptly, and when I opened the envelope a draft for three hundred dollars fell at my feet. The note accompanying the money said: 'I like the style of your letter. Enclose draft. Go to work.' The letter bound me to Cope for four long years and enabled me to endure immeasurable hardships and privations in the barren fossil fields of the West; and it has always been one of my joys of my life to have known intimately in the field and shop the greatest naturalist America has produced." That same year, Sternberg, Cope, and a small crew collected 18 specimens of dinosaurs from the Judith River badlands, which had been scouted by Hayden in 1855. This was Sternberg's first professional expedition. Sternberg was grateful to Cope, pledging to "make a collection of [the area's] wonderful fossils no matter what it might cost me in discomfort and danger." The *Topeka Daily Capital* later praised Sternberg as "the man who made fossil hunting a profession." Sternberg spent 13 of his 21 years with Cope in the field and, with Barnum Brown, collected more dinosaurs than anyone in history.

1876 Charles Darwin exclaims to Alfred Russel Wallace, "What progress Palaeontology has made during the past twenty years!"

1876 Edward Drinker Cope moves from Haddonfield, New Jersey, to nearby Philadelphia, Pennsylvania, where he buys adjacent buildings at 2100 and 2102 Pine Street. One building became his museum, and the other his home.

1876 Harry Ford Sinclair (1876–1956) is born in Benwood, West Virginia. After being trained as a pharmacist at the University of Kansas, in 1916 he formed Sinclair Oil and Refining Company by merging 11 small petroleum companies. He later sponsored several expeditions by AMNH to the American West to search for dinosaurs. Sinclair also owned horses that won the Kentucky Derby and, after being sentenced to prison later in his life, worked in the prison infirmary as a pharmacist.

1876 Othniel Marsh (Figure 9), armed with more than 40 different specimens of *Hesperornis*, establishes the most convincing link between birds and reptiles since *Archaeopteryx*. (Marsh referred to *Hesperornis* as a "carnivorous swimming Ostrich.") When Marsh subsequently published a monograph describing toothed birds in 1880, Thomas Huxley noted that Marsh's work "completed the series of transitional forms between birds and reptiles, and removed Mr. Darwin's proposition . . . from the region of hypothesis to that of demonstrable fact."

1876 The Centennial Exposition (the first official World's Fair in the United States) opens in Philadelphia, and a major attraction for the 10 million attendees is Benjamin Waterhouse Hawkins's 1868 mount of *Hadrosaurus*. (*Hadrosaurus* shared the spotlight with the world's largest steam engine and the torch of the soon-to-be-completed Statue of Liberty.) This mount was later placed in front of the new "castle" headquarters of the Smithsonian Institution in Washington, DC. (The reconstruction was later taken down after several years of weather-related damage.) Soon after Philadelphia's Centennial Exposition ended, plans were made to exhibit Edward Drinker Cope's fossils in a museum in Philadelphia's Fairmount Park. These plans never materialized, and most of Cope's collection remained in storage in the basement of Fairmount Park's Memorial Hall for the next two decades.

1876 The second paleontological expedition to what is now Montana includes three collectors who would later become famous fossil hunters: Charles H. Sternberg, J. C. Isaac, and Edward Drinker Cope.

1876 Thomas Huxley (Figure 11), who became known as "Darwin's Bulldog," visits Othniel Marsh in New Haven. While touring America, Huxley told crowds in New York and elsewhere that "it is, in fact, quite possible that all these more or less avi-form reptiles of the Mesozoic epoch are not terms in the series of progression from birds to reptiles at all, but simply the more or less modified descendants of Palaeozoic forms through which the transition was actually affected." Huxley placed *Archaeopteryx* on an evolutionary side-branch; he doubted that it resembled a stage in the reptile–bird transition.

1876 While exploring the Missouri Breaks in the aftermath of Custer's defeat at Little Bighorn, Edward Drinker Cope and Charles H. Sternberg (1850–1943)—assuming that "since every able-bodied Sioux would be with the braves under [Chief] Sitting Bull . . . there would be no danger for us"—find hundreds of bones that Cope uses to describe new species. Today, most of those species—which were collected from the Judith River Formation (Late Cretaceous)—are no longer recognized because they were based on fossils that were not particularly unique (*Monoclonius* was a notable exception). Cope and Sternberg used a stabilizing rice-based paste when excavating and shipping their fossils.

1876 Yale University opens the first building of YPM.

1877 Another fossil of *Archaeopteryx*—this one well-preserved and having a tooth-filled jaw—is given to Johann Dörr by a worker at a small quarry in Eichstätt, Germany, about nine miles from where the first *Archaeopteryx* was found in Solnhofen. This specimen (HMN 1880) was purchased in 1881 by Ernst Otto Häberlein (whose father, Karl, had found the first *Archaeopteryx*) for 140 marks. Häberlein reported the fossil in the journal *Leopoldina*, and hinted that it would be for sale. Since this specimen included impressions of feathers, as well as an attached skull, Häberlein asked for 36,000 marks ($108,000), and finally sold it in 1880 to industrialist Werner von Siemens for 20,000 marks. The so-called "Berlin specimen," which was described in 1884 by Wilhelm Dames (1843–1898), now resides in the Humboldt Museum für Naturkunde in Berlin. This specimen, which Dames in 1897 named *Archaeopteryx siemensi* to honor Siemens's generosity, has the best preserved and most complete feathers of all the specimens of *Archaeopteryx*.

1877 Arthur Lakes (1844–1917), a Denver teacher and part-time geologist, writes in his journal of finding vertebrae (embedded in sandstone) "so monstrous that I could hardly believe my eyes." Lakes then wrote to Othniel Marsh about his discovery of giant Jurassic dinosaur bones along the Front Range of the Rocky Mountains (now Dinosaur Ridge, where the Rocky Mountains meet the High Plains), outside of Morrison, Colorado: "A few days ago whilst taking a geological section and measurements and examining the rocks on the banks of Bear Creek near the little town of Morrison about 15 miles west of Denver, I discovered in company with a friend a Mr. [Henry] Beckwith of Connecticut some enormous bones apparently a vertebra and a humerus bone of some gigantic saurian in the upper Jurassic or lower Cretaceous at the base of (Hayden's) Cret. No. 1 Dakota group." Lakes enclosed sketches of the bones and geologic descriptions of where the bones were discovered. Lakes later sent Marsh several more letters describing "colossal" bones of "no less than six different animals if not different species." Such bones would have represented some of the first such fossils from this part of the Mesozoic of North America. When Marsh did not respond, Lakes sent 10 boxes of fossils weighing more than 1,500 pounds to Marsh and, assuming that Marsh was not interested, also sent some specimens to paleontologist Edward Drinker Cope at the Philadelphia Academy of Sciences. When Marsh learned that Cope had received the specimens, Marsh instructed his collector Benjamin Mudge, a former professor who was working for Marsh in Kansas, to investigate the area described by Lakes. The first published announcement of Lakes's discoveries appeared in an article by Marsh titled "Notice of a New and Gigantic Dinosaur" in the July 1, 1877, issue of *American Journal of Science*. Marsh named the 50-foot-long specimen found by Arthur Lakes *Titanosaurus montanus* ("giant mountain reptile," which surpassed "any land animal hitherto discovered"), but later changed the name to *Atlantosaurus* ("Atlas reptile") when Cope pointed out that the name *Titanosaurus* had already been used for a sauropod (now dubious) in India. (Marsh had used a footnote in his *Titanosaurus* paper to point out that *Laelaps*—one of Cope's favorite discoveries—had already been applied to a genus of mites, and changed the name of *Laelaps* to *Dryptosaurus*). The bones probably belong to what is now called *Apatosaurus,* but are too fragmentary to know for sure. Despite Marsh's renaming, Cope continued to use the name *Laelaps* for the rest of his career. The site where Lakes collected his fossils, now called Quarry 5 and Dinosaur Bone Quarry, is along the west side of Dinosaur Ridge at Alameda Parkway in Morrison, Colorado. Two decades later, Charles Knight painted his now-famous image of a fighting, active *Dryptosaurus*; that painting—one of the few pre-1960s images to present dinosaurs as active, agile animals (thereby anticipating the theories of John Ostrom in the 1960s)—presaged agile dinosaurs and now hangs in AMNH.

1877 *Diplodocus* is found by Samuel Wendell Williston (1851–1918) near Cañon City, Colorado. Seven years later, *Diplodocus* became the first sauropod for which a complete skull was described (by Othniel Marsh).

1877 Henry Osborn and some friends take a spur-of-the-moment trip to Wyoming to collect fossils, beginning what will be a distinguishing career for Osborn as a paleontologist.

1877 In a keynote speech given at the annual meeting of AAAS in Nashville, Othniel Marsh tells attendees, "To doubt evolution today is to doubt science, and science is only another name for Truth. . . . Birds have come down to us through the Dinosaurs. . . . The case amounts almost to a demonstration, if we compare, with

Dinosaurs, their contemporaries, the Mesozoic birds. The classes of Birds and Reptiles, as now living, are separated by a gulf so profound that a few years since it was cited by the opponents of evolution as the most important break in the animal series, and one which that doctrine could not bridge over. Since then, as Huxley has clearly shown, this gap has been virtually filled by the discovery of bird-like Reptiles and reptilian Birds. *Compsognathus* and *Archaeopteryx* of the Old World, and *Ichthyornis* and *Hesperornis* of the New, are the stepping stones by which the evolutionist of today leads the doubting brother across the shallow remnant of the gulf once thought impassible." The following day, the *Nashville Daily American* reprinted Marsh's lecture, calling his discoveries "more heroic and not less important than those of Cuvier" and "hardly second in importance to that of Darwin."

1877 In a letter to Othniel Marsh, Samuel Williston describes pasting "strips of strong paper on fractured bones before removing" them. That same year, in another letter to Marsh, Arthur Lakes described using "a strong coat of plaster of Paris on the outside of the bone to preserve it whilst the rest of the rock [is] being jarred by the hammer." By 1880, all of Marsh's collectors were using Lakes's technique for stabilizing bones. Although such techniques enabled paleontologists to better preserve bones, new technological advances have not significantly changed the hard work and detailed observations necessary to find fossils.

1877 In a two-paragraph article (that includes no illustrations) in *American Journal of Science* titled "Notice of New Dinosaurian Reptiles from the Jurassic Formation," Othniel Marsh claims that a gigantic dinosaur "is represented in the Yale Museum by a nearly complete skeleton in excellent preservation" but uses only a description of the vertebral column to name the 50-foot-long *Apatosaurus* ("deceptive lizard"). Similar fossils were named *Brontosaurus* by Marsh in 1879, and *Elosaurus* by others in 1902, and today *Brontosaurus* and *Elosaurus* are considered to be *Apatosaurus*. In 1879, Marsh published a sketch of the shoulder blade, pelvis, and a few vertebrae of *Apatosaurus* in "Principal Characters of American Jurassic Dinosaurs, Part II" in *American Journal of Science*. Marsh also claimed that Cope misnamed and inaccurately described several fossils, and that "conclusions based on such work will naturally be received with distrust by anatomists." The discoveries of *Apatosaurus* and *Allosaurus* at what today is Dinosaur Ridge (just outside Denver, Colorado) began the "dinosaur rush" to the American West.

1877 In his haste to publish his findings before Edward Drinker Cope, Othniel Marsh adorns *Brontosaurus* with a hybrid of two different skulls found miles from the headless skeleton. The authenticity of the skull described by Marsh was later challenged by Earl Douglass of the Carnegie Museum of Natural History (CMNH). The editor of *American Journal of Science* began publishing Marsh's papers in a special appendix after getting tired of resetting its type to accommodate Marsh's last-minute submissions.

1877 John Ryder, under the direction of Edward Drinker Cope, builds a life-size "over fifty feet in length" reconstruction of *Camarasaurus supremus* for display at a meeting of the APS. This reconstruction, which was the first reconstruction of a sauropod, was later exhibited at AMNH, but was not published until after Cope's death more than 35 years later. In 1877, Cope published *The Origin of the Fittest*, which described an "originative" and "directive" force, which he believed Darwin had failed to explain. Cope argued that natural selection might cull poorly adapted organisms, but he rejected the idea that variation among species was due to inheritance, and was instead acquired from the environment (i.e., that Lamarckism produces variation). In 1895, after losing more than $250,000 in a mining fraud, the religious Cope sold most of his dinosaur collection to Henry Osborn (another advocate of Lamarckism) of AMNH. Most of the *Camarasaurus* collection in AMNH (e.g., AMNH 467) was collected by Cope and his workers.

1877 Othniel Marsh (Figure 9) claims that all of the three-toed tracks studied by Edward Hitchcock in the Connecticut Valley had been left by dinosaurs, not birds.

1877 Othniel Marsh tries to buy the "Berlin specimen" of *Archaeopteryx* for YPM for $10,000, but his offer is rejected.

1877 Othniel Marsh uses a collection of ribs, vertebrae, teeth, and other bone fragments collected by Benjamin Mudge and others from the Morrison Formation of Garden Park, Colorado, to name and describe *Allosaurus* ("different lizard"), a large (28-foot-long) bipedal theropod that had a large skull and reduced forelimbs. Although this is the type-specimen of *Allosaurus* (YPM 1930), in 2010 a petition was submitted to the International Commission on Zoological Nomenclature (ICZN) to transfer the name *Allosaurus fragilis* to a more complete specimen. *Allosaurus* later became one of the best-known and most popular theropods in films and documentaries.

1877 Othniel Marsh, using fossils uncovered by Arthur Lakes in the Morrison Formation near Morrison, Colorado, describes the distinctive and small-brained *Stegosaurus* ("spiked lizard"). This specimen (*Stegosaurus armatus*, YPM 1850) was the first *Stegosaurus* found in North America and the first stegosaur found with preserved armored plates. The dinosaur's midline was adorned by a single row of plates, its throat was covered by hundreds of button-size, bony ossicles, and its tail was armed with large spikes that were often more than 3 feet long (one of Arthur Lakes's assistants at Como Bluff referred to the spikes as "devil's tails"). The tail spikes of *Stegosaurus* often show evidence of injury and healing, suggesting that they were damaged in use. Marsh noted that although *Stegosaurus* weighed more than 8,000 pounds, its brain weighed only 2.8 ounces. Soon after describing *Stegosaurus*, Marsh suggested that a cavity in the hip region might have housed a second brain that coordinated the movements of the back legs and torso. The 3-foot-tall plates of *Stegosaurus*, which were larger than plates in any other stegosaur, were probably used for defense, thermoregulation, display, and species recognition. *Stegosaurus*, which was up to 25 feet long, was the largest non-sauropod herbivore when the Morrison Formation was being deposited. Such specimens have been found at Como Bluff, Dinosaur National Monument (DNM), the Utah Field House of Natural History State Park Museum (UFHNH) Quarry, and elsewhere. Othniel Marsh proposed the Stegosauria to include large, extinct, plated reptiles from the Upper Jurassic near Morrison, Colorado. Soon after naming *Stegosaurus* and noting that it is "without parallel" and has "the smallest known brain of any known land vertebrate," Marsh wrote to BMNH's Richard Owen, "I think we are on the eve of some important discoveries." Today, the most popular species of *Stegosaurus* is *S. stenops*, the best specimens of which have been discovered at Garden Park near Cañon City, Colorado. *Stegosaurus*, which Marsh called "one of the most remarkable animals yet discovered," is the state fossil of Colorado,

1877 Paul Gervais (1816–1879) makes the first microstructural study of dinosaur eggs. Gervais concluded the eggs were structurally similar to those of turtles, but did not rule out dinosaurs.

1877 The first discoveries along the northern lee of an obscure, low-lying, east-west ridge—15 kilometers long and 1.5 kilometers wide—called Como Bluff (near Como Station on the Union Pacific Railroad) in southeastern Wyoming begin to revolutionize paleontology, procedures for collecting dinosaur fossils, and the public's interest in dinosaurs. Como Bluff, which was investigated intensively for the next 12 years by Yale's Othniel Marsh, exposes the Sundance, Morrison, and Cloverly Formations, and contains one of the greatest assemblages of small dinosaurs and Jurassic mammals ever discovered. Although the Como Bluff dinosaur specimens were not the first such discoveries in North America, they were the first large-scale discoveries anywhere; when Samuel Williston arrived at Como Bluff, he wrote to Marsh that the area was superior to the Morrison (Colorado) and Garden Park (Colorado) sites, adding that the bones were "magnificently preserved and scattered for *seven* miles and are by the ton. . . . The bones are very thick, well preserved and easy to get out." Today, dinosaur fossils from Como Bluff stock many famous museums, including YPM, AMNH, and NMNH. Discoveries at Como Bluff (named for spring-fed Lake Como at the western end of Como Bluff) prompted other museums to fund expeditions searching for comparable specimens to display in their galleries, but today there is relatively little paleontological work there. Como Bluff is included on the National Register of Historic Places (1973) and is a National Natural Landmark (1973).

1877 U.S. president Rutherford B. Hayes (1822–1893) presides at the opening of AMNH's first building. The museum later became famous for its research and collection of dinosaurs.

1877 Using vertebrae collected by Oramel Lucas (1845–1935) in 1877 near Cañon City, Colorado, Edward Drinker Cope names and shows the APS a restored skeleton of *Camarasaurus*. *Camarasaurus*, which was the first North American sauropod to be named, is one of the most common dinosaurs of the Morrison Formation. Cope's reconstruction was not published until 1921.

1877 While Benjamin Mudge and Arthur Lakes are excavating near Morrison, Colorado, Othniel Marsh receives a letter from "Harlow and Edwards" reporting their discovery of large fossil bones in Wyoming: "I write to announce to you the discovery not far from this place of a large number of fossils, supposed to be those of the Megatherium, although there is no one here sufficient of a geologist to state for a certainty. We have excavated one (1) partly, and know where there is several others that we have not, as yet, done any work upon. The formation in which they are found is that of the Tertiary Period. We are desirous of disposing of what fossils we have, and also, the secret of the others. We are working men and are not able to present them as a gift, and if we can sell the secret of the fossil bed, and procure work in excavating others we would like to do so. We have said nothing to any-one as yet. We measured one shoulder blade and found it to measure four feet eight inches 4 ft. 8 in. in length. One joint of the vertebrae measures two feet and one half 2 1/2 in circumference and ten inches (10) in length. As proof of our sincerity and truth, we will send you a few fossils, at what they cost us in time and money in unearthing. We would be pleased to hear from you, as you are well known as an enthusiastic geologist, and a man of means, both of which we are desirous of finding—more especially the latter. Hoping to hear from you very soon, before the snows of winter set in, We remain, Very respectfully Your Obedient Servants [Signed] Harlow and Edwards". The letter was mailed from Laramie, Wyoming, to conceal the location of their find—Como Bluff. Marsh urged "Harlow and Edwards" to ship the bones they collected. The two boxes of fossils were shipped on September 17—nearly two months after the original letter—because "Harlow and Edwards" did not have enough money to pay the freight charges ($26.95, which was equal to about two weeks' salary). "Harlow and Edwards" then wrote another letter to Marsh, noting that they had excavated another 1,500 pounds of fossils, adding, "Besides this, we have discovered the bed of two more animals which we judge to be of the same kind. But, which we have not done any work upon yet, and shall not until we hear from you and learn whether they will pay us for the labor and time which will be necessary to get them out in good shape." When Marsh received the boxes in mid-October, he promptly telegraphed "Harlow and Edwards" to "send rest [of fossils] with all small pieces" and included a check for $75 payable to "Harlow and Edwards" (which could not be cashed; see below). When "Harlow and Edwards" notified Marsh that "we are keeping our shipments of fossils to you as secret as possible as there are plenty of men looking for such things and if they could trace us they could find discoveries which we have already made and which we do not desire to have known," Marsh instructed Samuel Williston to leave Morrison, Colorado, immediately and go to Como, Wyoming. Williston told colleagues in Morrison that he was going to Oregon, and later told Marsh that "I arrived here a few hours ago . . . I found Messrs. "Harlow and ~~Reed~~ Edwards" (~~Collins~~ Reed and Collins). Mr. Collins is the station agent and Mr. Reed the section foreman. They used those names to conceal their identity with the fossils and you had better address them thus still. They have so far succeeded in keeping the matter very quiet, although it is going to be hard to still contrive it. I have seen a lot of bones that they have ready to ship and they tell me the bones extend for *seven* miles and are by the ton . . . I go out in the morning to find all I can and will write you tomorrow." The following day, Williston confirmed that "Cañon City [Garden Park] and Morrison are simply nowhere in comparison with this locality both as regards perfection, accessibility and quantity. . . . [The bones] are magnificently preserved and scattered for 6 or seven miles, and are by the ton. The shales are clean and exposed and it will be grandest place imaginable to hunt for small specimens. . . ." Thus, only 11 years after the founding of the Yale College Museum (now the Peabody Museum of Natural History) and barely a year after Custer's Last Stand at nearby Little Big Horn, the treasures of Como Bluff were discovered. Conditions for collectors during winter were harsh; Arthur Lakes later recounted huddling "at the bottom of a narrow pit 20 feet deep . . . fingers benumbed with cold . . . and snow blowing blindingly down and covering up a bone as fast as it is unearthed." Como Bluff would yield some of the most important discoveries about dinosaurs.

1877 While collecting plants along Oil Creek, school superintendent and amateur geologist Oramel W. Lucas (who later became a clergyman) discovers large fossilized sauropod bones in Upper Jurassic formations near Garden Park, Colorado (a few miles north of Cañon City, Colorado). This sauropod was named *Diplodocus longus* (YPM 1920) by Othniel Marsh the following year. Lucas sent some of the specimens to Edward Drinker Cope and Othniel Marsh; by doing so, Lucas fueled the famous feud between Cope and Marsh that became known as the "Bone War." When Marsh learned of Lucas's discoveries, he instructed Benjamin Mudge to leave Morrison and investigate the claims in Garden Park. In late summer, when Mudge reported to Marsh that the bones found by Cope's collectors were larger and more easily extracted than those found by Arthur Lakes in Morrison, Marsh instructed Mudge to start collecting in Garden Park. Marsh's agent Samuel Williston (1851–1918)—a Yale graduate and a former student of Mudge—later took charge of the Garden Park collections, but Marsh soon ordered his collectors to return to Morrison when Williston reported that Cope's crews were excavating the best specimens. After working for Marsh, Williston worked as a professor at the University of Kansas, where he helped train Barnum Brown.

1877 While touring the United States, Thomas Huxley observes, "There is no evidence that *Compsognathus* possessed feathers; but, if it did, it would be hard indeed to say whether it should be called a reptilian bird or an avian reptile."

1877 While working on the Judith River, Edward Drinker Cope and Charles Sternberg collect *Centrosaurus*. This specimen is now displayed in AMNH (AMNH 3999).

1877 While working for the Geological Survey of India, Richard Lydekker produces the first systematic description of dinosaurs on the Indian subcontinent.

1877 Writing in the *American Journal of Science*, Othniel Marsh describes—without any illustrations—a sauropod from Como Bluff that he names *Apatosaurus ajax* (YPM 1860). In another paper published two years later, Marsh described—this time, with illustrations—another sauropod that he found at Como Bluff in Wyoming. Marsh named this specimen (reconstructed from two skeletons) *Brontosaurus excelsus* (YPM 1980). *Brontosaurus* (i.e., *Apatosaurus*) later became famous, and subsequent depictions of this sauropod wading in marshes and swamps gave the Mesozoic world an image that would typify images of "the Age of Reptiles" for decades to come.

1878 Famed dinosaur-hunter Samuel Wendell Williston expresses his surprise that more dinosaur fossils had not found earlier in the American West because "in several of the localities I have observed acres literally strewn with fragments of bones, many of them extremely characteristic and so large as to tax the strength of a strong man to lift them. Three of the localities known to me are in the immediate vicinity, if not upon the actual townsites of thriving villages, and for years numerous fragments have been collected by . . . tourists and exhibited as fossil wood . . . I have no doubt that many hundreds of bones will eventually be exhumed."

1878 In a four-page paper, Edward Drinker Cope uses an incomplete skeleton collected the previous year by school superintendent Oramel Lucas from the Morrison Formation near Cañon City, Colorado, to describe the sauropod *Amphicoelias*. These fossils (AMNH 5777) were lost, but descriptions of the fossils suggest that *Amphicoelias* may have been the biggest dinosaur (the femur was 6 feet 4 inches long), weighing up to 100 tons and being more than 150 feet long.

1878 More than 30 virtually complete skeletons of *Iguanodon* are found by Jules Creteur and his coworkers in a coal mine more than 1,000 feet below in Bernissart, Belgium. Never before had so many intact individuals of a single genus of dinosaur been found in one place. Mine officials contacted Brussels's Musée Royal d'Histoire Naturelle de Belgique (now the Royal Belgian Institute of Natural Sciences), which sent Pierre-Joseph van Beneden (1809–1894) to examine the fossils. Beneden identified the fossils as *Iguanodon*. The dinosaurs were later described by mining engineer-turned-paleontologist Louis Antoine Marie Joseph Dollo (1857–1931) and his colleagues at the Royal Museum. Unlike many of his contemporaries, Dollo (French by birth but a citizen of Belgium) did not search for other sites or feel compelled to compete with others to tally new species. Instead, Dollo spent the rest of his professional life (more than 25 years) studying the more than 140 tons of specimens excavated at Bernissart; that work included 27 scientific papers about the Bernissart *Iguanodon*. Since Dollo had several entire skeletons with which to work, he could determine that the animals, unlike those designed by Owen and displayed in the Crystal Palace exhibit, were bipedal and moved about on their hind legs (as Thomas Huxley had suggested). Dollo also determined that the bony spike, which Gideon Mantell had assumed was a horn on the animal's nose (i.e., similar to a rhinoceros), was instead a claw at the tip of the thumb. (In the 1980s, an enlarged thumb-claw was also put on the foot of *Noasaurus, Baryonyx*, and *Megaraptor*.) These famous excavations remain the most extensive discoveries of *Iguanodon*. Many of the skeletons excavated by Dollo were larger than those discovered earlier in England, but one was classified as a new species, *Iguanodon mantelli* (now *Dollodon bampingi*). The largest of these skeletons is 32 feet long (from nose to tail) and 16 feet high. Since there were no young *Iguanodon* in the herd, Dollo concluded that the animals had traveled in non-familial groups. The presence of five kinds of crocodilians amongst the dinosaurs suggested to Dollo that the dinosaurs were scavenged by crocodiles. (Paul Sereno and other made similar claims involving "mega-crocs" more than 120 years later.) The Bernissart fossils, which were restored in 1882 (and displayed the following year), were the last major discoveries of dinosaurs in Europe before the shift of major discoveries to North America. Today, 11 of the *Iguanodon*—the first complete dinosaur skeletons ever discovered—are a permanent monument to Dollo's work as they glare at visitors to the Royal Belgian Institute of Natural Sciences in Brussels.

1878 On the first geologic map of Colorado's Grand Valley, A. C. Peale of the Hayden Survey refers to rocks immediately below the Dakota Sandstone as Jurassic shales and marls. This area yielded several iconic dinosaurs, including *Brachiosaurus* and *Apatosaurus*.

1878 Othniel Marsh (Figure 9) classifies dinosaurs into four orders: Stegosauria (armored dinosaurs), Ornithopoda (iguanodonts and duckbills), Sauropoda (giant marsh-dwelling brontosaurs), and Theropoda (carnivorous dinosaurs). Like previous systems for classifying dinosaurs, Marsh's system was fraught with problems, and a decade later was replaced by a system proposed by British paleontologist Harry Seeley that divided Dinosauria into two clades: Ornithischia and Saurischia. In 1878, Marsh also noted that Darwin's *On the Origin of Species* (1859) "changed the whole course of scientific thought . . . Darwin spoke the magic word—'Natural Selection,' and a new epoch in science began."

1878 The first year of excavations at Como Bluff yield some 30 tons of dinosaur fossils, including those of *Apatosaurus, Barosaurus, Laosaurus* (*Othnielosaurus*),

Camarasaurus, and *Stegosaurus*. Samuel Williston, who was one of Marsh's collectors, noted that most quarries at Como Bluff contained "remains of numerous individuals mingled together in the most inextricable confusion, and in every conceivable position, with connected limb bones standing nearly upright, connected vertebrae describing vertical curves, etc., precisely as though in some ancient mud holes these huge monsters had become mired and died, and succeeding generations had trodden their bones down, and then left their own to mingle with them." Union Pacific Railroad foreman William "Bill" Reed, another collector, wrote to a friend, "I wish you were here to see the bones roll out and they are beauties." The skeleton of *Apatosaurus excelsus* discovered by Reed in 1879 at Quarry #10—one of the most complete sauropods ever discovered—is today a centerpiece of YPM. Although this specimen is not significantly longer than the *A. louisae* discovered in 1909 by Earl Douglass at DNM, it weighed about 22,000 pounds less than *A. louisae*.

1878 To counter Marsh's quick access to publication in *American Journal of Science*, Cope buys controlling interest in *American Naturalist* for the quick publication of his manuscripts. Henry Osborn later loaned Cope money to sustain *American Naturalist*. Cope's investment was a poor one; several prominent scientists—including Asa Gray, James Dana, Alexander Agassiz, and, not surprisingly, Marsh—dissociated themselves from the journal.

1878 To promote interest in their work and raise the price of their discoveries, dinosaur hunters in Wyoming leak information about their discoveries to the *Laramie Daily Sentinel,* which publishes articles about their dinosaur bones.

1878 When fire destroys the school where Arthur Lakes works, Lakes begins collecting fossils to make a living. Lakes soon began working for Othniel Marsh, for whom he helped unearth *Allosaurus, Apatosaurus*, and *Stegosaurus* at Como Bluff and other locations. By 1880, Lakes had returned to teaching at what would become Colorado School of Mines. The library at the Colorado School of Mines is named for Lakes, the first person to collect a dinosaur fossil from the famed Morrison Formation.

1878 William Harlow "Bill" Reed (1848–1915) and William Carlin, working for Othniel Marsh, excavate the type-specimens of *Camarasaurus grandis* and *Dermodactylus* (the first Jurassic pterosaur from North America), and the type-specimen of the small ornithopod *Dryosaurus altus*. At about the same time, a crewmember discovered the first Jurassic mammal in North America, which Marsh named *Dryolestes priscus*.

1879 A cast of Benjamin Waterhouse Hawkins's *Hadrosaurus* for the 1876 Philadelphia Centenary is shipped from Philadelphia's ANS to Edinburgh's Royal Museum of Scotland, where it becomes the first dinosaur skeleton to be mounted in Europe. The model was dismantled during World War II to use its ironworks, and discarded in 1928. In 1883, fossils of *Iguanodon* became the second type of mounted dinosaur displayed in the world when they were displayed in Brussels.

1879 American paleontologist Othniel Marsh names *Coelurus fragilis* from a partial skeleton collected from Reed's Quarry 13 at Como Bluff, Wyoming. Although *Coelurus* ("hollow tail") is associated with one of the main divisions of theropods (i.e., Coelurosauria), *Coelurus* is poorly understood. The only species of *Coelurus* that is recognized as valid is *C. fragilis*, which was named for its hollow tail-vertebrae. This 8-foot-long specimen resides at YPM.

1879 Congress consolidates the four largest surveys of the American West into the USGS. Two years later, John Wesley Powell assumed control of the USGS and named Othniel Marsh its chief paleontologist.

1879 Edward Drinker Cope arrives at Como Bluff to inspect the fossil fields, but stays for only one day. Arthur Lakes, who met Cope, commented, "The monstrum horridum [terrible monster] Cope has been and gone and I must say that what I saw of him I liked very much, his manner is so affable and his conversation very agreeable. I only wish I could feel sure he had a sound reputation for honesty." Discoveries of fossils at Como Bluff soon peaked, including the discovery of the now-famous Quarry 10. This quarry yielded *Brontosaurus*, the most famous specimen collected at Como Bluff. This near-complete specimen is mounted at YPM.

1879 H. F. Hubbell collects a well-preserved *Allosaurus* at Como Bluff near Medicine Bow, Wyoming. This specimen later became part of Edward Drinker Cope's collection, which was sold to AMNH in 1899. Hubbell's records were lost, and the crates were not unpacked until 1903. When *Allosaurus* (AMNH 5753) was displayed in 1908 in AMNH, it was the first freestanding mount of a theropod. It later became famous when it was illustrated by artists such as Charles R. Knight. *Allosaurus*, which had meat-hook-like claws on its hands, is one of the most abundant theropods in the Morrison Formation; it accounts for more than two-thirds of all Morrison Formation theropods. The discovery of *Allosaurus* in Late Jurassic sediments of Portugal was the first confirmation of a dinosaur species on two different continents, and suggested that the North America and Europe were connected during the late Jurassic. *Allosaurus* is the state fossil of Utah.

1879 In a two-paragraph paper published in *American Journal of Science* (that is even shorter than the 1877 paper that described *Apatosaurus*), Othniel Marsh uses a pelvis and vertebrae to describe a more complete (but headless) specimen of a sauropod, which he names *Brontosaurus excelsus*. Marsh's paper, titled "Notice of New Jurassic Reptiles," included comments about the vertebrae and pelvis, but no illustrations. *Brontosaurus* was the largest dinosaur ever discovered at that time. Four years later, this 70-foot-long dinosaur was immortalized in Marsh's first reconstruction of a dinosaur, after which it became the most popular Jurassic dinosaur in America. However, in 1903, Elmer Riggs (1869–1963) of the Chicago Natural History Museum (now the Field Museum) noted that *Brontosaurus excelsus* was nearly identical to the earlier-named *Apatosaurus*, and reclassified *Brontosaurus excelsus* as *Apatosaurus excelsus*. (Samuel Williston of YPM had suspected this as early as 1891.) According to the rules of nomenclature, the name *Apatosaurus* had been published first and therefore had priority as the official name. *Brontosaurus* was therefore discarded from formal use, but remained popular with the public. When this dinosaur was mounted in YPM in 1905, it became the first sauropod displayed anywhere in the world. *Brontosaurus* later became a canonical sauropod that appeared in corporate logos (e.g., Sinclair Oil), movies (*Fantasia*), and elsewhere.

1879 Othniel Marsh names *Camptosaurus. Camptosaurus* was originally named "Camptotonus," but that name had already been used to name a cricket.

1879 Samuel Williston proposes that birds learned to fly from the ground up: "The wings must have first been used in running, next in leaping and descending from heights, and, finally, in soaring."

1879 The self-taught Bill Reed discovers Como Bluff's Quarry 13, which later yields "the greatest accumulation of stegosaurian remains ever discovered" and several species of the bipedal herbivore *Camptosaurus* ("flexible lizard"). Soon thereafter, at nearby Quarry 10, Reed and his crew excavated *Brontosaurus excelsus* (now *Apatosaurus excelsus*; YPM 1980), which today oversees visitors to YPM. In Quarry 12, workers found *Stegosaurus ungulatus*. In 1897, working near Quarry 9, Barnum Brown excavated *Diplodocus. Diplodocus* was the longest known dinosaur, but not the most massive (that title went to *Apatosaurus*). Reed, a former buffalo-hunter, collected more than 80 tons of specimens for the University of Wyoming; he also worked for CMNH and AMNH.

1879 While studying under Thomas Huxley, Othniel Marsh meets Charles Darwin. That same year, Henry Osborn met Huxley, of whom Osborn later in his life noted "by [Huxley's] way of living and by the unlimited personal sacrifices he made he taught me that we men of science must do our part in public education."

1879 With his finances dwindling, Cope makes his last purchases of important specimens from the field.

1880 After rejecting a job at AMNH, Henry Osborn joins the faculty at Princeton University and studies mammalian paleontology. Osborn, who was mentored by Edward Drinker Cope, later helped make AMNH one of the greatest museums in the world.

1880 Othniel Marsh (Figure 9) produces his famed monograph, *Odontornithes: A Monograph on the Extinct Toothed Birds of North America*. This book, which was published by the USGS, included 201 pages of text, 40 woodcuts, and 34 lithographic plates (some life-sized); it described, bone by bone, the nearly entire skeletons of five species of toothed birds: *Hesperornis regalis, H. crassipes, Ichthyornis dispar, I. victor*, and *Apatornis celer*. Richard Owen described the book as "the best contribution to Natural History since Cuvier!" In August, Charles Darwin wrote to Marsh claiming that Marsh's studies of "birds and the many fossil animals of North America has afforded the best support to the theory of evolution which has appeared within the last 20 years . . . I can say nothing stronger than this."

1880 While working in Algeria, George Le Mesle and Pierre Peron find the first dinosaur tracks in Africa.

1881 Como Bluff, now famous for its fossil beds, is overrun by bone hunters. Two years later, famed Como Bluff collector William Reed quit prospecting and returned to sheepherding. When his sheepherding venture failed, Reed worked as a night watchman for the Union Pacific Railroad, after which he became curator of the University of Wyoming Geological Museum. Today, the museum's exhibit hall is dominated by a 75-foot-long *Apatosaurus*.

1881 David Baldwin, a former collector for Othniel Marsh who was now collecting for Edward Drinker Cope, finds some hollow leg-bone fragments and other bones near Abiquiu in northern New Mexico. In 1889, after originally claiming that some of the bones were from *Coelurus* (and then *Tanystropheus*), Cope named the specimen *Coelophysis* ("hollow form," so named for the hollow bones of its limbs). *Coelophysis*, a small (10 feet long) carnivore, was one of the earliest known genera of dinosaurs from the Late Triassic. It has been featured in numerous films and exhibits, including *Walking with Dinosaurs, When Dinosaurs Roamed America, Dinosaurs Alive!,* and as "Spot" in the children's television series *Land of the Lost*.

1881 In western South Dakota, Jacob Wortman finds a well-preserved *Atlantosaurus* (now *Anatotitan copei*).

1881 Othniel Marsh first names the Ornithopoda. That same year, Marsh proposed the name "Theropoda" for meat-eating dinosaurs known at the time, including *Allosaurus, Megalosaurus*, and *Compsognathus*. This lineage appeared 230 million years ago (i.e., in the Triassic); birds, which are descendants of theropods, are classified as theropod dinosaurs. Theropods have been found on all continents.

1881 Thanks largely to the efforts of Richard Owen, BMNH (Figure 1) opens in London, but the museum's move from its former location is not completed until

1883. BMNH, which hosts more than 3.5 million visitors per year, transformed the public's expectations of museums.

1882 BMNH buys more than 500 fossils—many of them dinosaurs—from Rev. William D. Fox (1813–1881). Several of the dinosaurs in Fox's collection were later described by Richard Owen (e.g., *Polacanthus foxii*, *Hypsilophodon foxii*).

1882 British surgeon and amateur paleontologist John Whitaker Hulke describes, and Richard Owen names, *Polacanthus*, which has a "sacral shield" of bone over its back and hip area. Another such armored dinosaur was not discovered for almost 100 years.

1882 John Whitaker Hulke produces a skeletal reconstruction of *Hypsilophodon* (as part of a monograph for the Geological Society of London) and claims that its grasping hands enabled it to climb trees; for decades thereafter, *Hypsilophodon* was shown sitting on tree branches like a giant squirrel. Hulke also claimed that *Hypsilophodon* was armored; were this true, *Hypsilophodon* would have been the only known armored ornithopod. Subsequent studies in the 1970s refuted both of these claims. During his career, Hulke named 10 dinosaurs (5 of which remain valid).

1882 Othniel Marsh is appointed Vertebrate Paleontologist to the USGS, and is given $15,000 per year for his salary and those of his 37 field-collectors, 9 preparators, 8 scientific aids, and numerous artists. Marsh accepted $4,000 in salary, which he spent on research and field work. Marsh proposed to write 10 monographs, but finished only those about odontornithes and dinocerata.

1882 *Scientific American* publishes a restoration of *Stegosaurus*.

1882 While working as an assistant naturalist at the Royal Natural History Museum in Brussels, Louis Dollo publishes the first of 27 scientific papers describing the *Iguanodon* discovered in the Fosse Sainte-Barbe coal mine near Bernissart in southwest Belgium (he also published several papers about the geology of the site). These studies changed the views of *Iguanodon* derived from the earlier work of Richard Owen and Benjamin Waterhouse Hawkins at Crystal Palace; for example, Dollo recognized that *Iguanodon* had a spiked thumb on each hand (Owen and Hawkins claimed that the spike was on the skull). Dollo endorsed Joseph Leidy's claim that *Iguanodon* had a bipedal "kangaroo-like" stance, and even went so far as to break the dinosaurs' tails to get them to stand upright. This view was later modified to make *Iguanodon*'s body horizontal and its tail off the ground. Dollo argued that *Iguanodon* was a "reptilian giraffe because they were scaled animals as are typically all reptiles, and not furred animals as are the mammals. Moreover, like the giraffes they searched for their food among the leaves of trees." Dollo did no fieldwork of his own, instead devoting most of his career to analyzing *Iguanodon* discovered near Bernissart.

1882 While working for Edward Drinker Cope north of the Black Hills of South Dakota, R. S. Hill and J. I. Wortman (both of whom later worked at AMNH) find the medium-sized hadrosaur, *Anatotitan* ("large duck"). This specimen (AMNH 5730), and another (AMNH 5886) found in 1904 by rancher Oscar Hunter in what is now central Montana, are now displayed in the AMNH Hall of Ornithischian Dinosaurs. Hunter's fossils were obtained by Alfred Sensiba for a pistol, after which Barnum Brown bought the fossils from Sensiba for $250. Both skeletons were originally considered to be *Anantosaurus*. However, when subsequent research in 1990 showed that name to be invalid, both were renamed *Anatotitan*.

1882 While working near Neuquén in western Argentina, an Argentina army captain discovers fossilized bones that he sends to Argentinian paleontologist Florentino Ameghino (1854–1911). Ameghino, Argentina's "father of vertebrate paleontology," later determined that the fossils were from Upper Cretaceous dinosaurs. These were the first discoveries of dinosaurs in South America. The first dinosaur-tracks were not found in South America for more than 50 years.

1883 In an article titled "Principal Characters of American Jurassic Dinosaurs, Part IV: Restoration of *Brontosaurus*," published in *American Journal of Science*, Othniel Marsh publishes the first—and most famous—restoration of his *Brontosaurus* (now *Apatosaurus*) skeleton (Figure 12). Marsh lacked a head for the sauropod, so he guessed that it had a *Camarasaurus*-like skull with large, spoon-like teeth.[14] Marsh portrayed the sauropod as a slow, lumbering giant; as Marsh wrote, "a careful estimate of the size of *Brontosaurus*, as here restored, shows that when living the animal must have weighed more than twenty tons. The very small head and brain, and slender neural cord, indicate a stupid, slow-moving reptile. The beast was wholly without offensive or defensive weapons, or dermal armature. In habits, *Brontosaurus* was more or less amphibious, and its food was probably aquatic plants or other succulent vegetation." This lifelike restoration, which was the first published restoration of a North American sauropod, appeared in many textbooks, and was revised for Marsh's publication of "Restoration of *Triceratops*" in 1891. (Twelve of Marsh's 22 restorations were of dinosaurs.) Marsh did not disclose that he adorned the *Brontosaurus* with a square-snouted skull of what was later identified as *Camarasaurus* (another recently discovered sauropod) that had been found at another quarry four miles away from the other bones collected at Como Bluff. That skull had been collected from rocks older than those containing *Brontosaurus*, and had not been found with any *Brontosaurus* fossils. (Excavations at DNM in the 1970s revealed that *Brontosaurus* had a tapered head with a slender snout similar to that of *Diplodocus*.) Marsh's *Brontosaurus* (YPM 1980) skeleton, collected by William Reed from Quarry 10 at Como Bluff in 1879, now stands in YPM; it is 67 feet long, 16 feet high at the hips, and is estimated to have weighed nearly 25 tons,

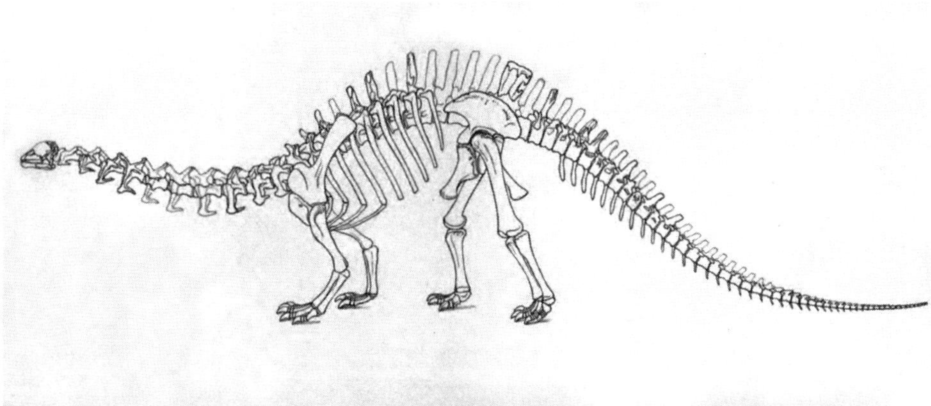

Figure 12 In 1883, famed fossil hunter Othniel Marsh (1831–1899; Figure 9) published the first reconstruction of *Brontosaurus* (now *Apatosaurus*). Marsh did not disclose that he placed the head of a different dinosaur (later identified as *Camarasaurus,* which was collected several miles from the *Apatosaurus* fossils) on his reconstruction. *Brontosaurus,* the largest known dinosaur at the time, became the most popular Jurassic dinosaur in America. The reconstruction shown here appeared as Plate I in Marsh's *The Dinosaurs of North America* (1886). Marsh's *Apatosaurus* dominates the Great Hall of the Yale Peabody Museum of Natural History, which was established by Marsh's uncle, George Peabody. (Marsh, O.C. 1883. "Principal characters of American Jurassic dinosaurs, Part IV: Restoration of *Brontosaurus,*" *American Journal of Science* 26 (Ser. 3), 81–85)

but its brain weighed less than a pound. Marsh's restoration of *Brontosaurus* influenced pictures and models of *Brontosaurus* for more than 60 years. Many of Marsh's other restorations—for example, *Triceratops* and *Stegosaurus* (1891), *Claosaurus* (later renamed *Edmontosaurus*) and *Ceratosaurus* (1892), *Anchisaurus* (1893)—were reviewed favorably by Richard Lydekker (1849–1915) in *Nature* in 1893, but Marsh protested Lydekker's use of Cope's names *Hypsirophus seeleyanus* (for Marsh's *Stegosaurus*) and *Agathaumas* (for Marsh's *Triceratops*).

1883 Othniel Marsh describes a "foetal" sauropod found with *Morosaurus* (*Camarasaurus*). That same year, M. P. Felch returned to Marsh's Garden Park Quarry and discovered a near-complete skeleton of *Allosaurus*. At the same quarry, Felch also found *Ceratosaurus*.

1883 Bill Reed resigns at Como Bluff, and work there begins to decline.

1883 Richard Owen retires from BMNH, the museum he helped found. By this time, Owen had become "the Grand Old Man of British Science" and the recipient of a £200 annual Civil Pension. During his career, Owen named 36 dinosaurs, seven (i.e., 19%) of which remain valid.

1883 When Othniel Marsh is elected president of NAS, Cope casts the only dissenting vote, noting, "I suspect that the majority will be sorry for it. Certainly they will be if they live long enough." After visiting Marsh's lab, Cope described Marsh as "more of a pretender than even I had supposed him to be."

1884 *Ammosaurus* is discovered in a quarry near Manchester, Connecticut, but only after the front half of the fossil had been made a part of Hop Brook Bridge. Marsh claimed that the 6-foot-long dinosaur "was one of the animals that are supposed to have made the footprints" in Connecticut.

1884 Charles H. Sternberg hires 23-year-old John Bell Hatcher to help him search for dinosaurs. Like Sternberg, Hatcher became a renowned fossil-hunter.

1884 Edward Drinker Cope (Figure 10) publishes 79 scientific papers. The following year, he published another 62 papers.

1884 J. L. Wortman, Henry Osborn's field-foreman and first hire at AMNH, discovers what became the type-specimen of *Trachodon mirabilis* (AMNH 5730) "surrounded by a natural cast of its epidermal impressions." Wortman destroyed most of the skin impressions when he excavated the fossil.

1884 Othniel Marsh names and describes *Ceratosaurus*, which he describes as a "ferocious biped" resembling a bird. The year before, Marsh had described *Brontosaurus* as "a stupid, slow-moving reptile" and *Diplodocus* as "passive."

1884 While studying coal seams in the Horseshoe Canyon Formation near the present-day town of Drumheller, Joseph B. Tyrrell of the Canadian Geological Survey collects the first dinosaur bones in Canada. These fossils, which were found at Red Deer River Valley, were named *Laelaps* by Edward Drinker Cope, and renamed *Dryptosaurus* by Lawrence M. Lambe (1849–1934) in 1903. Two years later, Henry Osborn renamed the fossils *Albertosaurus sarcophagus.* Subsequent excavations in the area, which spanned the next 17 years, were directed by Lambe (also of the Canadian Geological Survey). This area was later named Dinosaur Provincial Park (DPP). Twenty-five years later, Barnum Brown and Peter Kaisen (1870–1936) collected dinosaurs at this site, and it is from here in 1917 that Charles H. Sternberg collected the *Albertosaurus* displayed today at AMNH (AMNH 5664; AMNH bought the specimen from Sternberg in 1918 for $2,000). This specimen is the most complete advanced carnosaur known in North America. In nearby Drumheller, a famous sculpture of a giant green tyrannosaur guards a bridge over the Red Deer River.

1885 Richard Owen resigns as president of the Palaeontological Society, his last office. This ended Owen's public work and responsibilities. Owen published his last paper in 1888.

1885 The primitive ornithischian dinosaur *Camptosaurus* ("bent reptile") from Como Bluff is named. The specimen at AMNH (AMNH 6120)—collected by Peter Kaisen in 1905 at Bone Cabin Quarry in Wyoming—was first exhibited in 1908.

1886 British/Tasmanian geologist Charles Gould (1834–1895) first notes the possible connection between Chinese "dragon bones" and dinosaurs. Although Chang Qu wrote about dragon bones near 300 BCE in what is now China's Sichuan Province, these bones were not kept or adequately described.

1886 Edward Drinker Cope's investments in mining are nearly worthless, forcing him to start selling duplicates of his fossils. Cope tells his wife, "My entire future in a financial sense . . . depends on that collection so far as I can see now." Two years later, Cope's applications for jobs at AMNH and the Smithsonian were rejected.

1886 French scientist Camille Flammarion's (1842–1925) popular geology book, *Le Monde avant la Création de l'Homme* (*The World before the Creation of Man*), shows a bipedal *Iguanodon* on its back legs eating food from a fifth-floor apartment of a modern city (Figure 13). This image anticipated Godzilla-like poses of dinosaurs and monsters a century later. In the United States, a larger, similarly posed *Brontosaurus* was later shown beside skyscrapers in publications such as *American Century* and *New York World and Advertiser*. By the 1880s, dinosaurs had influenced popular imagination.

1886 Samuel Williston, an assistant in Othniel Marsh's lab at Yale, tells Cope that Marsh "has never been known to tell the truth when a falsehood would serve the purpose as well." Williston described Marsh as a liar, as someone who ordered him to destroy fossils rather than let Cope have them, as someone who never did "two consecutive, honest days' work in science," and who never "averaged more than one hour's work per day." Williston criticized Marsh's claims about fieldwork, noting, ". . . since 1875, when my personal relations with Professor Marsh began, he himself did no field work, his knowledge of the formations being derived from a few transient and hasty visits to the different fields where his collectors were at work. His references to the personal dangers encountered by hostile Indians is amusing in the extreme. . . ." Williston also claimed to have written Marsh's dinosaur papers, adding that Marsh "has no friends [at Yale] save those who do not know him well." Cope published Williston's letter in a newspaper article, after which Williston moved to the University of Kansas, where he taught geology and paleontology to Barnum Brown.

1886 When Congress again refuses to appropriate money for the publication of the concluding volume of what became known as "Cope's Bible" ("The Vertebrata of the Tertiary Formations of the West"), Cope writes to his wife, "We will have to live very cheaply for a year."

Figure 13 In 1886, French writer and scientist Camille Flammarion (1842–1925) imagined *Iguanodon* eating from a fifth-floor window in *A Prehistoric Monster in a Modern Town*. Flammarion's dinosaur had a Godzilla-like pose that was popularized a century later. The horn on the dinosaur's nose resembled that of early reconstructions of *Iguanodon* (e.g., Figure 4), despite the fact that Louis Dollo had shown it to be a thumb spike. (Private Collection/Archives Charmet/The Bridgeman Art Library)

1887 In his paper titled "On the Classification of the Fossil Animals Commonly Named Dinosauria," British geologist Harry Govier Seeley (1839–1909)—a friend of Edward Drinker Cope and a former assistant to Adam Sedgwick at Cambridge's Woodwardian Museum—argues, "Dinosauria has no existence as a natural group of animals, but includes two distinct types of animal . . . which show their descent from a common ancestry rather than their close affinity. These two orders of animals may be conveniently named the Ornithischia and the Saurischia." Seeley based his classification scheme on the geometry of the pelvic bone: The Saurischia were lizard-hipped dinosaurs (e.g., the theropods and sauropods, including *Coelophysis, Tyrannosaurus*, and *Diplodocus*), and the Ornithischia were bird-hipped dinosaurs (e.g., duckbills, plated and armored dinosaurs, and horned dinosaurs, including *Torosaurus* and *Triceratops*). Interestingly, birds arose from lizard-hipped dinosaurs.

1887 Near what is now Federal Boulevard in Denver, Colorado, George L. Cannon finds two bison-like horns. The following year, Othniel Marsh claimed that these horns were from *Bison alticornis*, a fossil buffalo. When John Bell Hatcher later sent Marsh a horn from what would be named *Triceratops*, Marsh realized that his "bison" was a dinosaur. Hatcher spent the next four years (1889–1892) collecting horned dinosaurs from the Lance Formation near Lusk, Wyoming.[15]

1887 The discovery of a near-complete, articulated *Stegosaurus* shows that the dinosaur had vertical spikes and a spined tail.

1888 John Bell Hatcher, a graduate of Yale and one of Marsh's collectors, is told by a cowboy from eastern Wyoming of finding a skull having "horns as long as a hoe handle and eye holes as big as your hat." Hatcher later discovered a horn protruding from the Late Cretaceous exposures near Lusk, Wyoming. After learning that the horn was attached to a giant skull, Hatcher sent the horn to Marsh, who suspected that it belonged to a new group of dinosaurs. Marsh hired Hatcher to collect the massive skull, which weighed more than a ton. The fossil was originally named *Ceratops montanus*; this specimen (YPM 1820) began several years of remarkable discoveries and collections of ceratopsians by Hatcher. Indeed, during the next three years, Hatcher discovered and collected more than 30 ceratopsian skulls and partial skeletons; some of the skulls weighed more than three tons. Although two of the specimens determined by Marsh to be separate species were assigned to the new genus *Torosaurus*, Marsh assigned all of the others to new species of *Triceratops*. The specimen originally named *Ceratops montanus* later became the type-specimen of *Triceratops horridus*. Today, *Triceratops* is one of the most popular dinosaurs.

1888 Spurred by Joseph Tyrrell's discovery of *Albertosaurus*, geologist Thomas Chesmer Weston (1832–1911) is sent by the Geological Survey of Canada (GSC) to Red Deer River Valley to collect dinosaur bones. Weston's boat sank a few hours after the trip began, forcing Weston to delay his expedition until 1889. During his trip down the river—which showed the feasibility of a boat-based expedition to collect dinosaur fossils—Weston and his colleagues discovered dinosaur bones between Deadlodge Canyon and the small village of Steveville.

1889 Edward Drinker Cope (Figure 10) receives a letter from John W. Noble (1831–1912), Secretary of the Interior, demanding that he return all of his vertebrate fossils collected in the West to the U.S. National Museum, on the grounds that they were government property. Although Cope had been a member of a government-funded survey when he collected many of the fossils, he had been an unpaid volunteer. The request had been recommended by John Wesley Powell, who had probably been encouraged by Othniel Marsh. Cope responded by telling Henry Osborn, "When a wrong is to be righted, the press is the best and most

Christian medium of doing it. It replaces the old time shotgun & bludgeon & is a great improvement."

1889 Edward Drinker Cope (Figure 10) succeeds Joseph Leidy as professor of zoology and comparative anatomy at the University of Pennsylvania. That same year, Cope described the type specimen of *Coelophysis bauri*, a carnivorous dinosaur that was 9 feet long, 3 feet tall, and lived during the Late Triassic. A vast deposit of *Coelophysis*, one of the earliest known dinosaurs, was found in 1947 at Ghost Ranch, New Mexico. Most of our knowledge about *Coelophysis* comes from the work of Edwin Colbert (1905–2001) at that site.

1889 George Cannon discovers the first remains of an ornithomimosaur near Denver, Colorado. The specimen was described by Othniel Marsh the following year and named *Ornithomimus velox* ("speedy bird mimic"). Two other species discovered by J. B. Hatcher in Montana were also described by Marsh. Later, Henry Osborn—noting the similarities of *Ornithomimus* with large running birds—suggested that one of the species (*Ornithomimus altus*) be renamed *Struthiomimus* ("ostrich mimic"). *Ornithomimus* later became a favorite dinosaur in several movies (e.g., *Fantasia* [1940], *The Valley of Gwangi* [1969]) and television shows.

1889 Othniel Marsh ends work at Como Bluff, Wyoming, his longest-lasting Jurassic quarry. In 12 years of work there, Marsh collected the largest group of Jurassic terrestrial vertebrates known up to that time, including 26 new species of dinosaurs (as well as six new species from near Morrison, and eight from Cañon City, Colorado). William Reed, the co-discoverer of Como Bluff, returned to the site with Wilbur Knight to work in 1895. Marsh's many quarries in Colorado and Wyoming ultimately produced more than 1,000 crates of fossils that included 21 new genera and 41 new species of dinosaurs. Subsequent expeditions to Como Bluff made important discoveries, but today only a few of the original quarries are readily identifiable.

1889 While working near Lusk, Wyoming, John Bell Hatcher excavates the first fossils of *Torosaurus*. Today, an image of *Torosaurus* adorns Hatcher's tombstone in Pittsburgh's Homewood Cemetery.

1889 Othniel Marsh names *Triceratops* ("three horn face"), the first horned dinosaur known from a complete skull. Many skulls of *Triceratops* have been found in the American West, thus explaining why *Triceratops* is common at natural history museums. Two years after naming *Triceratops*, Marsh named *Torosaurus*,[16] and later coined the name *Ceratopsidae* for the family of horned dinosaurs. In ensuing years, Marsh described 10 of the 16 proposed species of *Triceratops*. When Harry Govier Seeley learned of *Triceratops,* he noted, "After this, anything in skull structure is possible."

1889 While working along Piedmont Butte a few miles north of Rapid City, South Dakota, John Bell Hatcher discovers six tail-vertebrae of a long, slender sauropod that Othniel Marsh later named *Barosaurus* ("heavy lizard"; YPM 429). (Unlike most other collection-sites, the quarry in South Dakota was covered with grass and pine trees.) The rest of the specimen was collected in 1898 by Marsh's assistant George Wieland (1865–1953). Richard S. Lull described *Barosaurus* in 1919. The vertebrae collected by Hatcher are stored at YPM.

1890 After Congress ignores his 23,000-word allegation of fraud against Othniel Marsh, Edward Drinker Cope takes his feud with Marsh to the public in a meeting with *New York Herald* freelance reporter William Hosea Ballou. Cope charged Marsh with bribery, financial mismanagement, embezzlement, plagiarism, dereliction of duty, and more, and John Wesley Powell with financial mismanagement. Cope concluded his indictment by claiming that "[Marsh] is not a proper person to be the president of the National Academy of Sciences." The article began with an interview in which Cope labeled John Noble's demand that Cope return all of his vertebrate fossils collected in the West as "absurd" and "outrageous." (Cope claimed that he had "not more than a bushel of specimens belonging to the government and to those it is welcome.") Cope then described his use of $80,000 of personal funds for his collection and claimed that Noble's request was made so that "my work may be postponed until it has been done by Professor Marsh of Yale University." Cope also recited Marsh's technical errors, claimed that Powell had bought influence in Congress, and presented the scandalous letter written to him by Samuel Williston in 1886. Marsh, who responded with equally scandalous charges against Cope, later extracted a disclaimer from Williston. Cope followed the *New York Herald* article with a series of articles by Marsh's assistants in the *American Naturalist*, which were published in a small book that supplemented the newspaper articles. In these articles, one of Marsh's assistants claimed that Marsh wrote a glowing review of one of his own books on the Dinocerata and asked several people to sign it. When they refused, Marsh signed the review with the initials of the "lady typewriter." Marsh and Powell responded in the *New York Herald* that Cope's criticisms—most of which were wrong or not provable—were "the crowning work of his life." That was followed by Marsh's claim that Cope had erroneously placed the skull of *Elasmosaurus* on the animal's tail. Cope immediately tried to buy back the journals in which he'd published this error, and he paid for the republication of a new engraving of the corrected anatomy, but some remained with buyers. (Two decades later, Cope published another article in *Transactions of the American Philosophical Society* in which he corrected his mistake.) Marsh later noted that Cope's "wounded vanity received a shock from which it never recovered, and he has since been my bitter enemy." In response to Cope's charge that Marsh's work on horse evolution was plagiarized (from himself and Russian paleontologist Vladimir Kovalevsky [1842–1883]) and was "the most remarkable collection of errors of ignorance of anatomy and literature on the subject ever displayed," Marsh responded that Cope was a thief and that "Kovalevsky was at last stricken with remorse and ended his unfortunate career by blowing out his own brains. Cope still lives, unrepentant." The bitter feud between Marsh and Cope stimulated the public's growing interest in dinosaurs. In their celebrated careers, Cope published a staggering 1,400 publications and named 1,200 species; Marsh published 270 papers and named 500 species. By the end of the 1890s, Marsh and Cope had described 144 new species of dinosaurs, although many of these species were later disallowed.

1890 Marsh coins the term *Ceratopsia* to include the large, quadrupedal, frilled, horned dinosaurs of western North America.

1890 In the *American Journal of Science*, Othniel Marsh describes the long, slender *Barosaurus* ("heavy reptile"), which had been collected by Marsh and John Bell Hatcher (his assistant at the time) from the north slope of Piedmont Butte near Rapid City, South Dakota.

1890 The *Philadelphia Herald* publishes an entire page of denials by Othniel Marsh of charges made by Edward Drinker Cope. Marsh also pointed out several errors made by Cope. Although Cope and Marsh feuded until their deaths in 1897 and 1899, respectively, the "Bone War" was over in the West by the 1890s.

1890 When Congress drastically cuts the budget of USGS, John Wesley Powell fires Othniel Marsh. That same year, the Smithsonian claimed all of the fossils collected with funds from USGS.

1890 Henry Osborn joins the staff of AMNH. Initially, Osborn received no salary and paid AMNH $1,500 to do fieldwork. Osborn helped make AMNH one of the finest natural history museums in the world.

1891 After being hired jointly by Columbia University and AMNH, Henry Osborn establishes the Department of Vertebrate Paleontology at AMNH (AMNH president Morris Jesup had originally wanted to hire Othniel Marsh for the job.) Osborn hired preparators and artists such as Rudolph Weber, Barnum Brown, Erwin Christman, Walter Granger (who would later become Roy Chapman Andrews's assistant and the chief paleontologist during the AMNH's Central Asiatic Expeditions of the 1920s), and Charles Knight to develop collections, displays, and artwork to foster the public's interest in evolutionary history. Osborn, who hoped to educate "a very large class of inquisitive but wholly uninformed people," brought dinosaurs to AMNH while making it one of the best natural history museums in the world. Osborn, who subsidized AMNH's paleontology department with portions of his own salary, set the agenda for vertebrate paleontology in the United States.

1891 Edward Drinker Cope stops actively attacking Marsh and Powell.

1891 Harry Govier Seeley describes *Agrosaurus macgillivrayi*, the first dinosaur from Australia. The fossil was discovered along the continent's Queensland coast. The next Australian dinosaur (*Rhoetosaurus brownie*, named in 1926) was not found until 1924. This fossil was advertised as "Australia's largest fossil."

1891 Othniel Marsh revises his *Brontosaurus* with a skull collected 400 miles away in Cañon City, Colorado. By this time, as Ned Colbert later noted, Marsh was "almost literally being buried by tons of dinosaur bones."

1891 Othniel Marsh's restoration of *Stegosaurus* has a tail with four pairs of spines and a single row of plates along the midline. (A second row of plates may have been erased from the drawing before it was published.) Charles Knight, Arthur Lakes, and others used this restoration for their drawings. Interestingly, Marsh's diagrammatic restoration of *Stegosaurus* contradicted his 1887 (and 1896) written descriptions, which included a paired, double row of plates.

1891 Paleontologists John Bell Hatcher and William H. Utterback discover what will be the type specimen of the 10-foot-long herbivore *Thescelosaurus* ("marvelous lizard") in beds of the Late Cretaceous in Converse County, Wyoming. However, the fossils remained in crates for more than two decades, and were not described until 1913. While working along the eastern edge of the Bighorn Mountains in north-central Wyoming, Utterback later discovered *Diplodocus hayi*, which is now mounted at the Houston Museum of Science.

1891 The American Anthropometric Society acquires the 1,559-gram brain of Joseph Leidy, which it declared to be "unusually heavy and richly convoluted . . . the most interesting of all the specimens in the society's possession." The society later acquired the brains of Edward Drinker Cope (1,545 grams), John Wesley Powell (1,488 grams), and other prominent Americans. Cope was among the first people to will his body to science.

1891 The only known specimen of the horned dinosaur *Diceratops hatchery* is found in Wyoming. The poorly preserved specimen, which includes a six-foot skull, lacks the nose horn of its relative, *Triceratops*. *Diceratops*, which was named by Richard Swann Lull in 1905, is displayed at the Earth Science Museum in Provo, Utah.

1892 A drawing of the horned *Ceratosaurus* is the first published reconstruction of a North American carnivorous dinosaur based on a relatively complete specimen. That specimen had been collected in the Marsh Quarry in 1883–1884 near Cañon City, Colorado. Today, the type specimen of *Ceratosaurus nasicornis*, which is distinguished by the horn on its snout, resides at NMNH.

1892 Artist Carl Dahlgren (1841–1920) shows spear-wielding humans confronting *Stegosaurus* and *Amphicoelias*.

1892 Barnum Brown reports from Wyoming that the "Cope quarry [containing the *Diplodocus* specimen] is a veritable gold mine and I have been in bones up to my eyes." The Cope Quarry was opened in 1877 in strata deposited late in the Morrison Formation.

1892 Edward Drinker Cope uses a partial skull discovered in Alberta, Canada, to name "Laelaps incrassatus," but the name *Laelaps* had already been used to name a mite. Although Othniel Marsh had changed *Laelaps* to *Dryptosaurus* in 1877, Cope refused to recognize Marsh's change. In 1904, Lawrence Lambe changed "Laelaps incrassatus" to *Dryptosaurus incrassatus*, after which Henry Osborn—noting that *D. incrassatus* was based on generic teeth—included it in the genus *Albertosaurus* in 1905. More than 60 percent of all specimens of *Albertosaurus* have been found near Alberta's Red Deer River.

1892 Fifty years after inventing the clade Dinosauria, famed British anatomist Richard Owen dies "of old age" in London at age 88. Owen's knowledge of anatomy was unrivaled, and he worked tirelessly to oppose Darwinian evolution while helping create BMNH (Figure 1). Owen, who had accepted knighthood eight years earlier, had spent the past half-century at Sheen Lodge, an estate given him by Queen Elizabeth. Owen, who published more than 600 papers and more than a dozen book-length studies, was buried in London's St. Andrew Churchyard.

1892 Henry Osborn writes to U.S. senator William B. Allison (head of the Senate Appropriations Committee) criticizing the work of Othniel Marsh and the USGS. Eight years later, Osborn replaced Marsh as the USGS vertebrate paleontologist.

1892 In South Dakota, Edward Drinker Cope (Figure 10) finds two partial vertebrae having several pores that he uses to name *Manospondylus gigas* ("giant porous vertebrae"). In 1917, Henry Osborn recognized the similarity of *Manospondylus gigas* with *T. rex*, but did not immediately synonymize the two genera. When *Manospondylus gigas* was later synonymized with *T. rex*, *T. rex* remained the valid name because, according to ICZN, "the prevailing usage must be maintained [when] the senior synonym or homonym has not been used as a valid name after 1899." Thus, *T. rex* is a *nomen protectum* (i.e., protected name), whereas *Manospondylus gigas* is a *nomen oblitum* (i.e., forgotten name). In naming *M. gigas*, Cope had described (although with a different name) the first *T. rex* bone on record. Cope's fossils are housed at AMNH.

1892 James Erwin Culver's "Some Extinct Giants," published in *California Illustrated*, is the first popular article about dinosaurs to include restorations of dinosaurs. Although *Stegosaurus* had not yet been exhibited in public, Culver's artwork depicted *Hypsirophus* (*Stegosaurus*) facing spear-wielding humans.

1892 Liebig Continental Trading Cards produces "Prehistoric Animals in Different Ages," the first cigarette-cards to feature dinosaurs. A similar set of cards ("The Prehistoric World") was produced by Liebig in 1921. Cigarette cards (and tea cards) were used by tobacco and tea manufacturers to stiffen packages and advertise their brands. The most famous of these cards featured baseball star Honus Wagner; in 2007, that card sold for $2.8 million.

1892 Othniel Marsh names *Claosaurus annectens* from fossils collected by John Bell Hatcher in Montana. This dinosaur, later renamed *Edmontosaurus,* was the first hadrosaur to receive a skeletal restoration. In 1901, Marsh mounted this skeleton (YPM 2182) at YPM; this was the first essentially complete dinosaur skeleton to be mounted in the United States.

1892 The Marsh–Cope feud reaches the U.S. Senate as funding for irrigation plans proposed by John Wesley Powell (Director of USGS and an ally of Marsh) are being debated. Alabama Congressman Hilary A. Herbert (1834–1919), who was influenced by Harvard professor Alexander Agassiz (son of Louis Agassiz, founder of the Harvard Museum of Comparative Zoology), believed that scientific studies should be done with private funding and proposed that funds for paleontology be cut from the USGS budget. Herbert supported his arguments with references to Marsh's gilt-edged 1880 monograph about toothed birds, which he made synonymous with wasteful governmental spending. Congress acted quickly, and on July 20, 1892, Powell telegrammed Marsh, "Appropriation cut off. Please send your resignation at once." Marsh's 10 years with the USGS, and its funds, were over.

1892 YPM now contains more than 30 skulls of *Triceratops*, as well as two skulls of the larger, but more rare, *Torosaurus* collected by John Bell Hatcher.

1893 Barnum Brown enrolls at the fledging University of Kansas. There, Brown met his academic advisor Samuel Wendell Williston, "from whom I soon learned that I wanted to be a Paleontologist and Geologist." Brown's 66-year career in paleontology was a spectacular success; he is largely responsible for the vast collection of dinosaurs at AMNH.

1893 Franco-Belgian paleontologist Louis Dollo proposes what came to be known as "Dollo's Law," which states that evolution is not reversible. Dollo's law predicts that non-avian dinosaurs will not reappear in nature. This contradicted the now-discarded claim by Victorian geologist Charles Lyell that life and the environment develop cyclically and, as a result, that extinct organisms could reappear in the future. As Lyell had noted, "The huge iguanodon might reappear in the woods, and the ichthyosaur in the sea, while the pterodactyl might flit again through umbrageous groves of tree-ferns."

1893 Rev. Henry Neville Hutchinson's *Extinct Monsters: A Popular Account of Some of the Larger Forms of Ancient Animal Life* includes illustrations by Joseph Smit (an artist best-known for drawing birds) of *Stegosaurus, Brontosaurus*, and *Triceratops*. Smit's restoration of *Stegosaurus* had one row of plates and a sprawling, alligator-like stance. In the book's preface, Henry Woodward noted the inaccuracies in Benjamin Waterhouse Hawkins's restorations of dinosaurs at Crystal Palace.

1893 Richard Lydekker (1849–1915) describes the herbivorous titanosaur, *Argyrosaurus* ("silver lizard"). This 90-foot-long, 70-ton dinosaur, which lived in the Late Cretaceous, was the first dinosaur discovered in Argentina. Lydekker named 21 dinosaurs, 8 (i.e., 38%) of which remain valid.

1893 The Columbian Museum of Chicago is established at Chicago's Jackson Park. In 1905, the name of the museum was changed to the Field Museum of Natural History to honor its first major benefactor, Marshall Field. Today, the museum—now downtown along the shore of Lake Michigan—is known as the Field Museum.

1893 Unhappy with Othniel Marsh's policy of not allowing assistants to publish articles on their own, John Bell Hatcher leaves YPM and begins working as curator of vertebrate paleontology at Princeton University.

1894 American artist Charles R. Knight's (1874–1953) reconstruction of the mammal *Elotherium* attracts the attention of AMNH's Henry Osborn. Two years later, Osborn—who introduced Knight to Edward Drinker Cope—began including Knight's work in his publications. The subsequent collaborations of Osborn, Knight, and William Diller Matthew (1871–1930) contributed much to the public's growing fascination with dinosaurs.

1894 AMNH taxidermist John Rowley, a friend of Charles Knight, tells Knight that AMNH assistant curator Jacob Wortman is looking for an illustrator of prehistoric animals. Wortman was impressed with Knight's drawing of the pig-like *Elotherium*, and soon began offering more work to Knight. Wortman, like Henry Osborn, promoted dinosaurs as exhibit objects for "attracting the attention and interest of visitors" and donors to AMNH. Dinosaurs later became status symbols at many natural-history museums.

1894 Edward Drinker Cope conducts his final expedition to the American West.

1894 Elizabeth Gordon, a daughter of William Buckland, notes the growing importance of dinosaur tracks: "The variety and number of these impressions have created a new science, and Ichnology has taken a definite place as a branch of paleontological research."

1894 In *A Journey in Other Worlds*, John Jacob Astor IV (1864–1912) writes about "giant reptiles of prehistory" on Jupiter. Astor, the richest man in the world, went down with *Titanic*.

1894 Whitman Cross names the exposures near Morrison, Colorado, the Morrison Formation; this is the formation in which Arthur Lakes found his fossils. The Morrison Formation (from the Late Jurassic, between 155 and 148 million years ago) covers approximately 600,000 square miles of the western United States, ranging from New Mexico north to Canada, and from Idaho west to Nebraska. The original Morrison Formation included the "*Atlantosaurus* beds" named by Marsh in 1877.

1895 After more than a decade of working elsewhere, Charles H. Sternberg is rehired by Edward Drinker Cope to collect fossils in north Texas. Cope praised Sternberg, noting that "few men pursue a more useful life than yourself, and when the final account comes to be recorded you shall have no occasion to be ashamed of your record. I personally have the highest respect to your devotion to science."

1895 Arthur Lakes, the first person to collect a dinosaur fossil from the Morrison Formation, publishes his influential *Prospecting for Gold and Silver*.

1895 At a meeting of BAAS, Othniel Marsh describes the problems he has corrected regarding restorations of dinosaurs: "The dinosaurs seem to have suffered much from both their enemies and their friends. Many of them were destroyed and dismembered long ago by their natural enemies, but, more recently, their friends have done them further injustice by putting together their scattered remains and restoring them to supposed lifelike forms. . . . We now know from good evidence that both *Megalosaurus* and *Iguanodon* were bipedal, and to represent them as creeping, except in their extreme youth, would be almost as incongruous as to do this by the genus *Homo*."

1895 ICZN is founded "as adviser and arbiter for the zoological community by generating and disseminating information on the current use of the scientific names of animals." ICZN produces the International Code of Zoological Nomenclature, a set of rules for naming animals and resolving nomenclatural questions. (There is a separate code that regulates the naming of plants.)

1895 Ilona Nopcsa von Felsó-Szilvás (1883–1952), the younger sister of Hungarian paleontologist Franz Nopcsa (1877–1933), finds the first dinosaur bones in Hungary. Franz later discovered dinosaurs in Transylvania, England, and France, and was one of the first paleontologists to use histology to classify prehistoric life. Franz Nopcsa named 15 dinosaurs, four (i.e., 27%) of which remain valid.

1895 One year after his last fossil-hunting trip to the West, Edward Drinker Cope's deepening financial troubles force him to sell 10,000 of his fossils (comprising 463 species, mostly North American mammals) to AMNH for $32,149 (Cope had asked for $50,000). At the same time, Cope sold his ethnological artifacts to the University of Pennsylvania's Museum of Archaeology and Anthropology for $5,500. Two years later, Cope sold the rest of his fossils to AMNH for another $29,000. Because Philadelphia's ANS did not bid on Cope's treasures, Cope's fossils left his hometown. Cope used some of the money from the sales to hire Charles H. Sternberg to prospect for him.

1895 Samuel Wendell Williston, on his final expedition, works in Wyoming with Elmer Riggs, Barnum Brown, and others to excavate a skull of *Triceratops*. Williston's description of Brown as "the best man in the field that I have ever had" later helped Brown get a job at AMNH, where Brown became the greatest fossil-hunter of all time.

1895 The Carnegie Museum opens in Pittsburgh, Pennsylvania, as a library, music hall, art gallery, and natural history museum as "the gift of Andrew Carnegie to the People of Pittsburgh." Founder Andrew Carnegie (1835–1919; Figure 14)—who invited Othniel Marsh to the museum's dedication ceremony—considered the museum "one of the chief satisfactions of my life. [It] is my monument." Carnegie understood his debt to "the good of the people of Pittsburgh among whom I had made my fortune" and wanted to reward "their good by planting in their midst a potent influence for higher things." Carnegie promised, "I have $50,000 a year to spend [on the museum] forever." In 1898, William Holland—a scientist and ordained Presbyterian minister—was hired as the CMNH director; it was under Holland's leadership that CMNH discovered many of its famous dinosaurs. The Museum's first dinosaur-collecting expedition was mounted three years later, and its oldest specimens are *Diplodocus carnegii* and *Apatosaurus louisae*.

1896 Antoine Henri Becquerel (1852–1908), a professor of applied physics at the Muséum d'Histoire Naturelle in Paris, notices that photographic plates are exposed when they are placed near a mineral called "pitchblende." This observation of "strange radiations" emitted by uranium salts ultimately led to the discovery of radioactivity and radiometric dating, which helped establish that Earth is 4.55 billion years old—enough time for evolution by natural selection, as proposed by Charles Darwin, to produce life's diversity.

1896 Charles R. Knight (1874–1953) meets AMNH's Henry Osborn and William Diller Matthew. Osborn and Matthew became important advisors for Knight's work as an illustrator of dinosaurs and other animals. That same year, Knight began work on his famous painting *Dryptosauruses Fighting*. Matthew was a curator at AMNH from 1896 to 1925 and, as Knight later noted, was "for years was my

Figure 14 Scottish-American businessman and philanthropist Andrew Carnegie (1835–1919) sponsored much of the early work in paleontology in North America. Pittsburgh's Carnegie Museum of Natural History (established in 1895) houses a world-class collection of dinosaurs; its oldest specimens are *Apatosaurus louisae* and *Diplodocus carnegii*. Carnegie, a Social Darwinist, gave casts of "Dippy" (*D. carnegii*) to museums throughout the world. Although Carnegie funded many searches for (and displays of) dinosaurs, his will included no permanent provisions for paleontology. (Library of Congress)

chief consultant and advisor on matters of pose and difficult bone structure, and what unusual forms might indicate in the living animals." Knight, Osborn, and Matthew set out to mount fossil skeletons in AMNH. That same year (i.e., 1896), *Century Magazine* published an article by Osborn titled "Prehistoric Quadrupeds of the Rockies" that included nine paintings of extinct animals by Knight; this was one of the first accurate and modern introductions of prehistoric North American mammals. Knight's images of dinosaurs were produced with the assistance of Jacob Wortman (1856–1926) and appeared in newspapers, books, magazines (e.g., several covers of *Scientific American*), and schools, and were exhibited in AMNH; these paintings strongly shaped the public's views of prehistoric life. Although Knight never worked for any particular movie, his paintings inspired how dinosaurs were portrayed in movies (and elsewhere).

1896 Edward Drinker Cope notices Charles Knight's work in *Century Magazine*. The following year, Knight studied with Cope for two weeks, and Cope died three weeks later. Knight noted, "Under [Cope's] guidance, I felt that I had stepped back into an ancient world—filled with all sorts of bizarre and curious things, and in imagination I could picture quite distinctly just what these mighty beasts looked like as they walked or swam in search of food." Eight of Knight's restorations appeared in the November, 1897 issue of *Century Magazine* in an article by William Hosea Ballou titled "Strange Creatures of the Past: Gigantic Saurians of the Reptilian Age" and in a memorial written by Henry Osborn titled "A Great Naturalist: Edward Drinker Cope."

1896 French paleontologist Charles Depéret describes the first fossils from northwestern Madagascar. These fossils included *Dryptosaurus, Majungasaurus*, and a new species of *Titanosaurus*. An expedition dedicated to finding more of Madagascar's fossilized dinosaurs did not occur until 1993.

1896 Henry Osborn's paper titled "Prehistoric Quadrupeds of the Rockies" in *Century Illustrated Monthly Magazine* is the first AMNH publication to include work by artist Charles Knight. Throughout his career, Knight refrained from including graphic violence in his depictions of dinosaurs.

1896 In "On the Affinities and Classification of the Dinosaurian Reptiles," Othniel Marsh publishes restorations of "the twelve best known dinosaurs": *Anchisaurus, Brontosaurus, Camptosaurus, Laosaurus* (*Othnielosaurus*), *Stegosaurus, Triceratops, Ceratosaurus, Claosaurus, Compsognathus, Iguanodon, Hypsilophodon*, and *Scelidosaurus*. The first nine of these dinosaurs, all from North America, were described by Marsh. Marsh was a pioneer of skeletal reconstructions, but did not support displays of mounted dinosaurs; he had ridiculed the restorations displayed in London's Crystal Palace, and he refused to let Benjamin Waterhouse Hawkins examine his vertebrate fossils because he feared Hawkins would introduce errors in his restorations. In contrast, Henry Osborn made dinosaurs a showcase attraction at AMNH; he used mounted dinosaurs to increase the status of (and attendance at) AMNH as well as to raise money and promote his ideas about evolution.

1896 Barnum Brown accepts Jacob Wortman's (1856–1926) offer to work at AMNH. Famed paleontologist Henry Osborn later described Brown as "the most amazing collector I have ever known. He must be able to smell fossils. If he runs a test-trench through an exposure it will be right in the middle of the richest deposit. He never misses." Osborn later used Brown and Roy Chapman Andrews to make dinosaurs the marquee attraction at AMNH and a staple of pop culture. There were only five years between 1897 and 1942 (when he retired) that Brown did not participate in a major excavation. Brown became the greatest dinosaur-hunter in history; 57 of his major specimens remain on display in AMNH.[17]

1896 Jacob Wortman becomes the first public advocate for dinosaurs at AMNH when he urges Henry Osborn to obtain dinosaurs to promote the museum. The following year, Osborn began doing just that. Under Osborn's leadership, AMNH eventually gathered one of the greatest collections of dinosaurs in the world.

1896 John Bell Hatcher conceives of, plans, and obtains funding from AMNH trustee and railroad magnate John Pierpont Morgan (1837–1913) for three expeditions to Patagonia. When Hatcher discovered that plants and animals in Australia and Patagonia were similar, he concluded that the two landmasses were once connected by land. Hatcher's expeditions were described in *Princeton University Expeditions to Patagonia, 1896–1899.*

1896 Othniel Marsh publishes *Dinosaurs of North America*, a founding book of American paleontology. In his book, which includes 84 lithographic plates and more than 100 pages of text, Marsh lauded his *Brontosaurus* as "a nearly entire skeleton, by far the most complete of any of the Sauropoda yet discovered." Marsh's book summarized most of the 55 papers and books that Marsh had published about dinosaurs between 1872 and his death in 1899. (Marsh's publications named and described 80 new species of dinosaurs [two of which were lost to Leidy] that he classified into 34 new genera [of which 27 are still in use], 7 new suborders, and 3 new orders.) Although *Dinosaurs of North America* was hailed as "the foundation of dinosaur knowledge," Marsh noted that the monograph was not "as finished a product" as his book about toothed birds. Marsh had planned to write a series of monographs about dinosaurs, but *Dinosaurs of North America* is the only one that he completed.

1896 While in Wyoming, Barnum Brown finds his first dinosaur fossil, a skull of *Triceratops. Triceratops* had the heaviest skull of all land animals.

1897 AMNH's Henry Osborn introduces Charles Knight to Edward Drinker Cope. Cope mentored Knight, and Knight later described Cope as "the greatest conversationalist that ever graced the service of paleontology."

1897 One year after finishing his presidency of AAAS, 56-year-old Edward Drinker Cope (Figure 10) dies of kidney failure while on a cot in his home at 2102 Pine Street in Philadelphia, Pennsylvania. During his long career, Cope named 64 dinosaurs. According to artist Charles Knight, who had visited Cope a few weeks before Cope's death, the furnishings of Cope's home were "unique and completely dust covered. Never have I seen such a curious place. . . . The place was absolutely bare of furniture and hangings. No pictures, no curtains, nothing but petrified skeletons of extinct monsters. . . ." Cope had described famous animals such as *Camarasaurus, Coelophysis*, and other dinosaurs (although Cope's 1,400 published papers included many studies of dinosaurs, his most exhaustive publications were about mammals). Cope's passing was noted with an eight-page memorial in *American Naturalist* that did not mention Marsh, but recognized Cope as "a peer of Huxley and Owen." In an obituary prepared for NAS, Henry Osborn remembered Cope as "the most brilliant creative mind in comparative anatomy and evolution that America has produced" and "a man who loved speculation. . . . Marsh, with less breadth and less ability, nevertheless was a paleontologist of a very high order." *American Journal of Science* remembered Cope with a six-paragraph obituary, and Charles H. Sternberg—who learned of Cope's death while working for Cope in Texas—remembered Cope this way: "I had lost friends before, and had known what it was to bury my own dead, even my firstborn son, but I have never sorrowed more deeply than I did over the news [of Cope's death]. . . . One thing is certain—as long as science lasts, and men live to study the animals of the present and the past, Cope's name and work will be remembered and revered." Cope left his relatively small estate (later valued at $84,600, which was far less than the $250,000 he had inherited from his father) to his wife and daughter, and his brain, skeleton, and cremated remains to the Anthropometric Society, hoping that it would be made the type-specimen of *Homo sapiens* (but "not be placed on exhibition").[18] For many years, Cope had collected fossils much faster than he could describe them, and many excellent specimens were discovered in boxes after his death. Among these was the nearly complete skeleton of *Allosaurus* that is mounted in AMNH. Today, Cope's desk is in AMNH. In 1975, Cope's home in Philadelphia was made a National Historic Landmark.

1897 A book of tributes to Edward Drinker Cope praises his work as a geologist, ichthyologist, herpetologist, and mammalogist, but barely mentions his work with dinosaurs.

1897 AMNH begins work at Wyoming's Como Bluff. The next year, AMNH expanded its work there, after which Walter Granger discovered what would become the famed Bone Cabin Quarry. Over the next eight years, the site yielded fossils of 64 dinosaurs, including *Stegosaurus, Apatosaurus,* and *Diplodocus* (AMNH 223). *Diplodocus* was one of Barnum Brown's first discoveries as a professional paleontologist; as he later noted, "I was . . . fortunate in discovering a partial skeleton of . . . *Diplodocus*. This was the first dinosaur excavated by any American Museum expedition, and here I introduced the use of plaster of Paris in excavating fossils." The *Diplodocus longus* displayed today at AMNH (AMNH 969) was collected at Como Bluff's Bone Cabin Quarry by Peter Kaisen in 1903. (This quarry was named for a sheepherder's cabin built about 10 miles north of Como Bluff in the late 1800s; the foundation of the house was made of dinosaur bones found nearby.) *Diplodocus* is one of the three most-abundant sauropods in the Morrison Formation. Henry Osborn noted the discovery of *Diplodocus* this way in AMNH's annual report: "Thus has been inaugurated the second great division of the work, namely the history of the reptiles in North America." This excavation began AMNH's rise as one of the world's greatest showcases for dinosaurs.

1897 AMNH's Henry Osborn hires artist Charles Knight (1874–1953) as a consultant for portraying dinosaurs at AMNH, in hopes of decorating the museum with paintings "unsurpassed in any part of the world, illustrated by your masterly hand under my scientific and artistic direction." Knight, despite being legally blind from inherited astigmatism and cataracts for most of his life,[19] was the first artist to work with paleontologists to produce paintings of dinosaurs for AMNH. These popular murals were funded by AMNH trustees such as John Pierpont Morgan, Jr. (1867–1943), who paid $10,500 for some of the murals that continue to be exhibited at AMNH and other famous museums (e.g., NMNH, the Field Museum). Knight's work defined dinosaurs for the world; virtually every dinosaur book published in the first six decades of the 20th century included Knight's illustrations. Knight's work also influenced numerous filmmakers, including Willis "O'Bie" O'Brien (1886–1962), who first realized prehistoric animals on film (e.g., *The Lost World*, *King Kong*). *Godzilla* was also based partly on Knight's paintings. Evolutionary biologist Stephen Jay Gould (1941–2002), who featured one of Knight's paintings on the cover of his popular book *Bully for Brontosaurus* (1991), memorialized Knight this way: "I cannot think of a stronger influence ever wielded by a single man in such a broad domain of paleontology. . . . Not since the Lord himself showed his stuff to Ezekiel in the valley of dry bones had anyone shown such grace and skill in the reconstruction of animals from disarticulated skeletons. Charles R. Knight, the most celebrated of artists in the reanimation of fossils, painted all the canonical figures of dinosaurs that fire our fear and imagination to this day."

1897 Barnum Brown leads AMNH's first expedition to Wyoming's Como Bluff, a famed fossil-site explored earlier by Othniel Marsh's workers. During the 1897 season, AMNH spent $1,600 on its Como Bluff expedition.

1897 As relations between Jacob Wortman and Barnum Brown become increasingly tense, Wortman tells Henry Osborn that Brown had caused problems by leaving camp "to hang around a woman for 3 or 4 days in Medicine Bow. . . . He has not only made a fool of himself . . . but he has caused a great deal of unfavorable comment . . . I am completely disgusted with his actions and I doubt very much whether it will be expedient to keep him at any price . . . I don't want him with me anymore."

1897 Bill Reed starts working as assistant geologist and curator at the geology museum at the University of Wyoming. During the next three years, while working south of Laramie, Reed collected almost 10,000 specimens of *Barosaurus, Camarasaurus, Allosaurus*, *Stegosaurus*, and other animals.

1897 Charles Knight paints *Brontosaurus*; this image may still be the most recognizable image of a sauropod. In the background, *Diplodocus* grazed on the shore of a lake. Knight's painting, which was produced under the direction of Henry Osborn, was republished by William Diller Matthew in 1905. Both of Knight's animals were dull gray, an approach copied for more than 70 years (including by Rudolph Zallinger in his famous mural in YPM). Knight based his painting on a 1:16 model created from the mounted skeleton of *Brontosaurus*; this model was later used to provide the first scientifically calculated estimate of a sauropod's mass (38 tons).

1897 Charles Knight's restoration of Othniel Marsh's *Stegosaurus ungulates* appears in an article titled "Strange Creatures of the Past" in *Century Magazine*. Although Edward Drinker Cope had claimed that *Stegosaurus latus* had two rows of spinal plates, Knight's restoration—along with one published by paleontologist E. Ray Lankester (1847–1929) in *Extinct Animals* (1905)—had one row of spinal plates. (Arthur Conan Doyle referred to Lankester's restoration in *The Lost World* (1912) when Professor Challenger noted, "This is an excellent monograph by my gifted friend, Ray Lankester!") Knight's restoration was the model for *Stegosaurus* in the movie *King Kong* (1933); in that movie, *Stegosaurus* is the first dinosaur encountered by the explorers of Skull Island. In 1901, Knight produced the first restoration of *Stegosaurus* to have a paired double-row of plates.

1897 Henry Osborn hires Barnum Brown to join the staff of AMNH. Although Brown's discovery of *T. rex* in 1902 was his most famous discovery, Brown collected fossils for more than 60 years in the United States, Canada, Mexico, India, and Pakistan. When Brown arrived at AMNH, the museum had no fossil dinosaurs. Osborn expected Brown and his other employees to be committed to their work, as exemplified by this entry in Brown's journal on December 7, 1898: "Yesterday . . . I was notified by Prof. Osborn that arrangements had been made for me to go to South America. Four of the Dept. men packed up my kit and I took another with me home to pack up. Imagine getting an outfit together in three hours to go on a seven thousand mile journey, to be gone a year or more. Such is the life of a fossil man."

1897 Jacob Wortman, the first curator of Vertebrate Paleontology at CMNH, instructs his collectors to excavate dinosaur bones by surrounding them with plaster poured onto burlap. The plaster held the bones in place when it set.

1897 Lawrence Lambe of the GSC collects the first dinosaur bones from an area later known as Dinosaur Provincial Park. Lambe's work popularized dinosaurs with the public and helped produce the Golden Age of Dinosaurs in Alberta.

1897 Soon after studying under Edward Drinker Cope, Charles Knight completes his famous painting of *Laelaps* (which Marsh renamed *Dryptosaurus*, but which Cope refused to recognize) in a violent fight. Dinosaurs were not again seen as agile until the beginning in the 1960s, when the "dinosaur renaissance" revived the image of dinosaurs being active creatures. In 1897, Knight restored *Agathaumas*, a ceratopsid discovered in 1872 in Wyoming by Fielding Meek and H. M. Bannister while working for Cope. Although Knight's reconstruction was flawed because the dinosaur's fossils were incomplete, it was copied for years, including in *The Lost World* (1925). Knight's painting of *Dryptosaurus* is in AMNH.

1897 William Ballou publishes, as one of six figures, a restoration of the sauropod *Amphicoelias*, which had been prepared by Charles Knight under the supervision of Edward Drinker Cope. Knight's work shows four individuals, only two of which have their heads above water.

1898 After a disappointing first season at Como Bluff, AMNH returns to Wyoming to focus on excavating the Medicine Bow Anticline (uplift) just north of Como Bluff. (The following year, outlaw Butch Cassidy's gang robbed a train and dynamited a safe near here.) Upon discovering Bone Cabin Quarry, Jacob Wortman wrote to Henry Osborn, "If it is what I think . . . it is certainly the greatest bone bed in existence. . . . There are so many [specimens] I would not know where to begin to describe them. . . . I never saw bones thicker in all my life." By 1905, Bone Cabin Quarry yielded more than 500 bones from at least 50 individuals and a dozen taxa (mostly the sauropods *Diplodocus, Camarasaurus*, and *Apatosaurus*, and the non-sauropods included *Stegosaurus, Camptosaurus*, and *Ornitholestes*). These specimens, which weighed over 70 tons, filled 275 crates that were sent to AMNH aboard boxcars provided by J.P. Morgan. One of the most important specimens was *Ornitholestes hermanni*, a slender, 5-foot-long theropod having long arms for grasping prey (Henry Osborn described this specimen in 1903). Today, *O. hermanni* remains one of the rarest theropods of the Morrison Formation. When Charles Knight painted *Ornitholestes*, he showed it in an active, leaping pose, thus defying the prevailing assumptions of his time that dinosaurs were slow, plodding reptiles.

1898 AMNH's Henry Osborn (Figure 15) describes a "*Brontosaurus*" skeleton as *Camarasaurus*, apparently believing the two animals were the same. Osborn envisioned *Brontosaurus* "as a great, slow-moving animal-automaton."

1898 As his health worsens, Othniel Marsh arranges for his fossils to be transferred to Washington. When Marsh died the following year, he had already sent more than 30 tons of fossils, and after his death his executors shipped another 80 tons that filled five freight cars. During his life, Marsh produced hundreds of publications, including monographs about toothed birds (1880) and dinosaurs (1896); produced more than 200 plates for monographs that he never wrote; described 496 new species and 225 genera; and described 80 new dinosaurs. Marsh, who often claimed that evolution is "key to the mysteries of past life," was described by John Bell Hatcher as "the best of them all."

1898 Charles Knight paints a swamp-dwelling *Apatosaurus* for AMNH. This image became the model used by Marcel Delgado for *Apatosaurus* in the movie *King Kong* (1933).

1898 Chicago businessman Marshall Field hires Barnum Brown's friend Elmer Riggs to lead a dinosaur-collecting project for the recently established Columbia

Figure 15 American paleontologist and eugenicist Henry Fairfield Osborn (1857–1935) brought dinosaurs to the American Museum of Natural History (AMNH) while making it one of the world's greatest museums. Osborn, shown in this photo from 1898 beside the hind leg of *Diplodocus* at Como Bluff, Wyoming, hired now-legendary colleagues at AMNH, including Barnum Brown, Roy Chapman Andrews, and Charles Knight. Osborn, who was president of AMNH for 25 years, believed that humans originated in Asia, and he sent Andrews there in the 1920s to find human ancestors. Andrews found none but did find new dinosaurs that Osborn used as marquee attractions to lure visitors and donors to AMNH. Osborn, who in 1905 named *Tyrannosaurus rex,* helped make dinosaurs part of pop culture. (Private Collection/Ken Welsh/The Bridgeman Art Library)

Museum of Chicago (now the Field Museum). By 1900, Riggs—a protégé of Samuel Williston, his former professor at the University of Kansas—had collected the front limb of the stout sauropod *Camarasaurus* and what would be the type-specimen of the tallest sauropod then known, *Brachiosaurus*. Today, the wall-mounted,

16-foot-long juvenile *Camarasaurus* displayed at CMNH (and collected at DNM) is the most complete sauropod skeleton ever found in the Morrison Formation; it was also the first sauropod skeleton ever to be mounted from the bones of one individual. Despite the fame of *Brontosaurus* (i.e., *Apatosaurus*), *Camarasaurus* was the first of the giant dinosaurs to be known from fossils complete enough to make viable reconstructions. In 2013, an unopened crate in the Field Museum from Riggs's excavation in Wyoming yielded the first non-skull elements of the rare *Torvosaurus*.

1898 Edwin R. Lankester (1846–1929), a protégé of Thomas Huxley and friend of Arthur Conan Doyle, replaces Richard Owen as director of BMNH, and shifts the emphasis of the Museum from classification to evolution. This change in emphasis was manifest in Lankester's moving the life-size statue of Charles Darwin to the Museum's most prominent position, atop a stairwell overlooking the Great Hall. Lankester, who rejected Alfred Wallace's defense of the supernatural, wrote *Extinct Animals* (1905), which inspired (and provided the factual information for) the creatures in Sir Arthur Conan Doyle's *The Lost World* (1912). (Lankester was the model for—and is mentioned by—Professor George Edward Challenger in *The Lost World*.) In the movie version of Doyle's novel, the ape-men were portrayed by white actors in blackface, who traveled with monkeys and chimps.

1898 Henry Osborn of AMNH convinces J. P. Morgan to purchase Knight's original sculptures and paintings for AMNH. Osborn then produced a catalog that included reproductions of Charles Knight's artwork. That same year, Osborn—sensing problems created by basing the names of dinosaurs on small differences in individual bones—noted, "It is *a priori* improbable that so many different genera of gigantic Saurians of size coexisted. It is against the principles of evolution that closely similar types of equal size should occupy the same territory at the same time. . . . The evidence which has been brought forward to demonstrate such an exceptional condition is inadequate." Four years later, John Bell Hatcher had the same concern: "It does not seem at all improbable that some of the remains which have been referred to *Pleurocoelus, Astrodon*, or *Elosaurus* may in reality belong to the young of some of these genera of the large sauropoda (i.e., *Brontosaurus, Morosaurus*, and *Diplodocus*)." Osborn argued that it is improbable that many giant sauropods existed simultaneously, and suggested that all of their genera (e.g., *Atlantosaurus, Camarasaurus, Apatosaurus, Amphicoelias*, and *Brontosaurus*) are synonyms for *Camarasaurus*, which had been originally named by Edward Drinker Cope. Two years later, Osborn and Walter Granger divided the group into *Morosaurus* (established by Marsh in 1878; now *Camarasaurus*), *Brontosaurus* (now *Apatosaurus*), and *Diplodocus*. Osborn, an admirer of Edward Drinker Cope, later claimed, "Marsh's reptile work is as defective and full of faults as his mammal work." Size is no longer considered an adequate criterion to establish a species of dinosaur.

1898 Henry Osborn (Figure 15) of AMNH tells Barnum Brown, "I want you to go to Patagonia today with the Princeton expedition [organized by John Bell Hatcher]. The boat leaves at eleven; will you go?" Brown responded, "This is short notice, Professor Osborn, but I'll be on that boat." A year and a half later, Brown returned to AMNH with numerous excellent fossils, thereby establishing his reputation as the best fieldworker and fossil collector at AMNH.

1898 Juanita Horton (1898–1986) is born in Midland, Texas. When her family later moved to Los Angeles, Juanita met pioneering film director D. W. Griffith, who gave her the screen-name Bessie Love. In 1925, Love starred in *The Lost World*, in which she became the first woman to be harassed by a dinosaur; this role became a stereotype of movies featuring dinosaurs. (Her character, Paula White, was not in the novel.) After appearing in more than 131 movies and television episodes, Horton died in London in 1986.

1898 Social Darwinist and steel magnate Andrew Carnegie (1835–1919; Figure 14) reads a front-page headline in the December 11 issue of *New York Journal and Advertiser* proclaiming "Most Colossal Animal Ever on Earth Just Found Out West!" The story, which was accompanied by a drawing showing a 130-foot-long *Brontosaurus* found by Bill "Reeder" Reed (i.e., William Harlow Reed, a collector for Othniel Marsh at Como Bluff) peering into an 11th-floor window of the New York Life Building, claimed that when the dinosaur "walked, the earth trembled under its weight of 120,000 pounds, when it ate it filled a stomach big enough to hold three elephants, when it was angry its terrible roar could be heard ten miles, and when it stood up its height was equal to eleven stories of a sky-scraper." Intrigued by the discovery described in the mistake-riddled article, Carnegie sent a check for $10,000 and instructed newly-hired CMNH director William Jacob Holland (1848–1932) to "buy this for Pittsburgh." The self-confident and competitive Holland tried, but Reed telegrammed him that the "dinosaur [is] not for sale." (In fact, the fossil described in the article was still in the ground in Wyoming's Freezeout Hills, and was neither complete nor well-preserved.) The area where the fossil was discovered was soon filled with collectors, including those from the University of Wyoming, the University of Kansas, and the Field Museum of Natural History in Chicago. Carnegie funded an expedition to the western United States in which Jacob Wortman and his colleagues unearthed from an area now known as DNM an 84-foot-long skeleton of what was later named *Diplodocus carnegii*. This specimen, which was shipped to Pittsburgh in 130 crates, was the first dinosaur for CMNH. The Carnegie Institute funded other expeditions to search for dinosaur remains, and today the museum houses some of the world's most important dinosaur fossils, including one of the first *T. rex* remains ever found. A cast of the 150-million-year-old *Diplodocus* skeleton was unveiled at the BMNH in 1905, and two years later, the original *Diplodocus* was made a centerpiece of CMNH in Pittsburgh. Carnegie sent casts of "Dippy" to numerous European and American museums, including those in Berlin, Paris,

Vienna, Austria, Bologna, St. Petersburg, Buenos Aires, Madrid, and Mexico City. (The cast given to the Paleontological Museum in Munich in 1932 is not yet mounted.) These casts were the first dinosaurs seen by millions of people throughout the world, and *Diplodocus* ("double beam," referring to double-beamed chevron bones in the underside of its tail) is one of the best-known dinosaurs. Decades later, Gerhard Ernest Untermann (1898–1975) and his wife Billie (1906–1973) obtained the original molds for "Dippy" and created a fiberglass cast now visible along Highway 40 in Vernal, Utah. The worldwide distribution of "Dippy" made *dinosaur* a household word and soon prompted people to refer to CMNH as "the House That Dippy Built."

1898 The *New York Herald* describes the American Anthropometric Society—which housed the brains of Joseph Leidy, Edward Drinker Cope, and hundreds of other citizens—as "a grewsome [*sic*] organization." In the early 1900s, the society dissolved and passed its collection to the Wistar Institute of Anatomy and Biology, its parent institution.

1898 Walter Granger (1872–1941), who discovered Como Bluff the previous year, discovers *Brontosaurus* nine miles north of Medicine Bow, Wyoming. The fossil was displayed at AMNH in 1905. In 1921, Granger went to Zhoukoudian, China, and began excavations that later yielded "Peking Man" (*Homo erectus*).

1898 Lawrence M. Lambe becomes the first person to intentionally collect fossils of dinosaurs in what is now DPP in Alberta, Canada. Although Lambe did not recover any complete skeletons or skulls, his work notified others of the area's vast potential. Today, specimens from DPP are displayed in museums throughout the world.

1899 At the end of Wardon Allan Curtis's *The Monster of Lake LaMetrie*, soldiers kill an *Elasmosaurus* that had swam to the edge of Wyoming's Lake LaMetrie. After receiving a human brain, the dinosaur had been happy and social, but then killed McLennegan, its human friend. The dinosaur was then killed by infantrymen.

1899 Henry Osborn argues, "There is a traditional view that [large sauropods] were ponderous and sluggish. This view may apply in a measure to *Brontosaurus*. In the case of *Diplodocus*, it is certainly unsupported by facts."

1899 In one of his last publications, Othniel Marsh identifies the first dinosaur tracks in the Morrison Formation. Today, these tracks—which were found by George Wieland in the Black Hills of South Dakota—are in the basement of YPM. Most of the tracks were left by theropods.

1899 In the deed giving his estate to Yale University, Marsh describes his collection of fossils: "This is the most important and valuable of all, as it is very extensive, contains a very large number of type-specimens, many of them unique, and is widely known. . . . The collection was pronounced by Huxley, who examined it with care in 1876, to be surpassed by no other in the world. Darwin, in 1878, expressed a strong desire to visit America for the sole purpose of seeing this collection." Marsh described some of the best-known dinosaurs, including *Brontosaurus* (*Apatosaurus*), *Stegosaurus*, and *Allosaurus*.

1899 John Bell Hatcher collects the first fossils of a ceratopsian dinosaur. During the next four years, Hatcher collected more than 30 such specimens in eastern Wyoming; these fossils were used to describe many of the 16 species of *Triceratops*. John Ostrom (1928–2005) and *Archaeopteryx* expert Peter Wellnhofer (b. 1936) later suggested that there is only one species of *Triceratops*.

1899 John Bell Hatcher first proposes a map of a dinosaur quarry to show the relation of bones to each other and their surroundings. Today, such maps are created at all excavations and quarries.

1899 John Joly (1857–1933) uses calculations of the ocean's salinity to estimate that Earth is 90 million years old.

1899 Just before his death, Othniel Marsh lists his seven most important discoveries. Topping the list were his studies of fossil horses and toothed birds, followed by uintatheres, brontotheres, pterodactyls, and mosasaurs. Dinosaurs were listed last.

1899 On July 4, while working in Quarry 3 near Sheep Creek in Albany County, Wyoming, Arthur Coggeshall finds *Diplodocus carnegii*. This specimen, which Coggeshall suggested naming "the Star-Spangled Dinosaur," was formally described in 1901 and for many years was the longest (at 84 feet) dinosaur known; CMNH's exhibit hall had to be expanded to accommodate the fossil.

1899 Othniel Marsh, who had described such famous dinosaurs as *Triceratops, Stegosaurus*, and *Allosaurus*, dies of pneumonia at the age of 67 in New Haven, Connecticut. During his career, Marsh named 80 dinosaurs (more than anyone else in history), including iconic ones such as *Allosaurus* (1877), *Apatosaurus* (1877), *Barosaurus* (1890), *Diplodocus* (1878), *Stegosaurus* (1877), and *Triceratops* (1889). (Although Marsh's work with dinosaurs was impressive, it did not match the evolutionary significance of his studies of horses and toothed birds.) The epitaph on Marsh's tombstone in New Haven's Grove Street Cemetery (adjacent to Yale University, and near the grave of Benjamin Silliman) noted Marsh's accomplishments: "Eminent as an explorer, collector, and investigator in science. To Yale University He Gave His Services, His Collections, and His Estate." (Marsh gave another $30,000 to YPM, and $10,000 to the NAS; Yale sold the contents of

Marsh's home for $18,694.) *The American Journal of Science* included a 16-page obituary for Marsh, and in a tribute published in *Science*, longtime Cope assistant Jacob Wortman described Marsh's discoveries as "one of almost continual triumph in the bringing to light of new and strange forms of life that had inhabited the western hemisphere in the distant past. . . . The methods of collecting and preparing these fossils for study and exhibition which he has introduced in the course of his long experience form the basis very largely of all similar work in almost every paleontological laboratory of the world, and it is a matter of common remark that nearly all the noted collectors and preparateurs have received their training under his immediate influence." Marsh's fossils were appraised at over $1,000,000. The dinosaur *Othnielia* (1977) was named as a tribute to Marsh, as was *Marshosaurus bicentesmus* (1976). Like his benefactor George Peabody, Marsh was a lifelong bachelor. Marsh's 18-room home, now known as Marsh Hall, is a National Historic Landmark, and its grounds are known as the Marsh Botanical Garden. Just as the Marsh–Cope feud had destroyed Cope financially, so too did the feud hurt Marsh; he died with only $186 in his bank account. Marsh was replaced by Richard Swann Lull as director of the YPM. The Cope–Marsh feud changed the face of American paleontology; many modern paleontologists are intellectual descendants of Marsh and Cope.

1899 Samuel Williston recognizes that *Apatosaurus* and *Brontosaurus* are the same species of dinosaur. Four years later, the names were formally made synonyms. Although *Apatosaurus* retained priority because it was proposed before *Brontosaurus*, the public continued to recognize *Brontosaurus*.

1899 Thomas Weston discovers fossil-rich beds near Berry Creek and Deadlodge Canyon, both of which are in or near what now is DPP in Alberta, Canada.

1899 Wilbur C. Knight (1858–1903, a geology professor at the University of Wyoming) and Edward L. Lomax (a ticket agent for Union Pacific Railway) organize the "Wyoming Fossil Fields Expedition of July, 1899" that offers free railroad fare to scientists wanting to explore Wyoming's fossil fields. The promotion, whose goal was to help scientists "collect and create museums," attracted more than 70 geologists from areas ranging from California to North Carolina. Among these scientists was William Holland, who was sent to see the fieldwork and give lectures to fossil enthusiasts in Laramie.

1900 After enduring "the most discouraging, if not the most uncomfortable, period that I have known in seven years of collecting," Elmer Riggs (Figure 16) finds vertebrae and shoulder bones of *Camarasaurus* in what will later be named Colorado National Monument. Soon thereafter, Riggs and his assistant H. W. Menke (Figure 16) went west and, on the south side of a hill near Fruita that exposed the Morrison Formation, discovered what came to be known as *Brachiosaurus altithorax*

Figure 16 In 1900, near Grand Junction, Colorado, Elmer Riggs and H.W. Menke (Riggs's field assistant) found *Brachiosaurus* and *Apatosaurus*. In this photo, Riggs and Menke are shown preparing *Brachiosaurus* in the preparatory lab at the Field Museum. (Field Museum Library)

("deep-chest arm-lizard"). These fossils were shipped to the Field Museum in 37 crates; its femur was 6 feet, 10 inches long, and the intact animal almost 75 feet long and 40 feet tall. Newspapers announced the discovery with headlines such as "Bones of the Largest Known Animal Found" (*Chicago Sunday Tribune*), "The Monster of All Ages" (*Boston Journal*), and "Chicago Has the Largest Land Animal That Ever Lived" (*Chicago Sunday Times-Herald*). This fossil was described as "the largest of them all" and is today displayed at the Field Museum in Chicago, Illinois. In 1903, Riggs's discovery was named *Brachiosaurus*; its forelimbs were longer than its hind limbs, thereby giving *Brachiosaurus* proportions unique to Jurassic dinosaurs. Riggs suggested that, contrary to popular opinion, giant sauropods were not amphibious, but terrestrial. In 1938, the site of Riggs's discovery of *Brachiosaurus*—a hill now called Riggs Hill—was marked with concrete vertebrae and a plaque that misspelled the word *Brachiosaurus*. In 2000, after a stint in the Field Museum (and the arrival of *T. rex* "Sue"), *Brachiosaurus*—at one time the largest known dinosaur—was moved to Concourse B of United Airlines Terminal One at Chicago's O'Hare Airport. Riggs Hill is surrounded by housing developments just south of Fruita, Colorado. More than 30 species of dinosaurs—including the oldest known *Iguanodon*—have been found near this area.

1900 After repeated feuds with William Holland, Jacob Wortman—who had earlier hired Barnum Brown at AMNH, helped recruit Charles Knight at AMNH, and found the type specimen of *Diplodocus carnegii* for CMNH—is fired from CMNH.

Wortman soon gave up paleontology and, after a stint as a rancher in Nebraska, managed a drugstore in Brownsville, Texas, until his death there in 1926.

1900 CMNH's paleontology program is taken over by paleontologist John Bell Hatcher, a Yale graduate who had collected fossils for Othniel Marsh. That same year, in a note thanking Hatcher for sending a copy of Hatcher's *Diplodocus* monograph (which would be published the following year), Louis Dollo asked, "Have you ever thought about casting the bones of this Dinosaurian, so that one could mount them according to the restoration that you have just published? Or perhaps, to make a model of your restoration, since [its accuracy] is established by multiple specimens? This would be an element of exchange against a cast of an Iguanodon." Hatcher led a crew back to Wyoming, where they excavated a second specimen of *Diplodocus* from Quarry D (also known as Quarry 3), the same quarry that had yielded the first *Diplodocus* the previous year. On the same trip, Hatcher's crew unearthed an adult *Apatosaurus* (now displayed at the University of Wyoming) and a well-preserved skeleton of a juvenile *Apatosaurus*. While at CMNH, Hatcher named, helped display, and exported the famous *Diplodocus carnegii*.

1900 Elmer Riggs, an assistant curator at the Field Columbian Museum (now the Field Museum), receives a letter from Grand Junction, Colorado, dentist (and president of the Western Colorado Academy of Science) Stanton M. Bradbury, noting that "dinosaur bones have been known locally since 1885 and have been collected largely as curios by local ranchers." Intrigued, Riggs and his coworkers went to Grand Junction where, three months later, they found an articulated forelimb of *Morosaurus* (*Camarasaurus*) in the Morrison Formation near the eastern entrance of present-day Colorado National Monument. This site was called Quarry 12 (in an area that today is known as the Redlands), but its location was lost until 1988.

1900 Henry Osborn (Figure 15) succeeds Othniel Marsh (Figure 9) as vertebrate paleontologist with USGS. Osborn inherited Marsh's unfinished government-funded work. In 1924, Osborn became the senior vertebrate paleontologist with USGS.

1900 Hungarian paleontologist Franz (Ferenc) Baron Nopcsa (1877–1933) describes his first dinosaur, the hadrosaur *Limnosaurus* (now *Telmatosaurus*). Nopcsa was among the first paleobiologists interested in what dinosaurs were like when they were alive, and later became one of the first paleontologists to apply bone histology to studies and classifications of dinosaurs.

1900 John Bell Hatcher recruits Charles Whitney Gilmore (1874–1945) to work for CMNH. Three years later, Gilmore—who had helped AMNH excavate *Apatosaurus*—moved to NMNH to work on Othniel Marsh's fossils. Gilmore named several dinosaurs (e.g., *Alamosaurus, Brachyceratops, Thescelosaurus*) and wrote monographs about *Stegosaurus* (1914), *Apatosaurus* (1936), *Camarasaurus* (1925), and several theropods.

1900 Othniel Marsh and Edward Drinker Cope have named 144 (i.e., 40%) of the 359 named dinosaurs.[20] By 2010, their 144 names represented approximately 10 percent of the 1401 named species. However, only 23 (i.e., 29%) of Marsh's 80 named dinosaurs, and only 9 (i.e., 14%) of Cope's 64 names, remain valid.

1900 The late Edward Drinker Cope has described 1,115 of North America's 3,200 known fossil vertebrates.

1900 When William Reed quits his job with CMNH after feuds with several of his coworkers, famed field-paleontologist John Bell Hatcher tells Holland, "Reed is afflicted, as unfortunately so many partially educated and 'self-made' men are apt to be, with an exaggerated idea of their importance and the value of their attainments." Holland responded, "I feel greatly relieved to know that he is no longer on our force . . . I look to you to [find a good replacement]. We wish no more Reeds, no more Wortmans."

1900 Working in Weston County, Wyoming, Barnum Brown discovers a large carnivore that Henry Osborn names *Dynamosaurus imperiosus* (AMNH 5866). Osborn named *Dynamosaurus* in the same paper that he named *Tyrannosaurus*, but he later determined that *Dynamosaurus* and *Tyrannosaurus* were the same species. Because *Tyrannosaurus* was named a page earlier in the paper than *Dynamosaurus*, the name *Dynamosaurus* became an invalid name. Between 1900 and 1906, Barnum Brown found the first three recognized *T. rex* skeletons (BM-R7995, AMNH 5027, and CMNH 9380), and later found two more. *Tyrannosaurus* has appeared in countless shows, video games, and movies (e.g., *Barney & Friends, Dinosaur Train, Dino-Riders, Walking with Dinosaurs, When Dinosaurs Roamed America, Jurassic Fight Club*).

1901 A year after discovering *Brachiosaurus* near Grand Junction, Colorado, Elmer Riggs—an assistant curator of paleontology at the Field Museum—returns to Colorado to search for *Apatosaurus*. Riggs had a budget of $800 for his work. The 72-foot-long specimen he excavated at Quarry 15 (at what is now known as Dinosaur Hill just south of Fruita, Colorado), along with a *Triceratops* also found by Riggs, was mounted in 1908 in Chicago's Field Museum.[21] (Because these fossils were to be used for scientific reasons, railroads shipped the six tons of crates to Chicago for free.) Behind these specimens hang paintings by Charles Knight, who illustrated the legacy of Othniel Marsh, Edward Drinker Cope, and others. One painting depicts *Apatosaurus* in a Jurassic swamp, while another shows *Tyrannosaurus* glaring at *Triceratops*. The latter painting, which showed the body of *Tyrannosaurus* oriented horizontally (as opposed to the more-erect "Godzilla" pose), foreshadowed later ideas about *Tyrannosaurus*' posture. Indeed, almost all popular images of *T. rex* can be traced to Knight's two paintings of *T. rex*[22] (as well as to those by Neave Parker, Zdenêk Burian, and Rudolph Zallinger). We now know that *Apatosaurus* lived on dry land, not in swamps. In 1938, the site of Riggs's discovery of *Apatosaurus* was commemorated with a monument and bronze plaque paid for by Riggs, amateur paleontologist Al Look, and the Grand Junction Chamber of Commerce. The seven acres of land along the eastern side of Dinosaur Hill in Fruita, Colorado, were given by the Eugene Fletcher family to the Museum of Western Colorado.

1901 Andrew Carnegie sells his steel interests and has a fortune exceeding $350 million. Before he died in 1919, Carnegie had given away all but about $30 million. Much of this money supported CMNH and dinosaur excavations in the American West.

1901 Charles Knight, working with paleontologist Frederic A. Lucas, produces the first restoration of *Stegosaurus* unequivocally having a paired, double-row of spinal plates. Two years later, again working with Lucas, Knight sculpted a miniature model of *Stegosaurus* with a staggered, double row of plates. In the movies *The Lost World* (1925) and *King Kong* (1933), Knight used this model to produce the first, life-size sculptural restoration (made of papier-mâché) of *Stegosaurus*, which was displayed at the St. Louis World's Fair Exposition in 1904. *Stegosaurus* had a paired, double-row of plates. However, at the Sinclair exhibit at the Chicago World's Fair in 1933–1934, *Stegosaurus* had a staggered double-row of plates.

1901 German paleontologist Eberhard Fraas (1862–1915), who would famously collect dinosaur fossils from eastern Africa, visits Bone Cabin Quarry with AMNH's Henry Osborn.

1901 Henry Osborn makes the mounting of *Brontosaurus* from Nine Mile Crossing a top priority at AMNH. The 66.5-foot-long, 15-foot-tall (at the pelvis) specimen premiered in 1905 at a reception for more than 500 scientists at AMNH's Hall of Fossil Reptiles. This was the first permanently mounted sauropod skeleton in the world.

1901 Hugh Gibb, under the direction of Charles E. Beecher, mounts *Claosaurus annectens* at Yale University. This 30-foot-long and 15-foot-high skeleton, which had been collected a decade earlier by Hatcher and Sullins in Converse, Wyoming, was the first dinosaur skeleton to be mounted at Yale.

1901 John Bell Hatcher (1861–1904), then director of the newly created CMNH in Pittsburgh, describes and names *Diplodocus carnegii,* which becomes known at CMNH as "Dippy." Hatcher argued that giant sauropods such as *Diplodocus* being discovered in the American West "lived and flourished" there in environments similar to those of Brazil's Amazonas. Arthur Conan Doyle later situated his *The Lost World*, complete with *Iguanodon* (from Europe) and *Stegosaurus* (from North America), in a Brazilian forest. Hatcher's specimen, a near-complete skeleton collected by Jacob Wortman, was named in honor of

Hatcher's patron, Andrew Carnegie. This 84-foot-long specimen, made of bones from four individuals, was the second sauropod to be mounted. Casts of this specimen were later sent to museums in London, Berlin, Paris, Vienna, Madrid, St. Petersburg (now Leningrad), Bologna, La Plata, Mexico City, Munich, and elsewhere, making it the single most viewed skeleton of any animal in the world. Hatcher's *Diplodocus Marsh: Its Osteology, Taxonomy, and Probable Habits, with a Restoration of the Skeleton* (1901) described his study and display of *Diplodocus carnegii.*

1901 Newspapers begin to increase their coverage of dinosaur-related stories. For example, the word *dinosaur* did not appear in the *Washington Post* between 1877 and 1895, but then appeared an average of six times per year between 1901 and 1920. During the 1920s, *dinosaur* appeared an average of 16 times per year, and in the 1930s and average of 24 times per year.

1901 Samuel Williston notes that Othniel Marsh never could adequately distinguish *Allosaurus* from *Creosaurus*. (Today, these genera are probably synonyms.) This criticism would become increasingly common of Marsh, Cope, and others, thereby creating confusion about the true identity of many of the earliest discoveries.

1901 The gazelle-size *Ornitholestes* ("bird robber"; AMNH 619) is found near Bone Cabin Quarry in Wyoming. Henry Osborn described *Ornitholestes* in 1903, and then in more detail in 1916. Fossils of this theropod, one of the earliest birdlike dinosaurs, are rare.

1901 While hunting in Montana, William Temple Hornaday (1854–1937)—the director of the New York Zoological Park (now the Bronx Zoo)—discovers large fossilized bones. When he returned to New York, Hornaday showed photos of the bones to AMNH's Barnum Brown, who identified them as *Triceratops*. The following year, Brown and Richard Swann Lull (1867–1957) went to Montana and found *Triceratops* at the site photographed by Hornaday. Hornaday's discovery was the first such report from the Hell Creek Formation. (The Hell Creek Formation, which is from the Upper Cretaceous, is named for intensely studied exposures along Hell Creek, near Jordan, Montana. The formation occurs in Montana, Wyoming, South Dakota, and North Dakota.) Soon thereafter, Brown discovered *T. rex*, which would become the most widely recognized dinosaur in popular culture, and the only dinosaur primarily known by its full scientific name.

1901 Yale mounts what will later be named *Edmontosaurus*, one of the first skeletons of dinosaur bones to be mounted in the Western Hemisphere. The dinosaur was mounted in an erect, bipedal pose.

1902 After being scolded by Henry Osborn for sending a poorly preserved specimen to AMNH, Barnum "Mr. Bones" Brown (1873–1963; Figure 17)—the

Figure 17 In this scene from the early 1900s, Barnum Brown (in the white shirt in the lower half of the image) and his excavation team prepare to move plastered bones of *Tyrannosaurus* from Hell Creek, Montana. Brown, who discovered *T. rex* in 1902, is universally regarded as the greatest dinosaur hunter of the 20th century. (Bettmann/Corbis)

greatest dinosaur-hunter of the 20th century—informs Osborn that he's found a skull of *Sterrholophus* (*Triceratops*) "in fairly good condition." (Today, Brown's *Triceratops* and duckbill *Anatotitan* are displayed in the center of AMNH's Hall of Ornithischian Dinosaurs.) Osborn also told Brown that he had "struck what Marsh used to call the 'Holy Ground'; and there is every reason to think that by careful inquiry among the natives, by making friends wherever you can, and by energetic prospecting you may find something of real value." A few days later, writing from Hell Creek, Montana, Brown responded with one of the most famous letters in the history of paleontology: "Quarry No. 1 contains the femur, pubes, [partial] humerus, three vertebrae and two undetermined bones of a large Carnivorous Dinosaur not described by Marsh . . . I have never seen anything like it from the Cretaceous." This was the first interaction of humans with *Tyrannosaurus rex* ("King of the tyrant lizards"). That specimen (AMNH 973; CM 9380), with its 4-foot-skull and a foot-wide neck vertebrae, was 50 percent complete and described by Brown as "the find of the season so far for scientific importance." (Six years later, in 1908, Brown found another, better-preserved skeleton.) Brown's famous discovery was made at Max Sieber's ranch 16 miles northwest of Jordan, Montana. (Montana had earned statehood only 13 years before Brown's visit.) There, Brown found the

T. rex "before the cook's call for dinner," after which he and his workers had to dig a quarry 30 feet wide, 21 feet across, and 20 feet deep to extract the bones. The blocks of rocks, some of which weighed more than 3,500 pounds, were transported on horse-drawn wagons across more than 130 miles of rough, pothole-riddled road to Miles City, where they were shipped to AMNH. (Days earlier, Brown had noted that "the great drawback to this region is the distance to freight fossils. It is over a hundred and thirty miles to Miles, the only available point.") The block containing the *T. rex* pelvis alone weighed two tons, and AMNH spent $1,345 (excluding shipping) to collect *T. rex*. It would be more than two years before Osborn wrote two papers announcing and naming *T. rex*. In those papers, Osborn acknowledged that *T. rex* had been discovered by Brown, but did not list Brown as a coauthor. (There is no evidence that Brown was ever dissatisfied with Osborn's decision to not list Brown as coauthor of the papers announcing *T. rex*.) The type specimen of *T. rex* discovered by Brown in 1902 (AMNH 973, now CM 9380) was sold to Pittsburgh's CMNH in 1941 in response to fears that New York might be bombed; as Brown noted, "The type-specimen which had most of the limb bones preserved was sold to the Carnegie Museum after we made casts of the limbs in 1941; as we were afraid the Germans might bomb the American Museum in New York as a war measure, and we had hoped that at least one specimen would be preserved." (Before the war erupted, BMNH had stored many of its dinosaurs in the chalk caves of Kent.) *T. rex*, a bipedal carnivore that had a wide range in western North America, was among the last non-avian dinosaurs to exist, and later became a fixture in popular culture. In the 90 years after its discovery, scientists uncovered only eight other skeletons of *T. rex*, none of which was more than 60 percent complete. Today, Brown's fourth (and largest) *T. rex* glares at visitors to AMNH in New York. Many of Brown's expeditions were funded by Harry Sinclair (1876–1956), who incorporated an image of *Diplodocus* into his company's logo to emphasize corporate gigantism. In return, Brown—the "Father of the Dinosaurs"—wrote the dinosaur booklets that were distributed at Sinclair gas stations in the 1930s and 1940s. Brown, who was also the scientific consultant for the Disney movie *Fantasia* (1940, in which the killing of *Stegosaurus* by *Tyrannosaurus* is accompanied by dramatic music), was one of the greatest scientific celebrities of his day. He hoped *T. rex* would "inspire young boys when they grow up to search for the same or other fossils." Other tyrannosaurids include *Albertosaurus* (described by Henry Osborn in 1905), *Daspletosaurus* (described by Dale Russell in 1970), *Nanotyrannus* (described by Robert Bakker, Mike Williams, and Phil Currie in 1988), *Tarbosaurus* (described by Evgeny Maleev in 1955), and *Gorgosaurus* (described by William Diller Matthew and Barnum Brown in 1923).

1902 Barnum Brown (Figure 17) explores the Missouri Breaks, the Hell Creek region first explored for the United States by Lewis and Clark. These 66-million-year-old buttes and ridges of pastel-colored badlands yielded numerous skeletons of dinosaurs.

1902 Canadian paleontologist Lawrence Lambe (1863–1919) describes the hadrosaur *Stephanosaurus* and the armored *Stereocephalus. Stereocephalus* was later renamed *Euoplocephalus* when it was discovered that the name *Stereocephalus* had been used to name an insect. *Euoplocephalus* was the first articulated ankylosaurid to be discovered.

1902 Fishermen tell Russian army officials about giant bones along the border of Manchuria in northern China. The site—Yuliangzi, at which excavations began in 1915—produced a well-preserved skeleton of the flat-headed hadrosaur *Mandschurosaurus*. This fossil—the first dinosaur found in China—was named in 1935 and displayed in the Central Geological and Prospecting Museum in Leningrad (now St. Petersburg).

1902 Forty-year-old Earl Douglass (1862–1931) joins the staff of CMNH, soon after which he begins helping Andrew Carnegie search for something "big as a barn" for the museum. Carnegie allocated $5,000 for the work.

1902 Franz Nopcsa describes a dinosaur later named *Zalmoxes*, a small herbivore. Nopcsa attributed *Zalmoxes*' small size to insular dwarfism.

1902 John Bell Hatcher and Oliver P. Hay (1846–1930) suggest that most hadrosaurs are synonyms of *Trachodon*. This revision helped popularize *Trachodon* in the early 20th century.

1902 *Stegoceras,* the first of the "dome headed" dinosaurs, is named by Canadian paleontologist Lawrence Lambe. A better skeleton of this 6-foot-long, Late Cretaceous dinosaur was found in 1921 by George Sternberg in DPP, and announced by Charles Gilmore in 1924.

1902 The book *Leading American Men of Science* includes biographies of Edward Drinker Cope and Othniel Marsh, but does not mention their famous feud.

1902 The Field Columbian Museum hires Samuel Williston for an annual salary of $1,500.

1902 YPM's Charles Beecher (1856–1904) begins assembling the *Apatosaurus* skeleton that Reed had discovered at Como Bluff in 1879. Reed described the fossil as a "larger *Brontosaurus*."

1902–5 Barnum Brown's *T. rex* is prepared in the AMNH by Peter Kaisen, Richard Swann Lull, and Paul Miller. Brown believed that *T. rex* was "active and swift of movement when the occasion arose."

1903 Bill Reed returns to paleontology as curator of the University of Wyoming Geological Museum. Reed taught geology and paleontology there until his death in 1915.

1903 Charles Gilmore begins work at NMNH, where he enjoyed a productive, 42-year career studying dinosaurs.

1903 Elmer Riggs argues in *Geological Series of the Field Columbian Museum* that *Brontosaurus* is not distinct enough from *Apatosaurus* to warrant its own genus, adding "that [Marsh's] Apatosaur specimen is merely a young animal of the form represented in the adult of the Brontosaur specimen. . . . In fact, upon the one occasion that Professor Marsh compared these two genera he mentioned the similarity between . . . their respective types. In view of these facts the two genera may be regarded as synonyms. As the term 'Apatosaurus' has priority, 'Brontosaurus' will be regarded as a synonym." Riggs's paper making *Brontosaurus* a junior synonym of *Apatosaurus* rendered the name of best-known Jurassic dinosaur obsolete. The following year in *Science*, Henry Osborn praised the "novelty" of Riggs's one-paragraph proclamation, but continued to use the name *Brontosaurus*, as did many other paleontologists. The next year, AMNH mounted a specimen it named *Brontosaurus,* not *Apatosaurus*. Although paleontologists slowly began using *Apatosaurus*, the public preferred *Brontosaurus*.

1903 Elmer Riggs, who was trained by Samuel Williston and who had worked as a field-hand for Othniel Marsh, describes *Brachiosaurus altithorax* from 150-million-year-old Late Jurassic state of western Colorado. *Brachiosaurus,* whose front legs were longer than its back legs, was later also discovered in Africa and Europe. *Brachiosaurus,* which was found three years earlier, became a popular dinosaur that appeared in numerous movies, books, and paintings. Riggs's monograph about *Brachiosaurus* (1904) included a strong argument for a terrestrial lifestyle for sauropods (e.g., "if the foot structure of these animals indicates anything, it indicates specialization for terrestrial locomotion.").

1903 John Bell Hatcher describes *Haplocanthosaurus* collected from a site mined by Othniel Marsh's collectors near Cañon City, Colorado.

1903 *Allosaurus,* collected in 1879 by H. F. Hubbell for Edward Drinker Cope, is unpacked from its original crates. Preparators soon realized the importance of the specimen and began mounting the bones. This AMNH skeleton, which was displayed in 1908, now stands over a carcass of *Apatosaurus,* which was found in 1897 at Bone Quarry, Wyoming. The idea for the famous display was inspired by the discovery of *Allosaurus* teeth beside *Apatosaurus* vertebrae having grooves that may have been created by teeth or claws.

1903 While visiting Andrew Carnegie at Carnegie's Skibo Castle in Scotland, King Edward VII (1841–1910, son of Queen Victoria) sees a watercolor sketch of "Dippy" the *Diplodocus carnegii* and asks that Carnegie send a *Diplodocus* to BMNH ("We must have one of these in the British Museum. Do not fail to secure us a specimen."). Working under the direction of William Holland and Arthur Coggeshall (who found *Diplodocus carnegii* in 1899), Italian modelers began making molds of the skeleton and, two years later, these molds of 84.5-foot-long "Dippy" were reconstructed into a full-size *Diplodocus* in BMNH's Main Hall (Figure 18). (The molds cost £2,000 and were shipped to London in 36 crates.) King Edward VII's interest in "Dippy" sparked similar requests from governments throughout the world. As one poem noted, "The Crowned heads of Europe, All make an awful fuss, Over Uncle Andy, And his old *Diplodocus*." Anticipating requests from other dignitaries, the plasterers produced five extra sets of molds.

Figure 18 The Natural History Museum in London is famous for its dinosaur skeletons and ornate architecture, both of which are captured in this photo of a cast of *Diplodocus carnegii* that dominates the museum's Main Hall. This cast—"Dippy," an 84.5-foot-long *Diplodocus* given to the museum by American steel magnate Andrew Carnegie (Figure 14)—was unveiled in 1905, two years before it was displayed at Pittsburgh's still-under-construction Carnegie Museum of Natural History. "Dippy" became the figurehead of Carnegie's philanthropic efforts in natural history when Carnegie gave casts to several other museums, thereby making "Dippy" the most-viewed dinosaur fossil in the world. This photo, which was taken in 2006, shows a statue of Richard Owen (the museum's first director; Figure 5) atop the staircase overlooking the Main Hall. In 2009, the 200th anniversary of Charles Darwin's birth (and the 150th anniversary of the publication his *On the Origin of Species*), Owen's statue was replaced by one of Darwin. (Courtesy Randy Moore)

1903 Henry Osborn uses a partial skeleton and skull to describe *Ornitholestes hermanni* (AMNH 619), a slender, 5-foot-long theropod having a long tail and long arms for grasping prey. By this time, Osborn had 15 full-time preparators at AMNH; this was more than double that just three years earlier.

1904 Famed paleontologist John Bell Hatcher dies at age 43 of typhoid fever in Pittsburgh, Pennsylvania. Henry Osborn remembered Hatcher as "certainly our greatest collector, [and] probably the greatest on record . . . I shall always miss Hatcher." Holland, who praised Hatcher as "undoubtedly the foremost paleontological field collector in the world," never found a successor to Hatcher, and finally appointed himself to the position. Hatcher's grave remained unmarked for 91 years until members of the Society of Vertebrate Paleontology (SVP) bought him a tombstone adorned with an image of *Torosaurus*. Hatcher is buried beside his daughter Ruth in Pittsburgh's historic Homewood Cemetery.

1904 Fieldwork at Medicine Bow Anticline ends after seven years of excavations. During this time, Walter Granger and others removed tons of bones that later produced exhibits of *Stegosaurus, Ornitholestes, Apatosaurus, Camptosaurus*, and other dinosaurs in museums throughout the world.

1904 Henry Osborn describes the articulation of the manus (the wrist/hand segment) of *Diplodocus* with a paper that began with "my previous figures and descriptions of the manus are all incorrect."

1904 Lawrence Lambe uses a fragmented skull to describe *Centrosaurus. Centrosaurus,* which was common in what is now Alberta, Canada, had short horns over its eyes and a long horn on its nose.

1904 Life-size restorations of dinosaurs are seen by millions of visitors at the World's Fair in St. Louis.

1904 Newlywed Barnum Brown meets famed orator (and subsequent antievolution crusader) William Jennings Bryan at Arizona's Petrified Forest. After traveling with Bryan to the Grand Canyon and then to New York, Brown commented that it was a "privilege to hear this man speak." Brown later helped AMNH's Henry Osborn—who believed that evolution proved God's divine plan—try to counter Bryan's antievolution views by searching for fossils that could help document human origins from other primates. Osborn—who lamented the loss of "purity of race"—was an evolutionist, but rejected the popular prehistoric "ape man," instead believing that a human-like ancestor had preceded modern humans in Asia by some 30 million years.

1904 Robert Broom's description of the sauropod *Algoasaurus* stimulates paleontologists to begin hunting for more dinosaurs in Africa. In 1933, Broom's

The Coming of Man: Was It Accident or Design? claimed that evolution is driven by spirits whose ultimate goal was humans.

1904 The first cast of *Diplodocus carnegii* is erected temporarily at Western Pennsylvania Exposition Society in Pittsburgh, Pennsylvania. The next year, this cast was shipped in 36 crates to London, where it was unveiled at BMNH. John Bell Hatcher, who helped with the work, died before the restoration could be unveiled.

1905 A cast of *Diplodocus carnegii* from CMNH is unveiled at BMNH. The 84-foot-long, 15-foot-high dinosaur was a gift from Andrew Carnegie to the trustees of BMNH, one of which was King Edward VII. The dedication ceremony was attended by more than 200 people, including the 69-year-old Carnegie and BMNH director Ray Lankester, who commented that the size of the dinosaur was equaled only by the size of Carnegie's generosity. The gift was accepted by Lord Avebury (John Lubbock [1834–1913], a biologist and friend of Charles Darwin) on behalf of the museum's trustees. After the dinosaur was installed, attendance at BMNH became consistently larger than on any day since the museum had opened. The museum sold "*Diplodocus* vases," which featured the dinosaur's image; this was one of the first examples of dinosaur-related merchandising. Two years later, *Diplodocus carnegii* was displayed at CMNH.

1905 A mounted *Brontosaurus*—the first mounted sauropod—draws immense crowds to AMNH. Associate Curator William D. Matthew, who used dissections of alligators and other reptiles to understand the functioning of the dinosaur's joints, noted, "No museum has ever before attempted to mount so large a fossil skeleton." Although the mount was a composite of bones from several individuals, almost two-thirds of the skeleton came from one specimen (AMNH 460) discovered in 1898 by Walter Granger (and excavated in 1899) from Nine Mile Quarry (nine miles north of Medicine Bow) in Wyoming. Because workers who mounted *Apatosaurus* had no guidelines to follow, they did much of the mounting by trial and error. However, in most instances, they were accurate. (Elmer Riggs and Arthur Coggeshall were critical of the exhibit, and subsequent mountings of sauropods improved on the AMNH designs.) Other skeletons of *Apatosaurus* were soon displayed at Chicago's Field Museum, YPM, the University of Wyoming, and elsewhere. These displays, and the accompanying publicity, helped make *Apatosaurus* (*Brontosaurus*) a common word. (At that time, *Apatosaurus* was the largest known land-animal that ever lived.) Today's visitors to AMNH are greeted by this *Apatosaurus* at the entrance to the Saurischian Dinosaur Hall.

1905 AMNH paleontologist William Diller Matthew summarizes the belief that dinosaurs were stupid, plodding creatures when he notes "we can best regard the *Brontosaurus* as a great, slow-moving storehouse of organized matter directed chiefly or solely by instinct, and to a very limited degree, if at all, by conscious intelligence."

1905 English illustrator and butterfly expert Frederick W. Frohawk (1861–1946) criticizes the skeletal restoration of *Diplodocus,* claiming that its legs should be splayed out like those of an alligator.

1905 English paleontologist Arthur Smith Woodward (1864–1944) attributes a partial skeleton to what is now *Cetiosaurus*. Five years later, William Plane Pycraft (1868–1942) described *Cetiosaurus*—a long-legged, short-necked sauropod—as "an ancient British river dragon," while noting "it is difficult to imagine animals of this colossal size roaming about over what is now Great Britain."

1905 Ernest Rutherford (1874–1937), who would become known as the father of nuclear physics, proposes radiometric dating. This technique allowed paleontologists to determine Earth's age and when dinosaurs had lived.

1905 *Extinct Animals,* a popular paleontology book by E. Ray Lankester (a former student of Thomas Huxley), brings Marsh's *Diplodocus* to a fascinated public. Lankester's book included the bulky *Stegosaurus* that Charles Knight had published in 1897 in *Century Magazine*, and inspired Sir Arthur Conan Doyle's classic book *The Lost World* (1912), the first extensive treatment of dinosaurs in popular fiction and the book that established the "lost world" theme in popular fiction. The movie version of Doyle's book, which was released in 1925, inspired many subsequent films about lost worlds, including *King Kong* (1933), *Unknown Island* (1948), and *The Land Unknown* (1957).

1905 Henry Osborn (Figure 15) announces *T. rex*, "Tyrant King of the Dinosaurs," to the public with a full-page article on December 4 in the *New York Times*' first magazine section. The announcement added to Osborn's fame and debuted Barnum Brown as a scientific and public celebrity. Osborn claimed that *T. rex*, the first giant theropod to be discovered, was "the chief exterminator of *Triceratops*" and had teeth "pointed like daggers." On December 30, thanks to funding provided by American financier John Pierpont Morgan (1837–1913), a part of the *T. rex* skeleton was displayed publicly in the AMNH Hall of Dinosaurs; this showing, for the first time, presented *T. rex* to the public. *T. rex* was hailed as "the absolute warlord of the earth," "king of all kings in the domain of animal life," "a royal man-eater of the jungle," and "the most formidable fighting animal of which there is any record whatever." In the October 4, 1905, issue of *Bulletin of the American Museum of Natural History*, Osborn justified his naming of *T. rex*: "I propose to make this animal the type of the new genus, *Tyrannosaurus,* in reference to its size, which far exceeds that of any carnivorous land animals hitherto described. . . . This animal is in fact the *ne plus ultra* of the evolution of the large carnivorous dinosaurs: in brief it is entitled to the royal and high sounding group name which I have applied to it." An adult *T. rex*, one of the largest known land predators, weighed several tons, was more than 40 feet long, and was more than 12 feet tall at its hips. In its growing years, *T. rex* could gain as much as 1,000 pounds per year. *T. rex* became the most famous dinosaur; it was featured in countless books, games, toy sets, Sinclair ads, television shows, and movies (e.g., *Dinosaurus!, The Last Dinosaur, Night at the Museum, Ice Age, The Land Before Time, Jurassic Park, Jurassic Fight Club*).

1905 Henry R. Knipe's *Nebula to Man,* the first poem to mention *Stegosaurus,* is supplemented with Joseph Smit's third restoration of *Stegosaurus*. In 1912, Knipe's *Evolution in the Past* included Alice Woodward's (1862–1951) double-rowed *Stegosaurus*. Woodward's illustrations of dinosaurs often overshadowed her work with children's books.

1905 In a one-page note at the end of his paper describing *T. rex*, Henry Osborn describes *Albertosaurus,* a large carnivore discovered in Alberta, Canada, by Joseph B. Tyrrell several decades earlier. Most fossils of *Albertosaurus* have been found along the Red Deer River near Drumheller, Alberta, Canada. *A. sarcophagus* lived with *T. rex*, and *A. libratus* lived 10 million years earlier. A complete skeleton of *A. libratus* (formerly known as *Gorgosaurus libratus*) and several other dinosaurs collected along Red Deer River are displayed at RTM.

1905 John Bell Hatcher proposes that teeth of *Troodon* ("wounding tooth") belong to a dinosaur, not a lizard.

1905 Popular American newspaper cartoonist Winsor McCay (1867–1934), who in 1912 will produce the first animated cartoon featuring a dinosaur, includes a racing *Brontosaurus* skeleton in the comic strip *Dream of the Rarebit Fiend*. In this movie, McCay used the pen-name "Silas."

1905 Prehistoric animals first appear in moving pictures in *Prehistoric Peeps*. This movie came from the cartoon series of the same that appeared in England's *Punch* magazine.

1905 Richard Swann Lull, who designed many of YPM's exhibits after Othniel Marsh's death, restores *Diceratops* with muscular cheeks in the *American Journal of Science*. This was the first restoration to accurately depict horned dinosaurs. Several of Lull's models are displayed at YPM.

1905 When *Brontosaurus* is estimated to have weighed 38 tons, most people believe it needed water for buoyancy and support. The movies *The Lost World* (1925) and *King Kong* (1933) featured amphibious brontosaurs, and through the 1960s brontosaurs were usually restored in lakes or marshes.

1906 American engineer Daniel Barringer (1860–1929) suggests that a giant, 3,900-foot-diameter crater (now known as Barringer Crater, Meteor Crater, and Canyon Diablo Meteor Crater) in Arizona was produced by an iron-meteor impact. Barringer's claim was not fully confirmed until 1960 by astronomer Eugene Shoemaker (1928–1997). Barringer Crater, which was made a National Natural Landmark in 1967, was a source for meteorites that Claire Patterson used to determine Earth's age in 1956. A crater more than 100-times larger—Chicxulub in Mexico—was left by an impact that coincided with the extinction of dinosaurs.

1906 Charles Knight paints the first of his two famous confrontations of *T. rex* and *Triceratops*. The first of these hangs in AMNH, and the second (Figure 19)—painted two decades later—in the Field Museum. Knight's *T. rex* and *Triceratops* in the Field Museum, which triggered numerous "versus" restorations from other artists, is Knight's most famous and beloved dinosaur-painting of all time.

1906 The *New York Times* labels *Tyrannosaurus* "the prize fighter of antiquity" and the "last of the Great Reptiles and the King of Them All." The article closed by telling readers that because *T. rex* left no descendants, "an evolutionist would classify him as a leafless, flowerless branch on the tree of animal life."

1906 Henry Osborn publishes an article revising some of his earlier conclusions about *T. rex*. Osborn was puzzled by the short arms of the specimen, suggesting that they served as "a grasping organ for copulation." (This suggestion has since

Figure 19 Charles Knight's oil painting in 1927 of *Triceratops* confronting *Tyrannosaurus rex* is arguably the most famous single image of dinosaurs. This painting, which hangs in Chicago's Field Museum, is the standard image that portrays the Age of Dinosaurs; it established *Triceratops* and *T. rex* as enemies in public thought. Knight's *T. rex* in this painting foreshadowed the modern ideas of *T. rex*'s posture, with its backbone more parallel to the ground (as opposed to the more erect "Godzilla" pose of the *T. rex* in the background). Knight's images of dinosaurs were the basis of countless reconstructions of dinosaurs in magazines, books, advertisements, movies, and other art. (Field Museum Library/Getty Images)

been abandoned.) Although Osborn noted that Barnum Brown had "cooperated with me in all the details of describing and measurement," Osborn again listed himself as the sole author of the papers. Osborn wrote five papers about *T. rex*; the combined specimens provided all of the theropod's parts except its hands and tail. Osborn speculated—albeit incorrectly—that *T. rex* had three fingers (like *Allosaurus*).

1906 President Theodore Roosevelt signs into law "An Act for the Preservation of American Antiquities" (16 USC 431–433), better known as the American Antiquities Act of 1906. The law protected "historic landmarks, historic and prehistoric structures, and other objects of historic or scientific interest" as national monuments. The legislation was passed primarily to protect Native American sites in the southwestern United States, and some of the earliest areas it was used to protect included DNM, Petrified National Forest, and the Grand Canyon. The legislation was later invoked in legal proceedings involving dinosaurs and other fossils.

1906 Roy Chapman Andrews (1884–1960) begins work at AMNH as a janitor and an assistant in the Department of Mammalogy. Like Barnum Brown, Andrews became a protégé of Henry Osborn, who believed that Asia was "the incubating region for the land life of Europe and North America." Osborn's belief that humans originated in Asia later prompted him to send Andrews on the Central Asiatic Expeditions. Those expeditions failed to find any human ancestors, but did find numerous dinosaurs and dinosaur eggs. Andrews was primarily interested in modern mammals, but became famous for his expeditions that uncovered new dinosaurs.

1907 American physicist Bertram Boltwood (1870–1927) uses the radiometric decay of uranium to lead in a mineral from Connecticut to conclude that the mineral formed 410 million years ago. (Boltwood's estimate was later changed to 265 million years.) This was the first successful application of radiometric dating.

1907 AMNH mounts its famous skeleton of *Allosaurus* feeding on *Apatosaurus*.

1907 Barnum Brown (Figure 17) describes gastroliths (i.e., stomach stones; gizzard stones): "In 1900, while collecting fossils in Weston County, Wyoming, . . . I found a *Claosaurus* skeleton embedded in a hard concretionary sandstone. In chipping off the surplus stone three rounded well-worn pebbles were found near the fore legs, embedded in the same matrix. These specimens were preserved and the occurrence made note of at once, for similar stones had not been seen anywhere in the deposit. These pebbles are rounded and vary in size, the largest measuring nearly three inches across." In 1941, Brown described gastroliths in an ornithopod from Mongolia. Gastroliths presumably helped dinosaurs grind food, in the same way that chickens digest their food with gravel that they swallow.

1907 Charles H. Sternberg sells his *Hesperornis regalis*—a toothed Mesozoic loon-like bird collected two years earlier from Cretaceous-age chalk in Logan County, Kansas—to AMNH for $500. There, Barnum Brown prepared the specimen (AMNH 5100) for mounting, and it was displayed in 1909.

1907 While looking for minerals on a hill near what is now Tendaguru, Tanzania, engineer W. B. Sattler of the Lindi Prospecting Company finds a fossil bone. The bone eventually was given to German engineer Ernst Martin Werner Janensch (1878–1969), who took it to the Berlin Museum for Nature for examination. (Tendaguru is in eastern Africa, near the southern boundary of Tanzania and about 40 miles northwest of the port of Lindi). In 1909, Janensch and Edwin Hennig (1882–1977) directed the largest expeditionary team ever to search for more dinosaur fossils at this site. Among the 225,000 kg of bones that were collected was the 40-foot-long *Giraffatitan* (originally designated *Brachiosaurus brancai*), which now stands in the Berlin Natural History Museum. Hennig believed that *Diplodocus* was a juvenile *Brachiosaurus*.

1907 Hungarian paleontologist Franz Nopcsa (1877–1933) compares the movement of a boat's oars with the flapping of wings, and suggests that a theropod could have used flapping to run faster. According to Nopcsa, this flapping increased the theropods' running speed and eventually enabled flight.

1907 In an article titled "*Diplodocus* Restored: The Largest Creature That Ever Roamed the Earth" in the *Scientific American*, Charles Knight depicts *Diplodocus* as a nimble, rearing, grazing dinosaur. The dinosaur's torso is raised about 60 degrees above the horizontal, its neck high in the air, and its front feet raised to knee height.

1907 In John Bell Hatcher's *The Ceratopsia,* completed by Othniel Marsh's successor Richard Swann Lull at Yale and published three years after Hatcher's death, Henry Osborn notes, "I trust that this volume may prove to be a lasting monument to the rare and noble spirit of John Bell Hatcher." Lull, who worked at YPM for 50 years, was one of Osborn's first doctoral students at Columbia.

1907 Two years after its debut at BMNH, *Diplodocus carnegii* is unveiled at the newly created Dinosaur Hall at CMNH. In a letter to Elmer Riggs, Arthur Coggeshall claimed, "Everybody here admits that it beats the [AMNH] mount [by] a mile." By 1950, 10 complete skeletons of dinosaurs (and many other marine and flying reptiles from the Mesozoic) inhabited Dinosaur Hall. At the ceremony dedicating *Diplodocus,* Holland movingly remembered a colleague, noting, "I should be false to the promptings of my heart did I not allude tenderly and in terms of highest appreciation to the labors of Professor J. B. Hatcher, my learned colleague, who for fully eighteen months devoted himself to superintending the restoration of the object before us, but who unfortunately was stricken down in the midst of his activities by the hand of death, a premature victim to the hardships and exposures of those expeditions. . . . Hatcher was one of the most successful students of paleontology whom America has produced." Andrew Carnegie, from Skibo Castle in Scotland, helped Hatcher's widow Ann financially and noted "how difficult it will be to replace him." After assembling casts of *Diplodocus carnegii* throughout the world, Coggeshall worked as an administrator at several museums.

1908 American paleontologist and eugenicist Henry Osborn becomes president of AMNH upon the death of 77-year-old banker and philanthropist Morris K. Jesup (1830–1908), a founder of AMNH. Throughout his life, Jesup gave much money to AMNH and bequeathed it $1 million. In 1899, Jesup had purchased Edward Drinker Cope's fossils for AMNH. Osborn, the first scientist to head AMNH, established AMNH as a major research institution, supplanted Yale University as the institutional leader in vertebrate paleontology, displayed dinosaurs in "real-life" poses, and hired colleagues such as Charles Knight, Roy Andrews, Barnum Brown, Walter Granger, and other famous fossil-hunters to help the museum acquire fossils. Osborn was one of the most famous and influential scientists of his time, and he believed that the different "races" of humans resulted from the adaptive radiation of human ancestors from central Asia. Although his reputation suffered from his prejudices about the superiority of "Nordic races" (i.e., white Anglo-Saxon northern Europeans) and the inability of women to be good teachers (because they lacked "virile quality"), Osborn—who wrote more than 900 scientific and popular articles and books—remains the most famous president of AMNH.

1908 American paleontologist Oliver Hay (1846–1930) argues in *American Naturalist* that *Diplodocus* had a lizard-like posture (with its legs splayed out sideways from the body) rather than in the erect posture that W. J. Holland had restored in the CMNH specimen: "The mammal-like pose attributed to the Sauropods is one that is not required by their anatomy and one that is improbable." The following year, German paleontologist Gustav Tornier (1858–1938) published similar arguments, and soon thereafter artists such as Germany's Heinrich Harder (1858–1935) painted sauropods such as *Diplodocus* with splayed, reptilian legs. Holland's response in the May 1910 issue of the *American Naturalist* undermined each of the claims while branding Tornier a "closet naturalist" guided by his "brilliantly illuminated imagination." The controversy was settled in 1938, when Roland Bird discovered sauropod tracks just outside Glen Rose, Texas, showing that *Diplodocus*-like dinosaurs stood and walked, and that they did not creep or crawl.

1908 Barnum Brown, working at the ranch of John Willis (a hunting buddy of Teddy Roosevelt), notes that he "searched the badlands for a month without finding a thing. I had given up. Then I decided to give it one more day. Turned out to be my day." (Willis's ranch was near Big Dry Creek and approximately 30 miles east of his Hell Creek camp, in an area now submerged by the Fort Peck

Reservoir.) Brown made a "ten strike" when he discovered a second and better-preserved *T. rex,* which had 8-inch serrated teeth and a 4-foot skull. Brown knew the excavation would be expensive, but told Osborn, "I have seen nothing like it before . . . Magnificent specimen. This skull alone is worth the summer's work, for it is perfect . . . I am sure you will be more pleased with our new Tyrannosaurus when you see what a magnificent specimen it is. . . . We finished boxing it and moving it out of the quarry Saturday. This block weighed about [3000 lbs] while the lower jaws weight about [1000 lbs]." Unlike the earlier *T. rex* that Brown had discovered, this specimen included most of the skeleton, but was missing its forelimbs and hind limbs. Osborn wrote Brown, "Your letter . . . makes me feel like a prophet and the son of a prophet, as I felt instinctively that you would surely find a Tyrannosaurus this season . . . I congratulate you with all my heart on this splendid discovery. . . . I am keeping very quiet about his discovery because I do not want to see a rush into the country where you are working." The excavation, which cost $1,200 (approximately $25,000 today), was challenging—the heavy blocks had to be transported on wagons more than 100 miles on poor roads. As Brown noted, "Even after we got it back to the museum our troubles were far from over." The resulting *T. rex* (AMNH 5027)—and its exquisitely preserved 3,000-pound, 1.5-meter-long skull—is now mounted in AMNH. Osborn planned to mount Brown's two *T. rex* skeletons together over the remains of a hadrosaur, but the Cretaceous Dinosaur Hall was too small, and museum staff could not find a way to support the skeletons in the recommended poses. For more than 50 years after its original discovery, Brown's two skeletons at AMNH were the world's only known specimens of *T. rex.* In the 1990s, Brown's AMNH 5027 was re-mounted in a horizontal pose. This specimen can still be seen on the fourth floor of AMNH.

1908 Casts of *Diplodocus carnegii* are unveiled at national museums in Paris and Berlin.

1908 Charles H. Sternberg and his son, Charles M. Sternberg, discover the hadrosaur *Anatosaurus*, which includes a mitten-like layer of skin over the hand, thus giving it the appearance of a paddle.

1908 Eberhard Fraas publishes the first description of dinosaur fossils from what is now Tendaguru. That same year, the imperial German government protected the Tendaguru area. When Wilhelm Branca (1844–1928; director of Berlin's Museum of Friedrich-Wilhelm University) read Fraas's work, he began raising money for an expedition to search for more fossils at Tendaguru. These expeditions, which began the following year, hired hundreds of workers each year for the excavations, and shipped more than 200 tons of fossils to Germany.

1908 Elmer Riggs mounts a headless, truncated *Apatosaurus* in Hall 35 of the Field Museum. The specimen, which was remounted in Hall 38 in 1921, was not adorned with a skull until 1958. By about this time, the Field Museum had curtailed its excavations of giant sauropods because their display was becoming impractical. Unlike many paleontologists, Riggs argued that the giant sauropods lived on land; his interpretation was supported by structural analyses of sauropods done in 1971 by Robert Bakker and in 1975 by Walter Coombs.

1908 French geologist Marcellin Boule's (1861–1942) study of the first complete skeleton of a Neanderthal (from Chapelle-aux-Saints, France) erroneously gives Neanderthals a stooped, primitive posture. The skeleton's spine was deformed by osteoporosis, but Boule could not conceive that Neanderthals might have walked upright. Although Boule's work had little to do with evolution, it helped popularize the notion of "caveman," which, along with other compelling icons such as monkeys, apes, and other "missing links," infused the public's vocabulary about evolution.

1908 Friedrich von Huene's (1875–1969) influential *Triassic Dinosaurs of Europe* describes a variety of dinosaurs, including *Plateosaurus*. Six years later, von Huene published the follow-up, *Triassic Dinosaurs of the World.*

1908 Jules Verne's (1828–1905) *A Journey to the Center of the Earth* is made into a silent movie by Spanish filmmaker Segundo de Chomon. The original print of that film has since been lost.

1908 Oliver Hay begins a series of papers arguing that museums distinguish between casts and fossilized bones when displaying dinosaurs.

1908 The main building of the Colorado Museum of Natural History opens in Denver. The museum was later renamed the Denver Museum of Natural History.

1908 When the CMNH cast of *Diplodocus carnegii* is unveiled at the Museum d'Histoire Naturelle, France's president Armand Fallières proclaims, "Quelle queue!" Although the president meant to say "What a tail!", the press noted that the word *queue* also meant *penis*.

1908 While searching for *Triceratops* for BMNH near Lusk, Wyoming, the Sternbergs uncover a mummified corpse of a crestless, duck-billed dinosaur (described and named *Edmontosaurus annectens* by Henry Osborn in 1912). George Sternberg described the scene this way: "When I removed a rather large piece of sandstone . . . from over the breast I found, much to my surprise, a perfect cast of the skin impression beautifully preserved. Imagine the feeling that crept over me when I realized that here for the first time a skeleton of a dinosaur had been discovered wrapped in its skin. That was a sleepless night for me." This mummy remains one of the most complete dinosaurs available; it includes impressions of flesh, skin, and tendons. When AMNH's Henry Osborn learned of the Sternbergs' discovery, he paid them the then-astronomical sum of $2,000 and—despite their contracted work for BMNH—obtained the specimen for AMNH (AMNH 5060).

Newspapers announced the acquisition with headlines such as "Dinosaur Mummy 3,000,000 Years Old . . . Museum Now Has the Only Known Specimen in the World." Osborn later displayed Sternberg's specimen as it was found, "on its back, the head turned to one side, the forelegs stretched out, the hind ones doubled up close to the body" in a case labeled "Dinosaur Mummy." Despite 40 years of fieldwork that included many important discoveries, Charles H. Sternberg was elated, adding, "Shall I ever experience such joy as when I stood in the quarry for the first time?" He described the discovery as "the crowning specimen of my life's work," and—after receiving the money from AMNH—declared AMNH to be "by far the noblest [museum] in America and soon to be the greatest on earth, under the splendid management of Prof. Henry Fairfield Osborn." AMNH continues to describe Sternberg's mummified dinosaur as "one of the greatest discoveries in the history of paleontology."

1909 Arthur Conan Doyle (1859–1930) asks Arthur Smith Woodward (1864–1944) of BMNH to take him to the sites where Rev. Edward Tagart had discovered tracks of *Iguanodon*. Doyle displayed some of these tracks in his home. In Doyle's *The Lost World*, bombastic Professor Challenger (who was played by well-known actor Wallace Beery [1885–1949] and modeled after Ray Lankester, a close friend of Doyle's) and his three fellow British explorers find humans living with dinosaurs, as well as fresh tracks of a giant dinosaur; as Doyle wrote in Chapter 10, "'Look at this!' said [Lord John Roxton]. 'By George, this must be the train of the father of all birds!' An enormous three-toed track was imprinted in the soft mud before us. . . . If it were indeed a bird—and what animal could leave such a mark?—its foot was so much larger than ostrich's that its height upon the same scale must be enormous. . . . 'Waldean!' cried Challenger, in an ecstasy. 'I've seen them in the Waldean clay . . . Not a bird, my dear Roxton—not a bird.' 'A beast?' 'No; a reptile—a dinosaur. Nothing else could have left such a track. They puzzled a worthy Sussex doctor some ninety years ago; but who in the world could have hoped—hoped—to have seen a sight like that?'" The cover of a later edition of Doyle's *The Lost World*—an influential book that is seldom read today—was adorned with dinosaur tracks.

1909 At a meeting of the British Association for the Advancement of Science (BAAS), Arthur Smith Woodward cites tooth loss and "excess growth" in dinosaurs as evidence of "racial senility" that caused the dinosaurs to become extinct.

1909 Casts of *Diplodocus carnegii* from CMNH are unveiled at national museums in Bologna, Italy (Italian Museum of Paleontology), and Vienna, Austria (Kaiserlich-Königliches Naturhistorisches Hofmuseum, presented to Emperor Francis Joseph). The Vienna Natural History Museum also features an *Allosaurus* from the Cleveland-Lloyd Quarry in Utah, as well as 35-foot-long leg bones of an *Ultrasaurus* from Utah.

1909 CMNH's Earl Douglass (1862–1931) and local farmer George Goodrich find "a beautiful sight" consisting of "eight of the tail bones of a *Brontosaurus* in exact position" at the bottom of a ravine in Morrison Formation beds near Vernal, Utah. After searching unsuccessfully along the slopes above for the stratum from which the bone had come, Douglass decided to return to the area the following year to excavate more bones. While working in the "Carnegie Quarry" in what is now DNM, Douglass unearthed a near-complete skeleton; in 1915, William Holland named this large, well-preserved specimen (CM 3018) *Apatosaurus louisae* (the species epithet honors Andrew Carnegie's wife, Louise); it was later mounted at CMNH, where it remains on display today (beside *Allosaurus* found by Earl Douglass in 1913). Douglas Stewart—assistant director of CMNH—described Douglass' discovery in a letter to W. J. Holland: "I have found what bids fair now to be an almost complete skeleton of a Brontosaurus. I never saw anything that, on the surface, and so far as we have gone, had such promises of a whole thing. . . . We have exposed parts of the bones to the third or fourth vertebra anterior to the sacrum, and the bones are exactly in place, apparently except that the femur appears to have gone down several inches out of its place in the socket. . . .Of course, currents may have disturbed the thing before we get to the head but there is no indication as yet that any disturbance has taken place." In October, Douglass wrote to Stewart. "We evidently have the most complete of the huge Dinosaurs that ever was found. . . . So far as I know there is no such specimen of a large Dinosaur in any of the museums." The skull associated with *Apatosaurus* resembled that of *Diplodocus*, thereby contradicting the long and widely accepted belief that *Apatosaurus* had a skull resembling that of *Camarasaurus*. Douglass spent the next 15 years (13 for CMNH and two years at NMNH and the University of Utah) working at DNM; the 446 crates of fossils (weighing more than 350 tons) excavated there included *Stegosaurus, Dryosaurus, Camptosaurus, Apatosaurus* and the juvenile *Camarasaurus* displayed at CMNH, *Diplodocus* displayed at NMNH and the Denver Museum of Nature and Science, and the cast of *Barosaurus* displayed at AMNH. Between 1909 and 1924, DNM produced more complete dinosaur skeletons, well-preserved dinosaur skulls, and more juvenile dinosaurs than any other site; these fossils comprised the largest collection of Jurassic dinosaurs from a single location ever found in the Morrison Formation. More than 70 percent of the dinosaur fossils at DNM are sauropods (theropods account for less than 5%). These sauropods included the famous type specimen of *Apatosaurus ajax,* which was named by Othniel Marsh in 1877 after Ajax, a hero from Greek mythology. This was the most complete skeleton of *Apatosaurus,* but it was conspicuously displayed headless because a skull found 12 feet away was considered too small for the massive dinosaur. (Marsh's belief that brontosaurs had large, box-shaped heads was popularized in movies such as *King Kong*.) Seventeen years later, the skeleton was topped with an inaccurate skull, while the real head languished in the museum's basement. The mistake was rectified 40 years later (and 70 years after the specimen's excavation), when Wesleyan University physicist-turned-sauropod-expert John McIntosh (b. 1923) and CMNH paleontologist David S. Berman reunited the skeleton with its correct skull. Othniel Marsh left no notes about the head–body problem of his *Apatosaurus*. In 2013, Matthew Mossbrucker of the Morrison Natural History Museum in Morrison, Colorado, identified the first snout of *Apatosaurus ajax* (found in 2003) from a small roadside quarry near Denver that had been part of the "Bone War" of the 19th century.

1909 Fifty-nine-year-old Charles H. Sternberg chronicles his adventures as a dinosaur collector in his autobiography, *The Life of a Fossil Hunter*. Henry Osborn wrote in the book's introduction that fossil hunting "demands all sorts of hardships. . . . Hunters must be somewhat of an engineer, have a delicate touch, [and a] strong body to handle the work. Must be content with plain living and find his chief reward in the sense of discovery . . . [Sternberg's] is a career full of adventure, of self-sacrifice, worthy of lasting record and recognition by all lovers of nature." Sternberg's book, initially published by Henry Holt and Company in a limited edition of 500 copies, was described by *Chicago Herald* as "a curious union of scientific devotion and religious reverence that is as unusual as it is charming," and Boston's *Living Age* noted, "His [Sternberg's] name, as affixed to his specimens, is the only witness to his labors which will remain after him, except the work of three sons whom he has trained to follow in his footsteps; but he has been happy, and his single-hearted story is a book to renew our faith in man's capacity to work for pure delight in work." Sternberg and his three fossil-hunting sons—George, Levi, and Charles M.—discovered and sold dinosaur fossils to museums throughout the world.

1909 German paleontologist Werner Janensch (1878–1969) and other German paleontologists begin the Tendaguru Expedition to find more Jurassic dinosaurs in Africa. This project, the most ambitious paleontological expedition in history (500 collectors and their families), searched for dinosaurs and other fossils at Tendaguru in German East Africa (what is now Tanzania). The quarries stretched for more than three miles, and the size of the expedition exceeded that of any of the dinosaur expeditions to the American West. For example, in 1909 the Tendaguru expedition hired 170 local workers, the following year 400 workers, and in 1911, 500 workers. Between 1909 and 1912, the workers made 5,400 four-day marches carrying 235 tons of fossils from Tendaguru to the port of Lindi, from which the fossils were shipped to Berlin.[23] Many of these fossils remain unprepared. Among the discoveries was *Elaphrosaurus*, *Kentrosaurus* (a spiked relative of *Stegosaurus*), and *Brachiosaurus* ("arm lizard"; M.B.R. 2181), a giant from the Late Jurassic that was one of the largest of all herbivorous, quadrupedal sauropods (its head was almost 40 feet from the ground). *Brachiosaurus* ate more than 400 pounds of plants per day and, unlike most sauropods, had forelegs longer than its hind legs; this 39-foot-tall, 74-foot-long fossil now dominates the Humboldt Museum in Berlin. The success of the expedition (despite the area's dense vegetation) resulted in Janensch leading two similar expeditions the next year and in 1911. The fourth (and final) Tendaguru expedition was led by Hans Reck (1886–1937). During his career, Janensch named 10 dinosaurs, six of which remain valid.

1909 John L. Wagner, a rancher from Drumheller, Canada, visiting AMNH, tells Barnum Brown of dinosaur skeletons protruding from canyon walls along Red Deer River. Brown began exploring there the following year, and two years later was joined by Charles H. Sternberg and his three sons (like Brown, Sternberg was also from Kansas). The two groups enjoyed a strained, but relatively amicable, competition for the next three years, with Brown noting that although "we have at present no serious disputes . . . they have no regard for the ethics of bone digging." Brown collected *Corythosaurus, Monoclonius, Edmontia, Saurolophus,* and the dome-headed *Stegoceras,* and the Sternbergs collected the rhino-like *Styracosaurus, Chasmosaurus,* and *Gorgosaurus* (now considered to be *Albertosaurus*). By the end of 2009, more than 1,500 specimens of dinosaurs had been collected from the area.

1909 Soon after "the Great Flood of 1908" ripped through the Paluxy River, truant nine-year-old schoolboy George Adams discovers several large, three-toed theropod tracks in the 113-million-year-old limestone outcrops of the Paluxy River near Glen Rose, Texas (a small town about 60 miles southwest of Ft. Worth). Adams showed the tracks to Robert E. McDonald, his school principal, who recognized them as dinosaurs, and took all of the students in his school to see the tracks. The Cleburne (Texas) *Weekly Enterprise*—which reported the discovery as "a remarkable find"—described the tracks as being 4 feet 2 inches apart and "directly in line with each other, showing they were made by a large biped." Soon thereafter, Glen Rose merchants began selling postcards featuring the tracks, and the local newspaper reported, "People are pouring in from other counties to view them." After seeing sketches of the prints, the Smithsonian Institution's C. W. Gilmore (Custodian of the Reptilian Collection of the Division of Vertebrate Paleontology) attributed the prints to a bipedal dinosaur. The most famous of the Glen Rose tracks were left by *Pleurocoelus*, a 20-ton sauropod; others were left by *Acrocanthosaurus,* an early and smaller relative of *Tyrannosaurus rex*. (Fossils of *Pleurocoelus*, *Acrocanthosaurus*, and other dinosaurs—including *Tenontosaurus,* a species found nearby by seven-year-old Thad Williams—can be seen at the Fort Worth Museum of Science and History.) The Glen Rose tracks became famous when they were described in 1939 by Roland Bird; one of the original tracks (now known to have been made by *Acrocanthosaurus*) is mounted in the bandstand in the Glen Rose town square. Today, Glen Rose—"The Dinosaur Capital of Texas" and one of the first "dinosaur towns"—continues to market the many dinosaur tracks in the nearby Paluxy River.

1909 The first remains of the small stegosaur *Kentrosaurus* are found in Late Jurassic deposits in Tanzania. Many of these remains (some as long as 8 feet) were later lost.

1909 In Ambrose Bierce's (1842–1914) *For the Ahkoond,* an explorer reporting from the year 4591 in the western United States describes a meteor shower that triggers the reappearance of sauropods.

1910 A cast of *Diplodocus carnegii* from CMNH is unveiled at the Imperial Academy of Sciences in St. Petersburg. This cast is now displayed in Moscow.

1910 After Charles H. Sternberg and three coworkers excavate and lift a 3,500-pound block of fossils onto a wagon, Sternberg accepts "that the ancient Egyptians, with millions of laborers and endless tons of sand, could with nothing more than such simple tools have erected the pyramids."

1910 Arthur Smith Woodward finds fossils that he attributes to *Megalosaurus*, but in 1926 those fossils were reclassified as *Proceratosaurus* by Friedrich von Huene. This small (10-foot-long), Mid-Jurassic theropod is one of the earliest known relatives of tyrannosaurs.

1910 Barnum Brown and his assistant Peter Kaisen—with Henry Osborn's support—explore the Red Deer River, gathering their discoveries aboard a 12-foot by 30-foot barge. Kaisen reported that the "whole top of a hill is nothing but a mass of bone. There are four hind feet in sight, vertebrae, and a lot of limb bones." Brown and Kaisen sent eight crates of bones to AMNH, where the fossils were prepared, put on display for a short time, and then stored; their discoveries included the horned *Monoclonius,* the crested duck-billed *Corythosaurus,* the hadrosaur *Saurolophus,* the armored *Ankylosaurus,* and *Albertosaurus,* the smaller cousin of *Tyrannosaurus*. This began the Canadian Bone Rush. Although Brown mentioned his discovery in papers published in the *Bulletin of the Geological Society of America* in 1914, he never formally described the fossils or bone beds. Canadians, upset that their dinosaur skeletons were being shipped to the United States, competed with Brown by hiring Charles H. Sternberg. The religious Sternberg believed that fossils are testaments of God's work, and that studying fossils was a way to understand God. Sternberg and two of his sons (Charles and Levi) collected along the Red Deer River; their discoveries included duck-billed, carnivorous, armored, and a specimen of the spiked-collar ceratopsian *Styracosaurus*. The following year, Sternberg's other son, George (who had been working for Brown), terminated his contract with AMNH and joined his father and other brothers with the GSC.

1910 Barnum Brown (Figure 17) discovers a small (approximately 6 feet long) ceratopsian in Alberta, Canada's Red Deer Valley that is later named *Leptoceratops*. The specimen was described four years later. Although Brown's specimen was missing its skull, a later specimen collected by Charles M. Sternberg was complete. That same year, Brown found the type remains of *Hypacrosaurus* ("nearly the highest lizard") in Alberta.

1910 AMNH's Henry Osborn is now a leader in making paleoart acceptable for describing prehistoric animals.

1910 Charlie Moss (1896–1984) and his brother Grady of Glen Rose, Texas, find sauropod tracks near the Paluxy River. Moss, like other local residents, believed the bathtub-sized tracks had been made by ancient elephants. Several residents of Glen Rose began carving dinosaur tracks and "man tracks" that were sold to tourists visiting Glen Rose for its mineral water and sanitariums.

1910 Frank Taylor suggests that mountains had been formed by a "mighty creeping movement" of Earth's crust, and speculated that continents had been moved by tidal forces.

1910 Mount Holyoke College's Mignon Talbot (1869–1950) becomes the first woman to find and describe a dinosaur. Talbot's fossils of the theropod *Podokesaurus holyokensis* were lost when the college museum burned.

1910 Richard Markgraf (1856–1915) and Ernst Stromer von Reichenbach (1870–1952) of Munich's Museum of Natural History begin searching in Egypt for human ancestors, but instead discover a variety of Late Cretaceous dinosaurs; their discoveries included the carnivores *Carcharodontosaurus, Bahariasaurus, Aegyptosaurus,* and the bizarre theropod *Spinosaurus*, which had 6-foot-long bony spines along its back. Disputes among European countries that led to World War I delayed arrival of the fossils at Munich, and damages to the bones resulted in Stromer not publishing descriptions of the bones until the 1930s. The outbreak of World War I severed contact between Markgraf and Stromer. Markgraf died in poverty in 1915, and many of the specimens he and Stromer collected in Egypt were destroyed during an Allied bombing raid on Munich in 1944.

1910 The green-domed NMNH—a part of the Smithsonian Institution—opens on the National Mall in Washington, DC. By the end of the 1920s, the museum displayed several dinosaurs and other fossils collected by Charles Gilmore at DNM. Today, NMNH displays more than 40 dinosaurs, seven of which are type specimens. The specimens include one of the world's only public displays of the horned carnivore *Ceratosaurus,* which was collected in the Marsh Quarry in 1883–1884 near Cañon City, Colorado.

1910 The Sternbergs discover a nearly-intact second mummy of a hadrosaur in the Lance Creek Formation just down Canada's Red Deer River. The excavated specimen, which weighed almost five tons, was sold to the Senckenberg Museum in Frankfurt, Germany, where it is displayed today.

1910 William Holland and other members of CMNH go to the Imperial Academy of Sciences in St. Petersburg, Russia, to assemble and present a replica of "Dippy" the *Diplodocus carnegii* to Czar Nicholas II, after which they go to Spain to present a mold to King Alfonso XIII, and then to Argentina's National Museum to give a replica to President Roque Saenz Peña. The last remaining cast made in 1903 was unveiled in 1929 at the Mexican National Museum as a gift to President Emilio Portes Gil. Another replica would be made in 1935 for the Bayerische Staatssammlung für Paläontologie und Geologie (BSP) in Munich, Germany. "Dippy," which was distributed throughout the world, has been seen by more people than any other dinosaur fossil.

1910 Yale University professor Richard Swann Lull sculpts *Stegosaurus* carrying its tail above the ground. Like artist Charles Knight, Lull endowed ornithopods with cow-like cheeks.

1911 Barnum Brown visits Steveville, Alberta (Canada), with AMNH president Henry Osborn and soon begins making plans to collect more fossils from the area. When Charles H. Sternberg and his boys began collecting fossils in this area, Brown moved downriver to what is now DPP; there, he discovered the most abundant and diverse site for dinosaur bones he'd ever seen. From this site, Brown shipped more than 300 cases of fossils to AMNH.

1911 Charles Knight produces his first set of murals for AMNH.

1911 Geologist Arthur Holmes (1890–1965) uses Bertram Boltwood's findings about radioactive decay to determine that rocks from the Devonian are 370 million years old (at that time, the oldest claimed age of a rock) and that there is a strong correlation between dates obtained by uranium-lead dating and the relative stratigraphic age of the rock. Holmes gradually extended his estimates of Earth's age from 1,600 million years in 1944 to 3.45 billion years in 1956. By 1960, Holmes's geological timescale resembled that used today. As Holmes noted in 1964, "Earth has grown older much more rapidly than I have—from about 6,000 years when I was 10, to 4 or 5 billion years by the time I reached sixty."

1911 In *In the Wonderful Land of Dreams,* Winsor McCay includes a pink sauropod resembling what in the following year will be Gertie, the first dinosaur brought to life in a cartoon.

1911 U.S. president William Taft (1857–1930) creates Colorado National Monument in western Colorado. This monument housed some of the most famous dinosaurs of the Late Jurassic, including *Stegosaurus, Apatosaurus,* and *Allosaurus*.

1911 William Diller Matthew succeeds Henry Osborn as curator of the Department of Vertebrate Paleontology at AMNH.

1912 At the annual meeting of Germany's Geological Association, geologist Alfred Wegener (1880–1930) announces his theory of "continental displacement." Wegener's theory became popularly known as "drift," a derogatory term used by Wegener's critics to describe Wegener's idea. This theory, which proposed that 200 million years ago Earth had one giant continent (Pangaea) that broke and drifted apart, helped explain the distribution of organisms among present-day land masses. Wegener admitted that although his "theory in certain individual cases may still be uncertain, the totality of these points of correspondence constitutes an almost incontrovertible proof of the correctness of our belief that the Atlantic is to be regarded as an expanded rift. . . . It is just as if we were to refit the torn pieces of a newspaper by matching their edges and then check whether the lines of print run smoothly across. If they do, there is nothing left to conclude that the pieces were in fact joined in this way." Wegener's idea gathered little attention until 10 years later, when it was translated from his native German into other languages.

1912 Charles Roberts's *In the Morning of Time* describes a Cro-Magnon family cooking, inventing weapons, and fighting predators. *Diplodocus* fights *Ceratosaurus*, and *Triceratops* fights *Dinoceras* (a uintathere that lived 30 million years after the K-T extinction). *In the Morning of Time* was serialized in *London Magazine* near the same time that Arthur Conan Doyle's *The Lost World* was serialized in *Strand Magazine*.

1912 *Chicago Tribune* columnist Bert L. Taylor publishes a poem about the alleged two brains of *Stegosaurus*: "*Behold the mighty dinosaur, Famous in prehistoric lore,/ Not only for his power and strength, But for his intellectual length./ You will observe by these remains, The creature had two sets of brains—/ One in his head (the usual place), The other in his spinal base.*"

1912 Famed fossil-hunter Charles H. Sternberg begins working for the GSC, during which time he discovers several new genera of duckbill dinosaurs. Paleontologists have used evidence from trackways to conclude that hadrosaurs were quadrupedal, had a stiff tail that was carried off the ground (as a counterbalance to the trunk), and carried their spines nearly parallel to the ground.

1912 Henry Osborn (Figure 15) describes the mummified duckbill, *Edmontosaurus,* purchased for $2,000 from Charles H. Sternberg four years earlier: "After a natural death (in other words, not death by predators) the body lay exposed to the sun for a long time, perhaps on a sand bar or in a stream. The muscles and soft internal tissues became completely dried and shrunken while the skin, hardened and leathery, shrank around the limbs and was drawn down along the bones. In the stomach and abdominal areas the skin was drawn within the body cavity, while along the sides of the body and on the arms, it was formed into creases and folds. At some later date, the 'mummy' may have been caught in a sudden flood and carried downstream and rapidly buried in fine sand and clay. A cast, or impression, of the skin formed in the sand before the skin and other soft parts decayed. There is no remnant of the actual skin preserved, only its imprint." Osborn originally classified the specimen as *Trachodon*, but today it is classified as *Anatotitan* or *Edmontosaurus*. *Edmontosaurus,* which lived with *Triceratops* and *T. rex,* was one of the last non-avian dinosaurs.

1912 In Arthur Conan Doyle's novel *The Lost World* (Figure 20), a young Irish journalist accompanying an expedition to the Amazon basin finds a jungle plateau where dinosaurs live. (Doyle was also the creator of Sherlock Holmes.) Parts of *The Lost World* were first published in a supplement to the *Philadelphia Press* (in England, Doyle's novel was first serialized in London's *Strand Magazine*) before appearing as a book that featured *Iguanodon, Allosaurus, Megalosaurus, Ceratosaurus,* and *Stegosaurus* (taken from Charles Knight's *Stegosaurus* published 15 years earlier in *Century Magazine*). In Doyle's book, a fictional editor wonders where to send his reporter and notes, "The big blank spaces in

Figure 20 In 1912, Arthur Conan Doyle's (1859–1930) novel *The Lost World* was the first extensive treatment of dinosaurs in popular fiction. Doyle's book established the "lost world," a theme that has been repeated countless times in other books, movies, and stories. Dinosaurs, such as this *Stegosaurus* (which Doyle labeled "the Monster"), appeared only sporadically in *The Lost World*. The 1925 movie adaptation of Doyle's book was the first full-length movie to feature dinosaurs. (Doyle, Arthur Conan. 1912. *The Lost World*)

the map are all being filled in, there's no room for romance anywhere." Doyle, who was fascinated by the *Iguanodon* tracks discovered in Sussex, had casts of dinosaur tracks in his home, and described how travelers in *The Lost World*[24] found fresh footprints of a giant dinosaur. True to the thinking in 1912, dinosaurs of *The Lost World*—which appear only sporadically in Doyle's novel—were stupid and slow; herds of *Iguanodon* were raised as food for the primitive humans (dinosaurs were also raised for food in *Planet of Dinosaurs* [1978]). The 1925 film adaptation of Doyle's book was the first full-length motion picture to feature dinosaurs; both the book and film featured *Allosaurus* as the top predator. In the film, Willis O'Brien used pioneering "stop-motion animation" (also called "stop-motion photography") special effects that animated models frame by frame, thereby making dinosaurs into movie stars. *The Lost World* was serialized for radio by the British Broadcasting Corporation in the 1940s, and adapted to film in 1960, 1992, and 1998. Arthur Conan Doyle's premise—namely, the persistence today of plants and animals from the remote past—was the basis for countless subsequent sequels and imitations; for example, in 1995, Michael Crichton used Conan Doyle's title for his novel *The Lost World*, a sequel to *Jurassic Park* (1993).

1912 The Royal Ontario Museum (ROM) is created as a collection of smaller museums, and it opens two years later in Toronto. ROM began collecting dinosaurs in what would later be named Dinosaur Provincial Park in 1918, and its last major expedition there was in 1954.

1912 The seventh cast of *Diplodocus carnegii* from CMNH is unveiled by CMNH's William Holland and other dignitaries at the National Museum in La Plata, Argentina. The ceremony included a dinner at which dignitaries were served the hors d'oeuvre "Canapé Multimillionaire," the entrée "Grand pièce Diplodocus à la Holland," and similar treats named for Argentinian paleontologists attending the dinner. Ongoing requests from around the world for casts of *Diplodocus* prompted the *Pittsburgh Gazette Times* to describe CMNH as a "*Diplodocus* factory" and "one of Pittsburgh's busiest industries." At most of the ceremonies associated with the unveiling of *Diplodocus,* Holland was given medals, awards, and honorary degrees. The next year, Holland—an expert on butterflies and moths—described his adventures in *To the River Plate and Back*.

1912 The Sternbergs begin fieldwork near Drumheller, Canada. In the Horseshoe Canyon Formation near Alberta, Canada's Red Deer River, Charles M. and Levi Sternberg found a "wonderfully complete skeleton" that Lawrence Lambe (of the National Museum of Canada) later named *Trachodon marginatus*. (Lambe named several other dinosaurs discovered by the Sternbergs.) This specimen became the type specimen of *Edmontosaurus*. *Edmontosaurus,* which was mounted in the Victoria Memorial Museum, was named for the Edmonton Formation and was the first dinosaur to be mounted in a Canadian museum.

1912 While working in the Red Deer River area of Alberta, Canada, Barnum Brown excavates a spectacular skeleton "with skin impression[s]." As Brown noted, "The impression of the epidermis covering most of the body . . . although faint in places, where covered by masses of vegetable material, the general pattern is fairly well determined." Today, this specimen of *Corythosaurus* (AMNH 5240), along with a near-complete second specimen (AMNH 5338) excavated two years later by Brown and Peter Kaisen, stand in the AMNH Hall of Ornithischian Dinosaurs in their original death poses. The ossified tendons in the second specimen created a stiff backbone that would not have been compatible with the vertical reconstruction proposed earlier by Henry Osborn. After Brown's discovery, no more dinosaur mummies were reported for several decades.

1912 American cartoonist and animator Winsor McCay (1867/71–1934) produces *Gertie the Dinosaur*, the first animated cartoon to feature a dinosaur. The frisky, childlike Gertie was a *Brontosaurus* with humanlike feelings who did tricks like a trained elephant. The seven-minute cartoon began with a live-action sequence at AMNH. Since there were few movie theaters, McCay was part of the stage play; he stood beside the stage talking and holding up dialogue signs. After following all of Gertie's commands, McCay entered the cartoon through a slit in the screen, jumped on Gertie's back, and rode off into the prehistoric sunset. *Gertie the Dinosaur* was popular with audiences, and two years later, McCay produced another version of *Gertie the Dinosaur* that could be shown as film rather than as part of a stage show. *Brontosaurus* (*Apatosaurus*) ruled the early days of cinema, appearing prominently in *Gertie the Dinosaur, The Lost World* (1925), and *King Kong* (1933).

1912 Yale University's Richard Swann Lull names the sauropod *Anchisaurus. Anchisaurus* was a replacement name for *Amphisaurus,* which itself was a replacement name for Edward Hitchcock's *Megadactylus.*

1913 After staying upriver from Barnum Brown for a year, Charles H. Sternberg and his three sons move into what is now DPP. On the day they arrived, Charles M. Sternberg (1885–1981) discovered *Gorgosaurus*. Three years later, Levi and Charles H. Sternberg quit working for the GSC and moved to Alberta to work as freelance collectors. One of their prized fossils was lost when the boat transporting it to Europe was sunk by a German surface raider.

1913 After three years of preparation, Arthur Coggeshall and his coworkers erect *Apatosaurus* at CMNH. The dinosaur was 71.5 feet long and 15 feet tall at the arch of its back.

1913 Brazilian geologist Joviano Pacheco finds the first dinosaur tracks in Brazil in a quarried slab on a sidewalk in São Carlos.

1913 In *The Age of Earth*, Arthur Holmes outlines how radiometric dating could be used to construct an absolute geological scale for Earth's age. By 1927, Holmes knew that Earth's crust was at least 3–6 million years old. Soon thereafter, Holmes proposed a geologic time scale, noting, "It is perhaps a little indelicate to ask of our Mother Earth her age, but Science acknowledges no shame and from time to time has boldly attempted to wrest from her a secret which is proverbially well guarded."

1913 Charles Falkenbach and George Sternberg unveil *Saurolophus* at AMNH (AMNH 5220). The specimen, collected two years earlier by Barnum Brown along Canada's Red Deer River, was displayed exactly as it was found; that is, it was shown in relief by removing the rock while leaving the bones attached to the matrix.

1913 Charles W. Gilmore of NMNH describes *Thescelosaurus.* Since the fossils had been stored in crates for more than two decades, Gilmore named the small, neglected ornithopod *Thescelosaurus neglectus.* In 2000, *Thescelosaurus* attracted much attention when a specimen discovered in South Dakota in 1993 was interpreted as including a fossilized heart.

1913 In South Africa, paleontologist Robert Broom discovers *Euparkeria,* a small, bipedal, 230-million-year-old pseudosuchian ("false crocodile," a group of Triassic reptiles). Broom argued that the "reptilian" nature of this fossil was anatomically suited to be the ancestor of birds, thus creating the "pseudosuchian thecodont" (primitive archosaur) hypothesis of bird ancestry. This hypothesis was influential until the 1960s, when John Ostrom described *Deinonychus,* a small theropod anatomically similar to *Archaeopteryx. Euparkeria* was instrumental in Gerhard Heilmann's rejection of a direct link between birds and dinosaurs in his influential *The Origin of Birds* (1926).

1913 John Treadwell Nichols (1883–1958) establishes *Copeia,* a journal devoted to ichthyology and herpetology. The journal's name honors famed fossil-hunter Edward Drinker Cope (Figure 10).

1913 Lawrence Lambe describes *Styracosaurus* from DPP in Alberta, Canada. *Styracosaurus* was later often included in cartoons with people atop its back, using its horns like a steering wheel.

1913 Several members of the Calgary (Canada) Natural History Society form the Calgary Syndicate for Prehistoric Research to raise money for a paleontological museum.

1913 The International Zoological Congress, meeting in Monaco, adopts its "plenary powers decision" (Article 79): "When stability of nomenclature is threatened in an individual case, the strict application of the Code may under specified conditions be suspended" by ICZN. In other words, when a species has been given two different names, the authoritative name will be the name given first unless the later name has become popular and changing the name would threaten the "stability of nomenclature." This rule was passed a decade after the popular name of *Brontosaurus* became invalid because the name *Apatosaurus* was used earlier to describe the iconic sauropod. When Elmer Riggs in 1903 showed them to be the same, *Apatosaurus*—which was named first—replaced *Brontosaurus* as the official name of the iconic sauropod.

1913 The movie *Brute Force*, directed by famed motion-picture pioneer D. W. Griffith (1875–1948), shows primitive cavemen living with—and being threatened by—a life-size, animated *Ceratosaurus,* which was introduced by a title card describing it as "one of the perils of prehistoric apartment life." This first depiction of a dinosaur threatening humans, which helped establish the claim that dinosaurs lived with humans, contrasted sharply with good-natured Gertie from the previous year. Elsewhere in the movie, an alligator was adorned with fins to give it a menacing, prehistoric look; a small *Triceratops* was portrayed by a pig in a dinosaur outfit; and other dinosaurs were portrayed by decorated lizards or actors in dinosaur suits. *Brute Force*, which opened with the hero (Robert Harron) reading a book by Darwin about primitive humans, depicted a moving dinosaur and two tribes fighting to procure women. *Brute Force* was a sequel to *Man's Genesis* (1912, which included cavemen but no dinosaurs) and was the first movie to portray dinosaurs as villains. Clips of *Brute Force* (also promoted as *In Prehistoric Days, Wars of the Primal Tribes,* and *The Primitive Man*) were featured during the American Academy Awards show in 2000. Griffith became best known for his groundbreaking and controversial blockbuster *The Birth of a Nation* (1915), which was sympathetic to the Ku Klux Klan while negatively portraying African Americans. *Brute Force* was a popular film and among the best of the lizards-as-dinosaurs films; several of its scenes were used in later films such as *Teenage Caveman* (1958, in which a young Robert Vaughn starred with a modern-day haircut).

1913 While still exploring the Red Deer River area of Alberta, Barnum Brown finds a double-crested hadrosaur that in 1923 is named *Lambeosaurus* by Canadian paleontologist William A. Parks (1868–1936) in honor of famed paleontologist Lawrence Lambe.

1913–16 Gerhard Heilmann (1859–1946) begins publishing a series of articles titled "Our Present Knowledge about the Origin of Birds" in the *Journal of the Danish Ornithological Association*. In 1926, these edited papers became the basis for Heilman's influential *The Origin of Birds*.

1914 A cast of *Diplodocus carnegii* from CMNH is unveiled at Madrid's National Museum.

1914 As it begins its occupation of Belgium, Germany sends paleontologist Otto Jaeckel (1863–1929) to reopen the mine at Bernissart where the herd of *Iguanodon* had been discovered in the 1870s. However, the Allies again retook Belgium and no more skeletons were recovered. In 1921, the mine was abandoned and flooded.

1914 At a meeting of the Paleontological Society of America, William Holland suggests that the skull atop *Brontosaurus* is mismatched. As evidence, Holland noted (1) that there was nothing about the discovery of the skull (from the Carnegie Quarry, which is now the centerpiece of DNM) that Marsh placed atop *Brontosaurus* to suggest that it belonged there, (2) that a *Diplodocus*-like skull (CM 11162) was found only 12 feet from the skeleton of *Brontosaurus*, and (3) Richard Lull's discovery that the skull that Marsh placed atop *Brontosaurus* was found four *miles* from rest of the skeleton. Holland concluded, "Professor Marsh associated the skulls, which he had studied, with the remains of *Brontosaurus* as the result of a process of ratiocination rather than as the result of ocular evidence that the skull actually belonged with the skeleton." Nevertheless, Holland left his restoration of *Brontosaurus* (i.e., CMNH's *Apatosaurus louisae,* which was mounted in 1915) headless for almost two decades.

1914 Barnum Brown describes *Leptoceratops* and *Anchiceratops*. That same year, Brown discovered a nearly complete skeleton of an ostrich-like dinosaur. In 1917, Henry Osborn named this dinosaur *Struthiomimus* ("ostrich mimic").

1914 Barnum Brown (Figure 17) discovers the first known remains of *Dromaeosaurus* ("fast running lizard") during an AMNH expedition in what is now DPP in Alberta, Canada. *Dromaeosaurus* was a wolf-sized theropod from the Late Cretaceous which, like *Deinonychus*, had a sharp "sickle claw" on each foot. Although there are relatively few good fossils of *Dromaeosaurus,* skeletal casts (produced by RTM) appear in many museums. *Dromaeosaurus* also appeared in *Walking with Dinosaurs*, where it stalked a young *Torvosaurus* and tried to steal eggs from *Tyrannosaurus* (despite the fact that *Dromaeosaurus* did not live at the same time as either of these two genera). *Dromaeosaurus* was named in 1922 by Brown and William Diller Matthew.

1914 Barnum Brown finds the most complete dinosaur-skeleton he ever found. Brown described the articulated specimen, *Centrosaurus apertus* (AMNH 5351), as "surrounded by sandstone and ironstone matrix of such nature that much of the scientific nature of the skeleton would have been sacrificed by extracting it for a free mount. Consequently the skeleton [was] been worked out in relief and mounted as a plaque."

1914 C. F. Miller is appointed the first curator of what would later be the Sternberg Museum of Natural History at Fort Hays State University in Hays, Kansas. George

Sternberg helped develop the museum, which features several dinosaur fossils. The museum also displays the famous "fish within a fish" fossil (i.e., a 13-foot-long *Xiphactinus* which has inside it a 6-foot-long *Gillicus*), which Sternberg collected in 1852. The Sternberg family—father Charles and sons Levi, George, and Charlie—changed the face of American paleontology.

1914 C. W. Gilmore describes *Brachyceratops,* a small, curious horned dinosaur unearthed in Montana.

1914 Friedrich von Huene coins the name *Carnosauria* to include several large theropods having giant skulls and giant teeth.

1914 Lawrence Lambe shows that the hands of tyrannosaurids had only two digits. Lambe named several of Canada's most famous dinosaurs, including *Edmontosaurus* and *Styracosaurus*.

1914 Lawrence Lambe describes the short-horned *Chasmosaurus*. The frill of *Chasmosaurus* was decorated with large holes, suggesting that it was used as a display. That same year, Lambe also named *Gorgosaurus libratus,* which had been discovered the previous year by Charles H. Sternberg. Lambe argued that *Gorgosaurus* had short, two-fingered forelimbs. Dale Russell (b. 1937) later declared *Gorgosaurus* to be a junior synonym of *Albertosaurus,* but in 2003 Philip Currie (b. 1949) argued that *Gorgosaurus* and *Albertosaurus* be retained as separate genera until more is known about the dinosaurs.

1914 The second *Gertie the Dinosaur* cartoon by American cartoonist Winsor McCay depicts a cute, amicable *Apatosaurus* pet of McCay's that "laughs and cries, dances the tango, answers questions and obeys every command" (Figure 21). The film opened in AMNH, which was McCay's inspiration for creating Gertie. At one point in the cartoon, Gertie drank an entire lake, leaving behind a canyon-sized hole. *Gertie,* one of the first animated films, was made from more than 10,000 individual sketches and included an elongated, shallow skull (similar to that of *Diplodocus*) rather than a blunt, box-like skull (similar to that of *Camarasaurus*), as was presented in most other depictions of *Apatosaurus*. (The accuracy of this depiction was not shown until 1978.) Gertie, a Jurassic dinosaur, coexisted with wooly mammoths, despite the fact that they were separated by more than 100 million years; this portrayal of a generic "prehistoric world" was presented in many other movies, including *One Million B.C.* and *Fantasia*. Soon after the release of *Gertie the Dinosaur,* New York animator John Bray (1879–1978) created another version of *Gertie the Dinosaur,* as did several other moviemakers and writers. The following year (i.e., in 1915), Bray created *A Stone Age Adventure,* another animated film about dinosaurs. Gertie was the first animated cartoon character; ever since *Gertie the Dinosaur* (Figure 21), dinosaurs have been at the cutting-edge of movie animation.

Figure 21 *Gertie the Dinosaur* (1914) was the first animated cartoon to feature a dinosaur. The childlike Gertie did tricks and was often paired with locomotives, trucks, cars, and other vehicles to show the link between dinosaurs, machines, and the vast amounts of fossil fuels whose energy could be released by technology. (Box Office Attractions Company/Photofest)

1914 Werner Janensch describes *Brachiosaurus brancai*—one of the largest animals to ever walk on Earth—from Jurassic rocks of Africa. This specimen, which was later renamed *Giraffatitan brancai,* was the first near-complete brachiosaur to be discovered.

1914 While working near Steveville, Alberta, Canada, Barnum Brown finds a near-complete skeleton of *Struthiomimus altus* (AMNH 5339). The following year, this specimen—after being displayed by Peter Kaisen in a plaque mount as it was found in the rock—went on display in AMNH.

1914 William Holland, director of CMNH, writes to Earl Douglass that Othniel Marsh's *Apatosaurus* and its skull came from two widely separated sites, and that "other genera and species had skulls very much like *Diplodocus*, and I should not be at all surprised to find in the end that *Brontosaurus* and *Apatosaurus* and all the rest of them really belonged to the [Family] Diplodocidae, when the structure of their skull comes to be known, and that the separation into families of certain of these animals is in error. . . . The association of this skull material . . . is wholly arbitrary, and I doubt its correctness." Holland told the Paleontological Society of America that a *Diplodocus*-like skull was the correct skull for *Brontosaurus*.

(Earlier, Holland had expressed "doubt founded on observation, as to the correctness of Professor Marsh's surmise, which up to the present time has been unquestionably accepted. . . . The two skulls used by Marsh were found, one four miles from the rest of the skeleton, the other four hundred miles from it.") Indeed, a *Diplodocus*-like skull had been found 12 feet from the atlas of the *Brontosaurus*. Holland also noted that if the skull had belonged to *Brontosaurus*, it would have had to move upstream and move up through 8 feet of rock. Holland was right; the skull that Marsh had mounted on *Apatosaurus* was from *Camarasaurus,* not *Apatosaurus*. Nevertheless, when Holland had CMNH's *Brontosaurus* mounted, he left the skeleton headless. In 1934—two years after Holland's death—the CMNH *Brontosaurus* was adorned with a *Camarasaurus*-like head that Marsh had claimed was correct. Marsh's mistake long outlived Marsh; the incorrect skull remained on the CMNH *Brontosaurus* for the next 45 years.

1915 Andrew Carnegie (Figure 14), a business mogul who applied Darwinian ideas to his work, curtails funding for CMNH's paleontological expeditions to focus increasingly on world affairs.

1915 At the urging of paleontologists Earl Douglass and Charles Doolittle Walcott (1850–1927), the secretary of the Smithsonian and discoverer of the now-famous Burgess Shale, President Woodrow Wilson (1856–1924) uses a presidential decree to create and preserve DNM just east of Vernal, Utah, as an "American antiquity" (Figure 22). This fulfilled a wish that Douglass had written earlier in his diary: "I hope that the Government, for the benefit of science and the people, will uncover a large area, leave the bones and skeletons in relief and house them in. It would make one of the most astounding and instructive sights imaginable." DNM, includes a famous quarry whose centerpiece is a near-vertical riverbed more than 300 feet long and 50 feet wide that displays more than 1,000 fossils from 10 different species of dinosaurs. The U.S. Department of the Interior was given responsibility for the 80-acre site. The monument, which was later enlarged, includes some of the most spectacular and geologically interesting scenery in the United States. CMNH continued excavations at DNM until 1923, and a museum opened at the monument in 1958. In 1994, paleontologist Tony Fiorillo determined that material in DNM was probably deposited in a few years or less, indicating that the animals found there had probably lived at the same time. When Douglass mined the quarry, it was more than 150 feet longer and 80 feet higher than the "bone wall" at DNM today. Nevertheless, DNM remains the world's most impressive public collection of Jurassic dinosaurs. Outside DNM, a life-size *Stegosaurus* (from the Sinclair exhibit at the 1964–1965 World's Fair in New York) spans the equivalent of three parking spaces.

1915 Barnum Brown leaves what is now DPP. Although Brown returned several years later to prospect for gas and oil reserves, he never collected another dinosaur fossil from the area.

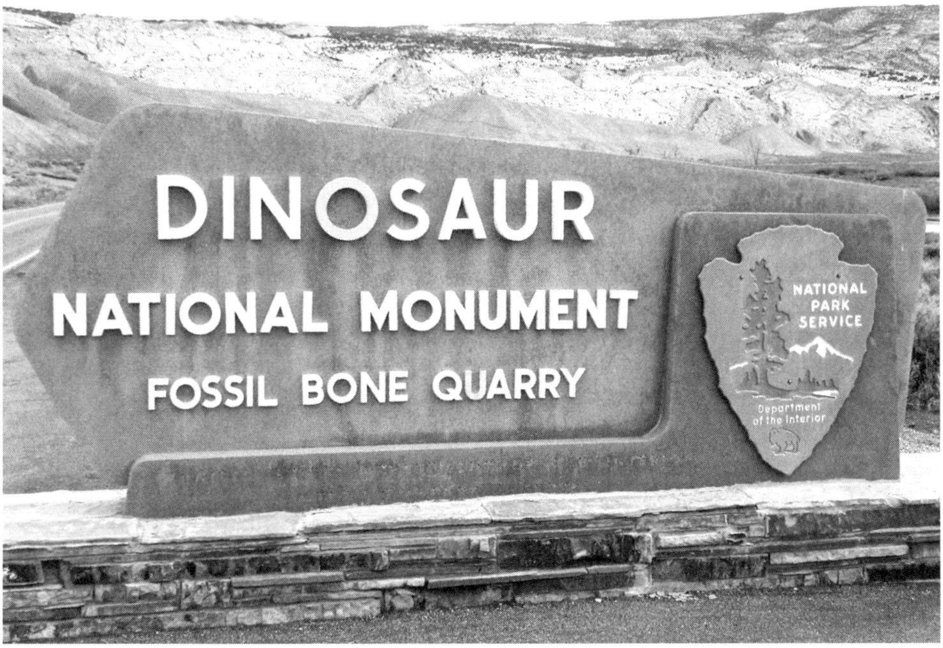

Figure 22 Dinosaur National Monument (DNM) along the Colorado–Utah border was created in 1915 by President Woodrow Wilson. The monument includes a 150-foot-long near-vertical in situ display of more than 1,000 dinosaur bones that have been carefully exposed from the rock face. Between 1909 and 1924, DNM produced more articulated skeletons and intact skulls of dinosaurs than any other site. Indeed, several crates shipped from DNM to the Carnegie Museum of Natural History remain unopened. DNM is the world's most impressive public display of Jurassic dinosaur fossils. In 1951, the Sierra Club began a successful five-year fight to block the federal government from building two dams at DNM. (Courtesy Randy Moore)

1915 Baron Nopcsa finds 10 small sauropods in western Romania and names them *Titanosaurus dacus*. In 1932, Friedrich von Huene renamed these fossils *Magyarosaurus dacus.*

1915 Charles Gilmore argues that *Stegosaurus* had alternating plates. Charles Knight's restoration of *Stegosaurus* in 1927, which incorporated Gilmore's ideas, is among Knight's 28 murals at the Field Museum. Rudolph Zallinger also included a *Stegosaurus* having a staggered, double-row of plates in his famed *The Age of Reptiles* (1947).

1915 Charles H. Sternberg excavates the hadrosaur *Hypacrosaurus* found by P. A. Bungert for the Canadian National Museum. In 1925, AMNH obtained the fossil. *Hypacrosaurus* remained an obscure genus until nests and eggs of *Hypacrosaurus* were found in the 1990s.

1915 Crowds flock to AMNH to see the first public exhibition of a complete and mounted *T. rex*. This specimen (AMNH 5027)—the second *T. rex* found by Barnum Brown in Montana—was mounted in an upright, Godzilla-like pose, and Osborn gave the fossil three fingers (i.e., like *Allosaurus*). Although paleontologists in the 1970s realized that *T. rex* oriented its body in a horizontal position, the AMNH exhibit was not changed until 1992. Osborn's display of *T. rex* at AMNH produced headlines across the United States, such as "The Prize Fighter of Antiquity" as people began imagining prehistoric life on Earth. This *T. rex* inspired and astounded visitors for the next 80 years.

1915 *Diplodocus longus* is mounted and exhibited at CMNH.

1915 German paleontologist Ernst Freiherr Stromer von Reichenbach (1870–1952) finds the long, shallow-snouted *Spinosaurus aegyptiacus* in Late Cretaceous rocks in Egypt. *Spinosaurus* was one of the largest known theropods. These and other fossils, which were later moved to Munich, were destroyed by Allied bombs in 1944 in another "extinction event" for dinosaurs.

1915 Irish American special-effects artist Willis O'Brien pioneers stop-action animation to re-create dinosaurs in the short film *The Dinosaur and the Missing Link: A Prehistoric Tragedy* (produced with a budget of $5,000) and, three years later, the 19-minute *The Ghost of Slumber Mountain* ("The Crowning Triumph of the Motion Picture!"). *The Ghost of Slumber Mountain* included a fight between two *Triceratops*, after which *Allosaurus* killed the survivor. Both films used clay models having hinged bodies that were incrementally adjusted for frame-by-frame filming. On a typical 10-hour workday, O'Brien completed about 500 separate frames, which accounted for about 30 seconds of on-screen movie. The first dinosaur brought to life by O'Brien was *Brontosaurus*. Several of O'Brien's early films were short films (e.g., *The Dinosaur and the Missing Link*) bought by (or prepared for) Thomas Edison's Conquest Pictures. Although O'Brien was not credited in advertisements for *The Ghost of Slumber Mountain,* his work on the film helped him secure a job with the movie version of *The Lost World,* which was O'Brien's first Hollywood feature. O'Brien's stop-action animations usually were based on 24-frames-per-second filming, the same principle used in most cartoons. O'Brien was the first person to understand the theatrical appeal of dinosaurs.

1915 New York pacifists organize what was later called the "Anti-Preparedness Committee" to keep the United States out of World War I. The Committee's mascot was Jingo the *Stegosaurus* ("All Armored Plate—No Brains").

1915 Richard Swann Lull's *Triassic Life of the Connecticut Valley* discusses the fossils in Triassic rocks of the Connecticut Valley. Lull reinterpreted Edward Hitchcock's fossilized trackways as being made by dinosaurs, not birds. A revised edition of Lull's book was published in 1953.

1915 Russian scientist Vladimir Obruchev's novel *Plutonia* features dinosaurs (e.g., *Triceratops,* which prompts "We're in the Cretaceous period now!") living underground in Siberia.

1915 William Diller Matthew's (1871–1930) *Dinosaurs* is the first book about dinosaurs for nonspecialists. Matthew's book included Charles Knight's 1897 painting of *Brontosaurus* and the first reconstruction of *Brachiosaurus*.

1915 William Harlow "Bill" Reed dies, and several years later, a flood destroys many of the fossils he had collected at the University of Wyoming. With a photograph and plaque in the geology building at the University of Wyoming, Reed's former students remembered him as a "faithful and useful man of science." A large *T. rex,* built in the early 1960s by geologist Samuel H. Knight (1892–1975), watches over the university's Geological Survey Building.

1915 William Holland describes *Apatosaurus louisae* and, based on its skull, again challenges Marsh's use of a *Camrasaurus*-like skull in his 1883 and 1891 reconstructions of "*Brontosaurus*." After noting the close proximity of the skull to the skeleton, their position in the same geological horizon, and the exact fit of the skull with the vertebra of CM 3018, Holland concluded that the *Diplodocus*-like skull was the true skull of *Apatosaurus*. AMNH's Osborn dared Holland to mount the new skull, but CMNH left its skeleton skull-less until the 1930s, when it was inaccurately crowned with a replica of the skull on Marsh's dinosaur.

1916 A German naval crew aboard SMS *Möwe* (a warship disguised as a cargo vessel) stops the Canadian Pacific Railway steamship SS *Mount Temple,* which was en route to BMNH in London via Brest, France. After killing four sailors, wounding several others, and taking the remaining passengers prisoners, the *Möwe* sank the *Mount Temple*. Among the cargo lost in the 14,435-foot-deep water were 22 crates of fossils, including several 75-million-year-old skeletons collected by the famous fossil-hunting family, the Sternbergs. The fossils included four partial hadrosaur skeletons and a near-complete skull of the horned dinosaur *Chasmosaurus. Mount Temple* sank in the general area of *Titanic*. The *Möwe*, which was later used as a German transport ship and renamed *Oldenburg,* was sunk in 1945 by Allied aircraft.

1916 Barnum Brown (Figure 17) goes to northwestern Montana and discovers numerous skeletons of duck-billed dinosaurs. Brown paid landholders $100 to collect the specimens. Paleontologist Jack Horner later described one of the new species discovered by Brown as *Gryposaurus latidens*.

1916 Oilman and former pharmacist Harry Sinclair borrows $20 million to combine 11 small petroleum companies to form Sinclair Oil & Refining Corporation, which later funds several dinosaur-hunting expeditions in the American West.

Sinclair's models of dinosaurs in the "Dinoland" exhibit at the 1964–1965 World's Fair in New York—viewed by 10 million visitors—fueled America's growing interest in ancient life. Some of Sinclair's models, which today are exhibited in Dinosaur Valley State Park (DVSP) in Glen Rose, Texas, were based on paintings by Charles Knight. Today, Sinclair's company—now known simply as the Sinclair Oil Corporation—is headquartered in Salt Lake City, Utah.

1916 Pioneering paleontologist Charles Gilmore of NMNH describes a layer of clams in Montana's Landslide Butte badlands. In 1986, Jack Horner retraced Gilmore's steps and discovered that the shell fragments were not from ancient clams, but instead were from dinosaur eggs.

1917 After 50 years of collecting fossils, Charles H. Sternberg laments, "I have seen my choicest treasures for forty years leave my hands forever, to add to the glories of museums I shall in all probability never see." In his privately published book *Hunting Dinosaurs in the Bad Lands of the Red Deer River, Alberta, Canada,* Sternberg described his life as a fossil collector, noting that his "whole object has been to give the information I have acquired through years of toil and hardship in the desolate fossil fields to the public, so they may realize something of the wonders of Nature, and the hope it may lead some of my readers to Nature's God, the Triune God we worship." The book included three chapters of science fiction in which Sternberg fell asleep, went back in time, and was joined with his daughter Maud (who had died several years before the book was published). Chapter XI of Sternberg's book included one of the first appearances of *T. rex* in fiction.

1917 Ellis Shuler of Southern Methodist University publishes the first studies of dinosaur tracks found at Glen Rose, Texas. Shuller's article, titled "Dinosaur Tracks in the Glen Rose limestone near Glen Rose, Texas," was published in *American Journal of Science*. Shuler reported prints 10 inches wide and 16 inches long (from the anterior end of the middle toe to the heel). Shuler named the footprint *Eubrontes (?) titanopelopatidus* ("the lime mud strider"); Shuler included the question mark because he was not sure of attributing the name *Eubrontes* to his discovery. Shuler's article attracted relatively little attention. Today, Glen Rose reaps more than $25 million per year from tourism, much of it related to the area's dinosaur tracks.

1917 Five years after moving to British Columbia to work as a mining engineer with his two sons, Arthur Lakes—the first person to collect a dinosaur fossil from the Morrison Formation—dies.

1917 Henry Osborn's *The Origin and Evolution of Life* argues that God set evolution in motion by changing the energetic interactions among organisms and the environment. That same year, Osborn named *Struthiomimus*, thus beginning the tradition that all species of this group be named after birds that they seemed to mimic (e.g., *Sinornithomimus* = "Chinese bird mimic"; *Gallimimus* = "chicken mimic"; *Anserimimus* = "goose mimic").

1917 Lawrence Lambe names *Edmontosaurus*, a 40-foot-long, four-ton duckbill dinosaur discovered in Canada. Unlike *Lambeosaurus, Edmontosaurus* did not have a crest on its head. That same year, Lawrence Lambe prepared the first reconstruction of *Gorgosaurus*.

1917 Two years before his death, Andrew Carnegie stops funding paleontology at CMNH and asks that the museum begin funding the work with money from a general fund. Although CMNH now owned more than 500 specimens of dinosaurs, some of the original crates from the Carnegie Quarry at DNM remain unstudied.

1917 Inventor Thomas Edison brings Willis O'Brien's *The Dinosaur and the Missing Link* to the public. The 19-minute film, which featured cavemen battling for Miss Araminta Rockface, showed anachronistic dinosaurs living with cave people. *The Dinosaur and the Missing Link* was a big hit, and O'Brien became one of the most famous animators of special effects involving dinosaurs, including the dinosaur sequences of *King Kong* (1933).

1917 While exploring the Big Bend region of Texas with Roland Bird, Barnum Brown finds a heavily armored nodosaur that is later named *Edmontonia rugosidens*. This specimen (AMNH 3076), from the Late Cretaceous, is now displayed at AMNH. It is named not after Edmonton, Canada, but instead after the Edmonton Formation (now the Horseshoe Canyon Formation) in which it was found.

1917 Winsor McCay releases *Gertie on Tour,* a sequel to his earlier *Gertie the Dinosaur.* This was one of the first sequels in movie history. That same year, O'Brien's film *The Birth of a Flivver* featured dinosaurs and cavemen. McCay, who made six more films after *Gertie,* died in 1934 of a cerebral embolism.

1917 YPM is demolished to make room for the Harkness Quadrangle dormitory complex. The expanded YPM reopened seven years later at its current location on Whitney Avenue.

1918 After consulting with Barnum Brown to obtain up-to-date information about dinosaurs, special-effects technician Willis O'Brien is hired by producer Herbert M. Dawley (1880–1970) to create *The Ghost of Slumber Mountain*. In this 19-minute film (cut down from an original length of 45 minutes), a character named Holmes (played by producer Herbert M. Dawley) has a dream in which he encounters the ghost of a mountain hermit named Mad Dick (played by O'Brien) that instructs him to peer through a time-viewing telescope atop Slumber Mountain. Apparently going back in time, Holmes sees *Tyrannosaurus* battling *Triceratops,* after which *Tyrannosaurus* breaks the time barrier and starts chasing Holmes. This was the theatrical premier of *T. rex*. In the end, the entire adventure turned out to be a dream. O'Brien was given three months to create the film, which was the most elaborate and last of his stop-motion short features. *The Ghost of Slumber Mountain* was a box-office success; although O'Brien was paid only $3,000, Dawley earned more than $100,000. (Dawley took credit for the animation, causing O'Brien to never again work with him). *The Ghost of Slumber Mountain* was the first movie featuring dinosaurs under the supervision of paleontologist Barnum Brown of AMNH.

1918 AMNH pays Charles H. Sternberg $2,000 for his *Albertosaurus* (AMNH 5664), which was mounted by Carl Sorenson and Peter Kaisen and unveiled in 1921. In 1923, Barnum Brown and W. D. Matthew named this specimen the type specimen of *Gorgosaurus,* but it is now considered a juvenile of *Albertosaurus*.

1918 Another flood of the Paluxy River destroys what Charlie Moss had claimed could be fossilized "footprints of apparent giant human beings." Today, the alleged "man tracks" beside dinosaur tracks near Glen Rose are used by many creationists and several museums to promote young-Earth creationism.

1918 Edgar Rice Burroughs's *The Land That Time Forgot* describes people "marooned in a lost land of Dawn Age monsters." This "lost world" was an uncharted island named Caprona in the South Seas. Caprona was inhabited by a variety of prehistoric creatures, including dinosaurs, cavemen, and Neanderthals. To protect themselves from the dinosaurs and other animals, American Bowen Tyler Jr. and the other marooned visitors (who had landed on Caprona after capturing a German U-boat) built Fort Dinosaur. *The Land That Time Forgot* featured several naked people, along with others in loincloths; some of the mechanical dinosaurs were full sized, but others were miniature or depicted as puppets. Burroughs, who created Tarzan and other "pulp" magazine heroes, included dinosaurs and other prehistoric animals in several of his stories. For example, 1914's *At the Earth's Core*, which appeared as a four-part serial in *All-Story Weekly*, depicted a "lost world" in a hollow Earth. In Burroughs's book, the hero (David Innes) drilled into Earth and found an inner world called Pellucidar inhabited by dinosaur-like animals. *At the Earth's Core* ("Guardians of the Gates of Hell!") was the first American science-fiction novel that featured prehistoric animals, some of which looked like dinosaurs. The book was so popular that Burroughs followed it with *Pellucidar*, another book describing a world inside Earth. When *At the Earth's Core* was made into a movie in 1976, prehistoric animals were portrayed by actors in costumes. Other volumes of the Caprona trilogy (also published in 1918) were *The People That Time Forgot* and *Out of Time's Abyss*.

1918 Samuel Williston dies in Chicago. Williston, with Benjamin Mudge, had discovered *Allosaurus* and *Diplodocus*.

1918 William Arthur Parks (1868–1936) begins organizing fossil-collecting trips to the American and Canadian West, which, with the help of Levi Sternberg, provided most of the dinosaur fossils for ROM. Parks, who named *Lambeosarus* (1923), *Parasaurolophus* (1923), *Dyoplosaurus* (1924), and *Arrhinoceratops* (1925), was the first director of ROM; he is honored by a large plaque near the front of the museum.

1919 Barnum Brown's article titled "Hunting Big Game of Other Days: A Boating Expedition in Search of Fossils in Alberta, Canada," published in *National Geographic,* describes his expedition's collection of "300 large cases . . . including twenty skulls and fourteen skeletons of large dinosaurs, besides many partial skeletons. The fossils represent many genera and species new to science." Today, 22 skeletons collected along the Red Deer River by Brown and his crews grace the halls of AMNH. The discovery of dinosaurs in Alberta produced what paleontologist Edwin Colbert called "the Canadian Dinosaur Rush."

1919 BMNH begins 11 years of expeditions to Tendaguru, but the results are disappointing.

1919 In New York, Louis and David Marx found Marx, a toy company that later became the largest toy-maker in the world. Marx made several popular dinosaur toys from 1955 to 1980. Although companies such as Ajax, Tim Mee, and Miller also made dinosaur toys, Marx was the iconic toymaker of the 20th century; in addition to dinosaur toys, Marx also produced Rock'em Sock'em Robots, Big Wheel tricycles, and toy soldiers.

1919 Lawrence Lambe dies suddenly and his work is taken over by Charles M. Sternberg (1885–1981), one of the greatest students of Canadian dinosaurs. During subsequent decades, Sternberg described species of *Triceratops, Monoclonius, Centrosaurus, Pachyrhinosaurus, Leptoceratops,* and *Montanoceratops*.

1919 Roy Chapman Andrews and Swedish geologist Johan Gunnar Andersson (1874–1976) decide to organize an expedition to Asia. These expeditions produced many important discoveries and launched Andrews to international fame.

1919 Sir Arthur Conan Doyle sells the movie rights of his novel *The Lost World* (1912) to British movie producer J. G. Wainwright for £500. Three years later, Wainwright sold these rights to Watterson R. Rothacker, who had hired Willis O'Brien for other projects. Rothacker had raised over $1 million to make a movie of Doyle's book. The 50 models in the movie (many of which were animated by O'Brien's stop-action techniques) were sculpted by Mexican artist Marcel Delgado (1901–1976), who based his models on paintings by Charles Knight. In some of his models, Delgado incorporated a rubber bladder to simulate the animals' breathing.

The Lost World, which was released in 1925, was one of the most expensive, influential, and elaborate movies made up to that time.

1919 Andrew Carnegie's death slows funding for dinosaur-hunting expeditions sponsored by CMNH. In the previous decade, the museum had amassed more than 300 tons of fossils, mostly from DNM. When CMNH's operations there ended in 1923, William Holland informed officials at NMNH of several more fossilized dinosaurs at the site, which were collected by a crew from NMNH. One of those specimens was another *Diplodocus,* which is displayed at the Smithsonian today. Between 1898 and 1911, Carnegie donated more than $250,000 to CMNH for the collection, preparation, and study of dinosaurs and other fossil vertebrates. However, Carnegie made no permanent provisions for paleontology when he died. Today, several crates of fossils from the American West remain unopened in the basement of CMNH.

1919 The type-specimen of *Barosaurus* is re-described by Richard Swann Lull. In 1968, Robert Bakker restored *Barosaurus* as a warm-blooded land-roving giant.

1919 William J. Holland names *Uintasaurus* from bones excavated at what would later be DNM. In 1958, Theodore White placed *Uintasaurus* with *Camarasaurus*.

1920 Advertisements for the dinosaur-based movie *Along the Moonbeam Trail* urge viewers to "See These 40-ton Monsters Fight to the Death!" All copies of the film were lost. *Along the Moonbeam Trail* included stop-motion dinosaurs from *Ghost of Slumber Mountain*.

1920 Charles Gilmore publishes the first monograph about *Ceratosaurus*. Gilmore later distributed several plaster dinosaur sculptures to museums, but these were not sold to the public. Gilmore's reconstructions were later criticized by Robert Bakker.

1920 Charles M. Sternberg completes a paper started by Lawrence Lambe, who had died the previous year. This paper, which described *Panoplosaurus,* was Sternberg's first scientific publication.

1920 German paleontologist Friedrich von Huene proposes the name Prosauropoda for a group of dinosaurs having long necks, small heads, and large bodies that were mostly herbivorous and lived on Earth from the Late Triassic through the Early Jurassic. Although the earliest known prosauropods were too specialized to be the ancestors of sauropods, the name *Prosauropoda* has remained. The first Triassic dinosaur (and the first sauropodomorph) to be discovered was *Thecodontosaurus* in 1836 (based on isolated teeth), and the best-known prosauropod is *Plateosaurus,* which was discovered in 1837 by Hermann von Meyer. There are more than 100 fragmentary or near-complete skeletons of *Plateosaurus*.

1920 In *Tarzan the Terrible,* Tarzan searches for his friend in Pal-ul-don, a lost world inhabited by *Triceratops*-like dinosaurs that were carnivores. Tarzan was captured by Germans and discovered the "missing link" between humans and apes. *Tarzan the Terrible* was the eighth book of the Tarzan series. In subsequent adventures of Tarzan, dinosaurs and other prehistoric animals often inhabited Pal-ul-don.

1920 Ruth Mason finds a bed of dinosaur bones on her family's ranch near Faith, South Dakota. This site later became the Ruth Mason Dinosaur Quarry.

1920 While at lunch, Roy Chapman Andrews proposes to Henry Osborn that AMNH explore the Gobi Desert for human ancestors. The resulting expeditions, which began in 1922 and concluded in 1930, were described as the "biggest scientific expedition ever to leave the United States." The expedition found no human ancestors, but produced spectacular discoveries of dinosaurs and other prehistoric creatures.

1920s Paleontologists begin shifting their work from dinosaurs to mammals. Dinosaurs would not again become the primary focus of American paleontologists until the "dinosaur renaissance" of the 1970s.

1921 Canadian mammalogist William Diller Matthew, who had been Henry Osborn's student at Columbia University, argues that continental drift, mountain building, and the diversification of mammals drove dinosaurs to extinction.

1921 Charles M. Sternberg finds what will later be named *Daspletosaurus* ("frightful flesh-eating reptile"). That same year, Sternberg attributed a partial skull that he discovered in the Judith River Formation in Alberta, Canada, to *Gorgosaurus*. In 1970, that specimen was fully described by Dale Russell, who made the specimen (a close relative and possible ancestor of *T. rex*) a new genus, *Daspletosaurus*.

1921 Friedrich von Huene and his colleagues open a quarry in a wooded valley near Trossingen. Thanks to money from AMNH (given in exchange for fossils), von Huene excavated several dinosaurs, including *Plateosaurus*. Von Huene participated in several other dinosaur-collecting expeditions, including those to Argentina (1923), South Africa (1924), and Brazil (1928).

1921 Henry Osborn and Charles Mook publish *Camarasaurus, Amphicoelias, and Other Sauropods of Cope*, a 141-page monograph that included a detailed, landmark description of *Camarasaurus*. The book also included the reconstruction of *Camarasaurus* from 1877 by John Ryder.

1921 Richard Lull describes *Nodosaurus* ("nodular reptile"), which was first mentioned by Othniel Marsh in 1889.

1921 The Field Museum opens in Chicago, Illinois. This museum was an outgrowth of Jackson Park's Field Columbian Museum, which was later rebuilt as the Museum of Science and Industry.

1921 The Second International Congress on Eugenics, presided over by Henry Osborn at AMNH, is guided by Osborn's hope that humans should "care for the race, even if the individual must suffer—this must be the keynote of our future." In 1932, AMNH—with Osborn serving as vice-president—hosted the Third International Congress on Eugenics.

1921 The silent animated film *The First Circus* features a "Stonehenge Circus 3009 years ago" that includes *Brontosaurus* and a trainer doing tricks for a spear-carrying crowd.

1921 While attending classes at the Otis Art Institute, Willis "O'Bie" O'Brien is impressed with the artwork of 19-year-old fellow-student Marcel Delgado. O'Brien later convinced Delgado to sculpt models for the movie *The Lost World*, for which Delgado was paid $75 per week (Delgado had been working at a grocery store for $18 per week). Delgado produced 50 models (from wire, cotton, sponges, and wax) that averaged 18 inches high and were based on the artwork of Charles Knight. The animation of Delgado's models by O'Brien for *The Lost World* took 14 months. Delgado's models (including those produced later for *King Kong*) were donated to Los Angeles's Museum of Arts and Science, where some were sealed between walls when a new wing was added to the museum. Today, the only two surviving models made by Delgado (a sauropod and *Pteranodon*) are displayed at the 18-room Forrest Ackerman Sci-Fi Mansion in Los Angeles, California. Ackerman (1916–2008) was the world's greatest private collector of memorabilia from science-fiction movies; in addition to Delgado's models, Ackerman's more than 400,000 specimens included King Kong's paw. Ackerman coined the genre nickname "sci-fi."

1921 While prospecting in Ethiopia, Barnum Brown comes within 100 miles of the Middle Awash drainage at Hadar, which in 1974 produced the then-earliest known human, "Lucy" (*Australopithecus afarensis*).

1922 A mummified anatosaur is found with seeds, twigs, fruit, and needles in its stomach, thus providing clues about its diet.

1922 After being translated into English, Alfred Wegener's ideas about continental drift are attacked by numerous scientists. At the annual meeting of the Geological Society of America in 1922, C. T. Chamberlin told attendees, "If we are to believe Wegener's hypothesis, we must forget everything which has been learned in the past 70 years and start all over again."

1922 After decades of little support for vertebrate paleontology at the Field Museum, Marshall Field III (1893–1956) gives money to support the department's work. This enabled Elmer Riggs to end his career working on expeditions in South America and Alberta, Canada.

1922 AMNH's president Henry Osborn (Figure 15), a protégé of Edward Drinker Cope and a neo-Lamarckian who believed that humanlike "Dawn Man" had few connections with primates, sends Roy Chapman Andrews (1884–1960) on the first of AMNH's legendary Central Asiatic Expeditions (1922–1930) to find hominid and other fossils. Andrews's expeditions, which were partly funded by the Dodge automobile company, were the first major scientific expedition to involve automobiles. Osborn believed that humans originated in the elevated regions of Asia ("the incubating region for the land life of Europe and North America") and was confident that Andrews would find fossils of human ancestors; as Osborn noted on the day of Andrews's departure "The fossils are there. I know they are. Go and find them." The expeditions' headquarters were in Peking (Beijing), and an advance caravan of 125 camels hauled in 4,000 gallons of gasoline, 3 tons of flour, 1.5 tons of rice, tents, water, spare parts, and other supplies. At some eroded U-shaped cliffs in Mongolia's Gobi Desert near Shabarakh Usu (also known as Shabarakh Usu and Bayn-Dzak) called "Flaming Cliffs" (named for how the setting sun seemed to turn the rocks into blowing embers), photographer J. B. Shackelford discovered a dinosaur egg. The following year, during a more extensive expedition, Andrews and his crew found 13 dinosaur eggs, with their ends pointing toward the center of the nest. These were the first dinosaur-eggs that the world had ever seen. One of the eggs was sold for $5,000 to Austin Colgate, who donated it to Colgate University (where it resides today). Many people who opposed the excavations claimed that the egg's sale was an attempt by AMNH to steal Mongolia's treasures for profit. Andrews and his coworkers also discovered numerous dinosaurs, the first of which was the primitive ceratopsian *Protoceratops andrewsi* (AMNH 6251), which AMNH's president Henry Osborn described as "a new chapter in the history of life upon the earth."[25] No bones of *Protoceratops* were directly associated with the eggs, and no embryos of *Protoceratops* were found in the eggs. In the Gobi Desert, Andrews stayed in a dark blue tent adorned with yellow bats, a sign of good luck in Mongolia. Andrews, who later became director of AMNH, was believed to be the inspiration for the movie character *Indiana Jones* (although George Lucas—Jones's inventor—denied it). Flaming Cliffs also yielded the medium-sized armored ankylosaur *Pinacosaurus*, the birdlike *Saurornithoides*, and *Velociraptor* ("fast-running robber"). The armor of *Pinacosaurus* consisted of individual bones fused to the skull during growth. Andrews later described Flaming Cliffs as "the most important deposit in Asia, if not the entire world." The eggs discovered by Andrews are displayed near *T. rex* in AMNH's Hall of Saurischian Dinosaurs. Andrews's expeditions did not find "Dawn Man," but they did uncover numerous dinosaurs, including many finds at "Flaming Cliffs." Andrews became a hero to a generation of would-be explorers.

1922 Animosity between William Jennings Bryan and Henry Osborn reaches the editorial pages of the *New York Times*. After Bryan wrote an invited editorial on "The Heretical Belief of Darwinism," Osborn replied that evolution does not necessarily conflict with religion. Osborn claimed that Bryan would see this if Bryan would "drop [his] books and narrow overspecialized thinking and go out into nature and observe it firsthand."

1922 At a party for magicians hosted by escape artist Harry Houdini (Erik Weisz, 1874–1926), Sir Arthur Conan Doyle treats guests to clips of the upcoming movie-version of his *The Lost World.* The next day, a headline in the *New York Times* reported "Dinosaurs Cavort in Film for Doyle." Doyle told Houdini, "The dinosaurs and other monsters have been constructed by pure cinema, but of the highest kind."

1922 By this time, CMNH's excavations at DNM have produced 446 crates of fossils weighing more than 350 tons. These excavations were difficult, and the crates could be shipped to Pittsburgh only after being hauled by four-horse wagons more than 60 miles to Dragon, Utah, and loaded onto boxcars of the now-defunct Uintah Railway.

1922 Edgar Rice Burroughs (1875–1950) publishes *At the Earth's Core,* the first of seven novels about the "lost world" of Pellucidar, which is populated by dinosaurs and other extinct animals of the past. *At the Earth's Core* was illustrated by Chicago artist J. Allen St. John (1872–1957), who later illustrated dinosaur-related articles in many magazines. In other books, Burroughs described other lost worlds (e.g., Caprona) inhabited by dinosaurs, and one of Burroughs's most famous characters, Tarzan, battled dinosaurs in lost worlds in *Tarzan the Terrible* (1920). Numerous subsequent movies and books featuring dinosaurs were based on "lost worlds."

1922 In the Outer Mongolian Gobi Desert, Peter Kaisen—accompanying Roy Chapman Andrews—discovers a crushed but complete skull and claw of *Velociraptor*, a 75- to 71-million-year-old bipedal, long-tailed, turkey-sized theropod. Two years later, Osburn's description made this specimen (AMNH 6515) the type-specimen of *Velociraptor mongoliensis,* the first dromaeosaurid to be described. Except for being smaller, *Velociraptor* resembled *Deinonychus*; each had long forelimbs with grasping hands, stiff tails, and a large sickle-shaped claw on the second toe of the foot that was used to kill prey. (One of the most famous specimens, found less than 60 miles from Flaming Cliffs and now displayed in the State Museum in Ulaanbaatar, Mongolia, shows *Velociraptor* locked in combat with the beak-faced *Protoceratops*.) Andrews, a former taxidermist, recognized the similarities of these dinosaurs and birds when he noted that "these dinosaurs are much too late in geological time to be ancestral to birds, but they do parallel them remarkably in their almost wing-like hands and lightly built skulls." Despite being shown with anatomical inaccuracies, *Velociraptor* became one of the most famous dinosaur genera when it was featured as a primary antagonist in the movie *Jurassic*

Park (1993). The model for *Velociraptor* in *Jurassic Park* and *The Lost World* was sculpted by artist Stan Winston.

1922 Near the Chinese village of Ning Chia Kou in the Shandong Province, Swedish geologist Johan Gunnar Andersson (1874–1960) and Chinese geologist Tan Xichou ("H.C. Tan") find what Carl Wiman later (1929) named *Helopus*. This was the first dinosaur to be named from China. Unlike most sauropods, *Helopus* had longer forelegs than hind legs. When it was discovered that *Helopus* was already the name of a bird, A. S. Romer renamed the Early Cretaceous sauropod *Euhelopus*.

1922 The Dominion Motion Picture Bureau (now the National Film Board of Canada) produces *Monsters of the Past,* one of the first films about collecting vertebrate fossils.

1922 William Parks describes *Parasaurolophus walker,* the first tube-skulled hadrosaur to be discovered. The following year, Swedish paleontologist Carl Wiman (1867–1944) noted that the crests of *P. walker* resembled a crumhorn (a musical wind-instrument), and speculated that they could have been used as a resonating chamber that would have enabled the dinosaurs to communicate over long distances.

1923 Although Charles Knight is best known for his paintings of dinosaurs, the U.S. Postal Service issues a 30¢ stamp featuring a drawing by Knight of a bison. That same year, Knight produced a famed mural for the AMNH's Hall of Man.

1923 Amateur fossil-collector Charles A. Matley (1866–1947) names the theropod *Lametasaurus* from tail bones and plating that he found two years earlier in India. A decade later, with German paleontologist von Huene, Matley identified numerous sauropods and theropods at Jabalpur, India, including *Indosaurus, Titanosaurus,* and *Antarctosaurus* (which had been identified earlier in South Africa).

1923 Buster Keaton rides an animated dinosaur in the comedy *The Three Ages*, a silent movie that tells three stories set in three different ages (prehistory, ancient Rome, and the Roaring Twenties). In one feature, Keaton rode a clay-animated brontosaur in homage to *Gertie*. In the previous year, the movie *Monsters of the Past* included stop-motion clay dinosaurs.

1923 Charles Gilmore (1874–1945) of NMNH begins working at DNM. Gilmore and his coworkers soon found the *Diplodocus* (USNM 10865) now displayed at NMNH. Later in 1923, the University of Utah began excavating near the monument's present-day parking lot; the *Allosaurus* skull that they found is displayed at the monument's quarry building.

1923 During AMNH's second expedition to Mongolia, paleontologist George Olsen of Roy Chapman Andrews's crew discovers dinosaur eggs at Flaming Cliffs.

Although Andrews was initially skeptical, he was soon convinced, as is evident from this conversation recounted later: "On July 13, 1923, George Olsen reported at dinner that he had found some fossil eggs. Inasmuch as the deposit was obviously Cretaceous and too early for large birds, we did not take his story very seriously. . . . Nevertheless, we were all curious enough to go with him to inspect his find. We saw a small sandstone ledge, beside which were lying three eggs partly broken. The brown striated shell was so egg-like that there could be no mistake. Walter Granger finally said, 'No dinosaur eggs have ever been found, but the reptiles probably did lay eggs. These must be dinosaur eggs. They can't be anything else.' . . . Finally we had to admit that we could make them out to be nothing else." (Granger was wrong—dinosaur eggs had been found earlier in France, but this was not widely known.) This discovery, and Andrews's growing fame, prompted *Time* magazine to feature Andrews on the cover of its October 29, 1923, issue. Andrews's first lecture about the expedition attracted 4,000 people hoping to get one of the 1,400 seats. Granger's most famous discoveries (*Velociraptor, Oviraptor,* and *Protoceratops*) were often attributed to Andrews.

1923 Earl Douglass writes a letter to Charles Walcott of the Smithsonian: "The number of visitors [at DNM] has increased from year to year and the popularity of the monument will undoubtedly grow rapidly in the future as people hear of it and learn where it is. The greater number come in the tourist season from June to October. Last August there were over 500 visitors to the quarry. They came from early morning to dark—in fact some have 'viewed the remains' by flash light. I hope—and there are thousands of others—that the government, for the benefit of science and the people, will uncover a large area, leave the bones and skeletons in relief and house them in. It would make one of the most outstanding and instructive sights imaginable." A decade later, workers from the Works Progress Administration (WPA) enlarged the quarry, and in 1958 a building was constructed so visitors could see bones in the quarry.

1923 F. H. McLearn first reports dinosaur tracks in British Columbia.

1923 German paleontologist Friedrich von Huene collects the prosauropod *Plateosaurus engelhardti*, the best-known relative of giant sauropods such as *Apatosaurus* and *Diplodocus*. This specimen (AMNH 6810) is now displayed at AMNH. Von Huene, who hoped that his research would reveal God's wonder, advocated Seeley's division of Dinosauria into Ornithischia and Saurischia.

1923 Henry Osborn describes *Pentaceratops* from New Mexico. That same year, Osborn's paper titled "Two Lower Cretaceous Dinosaurs of Mongolia" described the type specimen of the small, parrot-beaked, primitive ceratopsian *Psittacosaurus* ("parrot lizard"; AMNH 6253) and *Protiguanodon* (AMNH 6254) that had been found in Mongolia. The discovery of *Psittacosaurus mongoliensis* was the first major paleontological find of the Asiatic Expedition. *Psittacosaurus* and *Protiguanodon* were similar—so much so that Osborn noted, "In case the animals turn out to be the same, *Psittacosaurus* will have precedence." Osborn was prophetic: Both specimens were later determined to be *Psittacosaurus.* The 4-foot-long, 2-foot-tall *Psittacosaurus* had no horns, but is part of the same group as *Triceratops*.

1923 Roy Chapman Andrews publishes the popular book *All about Dinosaurs*. This book, which referenced *The Lost World*, inspired a generation of young paleontologists, including Jack Horner, Michael Novacek, and Philip Currie. As Currie recalled later, "after reading *All about Dinosaurs* . . . when I was 11 years old, I knew I wanted to be a dinosaur hunter." Currie helped found Alberta's Royal Tyrrell Museum of Paleontology (RTM).

1923 The film *Adam's Rib* includes a dinosaur skeleton.

1923 *Tyrannosaurus* appears in the films *Monsters of the Past* (in which Virginia May sculpts a clay tyrannosaur) and *Evolution,* which incorrectly identifies *Iguanodon* as *Tyrannosaurus*. Although *T. rex* was not in the book *The Lost World* (1912), it was in the 1925 film.

1923 While working at Flaming Cliffs in Mongolia, George Olsen collects the type specimen of the toothless *Oviraptor philoceratops* (AMNH 6517). Because the near-complete skeleton was found atop a nest of dinosaur eggs, AMNH's Henry Osborn concluded that *Oviraptor* had been "overtaken by a sandstorm in the very act of robbing the dinosaur egg nest" (*Oviraptor* means "egg robber"). However, subsequent evidence collected by a joint expedition of AMNH and the Mongolian Academy of Sciences showed that the nests and eggs originally attributed to *Protoceratops* were actually those of *Oviraptor,* and that *Oviraptor* was incubating its eggs and/or protecting its nest.

1923 While working at what would become DNM, a crew from the University of Utah unearths the *Allosaurus* skull that is now displayed at the monument. That same year, sauropod expert Charles Gilmore discovered the *Diplodocus* (USNM 10865) that is now displayed at NMNH (Figure 23). Other specimens excavated from the Carnegie Quarry at DNM include the long-necked *Barosaurus* displayed at AMNH (a cast is displayed), *Diplodocus* displayed at the Denver Museum of Nature and Science, *Apatosaurus* and the juvenile *Camarasaurus* displayed at CMNH, and *Diplodocus* displayed at NMNH. Parts of *Barosaurus* were spread between the institutions that helped with the excavation: the University of Utah (which got the ribs, pelvis, most of the tail, and some vertebrae), CMNH (which received tail vertebrae), and NMNH (which got neck vertebrae, some back vertebrae, and the shoulder girdle). In 1929, Barnum Brown brought all of the specimens together at AMNH. The AMNH's *Barosaurus* now appears in a rearing-up pose in the lobby of Theodore Roosevelt Hall defending itself and its baby against a menacing *Allosaurus*.

Figure 23 In 1900, American paleontologist Charles Whitney Gilmore (1874–1945) was recruited by John Bell Hatcher (1861–1904) to work at the Carnegie Museum of Natural History. Three years later, Gilmore moved to what is now the National Museum of Natural History (NMNH), where he spent the rest of his career. Gilmore named several dinosaurs in North America and Mongolia, including *Brachyceratops*, *Alamosaurus*, and *Styracosaurus*, and in 1923 found the *Diplodocus* displayed in NMNH. In this photo from 1924, Gilmore examines cervical vertebrae of *Diplodocus*. (Library of Congress)

1924 After BMNH seeks support "to obtain the bones of *Gigantosaurus*, a Dinosaurian reptile of stupendous size, which was discovered by the Germans a few years before the war," another dinosaur-collecting expedition to Tendaguru, Tanzania, is led by University of Manitoba paleontologist William Edmund Cutler. The expedition uncovered numerous ornithischians (e.g., *Dryosaurus, Kentrosaurus*), sauropods (*Barosaurus, Brachiosaurus, Dicraeosaurus*), and theropods (*Ceratosaurus, Megalosaurus, Labrosaurus*), as well as dinosaur eggs. The 1924 expedition included the young Kenyan-born paleontologist Louis S. B. Leakey (1903–1972), who later discovered human ancestors in Africa. Cutler died of blackwater fever in a hut in Lindi the following year.

1924 *Anatosaurus* is created to absorb several of the older species of hadrosaurs, including *Claosaurus annectens, Diclonius mirabilis* (which later became *Trachodon mirabilis*), and *Trachodon longiceps*. These later became *Edmontosaurus*.

1924 Arthur Brown, manager of Durham Downs sheep station in Queensland, Australia, finds *Rhoetosaurus,* one of the earliest sauropods.

1924 Earl Douglass and his coworkers have shipped more than 300 tons of fossils from DNM. Today at the Monument, visitors can see more than 1,500 bones (from more than 400 individuals) exposed *in situ* on a 150-foot-long, 50-foot-high face of sandstone. Unlike at most other museums, visitors to DNM are encouraged to touch the bones.

1924 Earl Douglass retires at CMNH and starts working for the University of Utah. Douglass later worked as a consultant for oil companies.

1924 George Patterson discovers the type specimen of *Edmontonia,* which was named *E. rugosidens* by Charles M. Sternberg.

1924 In an article titled "The Discovery of an Unknown Continent" in *Natural History* magazine, as well as in others that appeared in several U.S. newspapers, Henry Osborn uses the name *Oviraptor djadochtari* to refer to the specimen (later named *Velociraptor*) discovered almost two years earlier in Mongolia. (The species epithet *djadochtari* refers to the khaaochta Formation in which the specimen was found.) Because *Oviraptor* was not accompanied by a formal description or published in a scientific journal, *Oviraptor* is not considered valid name (i.e., *Velociraptor* is the proper name). Osborn claimed that *Velociraptor* was an "alert, swift-moving carnivorous dinosaur." *Velociraptor* remained relatively obscure until 1993, when it became famous after starring as the vindictive, matriarchal villain in the movie *Jurassic Park*. In real life, *Velociraptor* was only about 6 feet long.

1924 Richard Swan Lull proposes that the extinction of dinosaurs resulted from "racial senility."

1924 The expanded YPM reopens at Yale University, with its two-story Great Hall of Dinosaurs designed especially for Othniel Marsh's dinosaurs. The most prominent dinosaurs displayed in the gothic-designed building would be *Stegosaurus* and *Apatosaurus* (known originally as *Brontosaurus,* or "thunder lizard"), which was completed in 1931 after six years of work. This specimen, which was the first mounted skeleton of a huge dinosaur, originally (and mistakenly) was adorned with the bulkier skull of *Camarasaurus,* but was corrected with the 1931 restoration. Today, YPM features the type specimen of *Deinonychus* (YPM 5205), which was found in Montana in 1964 by Yale paleontologist John Ostrom. In front of the museum is the first full-scale reproduction of *Torosaurus latus*; this bronze sculpture, which is mounted on a 13-foot-tall, 70-ton foundation of Stony Creek granite (the same granite forming the base of the Statue of Liberty), is 21-foot long, 9-foot tall, and weighs more than 7,300 pounds. It was created in 2005 by Michael Anderson.

1924 The first intact dinosaur bones are discovered in Australia. Bone fragments suspected to be from dinosaurs had been discovered there in 1844.

1924 William Holland again argues that *Apatosaurus* should have a *Diplodocus*-like skull. Holland's claim was ignored until it was resurrected by John S. McIntosh (of Wesleyan University) and David Berman (of CMNH).

1924 The University of Utah ends its excavations at DNM.

1925 Arthur Conan Doyle's *The Lost World* is made into a movie by director Henry O. Hoyt, and invites viewers to "see gigantic pre-historic monsters clash with modern lovers." The silent movie, which cost more than $1 million to make and which employed more than 100 technicians, used models (each 16–20 inches tall filmed on sets 6 feet long) made by 20-year-old art student Marcel Delgado, who was paid $300 per month for his work. (The movie's spectacular stampede of dinosaurs was filmed on a set 150 feet long.) *The Lost World* was one of the first movies to feature young dinosaurs; at one point, a female *Triceratops* licks its baby after fending off a carnivore. Delgado's models, which have been lost, were animated by Willis H. O'Brien. Unlike in other movies of the time, O'Brien's villainous dinosaurs bled—O'Brien used chocolate syrup to simulate blood (in black-and-white, the syrup looked like blood), and their drool was depicted with shellac. O'Brien based his stop-action models for *The Lost World* on paintings by Charles Knight displayed at AMNH. *The Lost World* was the first feature-length movie to showcase stop-motion dinosaurs, and ended with a giant sauropod destroying London; this use of a rampaging dinosaur in a modern city would be repeated often in subsequent movies. The most menacing of the movie's dinosaurs was the theropod *Allosaurus. T. rex* appeared only once in *The Lost World,* when it killed a horned "*Agathaumas*" (also based on a painting by Charles Knight, and now a dubious genus of large ceratopsids; it is probably *Triceratops*). Many people who saw the movie believed its dinosaurs were real. In *The Lost World,* actress Bessie Love (Figure 24) became the first woman to be threatened by a dinosaur; this theme was also repeated often in subsequent dinosaur-related movies. Doyle showed a segment of movie version of *The Lost World* at a meeting of the Society of American Magicians; the following day, the *New York Times* reported that if the dinosaurs were not real, "they were masterpieces." *The Lost World* later became the first in-flight movie when it was shown on Imperial Airways' flights between Paris and London; the movie's erupting volcanoes, prehistoric lost world, and monster loose in a city were original in 1925, but later became dinosaur movie clichés (along with mythic caves and "rivers of time").

1925 Bessie Love, the first actress to be harassed by a dinosaur in movies (i.e., in *The Lost World*), appears in advertisements selling "Twistum, the Baby Dinosaur." This toy, manufactured by the Twistum Toy Factory, was the first dinosaur toy sold to help promote a dinosaur-related movie. Such marketing ploys later became common.

1925 Charles Gilmore (Figure 23) describes a nearly complete, articulated juvenile of *Camarasaurus* (CM 11228) found at DNM. Gilmore's discovery questioned claims

Figure 24 In 1925, American actress Bessie Love (born Juanita Horton; 1898–1986)—starring as Paula White in the movie *The Lost World*—became the first woman to be harassed by a dinosaur. Although the movie was an adaptation of Arthur Conan Doyle's book *The Lost World* from 13 years earlier, Love's character—the daughter of an earlier explorer who searched for the Lost Plateau and never returned—was not in Doyle's book. The *Apatosaurus* that threatened Love was sculpted by Marcel Delgado, who based his models on paintings by Charles Knight (e.g., Figure 19). Similar scenes of women being harassed by dinosaurs became a staple of movies featuring dinosaurs. (First National Pictures Inc./Photofest)

by Oliver Hay that sauropods had a lizard-like posture (with their legs splayed out sideways from the body) rather than an erect posture proposed by William Bell Hatcher. A cast of Gilmore's 15-foot-long specimen, one of the most famous skeletons of a sauropod in the world, is displayed at CMNH.

1925 During another expedition to Mongolia's Flaming Cliffs, Roy Chapman Andrews films the excavations of more dinosaur eggs. These expeditions skyrocketed Andrews to fame and helped him raise funds for more expeditions. This expedition was the most ambitious ever conducted by AMNH up to that time: It included 40 people, two trucks, five touring cars, and 125 camels.

1925 In "Dinosaur Extinction" published in *American Naturalist,* Yale University paleontologist George Wieland argues that dinosaurs became extinct because of "racial senility" and egg-robbing snakes and lizards. Charles Immanuel Forsyth Major (1843–1923) and Edward Drinker Cope had earlier speculated that small mammals ate dinosaurs' eggs, but Wieland believed that mammals were too weak to puncture the eggs. Wieland also speculated that dinosaurs provided some parental care and that the egg-diet of some reptiles led to gigantism. In 1994, paleontologist James Kirkland suggested small crocodilians such as *Fruitachampsa* may have eaten dinosaur eggs and babies, and in 2010, paleontologists announced the discovery of *Sanajeh,* a Late Cretaceous snake that may have eaten sauropods' eggs. However, there is no evidence that dinosaurs' extinction was due to egg-eaters.

1925 In the animated film *Felix the Cat Trifles with Time,* Felix goes back in time and hops on the back of a brontosaur as he tries to cross a pond; this and other features of the film appeared repeatedly in future animations about dinosaurs.

1925 Messmore and Damon's 48-foot-long, 9-foot-tall, 2-ton *Apatosaurus*, the world's first robotic dinosaur, is announced in *Iron Trade* magazine. The dinosaur, which could wag its tail, "breathe," and lift its neck, was used to lure visitors to Coney Island, after which it toured the country before gaining even more fame in the "The World a Million Years Ago" exhibit at the 1933 Chicago's Century of Progress. At this exhibit, visitors entered a globe-shaped rotunda and were then taken on a conveyor belt through geological history, where they saw "life-like ancient monsters that move and roar."

1925 Now at the height of his fame, artist Charles Knight has more pictures and square feet of canvas in AMNH than any other artist in any other major museum in the world. His most influential painting there—and one of the most admired restorations of dinosaurs of all time—shows *T. rex* confronting *Triceratops* (this painting is now displayed in the museum's "Life through Time" exhibit). Although modern paleontologists can quickly spot its errors (e.g., the eyes of *T. rex* are misplaced, its hands have three fingers instead of two, and it has an overly long tail that drags the ground), these mistakes were not the fault of Knight; when he painted this and other paintings of dinosaurs, he faithfully depicted what was then known or believed about the animals and their environment. Knight's images of *T. rex* (Figure 19) were presented as *the* images of *T. rex* in books, movies, comic strips, merchandise, and elsewhere.

1925 William A. Parks (1868–1936) describes *Arrhinoceratops* ("no nose-on face"), a frilled ceratopsid from Alberta, Canada.

1926 Danish artist and paleontologist Gerhard Heilmann's (1859–1946) influential *The Origin of Birds* rejects dinosaurs as birds' ancestors, instead ascribing their similarities to convergence (resulting from adaptations caused by bipedalism) and claiming that birds evolved from pseudosuchians (a group of Triassic thecodonts). Heilmann, a proponent of Dollo's Law of Irreversibility (i.e., that once a trait is lost during evolution, it can't be regained), did not believe that clavicles (i.e., collarbones) had evolved in birds after being lost in dinosaurs. (Birds have wishbones, which are clavicles that fuse during embryonic development and brace the shoulders against strains of flight.) Although Heilmann noted the similarities of birds and coelurosaurs (e.g., long hind limbs, long tails, long necks, hollow bones), he also pointed out—according to what was known at the time—that dinosaurs lacked wishbones and clavicles. Heilmann's conclusions were accepted by most biologists for several decades, despite the discovery in 1936 of a clavicle in *Segisaurus,* a primitive theropod. If dinosaurs had lost their collarbones, how could they be ancestors of animals that still had a wishbone made of collarbones? As Heilmann noted, the absence of clavicles in coelurosaurs "would in itself be sufficient to prove that these saurians could not possibly be ancestors of birds." Heilmann's claim that dinosaurs and birds were "cousins" having a common ancestor was favored for several decades after the publication of his acclaimed book. Heilmann was unaware that a small dinosaur had been discovered in Mongolia that had a wishbone almost identical in shape to that of *Archaeopteryx;* this wishbone had been misinterpreted as different bones. In 1926, Heilmann had also suggested that *Iguanodon*'s typical stance was a running position with its neck and tail horizontal. Later, British paleontologist Peter M. Galton (b. 1942) concluded that *Iguanodon* carried its forelimbs closer to the ground than depicted in Louis Dollo's models.

1926 In the comedy film *The Savage,* a scientist finds a giant brontosaur.

1926 The German Air Service chooses *The Lost World* as its first in-flight movie.

1926 The slender, medium-sized (approximately 450 pounds) theropod *Elaphrosaurus,* which had been found a few years earlier in Tendaguru (what is now Tanzania) and named in 1920, is mounted in Berlin's Museum of Natural History at Humboldt University. This museum also houses one of the largest mounted dinosaurs, *Brachiosaurus.*

1926 To get out of debt, Charles Knight accepts $139,000 (over four years) to produce 28 paintings showing life's history for the Field Museum of Natural History's Hall 38. (Hall 38 was commonly called "Dinosaur Hall," despite the fact that most of its fossils were not dinosaurs.) Knight considered these paintings to be his best work. Knight's friend (and AMNH curator of vertebrate paleontology) Edwin Harris Colbert said that "Knight's restorations of extinct animals are great not only because of his inherent abilities as an artist but also because of his readiness to work with scientists . . . his was a constant quest for truth in art and science . . . he had so much imagination that he could project himself back in time and feel that he was on a cliff with one of those monsters." The most famous of these paintings (Figure 19) shows a side view of *Tyrannosaurus* (horizontal and with its tail off the ground) confronting an oncoming *Triceratops*. Knight's paintings at the Field Museum were the source of the *T. rex* (and other dinosaurs) designed by Indiana model-maker P. G. Alen for Sinclair Oil's "Sinclair Dinosaur Exhibit" at the 1933–1934 "A Century of Progress" World's Fair in Chicago. Knight's work defined dinosaurs for the world; few publications in the first half of the 20th century did not include Knight's work.

1926 Werner Janensch describes the first dinosaur found in Mexico, a ceratopsian collected in the "Soledad Beds" of Coahuila.

1926 While excavating near Alberta, Canada's Red Deer River, Charlie H. Sternberg—known in Canada as "Mr. Dinosaur"—finds a skeleton of *Thescelosaurus,* one of the latest dinosaurs of the Late Cretaceous. Soon thereafter, the fossils were described and named by Canadian paleontologist William Parks (1868–1939). Before his death, Sternberg—who discovered 17 new dinosaurs—was featured in a movie produced by Canada's National Film Board named *Charlie.*

1927 Danish paleontologist Gerhard Heilmann depicts the animals that made the Connecticut Valley dinosaur tracks as coelurosaurs, ornithopods, and prosauropods.

1927 Friedrich von Huene names *Cetiosauriscus* ("whale-lizard-like"), a Jurassic sauropod. Von Huene named more dinosaurs (71, of which 18 remain valid) in the early 20th century than anyone else in Europe.

1927 In *Glen Rose and Geo. P. Snyder: A Texas Town,* Elna Martin describes "Dinosaur," the first visitor to Glen Rose. By the end of the 1930s, local newspapers began to use the area's dinosaur tracks to lure tourists with headlines such as "Glen Rose Claims Original Tourists of 10,000,000 B.C—Noted Resort Town Says Even Dinosaurs Chose It as Vacation Spot."

1927 In Utah, Barnum Brown finds *Sauropelta,* which today is featured prominently in AMNH's Hall of Ornithischian Dinosaurs. Like many of his discoveries, *Sauropelta* was never formally described by Brown. While collecting in Montana's Cloverly Formation on the same expedition, Brown also collected the turkey-sized "pygmy dinosaur" (named *Microvenator* by John Ostrom in 1969) and, in 1931, "Daptosaurus," a 2.5-meter-long theropod. Brown wrote a paper about "Daptosaurus," but did not publish descriptions of any of these specimens. In 1969, John Ostrom named and described the 11-foot-long, 160-pound theropod that Brown had named "Daptosaurus" as *Deinonychus* ("terrible claw"), noting that its wrists had crescent-shaped bones (just as in birds' wings, which enable them to swivel their wings). *Deinonychus* remains one of the most important and stunning dinosaurs discovered in the 20th century, for it rekindled the bird–dinosaur connection suggested earlier by Thomas Huxley and others.

1927 Messmore and Damon of New York begin work on "The World a Million Years Ago," an exhibit that will debut at the Chicago World's Fair in 1933. The exhibit featured animated dinosaurs and other prehistoric animals, as well as a booklet showing *T. rex* threatening the Statue of Liberty. This free, walk-through exhibit of life-size dinosaur models (some animated with internal motors) was a short walk from the Field Museum. The fair's Sinclair exhibit was seen by 16 million people; it also featured a giant red dome that housed "The World a Million Years Ago," in which visitors on a moving concourse met dinosaurs (e.g. *Triceratops, Apatosaurus*), prehistoric mammals, and early humans. Various versions of this exhibit, produced by the Messmore and Damon Company of New York, toured the United States, Japan, and Paris until 1972. In a famous exhibit and parade in New York City, *Brontosaurus* was adorned with a Nazi swastika to represent the war-monster Hitler. Only two years earlier, *The Lost World* had featured a rampaging sauropod destroying London.

1927 Occasional excavations begin at what later becomes known as the Cleveland-Lloyd Dinosaur Quarry about 30 miles south of Price, Utah. The following year, more excavations at the site were conducted by teams from the University of Utah; this work, led by geologist Golden York, continued for three years. York removed more than 500 bones, and the famous quarry later yielded more than 10,000 bones, including more than 45 specimens of *Allosaurus*, *Camarasaurus,* and *Stegosaurus*. Fossils from the Cleveland-Lloyd Quarry (especially *Allosaurus*, the most common predator of the Late Cretaceous) are displayed in more than 65 museums throughout the world. In 1967, the Cleveland-Lloyd Quarry was designated a National Natural Landmark. The quarry was renovated and expanded in 2006, and today it is managed by the U.S. Bureau of Land Management.

1927 The first remains of what will be named *Carcharodontosaurus* are found in Egypt. These remains, which were originally named *Megalosaurus,* were reclassified in 1931 as *Carcharodontosaurus* by Ernst Stromer, but in 1944 were destroyed during World War II. "New" fossils of *Carcharodontosaurus* were found in 1995 in southeastern Morocco by Paul Sereno. *Carcharodontosaurus*, which Sereno claimed in *National Geographic* was "perhaps the largest carnivore that ever walked the earth," was believed to have the longest skull of any theropod until the discovery of *Giganotosaurus*.

1927 The *Journal of Paleontology* begins publication. This is the oldest paleontological periodical in North America.

1927 William D. Matthew establishes the Department of Paleontology at the University of California, Berkeley. At the time, this was the only separate university department in the world devoted to paleontology. In the 1970s, the University of California, Berkeley—with Stockton State College—became the first college to offer courses about dinosaurs.

1927 Yang Zhongjian (1897–1979), known in the West as Chung Chien "C.C." Young, becomes China's first professional vertebrate paleontologist.

1928 Charles H. Sternberg is granted permission to explore Baja, California. This was his last field-expedition.

1928 Five-year-old Jack McIntosh visits the Carnegie Museum and falls in love with dinosaurs. McIntosh later became a physicist and expert about sauropods.

1928 Flamboyant Hungarian aristocrat and paleontologist Franz Nopcsa von Felsó-Szilvás (1877–1933), who in 1924 led a British expedition searching for dinosaurs in Tendaguru, names *Scolosaurus cutleri. S. cutleri* was an armored dinosaur found in 1914 by homesteader and freelance paleontologist William Cutler (1878–1925), who sold the specimen to BMNH the next year. Nopcsa's "Island Rule"—that is, limited resources on most islands reduce the size of animals over generations, thereby producing a localized form of dwarfism—later became widely accepted. In addition to paleontological discoveries, Nopcsa's colorful life included espionage and a plot to become the ruler of Albania, which he planned to finance "by marrying a wealthy American heiress aspiring to royalty, a step which under other circumstances I would have been loath to take." In 1933, Nopcsa—a founder of Albanian studies—wrote a note admitting "a complete breakdown of my nervous system." He then murdered his secretary and lover, after which he killed himself with a gunshot to his head.

1928 Following a request from Mexico's president Emilio Portes Gil, negotiations begin with Mexico for the final cast of CMNH's *Diplodocus carnegii*. Two years later, a replica of "Dippy" was unveiled at the National Museum of Mexico.

1928 Friedrich von Huene and his coworkers discover the first major assemblage of Late Triassic tetrapods from South America. Von Huene's success prompted several Harvard scientists to explore in 1936 the Santa Maria region, where they found *Staurikosaurus pricei,* one of the earliest known dinosaurs.

1928 Harry Sinclair, whose oil company funded several dinosaur excavations in the American West, appears on the cover of the April 9, 1928, issue of *Time* magazine. The article accompanying the cover described Sinclair's upcoming trial for criminal conspiracy to defraud the United States as part of the larger "Teapot Dome" scandal. (The previous year, the U.S. Supreme Court had ruled that the Sinclair Oil lease had been obtained illegally.) Sinclair was acquitted, but convicted of obstructing justice (he had hired private detectives to shadow jurors). When Sinclair was found in contempt of the U.S. Senate after refusing to answer questions about his campaign contributions, he was imprisoned for six months. Sinclair worked as the prison pharmacist (he had a degree from the University of Kansas School of Pharmacy). During legal proceedings associated with the "Teapot Dome" scandal, Secretary of the Interior Albert B. Fall was convicted of bribery, fined $100,000, and sentenced to a year in prison; this made Fall the first member of a Presidential Cabinet ever to go to prison for his actions while in office.

1928 J. B. White finds the first fossil remains of dinosaurs in Nebraska. The next dinosaur would not be found in Nebraska until 1981 by eight-year-old Aaron Thornburg. Despite the relative scarcity of dinosaurs in the state, the University of Nebraska's State Museum has an excellent display of dinosaurs, including *Allosaurus* and *Chasmosaurus*.

1928 NMNH's Charles Gilmore finds eggshells while excavating dinosaurs in Montana, but does not publish the find. A few years later, Gilmore found seven duckbills in one pit, but did not realize that they were juveniles in a nest. While at NMNH, Gilmore led or participated in 16 major fossil-collecting expeditions.

1928 Officials at the University of Utah hear reports from local cowboys and sheepherders of fossils south of Price, Utah. After investigating, the university began excavating fossils at what became known as the Cleveland-Lloyd Dinosaur Quarry.

1928 Pathologist Harry Marshall of the University of Virginia claims that dinosaurs became extinct because of reduced levels of ultraviolet light resulting from clouds of volcanic dust in the atmosphere. The resulting lack of vitamin D killed the dinosaurs by causing rickets. Marshall also suggested dinosaurs' extinction may have been hastened by migrations, cooling climates, and dinosaurs' huge sizes, sluggish habits, and stupidity. Subsequent explanations for dinosaurs' extinction included starvation, overeating, poisoning, blindness, constipation, meteorites, mass suicide, parasitic infections, shrinking brains, and slipped discs.

1929 Arthur Holmes proposes that convective currents in a fluid layer beneath Earth's crust (produced by heat released from radioactive decay) move the continents. This was the first plausible mechanism for how continents could move.

1929 Barnum Brown meets with officials from the U.S. Congress and the National Park Service to discuss plans for a museum at DNM. Today, Vernal, Utah, is one of several "dinosaur towns" of the United States; its seven roadside dinosaurs greet thousands of visitors on their way to DNM each year.

1929 British paleontologists, working in Kindope (just north of the Tendaguru), Tanzania, discover the plated stegosaur *Kentrosaurus*.

1929 Carl Wiman names *Helopus,* a 50-foot-long, 15-ton sauropod that lived in what is now the Shandon Province of China during the Early Cretaceous. In 1956, this name was changed to *Euhelopus* after it was discovered that *Helopus* was already used to name to a bird. *Euhelopus* was the first dinosaur to be named from China.

1929 Estonian paleontologist Alexander Audova rejects racial senility as the cause of dinosaur extinction and instead suggests that dinosaurs disappeared because of environmental changes.

1929 Excavations continue at the Cleveland-Lloyd Dinosaur Quarry in central Utah, and the first major discoveries begin a decade later with the arrival of Princeton student (and future state-geologist) William Lee Stokes. By 2005, the quarry had yielded a dinosaur egg and more than 12,000 bones from at least 70 individuals (more than 40 of which were the predator *Allosaurus fragilis*) and a dozen kinds of dinosaurs. The Cleveland-Lloyd Dinosaur Quarry has produced specimens (or casts) of *Allosaurus* for more than 65 museums. Four skeletons from the Cleveland-Lloyd Quarry—*Stegosaurus, Camarasaurus, Allosaurus,* and *Camptosaurus*—are displayed prominently at the College of Eastern Utah Prehistoric Museum in Price, Utah. Many specimens at this museum, which developed from a classroom in which local experts taught geology to local residents, are the actual fossilized bones of the dinosaurs (e.g., 70% of the *Allosaurus* is petrified bone) from the Cleveland-Lloyd Dinosaur Quarry.

1929 Following the discovery of "Peking Man" (*Homo erectus*) at Zhoukoudian, C. C. Young establishes the Cenozoic Research Laboratory as part of the Geological Survey of China. By 1960, this lab—the first federal agency for paleontological research and fieldwork in China—had become the Institute of Vertebrate Paleontology and Paleoanthropology (IVPP), which is one of the most famous dinosaur facilities in the world. During his career, Young named 28 dinosaurs, 11 (i.e., 39%) of which remain valid. Today, IVPP has more than 150 staff and 200,000 specimens.

1929 Friedrich von Huene provides the first reconstruction of a titanosaur. That same year, Huene depicted two fighting titanosaurs, with one rearing on its hind legs.

1929 Henry Osborn, who wove his religious beliefs into his views of evolution, uses *T. rex* to argue that an unknown force, and not natural selection, produces Lamarckian "orthogenetic" evolution.

1929 John Taine's (a pseudonym for Eric Temple Bell) "The Greatest Adventure" describes Eric Lane's expedition to Antarctica, where he finds oil and dinosaurs living in a "lost world." The creatures were "bad copies" of dinosaurs created when "nature must have been drunk, drugged, or asleep."

1929 Kodascope Libraries produces an abridged print of *The Lost World* movie that is more stable than the original. This hour-long version, which deleted 45 minutes from the original film, is what is usually seen today.

1929 Richard Lull mounts the horned dinosaur *Centrosaurus* in YPM.

1929 Teilhard de Chardin and C. C. Young report the first dinosaur tracks in China (from China's Shanxi Province).

1929 Werner Janensch describes *Dicraeosaurus* ("forked reptile") from the Late Jurassic of Tendaguru.

1929 Working near Cameron, Arizona, Barnum Brown discovers "300 dinosaur tracks . . . representing four different species of Triassic dinosaurs [which will] aid us greatly in determining posture and foot structure of early dinosaurs." Later, in southern Utah, Brown found specimens of *Apatosaurus* and *Diplodocus*, neither of which was ever excavated.

1930 A cast of *Diplodocus carnegii* from CMNH is unveiled at Mexico City's National Museum of Natural History.

1930 Charles M. Sternberg reports dinosaur tracks in the Peace River Canyon in British Columbia. In the 1970s, Philip Currie excavated 100 (and documented another 1,600) of the prints before the canyon was dammed and flooded by the power company B.C. Hydro.

1930 Dinosaur bones are discovered in the Xonbuly District of Laos. These bones are now displayed in the Dinosaur Exhibition Hall in Khanthabouly.

1930 Ed Colbert becomes Henry Osborn's assistant at AMNH. In Colbert's autobiography, which was published five decades later, Colbert tactfully described his five years with Osborn as "a time filled with experiences and incidents. . . . Professor Osborn was quite aware of his eminence. . . . He would often be seen striding along with two or three people in his wake, like a majestic ocean liner accompanied by several gulls."

1930 Extensive dinosaur tracks of "the Dinosaur Freeway" are uncovered in 100-million-year-old sediments along the east side of what later became known as Dinosaur Ridge just outside of Denver, Colorado. Beginning in 1877, this area—which includes Arthur Lakes's famous Quarry 5—yielded fossils of iconic dinosaurs such as *Diplodocus, Stegosaurus,* and *Apatosaurus*. A road-cut further north along Interstate 70 exposes the Morrison Formation. The nearby Morrison Natural History Museum includes other dinosaur fossils, including a skull of *T. rex*.

1930 In his novel *Tarzan at the Earth's Core*, Edgar Rice Burroughs describes how Tarzan, "the Lord of the Jungle," goes into the Earth to save David Innes and others from Pellucidar. The prehistoric creatures that Tarzan encountered included a *Stegosaurus* that could lower its dorsal plates and glide through the air. Carnivores ate all of the herbivores before eating each other to extinction. Virtually all superheroes of newspapers' comic pages (e.g., Superman, Captain Marvel, and Spider Man) would encounter—and usually overcome—dinosaurs.

1930 Roy Chapman Andrews leads his fifth (and final) Central Asiatic Expedition. These expeditions were administered from a house near what is now Tiananmen Square in Beijing. Andrews wanted to return, but uncooperative governmental officials, bandits, and disrespectful soldiers made such expeditions too dangerous. Andrews's discoveries revolutionized paleontology. The total price for the various Central Asiatic Expeditions—the last (1930) of which was joined by Jesuit priest and paleontologist Pierre Teilhard de Chardin—exceeded $1,000,000, a gigantic sum of money at the time. Although Henry Osborn had hoped to find a humanlike "missing link," no human fossils were found during any of the expeditions. Mongolia's conversion to communism resulted in AMNH not returning to Mongolia until 1990, when Mongolia shifted to democracy.

1930 Sinclair Oil and Refining Company begins its first dinosaur-based advertising campaign in 104 newspapers and five national magazines. These ads showcased dinosaurs such as *Tyrannosaurus, Brontosaurus,* and *Triceratops*, and promoted Sinclair's Wellsville oils. The ad campaign was effective, and one dinosaur was especially popular: "Dino" the *Brontosaurus* (*Apatosaurus*). "Dino" was registered as Sinclair's company trademark two years later. As Harry Sinclair noted, "The public equated him ["Dino"] with power, endurance and stamina, the prime qualities of Sinclair products." Today, "Dino" continues to dominate Sinclair's corporate logo, one of the most recognized corporate logos in the world. Sinclair's marketers knew that their crude oils had formed hundreds of millions of years ago, and dubiously claimed in "the Sinclair Law of Lubrication" that the oils were "mellowing and filtering in the earth even before dinosaurs roamed America."

1931 After six years of work, YPM displays the first mounted skeleton of *Brontosaurus*. Although the name *Brontosaurus* was replaced by *Apatosaurus,* the name *Brontosaurus* remains well-known.

1931 AMNH's Charles J. Long begins mounting *Stegosaurus*. Barnum Brown described this reconstruction the following year in a four-page article titled "A Spine-Armored Saurian of the Past" in *Natural History*. Although adult *Stegosaurus* weighed more than four tons, its brain weighed less than three ounces.

1931 Barnum Brown finds a curious fossil along Beauvais Creek west of Little Bighorn Battlefield (near Bridges, Montana) that is "encased in lime [and] difficult to prepare." The hollow-boned fossil had wrists with crescent-shaped bones (similar to those of a bird) that allowed its grasping hand to swivel. Brown wrote a paper describing and naming the fossil "Daptosaurus," but never published the paper. Nevertheless, Brown's discovery helped trigger the "dinosaur renaissance" when Brown—just before his death—suggested that student John Ostrom study "Daptosaurus." Ostrom returned to the Cloverly Formation in south-central Montana and found additional specimens of the carnivore, which he named *Deinonychus*. With these more-complete skeletons, Ostrom documented that birds evolved from—and are the only extant lineage of—dinosaurs.

1931 Henry Osborn writes *Cope: Master Naturalist*, which was financed largely by Cope's surviving family-members. *Cope: Master Naturalist,* which was meant to rescue Cope's reputation following his feud with Marsh, prompted Charles Schuchert (a Yale paleontologist and ally of Marsh) to publish *O.C. Marsh: Pioneer in Paleontology* (1940). Osborn admitted that Cope's "hurried publication led him to many errors of interpretation and nomenclature," whereas "the brevity and clarity of Marsh's descriptions and their uniform appearance in the *American Journal of Science* led to the immediate reception of his works abroad." Osborn also acknowledged, "Men of science are often carried away by personal ambition and by a fierce spirit of competition with their rivals, as well as by a sense of scientific power, to employ methods which are wholly unworthy of the true spirit of American scientific research."

1931 Henry Osborn notes that fossils of toothed birds were "the most important single paleontological discovery" of Othniel Marsh's life.

1931 Henry Osborn (Figure 15) claims, "The height attained by the leaping [*Laelaps* restored by Edward Drinker Cope and Charles Knight] is now considered too great."

1931 Sauropod tracks are discovered eight miles east and north of Kenton, Oklahoma, along with the remains of a 70-foot-long *Apatosaurus*. These sites were later excavated by geology professor J. W. Stovall of the University of Oklahoma.

1931 *The Mystery of Life,* a documentary narrated by John Scopes's defender Clarence Darrow, includes clips from *The Ghost of Slumber Mountain*.

1931 The never-completed movie *Creation* describes a submarine crew shipwrecked on a volcanic island inhabited by dinosaurs. In the movie, an adult *Triceratops* chases crewmembers after they shoot her baby. The models for *Creation,* some of which appeared in *King Kong* (1933), were sculpted by Marcel Delgado, who created the models used in *King Kong, The Lost World, The Beast of Hollow Mountain,* and other movies.

1931 The U.S. National Park Service asks Barnum Brown for advice about creating a public display of dinosaur bones at DNM. While visiting the monument, Brown learned of giant dinosaur bones on the ranch of Barker Howe at the base of the Bighorn Mountains in Wyoming. Thanks to financial help from the Sinclair Oil Corporation, Brown began excavations at Howe Quarry in 1934. There, Brown saw "a veritable herd of dinosaurs, their skeletal remains crossed, crisscrossed and interlocked in a confused and almost inextricable manner. . . . Never have I seen such a thirsty lot of dinosaurs." Brown eventually collected more than 4,000 bones and bone fragments from more than 20 specimens such as *Camptosaurus, Diplodocus, Camarasaurus,* and *Barosaurus*. He believed the dinosaurs had died of thirst as they "became more and more concentrated in the remaining pools as they were pushed together in huge herds. As the water receded, weaker dinosaurs were trampled and their bones scattered on the borders of the pool, the larger ones huddled closer and closer together as they made their last futile stand against fate." This scene depicting thirsty, dying dinosaurs became famous when it was incorporated into the Walt Disney movie *Fantasia* (1940).

1932 American cartoonist and amateur paleontologist Vincent Trout Hamlin (1900–1993) creates the syndicated comic strip *Alley Oop,* in which the strip's title character (Alley Oop, a time-travelling caveman) often rode Dinny, his faithful sauropod-like pet. Oop and Dinny were the model for many roadside dinosaurs built through the United States. Hamlin's popular and influential cartoon strip—which was commemorated in 1995 on 32¢ U.S. postage stamps—was a staple of the "funny pages" in hundreds of newspapers for decades. *Alley Oop* established prehistory as a theme in comic books and was published for more than 50 years; it pioneered the many caveman comics that would come later.

1932 Barnum Brown invents the Marsh Pick, a short-handled pick "especially suited for dinosaur hunters."

1932 Barnum Brown visits Wyoming while taking a break from excavations in Montana. While near Greybull, residents show Brown some bones. Two years later, Brown returned to the area and began excavating what became Howe Quarry, one of the most valuable deposits of dinosaurs in the American West.

1932 Charles H. Sternberg describes fossil collecting in Canada with a five-chapter addendum to the new edition of *Hunting Dinosaurs in the Badlands of the Red Deer River*. Sternberg reflected on his life's work: "No matter what the common herd may say about me, I have done my humble part toward building up the great science of paleontology. I shall perish, but my fossils will last as long as the museums that have secured them. . . . My own body will crumble in dust, my soul return to the God who gave it, but the works of His hands, those animals of other days, will give joy and pleasure to generations yet unborn."

1932 Charles M. Sternberg describes the theropod *Macrophalangia* ("large toes"), which he had collected from Red Deer River in Alberta, Canada. That same year, Sternberg used fragments of a small carnivore from Alberta to name *Stenonychosaurus inequalis*. In 1969, and again in 1982, Dale Russell used specimens collected in DPP to provide more-complete descriptions of *Stenonychosaurus.* After producing a model of the dinosaur, Russell and his colleague R. Séguin wondered how descendants of *Stenonychosaurus* might have looked if dinosaurs had not become extinct. Their answer was a bipedal, humanlike creature called a "dinosauroid."

1932 Friedrich von Huene finds *Magyarosaurus,* a dwarf (6-meter long) Romanian sauropod from the Late Cretaceous.

1932 Henry Osborn (Figure 15) retires after 25 years as president of AMNH. Under Osborn's leadership, AMNH became a world leader in paleontology.

1932 In *The New Conquest of Central Asia,* Roy Chapman Andrews chronicles the Central Asiatic Expeditions and describes his view of fieldwork: "I do not believe in hardships, if they can be avoided, for they lessen effectiveness; they are a great nuisance. . . . Neither do I believe in adventures. Most of them can be eliminated by foresight and organization."

1932 Sinclair Oil Company, which had used dinosaurs in their advertisements since 1930 (to relate the age of oil with the ancient creatures), registers "Dino"—an *Apatosaurus*—as its trademark. Sinclair used Dino "to impress on your mind the tremendous age of the crude oils from which Sinclair Motor Oils are made. By and large, the oldest crudes make the finest lubricants." Sinclair Oil's advertisements in the *Literary Digest* noted that "Down where the heat of earth's internal fires stands at 100 degrees—a mile and a quarter below the surface of Oklahoma—lies a treasure trove, the Cambro-Ordovician oil pool. Millions of years have passed since Nature formed that pool—ages which saw the rise and fall of the dinosaurs—ages which played a priceless part in the mellowing and filtering of this remarkable crude." Between 1934 and 1938, Sinclair funded many dinosaur-hunting expeditions, including several led by Barnum Brown.

1932 Sinclair Oil hires P. G. Alen to produce a giant exhibit for the Century of Progress Exhibition in Chicago. Alen, who had helped create papier-mâché dinosaurs for several movies, based his models on the artwork of Charles Knight. The success of the Sinclair exhibit produced the first corporate-sponsored geological educational materials for schools and libraries.

1932 William Holland dies at age 84 from a stroke, 10 years after retiring as director of CMNH. Today, Holland Hall—a residence hall at the University of Pittsburgh (where Holland served as chancellor from 1891 to 1901)—is named in his honor. Holland had long opposed mounting a camarasaurid skull atop *Apatosaurus*. Nevertheless, in the same month that Holland died and just prior to a visit by noted paleontologist C. W. Gilmore to CMNH to study *Apatosaurus*, a mold of a *Camarasaurus* skull was mounted atop CMNH's skeleton of *Apatosaurus louisae* (CM 3018).

1932 The Evolution Protest Movement is established in London to counter the growing acceptance of evolution. That organization, now called the Creation Science Movement, is the oldest antievolution organization in the world. The Creation Science Movement sponsors the Genesis Expo ("the biggest creation museum in the United Kingdom"), which claims that humans lived with dinosaurs.

1933 A well-preserved skeleton of the theropod *Albertosaurus* collected in 1913 on the Red Deer River in Alberta, Canada, is traded to the Smithsonian for *Barosaurus*. This *Albertosaurus* is now displayed in the Smithsonian's dinosaur halls.

1933 Alfred S. Romer suggests that cranial crests of *Parasaurolophus* functioned as snorkels or air tanks. In 1975, James Hopson argued that the crests were visual signals, and others later suggested that the crests were used for thermoregulation and as resonating chambers for communicating. Paleontologist David Weishampel determined that sounds made by *Parasaurolophus* ranged from G two octaves below middle-C to B below middle-C.

1933 AMNH's Barnum Brown collects the only known specimen of the small *Microvenator celer* (AMNH 3041) from the Early Cretaceous Cloverly Formation in southeastern Montana. Brown found several large, serrated teeth beside the skeleton, and named the specimen *Megadontosaurus* ("large teeth"). However, this name never became official because Brown did not publish a formal description of the fossil. In the 1960s, Yale's John Ostrom's work in the Cloverly Formation discovered that the large teeth found by Brown resembled those of *Deinonychus*. Ostrom described Brown's discovery as *Microvenator*.

1933 Barnum Brown hires Roland T. Bird (1899–1978) to conduct excavations and field explorations for AMNH. Despite having no college degree or formal training in paleontology, Bird became a famous collector of dinosaur tracks.

1933 Carlton Nash of South Hadley, Massachusetts, discovers a dinosaur footprint and later buys the 1.75-acre parcel containing the print. In 1939, Nash opened Nash Dino Land, an attraction visited by Barnum Brown, Roland Bird, Roy Chapman Andrews, and other famous dinosaur-hunters. Today, the site is known as Nash Dinosaur Track Site and Rock Shop.

1933 George Messmore is awarded a patent for the head- and neck-movement device (consisting of seven motors) he invented for his and Joseph Damon's robotic life-size *Apatosaurus,* the first moving dinosaur ever built. Messmore and Damon referred to their *Apatosaurus* as an "amphibious Dinosaurus Brontosaurus." Messmore's patent was noted in *Popular Science Monthly* with the headline "Hidden Motors Give Life to Prehistoric Monsters." Advertisements proclaimed that the "mechanical monsters swing their heads, roll their eyes, breathe, snarl, roar, and grunt in realistic fashion. A complicated mass of cogs, wheels, bellows, and silent motors produces the life-like sounds and motions."

1933 George Sternberg is named curator of museums at Fort Hays State University (Kansas). Sternberg held this job for 22 years.

1933 In the comic strip *Alley Oop,* Oop finds proof that dinosaurs lived with humans.

1933 In *The Land of Terror,* Doc Savage and other explorers go to Thunder Island (an uncharted island in the Pacific), where they encounter dinosaurs while searching for a magic solvent in a volcano.

1933 Metal brontosaur toys are sold at the Chicago World's Fair; these are some of the earliest dinosaur toys every produced. The following year, similar toys were sold at the Messmore and Damon "The World a Million Years Ago" pavilion and to promote the popular radio show *Og, Son of Fire*.

1933 More than 80 years after Victorian anatomist Richard Owen and Benjamin Waterhouse Hawkins produced their dinosaurs at Sydenham, dinosaurs come alive at the Chicago World's Fair. Sinclair Oil's seven life-size dinosaurs and a giant dinosaur-based diorama were the main attractions. At the exhibit, a moving "Magic Skyway" took visitors back through time, where they encountered a 50-foot-long *Brontosaurus* (having a 16-foot-long neck) and a faceoff between *T. rex* and *Triceratops*;[26] *T. rex* and *Triceratops* lunged at each other "by means of eleven automatically controlled electric motors." The exhibit by Sinclair—one of the first corporations to market their products with dinosaurs—was the first outdoor re-creation of a portion of prehistoric Earth. The exhibit's two-ton *Brontosaurus* was the largest animated animal in the world, as well as the most popular animal in Sinclair's exhibit. More than 15 million people toured the Sinclair exhibit to see *Brontosaurus, Stegosaurus, Tyrannosaurus, Triceratops, Protoceratops,* and *Trachodon,* which were based on paintings by Charles Knight. The exhibit's popularity—the first outdoor display to interpret dinosaurs in their native environment—prompted Sinclair to begin producing small rubber dinosaurs to promote its products and gas stations, as well as Sinclair's production of the first co-sponsored geology curricular materials for schools and libraries. The production of these materials was supervised by Barnum Brown of the AMNH.

1933 Owners of the now-gone Creston General Store in Creston, South Dakota, use wood, concrete, and scrap metal to build the first roadside dinosaur to attract business (Figure 25). In 1998, the dinosaur, which is 60 feet long and 20 feet high, was restored by students from the South Dakota School of Mines and Technology.

1933 Residents of Somervell County (which includes Glen Rose, Texas) excavate a theropod track at the Hendrix-Ramfield Crossing of the Paluxy River. Today, that track is exhibited on the Glen Rose courthouse square, where it attracts curious tourists. Dinosaur tracks have been found in more than 50 locations near Glen Rose, with the most famous ones being from the bed of the Paluxy River.

1933 Richard Swann Lull sketches the evolution of horned dinosaurs. This was the first diagrammatic representation of dinosaurian evolution. That same year, Lull likened the existence and disappearance of dinosaurs to the growth, expansion, and extinction of great civilizations.

1933 RKO's classic film *King Kong* features Kong fighting *Tyrannosaurus rex* on Skull Island, the most famous "lost world" island. The movie (which was originally

Figure 25 This brontosaur, erected in 1933 to lure customers to the now-gone General Store in Creston, South Dakota, was the first roadside-dinosaur. Today, this dinosaur—renovated in 1998—stands alone in a field a few miles southeast of Rapid City, South Dakota. (Courtesy Randy Moore)

titled *The Giant Terror Gorilla*) premiered with a live show including "native" dancers at Grauman's Chinese Theatre in Los Angeles, and featured *Elasmosaurus, Brontosaurus,* and *Stegosaurus* (which was the first dinosaur to appear after the crew began pursuing Kong). The original plan for *King Kong* was based on Kong being played by a gorilla, and dinosaurs by decorated lizards. Sculptor Marcel Delgado modeled *T. rex* directly from the paintings of Charles Knight, and special-effects artist Willis O'Brien brought the animals to life by photographing the models one frame at a time, moving each model a fraction of an inch between shots. (O'Brien allegedly referred to the dinosaur as *Allosaurus*; the script by Delos Lovelace referred to the dinosaur as a large carnivore.) In the movie, Kong kills *T. rex* with punches and wrestling moves, and then wards off pterosaurs trying to take Ann Darrow (played by actress Fay Wray [1907–2004]). In the meantime, Wray is menaced by *Elasmosaurus* and *Tyrannosaurus*. Near the end of the movie, in one of the most famous scenes in the history of cinema, Kong climbs New York City's Empire State Building and swats biplanes before being shot and falling to his death. Although the depiction of *T. rex* in *King Kong* is no longer accurate (e.g., *T. rex* carried its body horizontal to the ground, it had only two fingers, and its tail did not drag on the ground), it remains the classic image for many people. An advertisement for *King Kong,* which opened on O'Brien's birthday, showed

Kong confronting a surprisingly aggressive *Stegosaurus,* but that scene was not included in the movie. The battle between Kong and *T. rex* remains one of the most impressive scenes ever to appear in movies. (The movie's censored scenes included Kong crushing a native, Kong biting a man to death, and Kong smelling his finger after tickling Wray.) The success of *King Kong* prompted the release of the less ambitious sequel, *Son of Kong,* the same year. In that sequel, Kong's albino son Kiko fights prehistoric monsters (some left over from the original *King Kong*) while protecting humans; the movie ends with the island sinking beneath the waves. A remake of *King Kong* in 1976 included no dinosaurs.

1933 Roy Chapman Andrews publishes his third and final article about the Central Asiatic Expeditions in *National Geographic.*

1933 Seven miles east of Medicine Bow, Wyoming, on U.S. Highway 30 ("the Lincoln Highway"), the Boylan family—Grayce, her husband Thomas, and their son Edward—complete Fossil Cabin (Figure 26), a small (29 feet by 19 feet) home made of 5,796 mortared-together dinosaur bones from nearby Como Bluff.

Figure 26 Fossil Cabin along the Lincoln Highway outside of Medicine Bow, Wyoming, is made of 5,796 dinosaur bones collected from nearby Como Bluff. The small cabin, completed in 1933, was built by Grayce and Thomas Boylan to attract customers to their gas station. In 2008, Fossil Cabin was added to the National Register of Historic Places. (Courtesy Randy Moore)

The home, which is by the Como Bluff "Dinosaur Graveyard" historical marker, helped attract visitors to Boylan's nearby gas station. (One visitor, Walt Disney, helped Grayce install the floor of the cabin.) In 1938, the cabin was named "Oldest Cabin in the World" by Robert Ripley of "Ripley's Believe It or Not." Boylan originally called the building "the Como Bluff Dinosaurium" and "the World's Oldest Cabin," but after Ripley mentioned it, the cabin was described as the "Creation Museum" and "the Building That Used to Walk." In 2008, the cabin was added to the National Register of Historic Places.

1933 The first dinosaur-tracks are reported from Australia.

1933 The U.S. Postal Service issues 1¢, 3¢, and 50¢ stamps commemorating Chicago's "Century of Progress" Exhibition, at which a star attraction is the Sinclair Exhibition featuring life-size dinosaurs. Despite the popularity of the Sinclair Exhibit, none of these stamps featured dinosaurs.

1933 Three years after its discovery in Arizona, Charles Camp describes the small theropod *Segisaurus* ("reptile from Segi Canyon").

1933 When 13-year-old Ray Harryhausen sees *King Kong* at Los Angeles's famous Grauman's Chinese Theater, "this gigantic ape . . . simply overwhelmed me." Harryhausen, a protégé of Willis O'Brien, became a famous special-effects technician in several subsequent films featuring dinosaurs.

1934 Alex Raymond's comic strip *Flash Gordon* debuts in newspapers. The popular strip often featured battles with dinosaur-like creatures, and in 1935 was expanded into a radio series. In *Flash Gordon* movies, decorated iguanas substituted for dinosaurs.

1934 Barnum Brown sends 145 crates of fossils from Montana to AMNH. That same year, Brown began a two-month, 20,000-mile aerial survey of the southwestern United States aboard "Diplodocus," a four-passenger airplane. The project, funded by Sinclair Oil and Refining Company, was meant to explore Mesozoic outcrops from Alberta to Mexico, and was the first such aerial survey ever conducted. That same year, Brown—who had recently been sued by a spurned lover—began supervising production of *The Sinclair Dinosaur Book*. This popular pamphlet included artwork by James E. Allen that was based on dinosaurs displayed at the "A Century of Progress" World's Fair, which had opened the previous year in Chicago. Allen's paintings of dinosaurs, which were inspired by the work of Charles Knight,[27] later appeared on many stamps and booklets produced by Sinclair, as well as in magazines, books, and other products.

1934 CMNH makes another cast of "Dippy" the *Diplodocus carnegii* and sends it to BSP in Munich, Germany, in exchange for a giant collection (mostly invertebrates) of European fossils. The replica was not assembled, and records were lost during World War II. In 1977, the 36 crates containing the original replica were discovered in the museum's storerooms, but the cast was never assembled.

1934 Dhirendra Kishore Chakravarthy becomes the first Indian to name a dinosaur when he names the stegosaur *Brachypodosaurus* ("short-footed lizard") from the Late Cretaceous of central India.

1934 In John Taine's *Before the Dawn,* scientists study Belshazzar, a theropod with human and dinosaur traits. Belshazzar later confronted his nemesis, a dinosaur named Satan.

1934 New York judge Peter Schmuck dismisses the lawsuit by Messmore and Damon, Inc., against show-producer Earl Carroll for infringing on their patent of "Dinah," a large *Apatosaurus*. Carroll, who claimed his model "Dinny" was different from "Dinah," brought his model to the courthouse, but it would not fit through the door.

1934 *Og, Son of Fire,* a radio series featuring dinosaurs and cavemen, premiers on CBS.

1934 Paul Domke (1885–1981) of Ossineke, Michigan, begins construction of what became Dinosaur Gardens Prehistoric Zoo. The last dinosaur there was finished in 1974. Visitors tour the 40-acre park along a mile-long trail that features cavemen and dinosaurs. When visitors climb inside the 60,000-pound, 85-foot-long *Brontosaurus,* they encounter the dinosaur's heart and Jesus ("The Greatest Heart That Ever Lived"). Elsewhere in the park is a 30-foot-tall Jesus, erected as a testimonial to Domke's belief "that Christ was the master planner of an Earth that included dinosaurs. . . . The seven days in which God created the world lasted several million years, a period during which dinosaurs evolved."

1934 Roy Chapman Andrews is named director of AMNH. In 1941, he was pressured to resign, thereby ending his 35-year career at AMNH.

1934 Sinclair Oil and Gas Company begins issuing dinosaur booklets and stamps at its gas stations. The first booklet, which was titled *The Sinclair Dinosaur Book,* was used in schools throughout the United States. The 12-page, full-color booklet (which was announced on the *Sinclair Minstrel Program* on radio) featured dinosaurs such as *Brontosaurus, Tyrannosaurus,* and *Triceratops*. Before the promotion ended, the company had given away almost 50 million stamps, and sales were up significantly; people had to visit gas stations 12 times to collect all of the stamps needed to fill the booklet. That same year, Sinclair Oil began a four-year sponsorship of Barnum Brown's fossil-collecting expeditions. *The Sinclair*

Dinosaur Book later became a highly collectible item; in 2014, it was selling (with a complete set of stamps) for approximately $40.

1934 When the Chicago World's Fair closes, several of the dinosaur models from the exhibit (which had been upgraded during the fair's final year) begin touring the country.

1934 While working near Greybull, Wyoming, at the base of the Bighorn Mountains, Barnum Brown discovers a vast "dinosaur graveyard" that he names Howe Quarry (after Barker Howe, the owner of the land). Brown and his crew—funded by the Sinclair Oil Company—began excavations at the Howe ranch; this was the final excavation of large sauropods by AMNH. The site drew crowds of tourists, whom Howe charged 10¢ per person to watch the excavation. (Brown, who "did not feel that it was an ethical procedure to invite the pubic to see this work and then charge admission," used money from Sinclair Oil to pay the entrance fee of each visitor.) Roland Bird produced a map of Howe Quarry for Brown. Howe Quarry yielded more than 4,000 bones of nearly 20 individuals, including *Diplodocus, Barosaurus, Camarasaurus, Allosaurus,* and *Camptosaurus*, which Brown shipped in 145 crates to AMNH. These crates, which were stored in a courtyard for decades, later triggered a lawsuit by Howe, who believed that they were "to the best information worth $25,000." Brown believed that dinosaurs at Howe Quarry had gathered seeking water, and were later killed by a drought; as Brown later told Bird, "This site was probably an old water hole left over from a much larger body of water that dried up gradually. The bones you see here were those of sauropods . . . the big fellows who walked on all fours . . . creatures whose very size made them largely dependent on an aquatic environment. In all probability they were too heavy to support their enormous weight on their legs for much of the time; they floated or wallowed around along the shoreline, feeding on plant life along the banks. The drying up of the water drove many dinosaurs to this one spot, where they died, standing on the trampled bodies of dinosaurs who had died here before them. The water hole may have been refilled from time to time, and the process repeated. Rushing waters, bringing in silt and sand, buried everything. Eventually the bones fossilized, and they have remained in this condition for the last eighty or ninety million years." In 1940, Brown's imagined scene was popularized in Walt Disney's *Fantasia,* in which the dinosaurs turned into oil. Among paleontologists, Brown—a consultant during the production of *Fantasia*—is almost universally recognized as the greatest fossil-hunter of all time; more than 30 of his discoveries remain on display at AMNH today. Dozens of crates of Brown's fossils remain unopened at the AMNH.

1934 William Elgin Swinton writes *The Dinosaurs: A Short History of a Great Group of Reptiles*, the first scientific book about dinosaurs intended for the general public. Swinton based his illustrations on two-dimensional models created by Vernon

Edwards for BMNH. These eight models, which were among the earliest examples of dinosaur-related merchandise, included *Hypsilophodon* in a tree and a large sauropod almost submerged in water. The only remaining complete set of Edwards's models is housed in the Hunterian Museum at the University of Glasgow.

1935 A Social Credit government led by radio evangelist and politician William "Bible Bill" Aberhart (1878–1943) begins eight years of governing dinosaur-rich Alberta, Canada. Despite the abundance of dinosaurs in Alberta, Aberhart provided relatively little money for dinosaur-related projects. During one of his many radio broadcasts, Aberhart claimed that paleontologists were using dinosaur bones made of plaster of Paris to confuse religious believers. Charles Sternberg responded, "How clever we were."

1935 A hadrosaur found in southern Sakhalin, and described as *Nipponosaurus sachalinensis,* is the first dinosaur to be found in Japan. Sakhalin became part of Russia after World War II. The first dinosaur fossil found within today's Japan was a part of a humerus of a sauropod found in 1978 in the Upper Cretaceous of Iwate Prefecture.

1935 Barnum Brown describes the Howe Quarry in the June issue of *Natural History*. The most abundant animals were *Barosaurus* and *Morosaurus* (now known as a juvenile of *Camarasaurus*). The ornithopod *Camptosaurus* was also present, but *Stegosaurus*—common at many Morrison Formation sites—was not.

1935 Charles Knight publishes *Before the Dawn of History*, his first book. Some of the book's illustrations were reproductions of Knight's earlier work. Other books written or illustrated by Knight included *Life through the Ages* (1946) and *Prehistoric Man: The Great Adventure* (1949). Today, Knight's work is exhibited at some of the world's greatest museums, including AMNH, Chicago's Field Museum, and the Natural History Museum of Los Angeles County. A concrete version of Knight's *Stegosaurus* looms above the entrance to the National Zoo's Reptile House in Washington, DC.

1935 Czech paleoartist Zdenêk Burian (1905–1981), who had toiled as a construction worker to pay his art-school expenses, completes his first restorations of prehistoric animals. Burian became the most prolific paleoartist of all time; during the next 46 years, he restored nearly 500 ancient animals and landscapes. In 1965, Burian's art inspired a series of Polish postage stamps. With few exceptions, Burian portrayed dinosaurs as tail-dragging and lethargic.

1935 Geologist J. W. Stovall (1891–1953), who had studied with A. S. Romer and R. S. Lull, begins excavations near Kenton, Oklahoma, that eventually produce *Apatosaurus* and the only known specimen of the theropod *Saurophaganax*.

During the next several years, Stovall and his crews also uncovered *Camptosaurus, Diplodocus, Stegosaurus,* and a baby *Camarasaurus*.

1935 Henry Osborn, who had served as president of AMNH for 25 years, dies of a heart attack in his home in Garrison-on-Hudson, New York. Edwin Colbert, one of Osborn's protégés, described Osborn as "a larger than life figure in the field of vertebrate paleontology." Soon thereafter, in the *New York Sun*, artist Charles Knight published a tribute to his recently deceased "second father." Today, a memorial to Osborn in AMNH notes that "under his leadership, the museum attained great eminence in science and education."

1935 Schoolgirl Betty Jo Riddenoure, Lawrence "Cuddle" Riddenoure, and Ralph Owen discover more than 1,300 sauropod tracks near what local residents called "Elephant Crossing" in Purgatoire Valley south of La Junta, Colorado. These tracks—one of the largest and most spectacular track sites in the world—were described briefly by John MacLary three years later in *Scientific American*; the remoteness of the tracks is why they were overlooked long after their discovery. Most of the eroding tracks—at what is now called "Picketwire Canyonlands Dinosaur Tracksite"—are preserved in limestone. The 139-million-year-old Purgatoire tracks include about 100 different trackways, 40 percent of which were made by *Apatosaurus* (and several others by *Allosaurus*). These tracks, which comprise the longest mapped footprints of dinosaurs in North America, include the first identified footprints of sauropods.

1935 Levi Sternberg begins using "quarry stakes" to mark promising areas of DPP for prospecting. Some of those stakes remain visible today.

1935 Raymond L. Ditmars's (1876–1942) and Helene Carter's *The Book of Prehistoric Animals* is one of the first books about dinosaurs for children.

1935 Several dinosaurs, including "Dinny the Dinosaur" (107 feet long, 32 feet tall, and weighing 120 tons), whose construction was supervised by Charles M. Sternberg and Barnum Brown, are displayed at the new Prehistoric Park at the Calgary Zoo. The park offered an image of Western Canada as it might have looked in the Mesozoic Era. In 1959, Dinny became the zoo's mascot, and in 1987, the zoo was made an Alberta Historical Site. The Prehistoric Park, which includes specimens of *Albertosaurus, Allosaurus, Apatosaurus, Protoceratops,* and other dinosaurs, was renovated in 1956 and 2010.

1935 Sinclair Oil embarks on its most successful promotion when it issues albums that could be filled with 24 full-color dinosaur stamps given away by Sinclair dealers "to any boy or girl who asks for them, provided he or she is accompanied by an adult." The first three-stamp set included *Tyrannosaurus, Ornitholestes,* and

Parasaurolophus. The subsequent seven sets included *Brontosaurus, Stegosaurus, Allosaurus, Triceratops,* and other dinosaurs. These albums and stamps, whose designs were supervised by AMNH's Barnum Brown, were distributed to dealers two days after a single radio-broadcast advertising the offer. In 1938, Sinclair issued 24 different stamps for their albums that included *Psittacosaurus, Iguanodon, Troodon,* and *Camarasaurus*. Before the promotion ended, Sinclair had distributed four million albums and 48 million stamps, as well as countless books, signs, oil cans, maps, and matchbooks featuring dinosaurs.

1935 The Historic Sites Act of 1935 (16 USC 461) authorizes the National Parks Service and the U.S. Secretary of the Interior to organize and protect monuments, parks, and historic sites. This act was the first explicit recognition of historic preservation as a governmental responsibility (which had been hinted at in the Antiquities Act of 1906). Today, the Historic Sites Act protects a variety of dinosaur-related sites, including Como Bluff (WY), Cleveland-Lloyd Dinosaur Quarry (UT), DVSP (TX), and Ghost Ranch (NM).

1936 AMNH's Peter Kaisen, who collected dinosaurs for Barnum Brown and led expeditions to Alaska and Mongolia, dies.

1936 Charles Camp discovers *Segisaurus* in Early Jurassic rocks in Arizona.

1936 Charles Whitney Gilmore's (1874–1945; Figure 23) monograph *Apatosaurus* includes the first accurate postcranial anatomy of this famous sauropod.

1936 French geologist Josué Heilmann Hoffet (1901–1945) reports the first dinosaurs from Laos. These bones from a sauropod and a hadrosaur were described in 1942.

1936 London's Crystal Palace is destroyed by fire. Today, Benjamin Waterhouse Hawkins's sculptures of several of Richard Owen's dinosaurs can be seen in Sydenham (just south of London, and restored in 2002) and remain the best-known works of their designers.

1936 The Carter County Museum is chartered in Ekalaka, Montana, to display fossils collected by Marshall Lambert. The displays include the duck-billed *Anatotitan,* a skull of *Triceratops,* and a small tyrannosaur called *Nanotyrannus lancensis,* all of which were found in local exposures of the Hell Creek Formation. The Carter County Museum was the first of many dinosaur museums in Montana that were linked by the Montana Dinosaur Trail in 2004. The badlands of eastern Montana have the highest concentration of exposed dinosaur fossils in the world.

1936 The first dinosaur-tracks are reported from South America.

1936 The hilltop tourist-attraction Dinosaur Park is dedicated in Rapid City, South Dakota, to "perpetuate the facts of history" and to give people "a fair idea as to the appearance, size, and characteristics of our earliest known inhabitants" (Figure 27). Land for the park was donated to the city by private owners of the land. The park, built with funds (Project #960) from WPA, included five dinosaurs built by Frank Lockhart and George McGraw, and designed by businessman and sculptor Emmet A. Sullivan (1887–1970). (Sullivan, who was inspired by Charles Knight's murals at the Field Museum, also designed the 80-foot-long *Apatosaurus* that attracts more than 1.5 million visitors per year to eclectic Wall Drug in nearby Wall, South Dakota; Figure P-1). The original dinosaurs at the Rapid City park were *Trachodon* (now *Anatotitan*; 33 feet long, 17.5 feet high), *Brontosaurus* (now *Apatosaurus;* 80 feet long, 28 feet high), *Stegosaurus* (11 feet long, 7 feet high), and *Tyrannosaurus* (35 feet long, 16 feet high) confronting *Triceratops* (27 feet long, 11 feet high). Smaller models of *Dimetrodon* (the largest known carnivore of its day) and *Protoceratops* were later displayed across the street at the park's gift shop. Most of the dinosaurs at Dinosaur Park were based on fossils found in South Dakota and the western United States, and can be traced to Charles Knight's paintings at the Field Museum. Since the dinosaurs were built in the 1930s, they reflected the ideas

Figure 27 In 1936, Dinosaur Park—the first "dinosaur theme park" in the United States to feature life-size dinosaurs—opened atop a hill in Rapid City, South Dakota, just a few miles from the first roadside dinosaur in Creston, South Dakota (Figure 25). This photo features the park's 80-foot-long, 28-foot-high *Apatosaurus*. The year after Dinosaur Park opened, famed artist Charles Knight described its dinosaurs as "awful." (Courtesy Randy Moore)

of that decade, including the fact that many of the dinosaurs were dragging their tails. In 1990, Rapid City's Dinosaur Park—the first "dinosaur theme park" in the United States to feature life-size animals—was added to the National Register of Historic Places. Similar "dinosaur parks" soon began appearing throughout the United States, and today they promote businesses ranging from miniature golf to fundamentalist religion. One of the most unusual of these parks is "Professor Cline's Dinosaur Kingdom" in Natural Bridge, Virginia, which depicts a lost valley of dinosaurs discovered in 1863 by Union soldiers. The park shows how the soldiers' attempts to use the dinosaurs to fight Confederates go horribly wrong, and Union forces are eaten and defeated by the dinosaurs.

1936 The success of Sinclair's dinosaur-based advertisements prompts the company to start featuring Dino the *Brontosaurus* in humorous ads that include humans. By this time, Dino was depicted with a smooth Hogarth curve to balance its head and tail. The following year, Sinclair—now aware that their *Brontosaurus* had become a powerful symbol—began using Dino to promote its premium gasoline, urging motorists to "put new dinosaur power in your engine!"

1937 Barnum Brown (Figure 17) uses hadrosaur tracks found in the Charles G. States Mine near Cedaredge, Colorado, to estimate the dinosaur's speed to be 30 feet per second. Roland Bird assisted Brown with the excavation. Brown described these tracks the following year in an article in *Natural History* titled "The Mystery Dinosaur." In 1981, dinosaur specialist Tony Thulborn discovered that Brown had missed a track, and that the dinosaur's speed was closer to 15 feet per second. Today, these hadrosaur tracks (AMNH 3650) are displayed outside the AMNH Hall of Ornithischian Dinosaurs.

1937 Barnum Brown writes an article promoting DNM, but the public display of dinosaur bones there will not open for another 21 years.

1937 Charles M. Sternberg names *Parksosaurus* ("Parks' lizard') in honor of William Arthur Parks, the first director of ROM. *Parksosaurus* was an 8-foot-long, small, fast, bipedal herbivore from the Late Cretaceous. Some paleontologists argue that *Parksosaurus* is *Thescelosaurus*.

1937 Famed artist Charles Knight describes the Rapid City, South Dakota, dinosaurs (Figure 27) as "atrocities," "awful," "dreadful," "miserable," and "an insult to the scientific and artistic intelligence of [South Dakota's] citizens."

1937 Five years after beginning work on its first dinosaur, Calgary's Prehistoric Park opens. The park, which originally featured the 35-foot-tall "Dinny," the *Brontosaurus*, was renovated in the 1970s and 1980s, and today includes more than 25 life-size models of dinosaurs.

1937 Geologist Alexander Du Toit (1878–1948), an early supporter of Alfred Wegener's ideas about continental drift, proposes that Pangaea broke into Laurasia (a northern supercontinent) and Gondwanaland (a southern supercontinent).

1937 After seven years of work, and just two years before the start of World War II (and one year after the Berlin Olympics), the type specimen of *Brachiosaurus brancai* (HMNSII) is unveiled at Berlin's Humboldt Museum in front of swastika banners. A paper describing the skeletal mount was not published until 1950 by Werner Janensch.

1937 Near Garden Park, Colorado, local teacher Frederick Kessler finds and excavates (with the help of the Denver Museum of Nature & Science) a nearly complete skeleton of *Stegosaurus*. This specimen was displayed the following year at the Denver Museum of Nature and Science.

1937 On the northwest side of what would soon be named Riggs Hill near Fuita, Colorado, teacher and graduate student Edward Holt discovers partial skeletons of *Allosaurus* and *Stegosaurus*. Holt hoped that it would become a research site, so he fenced it and left the bones in the ground. Vandals later destroyed the site, and by 1960, the bones were gone.

1937 While on another expedition for the Sinclair Oil and Refining Company, Barnum Brown and Roland Bird find tracks of a "Mystery Trackmaker" in the Chesterfield Mine near Sego, Utah. To attract publicity and funding, Brown sensationalized the discovery; headlines described the tracks as the "World's Largest Footprint."

1938 A stone marker and plaque commemorating Elmer Riggs's discovery of *Brachiosaurus* in 1900 are dedicated at the original site of Quarry 13: "Brachyosaurus, the giant dinosaur, first known skeleton discovered in 1900 by Elmer S. Riggs and preserved in Field Museum, Chicago" (Figure 28). The plaque misspelled the word *Brachiosaurus*. At the dedication ceremony, the area was renamed Riggs Hill. That same year in nearby Fruita, Colorado, a plaque commemorating Riggs's discovery of *Apatosaurus* was dedicated at what became known as Dinosaur Hill: "Discovered here, 1901, a giant Apatasaurus skeleton by Elmer S. Riggs, now mounted in Field Museum, Chicago." Both plaques misspelled the name of the dinosaur. Riggs attended both of the ceremonies.

1938 AMNH unveils *Tenontosaurus tilletti,* a primitive relative of the duck-billed dinosaurs from the Early Cretaceous. Although Barnum Brown had collected some *Tenontosaurus* vertebrae (AMNH 5854) during his 1903 trip to south-central Montana, the specimen displayed today at AMNH (AMNH 3034) was collected by Brown and Peter Kaisen in 1932.

Figure 28 This plaque on Riggs Hill near Grand Junction, Colorado, marks the site where, in 1900, Elmer Riggs (Figure 16) discovered *Brachiosaurus altithorax*. *Brachiosaurus* is misspelled on the plaque. In nearby Fruita, Colorado, another plaque designates where Riggs found *Apatosaurus*. That plaque, like the one for *Brachiosaurus*, misspells the dinosaur's name. You can see casts of Riggs's *Brachiosaurus* outside of Chicago's Field Museum and in Concourse B of Chicago's O'Hare International Airport. (Courtesy Randy Moore)

1938 CMNH mounts the *Allosaurus fragilis* collected by Earl Douglass in 1913–1915. In 1976, James Madsen's monograph *Allosaurus fragilis: A Revised Osteology* became the definitive work on the genus.

1938 DNM expands to its present size of 330 square miles, including the canyons formed by the Yampa and Green Rivers.

1938 In Canada, Prime Minister R. B. Bennett (1870–1947) unveils a giant *Brontosaurus* that becomes popular as "Dinny."

1938 In his *Natural History* article titled "The Mystery Dinosaur," Barnum Brown claims that he—in 1897—was the first to use jackets of plaster to collect vertebrate fossils. However, others had used similar techniques previously. For example, while working in Montana in 1876, Edward Drinker Cope and Charles H. Sternberg removed fossils by covering them with strips of burlap and cloth soaked in rice paste (this paste later hardened into a protective coating). Similarly, S. W. Williston in 1877 suggested protecting fossil bones with paper soaked in flour paste, and, that same year, Arthur Lakes told Othniel Marsh that he protected bones with coatings of plaster of Paris.

1938 In the classic comedy film *Bringing Up Baby*, paleontologist David Huxley (alias "Dr. Bone," played by Cary Grant) of the Stuyvesant Museum of Natural History seeks an "intercostal clavicle" (a bone that does not exist) to complete a skeleton of *Brontosaurus*. Throughout the movie, Grant is intrigued and annoyed by young Katherine Hepburn, as well as by a dog who has stolen the bone required to complete his museum's mounted skeleton of *Brontosaurus*. The movie's dinosaur skeletons later appeared in a variety of other features, including *The Beast from 20,000 Fathoms*.

1938 Sinclair Oil and Refining Company's advertisements include the same text as their first dinosaur-based ads from the 1930s, but feature only *Brontosaurus*. Sinclair's dinosaur-themed advertisements again increased sales and stimulated the public's interest in dinosaurs.

1938 When the Denver Museum of Natural History acquires a large *Diplodocus* skeleton, attendance increases from 300,000 to 800,000 visitors per year.

1938 While herding cattle on the family ranch near Ekalaka, Montana, William Winkley (1900–1976) finds the first fossil of *Pachycephalosaurus*.

1938 While in Gallup, New Mexico, Roland Bird (1899–1978) hears of "strange tracks" at a trading post; these tracks were dinosaur tracks carved by George Adams and Jim Ryals that were 15 inches long. (Both Adams and Ryals probably had economic rather than antievolutionary motives for their carvings, which were made during the Great Depression when jobs were scarce.) Adams and Ryals lived near Glen Rose, Texas; when Bird realized Glen Rose's relation to the Lower Cretaceous, Bird went to Glen Rose and (with the help of Earnest Tolbert "Bull" Adams[28]) found similar tracks within an hour after arriving. When Bird examined the now-famous tracks in the Paluxy River, he was "suddenly filled with a wild thrill that I hardly dared to accept. . . . A full twelve feet away was another . . . gigantic footprint. . . . There come rare moments in the lives of all of us when we see things we do not actually believe." Bird noted, "I've been around dinosaur remains long enough to know I have made a find." He was right; as he later wrote in his book *Bones for Barnum Brown: Adventures of a Dinosaur Hunter* (1985), "As for the great plantigrade sauropod tracks, the Paluxy tracks were the first known in the world." The SVP named Bird's discoveries—funded in part by Sinclair Oil, AMNH, the Texas Statewide Paleontological Survey, and the University of Texas—its "Discovery of the Year." Bird's interpretation of the 23 trackways—all going in the same direction—at Davenport Ranch (along West Verde Creek on the ranch of Mrs. Cleora Davenport, about 13 miles south of the town of Bandera) made him the first person to suggest that sauropods lived in herds. Bird also concluded that the dinosaurs walked on land, but this conclusion was largely overlooked for many years. Bird seldom found tail drags with dinosaur tracks, prompting him to conclude "most of my dinosaurs run high-tailed. . . . Perhaps scientists of future ages may be able to settle" the question of why dinosaur trackways were not associated with tail-drags. At Bandera and Glen Rose, Bird and his crew of 10–12 WPA workers excavated 1,200 pieces comprising eight slabs containing 49 prints and weighing 40 tons. (Volunteers who helped Bird's crew, none of whom Bird described as a "shovel-leaner," were paid $1.25 per day for their help.) At Glen Rose, Bird excavated 87 feet of tracks, and each block was about 18 inches deep. To excavate the tracks, Bird loaded the seven tons of fossils into his car, drove them 13 miles to the railroad station at Walnut Springs, and shipped them to AMNH (and another 3.5 tons to the University of Texas). Slabs were also sent to several schools (e.g., Baylor University, Brooklyn College, and Southern Methodist University[29]) and museums (e.g., NMNH), whose only cost was paying the shipping costs to get the slabs. Today, the most famous exhibits of the 107-million-year-old dinosaur tracks from the "Main Track Site" (i.e., so-named because it contains the footprints most easily seen by casual observers) are at AMNH (AMNH 3065) and the Texas Natural Science Center (a part of the Texas Memorial Museum; established in 1935; TMM 40638–1). Although many of the tracks viewed by Bird have been destroyed by erosion, others remain visible along the Paluxy River in DVSP just outside of Glen Rose.[30]

1939 AMNH's Jurassic Dinosaur Hall, curated by Barnum Brown, opens. The hall's many dinosaurs included *Apatosaurus, Stegosaurus,* and *Allosaurus*.

1939 Chinese geologist C. C. Young and American geologist Charles Lewis Camp (1893–1975) name *Omeisaurus,* a sauropod found three years earlier during an expedition to China's Sichuan Province. The sauropod was named in honor of Mt. Omei, a nearby sacred mountain. Young—who had attended school with Mao Tse-tung—had worked at Beijing's IVPP, which had sponsored the work of French priest Pierre Teilhard de Chardin (1881–1955) and Canadian surgeon Davidson Black (1884–1934). The quarry administered by the institute was Chicken Bone Hill, a site that in 1926 became famous as the source of "Peking Man" (*Homo erectus*).

1939 In his article titled "Thunder in His Footsteps" in *National History*, fossil-hunter Roland Bird describes the Paluxy Riverbed fossils, a series of impressions suggested by some as evidence that humans co-occurred with the dinosaurs (rather than arising millions of years after the non-avian dinosaurs went extinct). Bird described dinosaur tracks and mentioned humanlike footprints having a "questionable origin," and described a "mystery track" as "something about

15 inches long, with a curious elongated heel." When creationists began claiming that Bird's "mystery track" (and other tracks) had been made by humans, Bird responded that the tracks "are nothing more than simple depressions or eroded flaws in a rock surface with an accidental (and very slight) resemblance of human footprints . . . in the minds of these beholders." Bird later noted that "it was ridiculous to think they were human footprints," and in 1986, John Morris (Henry Morris's son) declared the footprints "at best, ambiguous and unusable as an antievolutionary argument at the present time." However, other young-Earth creationists continued to use the footprints to promote their claims that humans lived with dinosaurs a few thousand years ago. Indeed, Carl Baugh's Creation Evidence Museum of Texas (located on the banks of the Paluxy River) claims to have fossils showing a human footprint imposed on a dinosaur footprint. Although Bird left undocumented thousands of other dinosaur footprints from several Paluxy River sites, his articles helped popularize the study of dinosaur tracks (Figure 29). Bird was among the first to understand that dinosaur tracks represent fossilized behavior that offers insights into dinosaurs' social lives.

Figure 29 To excavate the dinosaur tracks near Glen Rose, Texas, Roland Bird and his crew had to divert part of the Paluxy River and remove overlying water, debris, and sediment. In this photo, Bird sits with his feet in one of the area's tracks left by a sauropod more than 110 million years ago. (Somervell County Heritage Center)

1939 In *The Wizard of Oz*, Tin Man asks, "What if it was a Brontosaurus?" and the Cowardly Lion replies, "I'd teach him who was king of the forest!"

1939 Princeton University begins excavations at what becomes known as the Cleveland-Lloyd Dinosaur Quarry. The operation was partially funded by Princeton law graduate (and Philadelphia businessman) Malcolm Lloyd Jr., who wanted a dinosaur skeleton for Princeton's geology museum. Lloyd's support, and the quarry's proximity to the town of Cleveland, Utah, produced the quarry's famous name. In 1941, Lloyd made his only visit to the site. More than 75 percent of the bones found at the site come from carnivores (especially *Allosaurus fragilis,* the top predator of the Middle Jurassic). Bones found at the Cleveland-Lloyd Quarry are black because of staining by manganese oxide in the area's groundwater.

1939 The Raymond M. Alf Museum of Paleontology is founded in Claremont, California. Today, this museum—the only paleontology museum in the United States on a high school campus—displays skulls and skeletons of *Centrosaurus, Tyrannosaurus* and *Allosaurus,* and dinosaur eggs.

1940 Barnum Brown and E. M. Schlaikjer publish an influential monograph about *Protoceratops*. That same year, Brown lectured at the New York Academy of Science about "The Methods of Walt Disney Productions" and Disney's making of the movie *Fantasia*. In that movie (which opened on Broadway and was a box-office failure when it was first released), the dinosaurs were scientifically accurate and based on prevailing knowledge at the time. All of the dinosaurs in the movie actually existed; there were no imaginary dinosaurs or bizarre hybrids that typified many other films involving dinosaurs.

1940 CMNH exhibits *Camptosaurus* found at DNM in 1922. That same year, CMNH mounted the fast, bipedal herbivore *Dryosaurus,* which was collected by Earl Douglass in 1910.

1940 German bombs destroy Britain's Bristol City Museum, including several fossils collected by Mary Anning, the type specimens of *Plesiosaurus,* and some of the fossilized bones of *Thecodontosaurus,* a prosauropod described by Samuel Stutchbury in 1836. (*Thecodontosaurus* was the fourth dinosaur to be named, and was the first dinosaur to be named from the Triassic.) Some bones of *Thecodontosaurus,* which were stored in a tunnel in Avon Gorge, were later used to reconstruct a display for the museum; parts of this specimen show smoke damage from the bombing. During World War II, bombs destroyed or damaged 17 dinosaur fossils (including several type specimens) in European museums. In 2013, paleontologists in Bristol, England (which was heavily bombed during World War II) unveiled a life-size replica of *Thecodontosaurus.*

1940 In Robert Moore Williams's *Jongor of Lost Land,* the parents of Jongor are killed by pterodactyls, and Jongor must fight off dinosaurs and other creatures to survive. Ann Hunter tries to find her brother in the Lost Land. When Ann is captured by apelike Muros and prepared for sacrifice, Jongor rides a theropod to her rescue.

1940 In the movie *Granite Hotel,* humans use a giant sauropod to fight fires.

1940 Raymond Sternberg, a son of Charles M. Sternberg (1885–1981, and grandson of Charles H. Sternberg), discovers two toothless jaws and a wishbone. Because Sternberg believed wishbones were not present in dinosaurs, he described the fossil as the bird *Caenagnathus* ("recent bird"). The subsequent discovery of more fossils in Mongolia reclassified the fossil as a member of the *Oviraptor* group of dinosaurs.

1940 Roland Bird (Figure 29) writes to Barnum Brown "the bed of the Paluxy River seems to be a regular footprint treasure cove." That same year, Bird—thanks to financial support from AMNH, Sinclair Oil and Gas, the State of Texas, and the federal government—began excavating dinosaur tracks at Glen Rose. (Bird had wanted to excavate the tracks he first saw at the Davenport ranch, but when "Mrs. Davenport proved to be as big a problem as any in the field," he extracted those from what today is Dinosaur Valley State Park.) Bird's most famous excavation there—sauropod trackway "S2" along the west side of the Paluxy River—was removed in several pieces and today is exhibited in AMNH. (Along the opposite side of the river are the "Ozark Trails," so named because the first accurate map of the site was made by a team from Ozark Bible College in the late 1970s.) The tracks excavated by Bird provided further evidence that sauropods were erect with their legs directly beneath their bodies (rather than splayed out like lizards). Bird hoped that the area surrounding the dinosaur tracks in Glen Rose, Texas, would be made a National Monument, but it ended up as a state park. Today, there are more than 50 known sites of dinosaur tracks in Texas, several of which were discovered by Roland Bird.

1940 Soviet paleontologist and science fiction writer Ivan Antonovich Yefremov (1908–1972) coins the word *taphonomy* to refer to fossilization and the biases inherent in the fossil record. However, the term was not widely used until the 1960s, when University of Chicago paleontologist E. C. Olson introduced the term to books and papers written in English. In 1952, Yefremov received the Stalin Prize for his work involving taphonomy.

1940 The movie *One Million B.C.,* a remake of D. W. Griffith's *Man's Genesis* (1912), features humans living with dinosaurs, as does its remake *One Million Years B.C.* (1966) and *When Dinosaurs Ruled the Earth* (1970). *One Million B.C.* established many of the conventions that persisted in dinosaur-themed movies for several decades, and was one of the first films in which some of the dinosaurs were depicted by actors in rubber suits (e.g., stuntman Paul Stader, who had been an action double for Johnny Weissmuller in several *Tarzan* movies, appeared in a *T. rex* costume). Other prehistoric creatures in *One Million B.C.* included lizards, armadillos, pigs, and alligators decorated with horns and frills (e.g., *Triceratops* was a pig adorned with rubber appendages, and *Dimetrodon* was a small alligator wearing a rubber dorsal fin) filmed in slow motion on miniature sets. Except for the opening scenes, *One Million B.C.* included no dialogue. *One Million B.C.,* which was released as *Man and His Mate* and *Cave Man* in some foreign countries, starred Victor Mature (in his second movie-role, as the caveman "Tumak") and Carole Landis (as "Loana"), and was produced by Hal Roach, who was known for his comedies (e.g., *Our Gang*, *Laurel and Hardy*). Subsequent movies that used decorated lizards to depict dinosaurs included *King Dinosaur* (1955), *A Journey to the Center of the Earth* (1959), and *The Lost World* (1960). Footage from *One Million B.C.*—"So Amazing You Won't Believe Your Eyes!"—later appeared in a host of similar movies, including *Tarzan's Desert Mystery* (1943), *Two Lost Worlds* (1950), *Untamed Women* (1952), *The Lost Planet* (1953), *King Dinosaur* (1955), *Teenage Caveman* (1958), and *Attack of the B-Movie Monster* (1989). When shown on television, the segment featuring *Triceratops* near the beginning of *One Million B.C.* is often cut to make time for commercials.

1940 Thirty-four paleontologists meeting at Harvard University establish the SVP. The society's original members included George Gaylord Simpson, Charles Gilmore, A. S. Romer, and Edwin Colbert. The society's logo (featuring vertebrae of *Eryops* and a Marsh Pick) was designed by Margaret Colbert (1911–2007), the wife of Edwin Colbert and daughter of William Diller Matthew. Today, SVP annually awards the Edwin H. and Margaret M. Colbert Prize for the best poster presented by a student at the society's annual meeting. In 2012, the society had more than 2,400 members.

1940 Walt Disney Studios consults with prominent scientists Julian S. Huxley (1887–1975), Barnum Brown, and Roy Chapman Andrews to produce the cartoon *Fantasia,* which depicts the evolution of life on Earth up through the extinction of dinosaurs. The movie included a 20-minute sequence about dinosaurs, which attributed their disappearance to drought (as Brown had suggested for dinosaurs he discovered at the Howe Ranch Quarry in Wyoming). The plot, part of which was an evolutionary sequence titled "March of Life" that was set to the Philadelphia Symphony's rendition of Igor Stravinsky's *Rite of Spring,* culminates with a fight between *Stegosaurus* and *Tyrannosaurus*. (*Tyrannosaurus* had only two fingers, but Walt Disney insisted that they have three because "I really think [*T. rex*] looks better with three.") Although *Fantasia* depicted most dinosaurs according to the thinking of the times (i.e., as giant, lumbering animals), *Stegosaurus* and *Tyrannosaurus*—both of which appeared in the cartoon—never saw each other in real life, for they were separated by more than 80 million years. Similarly, and despite depictions to the contrary, *Apatosaurus* never saw *T. rex*; the time between these two iconic dinosaurs was approximately the same as between *T. rex* and the first humans (65 million years). *Fantasia* was one of the first films with "Superscope and Full Stereophonic Sound." Although *Fantasia*, which won an Honorary Oscar in 1942, was a cartoon, no one had ever seen anything like it. As American biologist Stephen Gould noted later, "I remember well my first exposure to [dinosaurs' extinction] at age five: the dinosaurs of Disney's *Fantasia* panting to their deaths across a desiccating landscape to the tune of Stravinsky's *Rite of Spring*." The models used in *Fantasia*'s battle between *Tyrannosaurs* and *Stegosaurus* are now part of "Primeval World" at Disneyland.

1941 CMNH trustee and biologist George H. Clapp (1859–1949) purchases the type specimen of *T. rex* from AMNH for $7,000 that was "to be set aside as an endowment for the Department of Paleontology, the interest to be used for expeditions and the purchase of specimens." (AMNH approached CMNH after Yale University did not respond to its query.) As Barnum Brown noted, "The transaction was accomplished because the American Museum was afraid that German airships might bomb [AMNH] and destroy the second *Tyrannosaurus rex* skeleton [AMNH 5027] now mounted here and that at least one specimen might be preserved." The specimen was displayed at CMNH the next year. In 1915, the University of Pittsburgh named its biology building Clapp Hall in Clapp's honor.

1941 E. T. Hall Jr. excavates some of the first vertebrate fossils collected in the Western Hemisphere. Those fossils, one of which was part of an upper jaw of the extinct mammal *Phenacodus,* were originally collected by Indians in the Lower Eocene beds of northwestern New Mexico in 700–900 A.D. and taken to a pit house, the ruins of which Hall excavated. The sediments that originally housed the fossils were described in 1874 by Edward Drinker Cope.

1941 In *How to Become Extinct,* satirist Will Cuppy (1884–1949) claims, "The Age of Reptiles ended because it had gone on long enough and it was all a mistake in the first place. A better day was already dawning at the close of the Mesozoic Era. There were some little warm-blooded animals around which had been stealing and eating the eggs of the Dinosaurs, and they were gradually learning to steal other things, too. Civilization was just around the corner."

1941 Isaac Asimov writes "Big Game," a 1,000-word short story in which an inventor named Hornby builds a time machine to go back to the Late Mesozoic and hunt dinosaurs to extinction for sport. The explorers then killed themselves. Asimov's short story, which was repeatedly rejected by publishers, was finally published in 1974.

1941 Nine years after receiving his only formal academic degree (an honorary doctorate from Middlebury College in Vermont), AMNH's famed fossil-collector Walter Granger dies of heart failure while on an expedition in Lusk, Wyoming. Granger, one of the first vertebrate paleontologists at AMNH, was the chief paleontologist in Roy Chapman Andrews's Central Asiatic Expeditions in the 1920s. Granger's ashes were spread on his mother's grave in the Pleasant View Cemetery in his hometown of Middletown Springs, Vermont. At AMNH, where his colleague George Gaylord Simpson lauded Granger as "the greatest collector of fossil vertebrates that ever lived," the Asiatic Hall of Fossils was renamed the Walter Granger Memorial Hall.

1941 Paleontologist C. C. Young names *Lufengosaurus hueni* (IVPP V15), a 20-foot-long prosauropod that lived in the Early and Middle Jurassic in what is now southwestern China. Young named the dinosaur after his teacher Friedrich von Huene. *Lufengosaurus* had been collected several years earlier near the city of Lufeng in the province of Yunnan, China. When *Lufengosaurus hueni* became the first complete dinosaur skeleton to be displayed and mounted in China, government officials issued a postage stamp (8 yuan) in 1958 to commemorate the event. This was the first postage stamp anywhere in the world to feature a dinosaur.

1941 Roland Bird's article titled "A Dinosaur Walks into the Museum" in *Natural History* shows three-year-old Tommy Pendley bathing in the 18 gallons of water needed to fill a sauropod track near Glen Rose, Texas (Figure 30). That same year,

Figure 30 In this photo taken by famed dinosaur hunter Roland Bird (Figure 29), three-year-old Tommy Pendley, Jr., plays in the hind-foot track left by a sauropod along the Paluxy River near Glen Rose, Texas. (Tommy's father was a member of Bird's excavation crew.) Bird described Tommy as "the first human in all history to have a dinosaur footprint for a bathtub." (Somervell County Heritage Center)

several of the dinosaur tracks excavated from the famous "S2" trackway by Bird near Glen Rose, Texas, were displayed in a small building adjacent to the Texas Memorial Museum in Austin, Texas. However, because the climate of this room was not controlled, the tracks began to deteriorate. In 2012, the museum began restoring Bird's famous tracks.

1941 Zdenêk Burian paints three snorkeling *Brontosaurus,* two of which were in a lake with only their heads and upper necks visible above water. Burian's influential and often-reproduced painting resembled Charles Knight's 1897 drawing of *Amphicoelias*.

1942 AMNH artist George Geselschap paints the first group of *Stegosaurus*.

1942 Arthur G. Nord of Utah's Ashley National Forest, concerned that the area's fossil treasures are being plundered by out-of-state collectors, begins planning what came to be known as the Utah Field House of Natural History State Park Museum (UFHNH). Three years later, the Utah legislature approved the museum, which opened in 1948. Today, a complete *Diplodocus* skeleton dominates the main hall, with its head peering down on the admissions counter.

1942 Charles Knight prepares a series of pictures published in *National Geographic* to show the history of life. Knight's father worked at a bank owned by the family of J. P. Morgan, who sponsored some of Knight's murals. Knight's article titled "Parade of Life through the Ages" was the first time Knight's paintings of dinosaurs appeared in color. The paintings included two battling *Tyrannosaurus, Protoceratops* positioned over a clutch of eggs (with its mate standing nearby), *Parasaurolophus, Styracosaurus,* and a group of *Diplodocus* in a swamp. Knight described *Diplodocus* as "harmless and stupid, the grotesque *Diplodocus* asked only to be left in peace." Knight also painted for the Field Museum a mural depicting *Tyrannosaurus* confronting *Triceratops* (Figure 19), arguably the most famous single image to ever feature dinosaurs; as the Field Museum's Alexander Sherman noted, Knight's image "is so well-loved that it has become the standard encounter for portraying the age of dinosaurs." The subsequent fame of this painting forever established these two dinosaurs as enemies in popular thought. Robert Bakker (b. 1945) later described Knight's *Tyrannosaurus–Triceratops* rivalry this way: "No matchup between predator and prey has ever been more dramatic. It's somehow fitting that those two massive antagonists lived out their co-evolutionary belligerence through the very last days of the very last epoch of the Age of Dinosaurs." Knight worked with many paleontologists, including Henry Osborn, Edward Drinker Cope, and William Diller Matthew. Right or wrong, Knight's dinosaurs were the ones that the public believed were real.

1942 David Dunkle of the Cleveland Museum of Natural History finds a skull that he classifies as *Gorgosaurus* ("fierce lizard"), a new species. The skull, which was discovered near Ekalaka, Montana, was later reclassified as *Albertosaurus*, but in the 1980s another reevaluation classified the fossil as a new species, *Nanotyrannus* ("tiny tyrant"). This fossil, which is displayed in Cleveland's Museum of Natural History, is the only known fossil of *Nanotyrannus*. (This museum also hosted the first life-size model of *Stegosaurus*.) Some paleontologists claim that *Nanotyrannus* is a species of *Tyrannosaurus*, and others argue that it is a juvenile of *T. rex*.

1942 Elmer Riggs retires from the Field Museum and returns to Kansas, where he died in 1963 at age 94.

1942 Geologist Edward Lee Holt finds fossils of *Apatosaurus* near Green River, Utah. In 1958, the Field Museum obtained the specimens and used them to complete its headless specimen of *Apatosaurus*.

1942 In Dave Fleischer's cartoon *The Arctic Giant*, much of New York is destroyed as Superman fights a 250-foot-tall tyrannosaur that has been frozen in Siberia. This theme—that is, humans fighting dinosaurs—was previewed at the end of *The Lost World* (1925) and became common in many dinosaur-themed movies.

1942 Paleontologists Richard Swann Lull and Nelda Wright reevaluate the hadrosaurs and propose the new genus *Anatosaurus* ("duck lizard") to encompass several of the existing species. Two species at AMNH were reclassified as *A. copei*, and *A. saskatchewanensis* became the new name for the skull and skeleton collected by Charles M. Sternberg in 1921 (and described by him in 1926 as *Thespesius*). *Anatosaurus* was later shown to be *Edmontosaurus*.

1942 The annual report of AMNH notes that 69-year-old "Dr. Barnum Brown retired from active duty as of July 1, 1942, and was given the title of Curator Emeritus. . . . In the history of the science of Vertebrate Paleontology there has never been a more energetic or more successful collector than Dr. Brown. . . . In fact, the present position that this Museum holds as a repository for dinosaurian fossils as well as the absolutely unequalled display of these animals in our two Dinosaur Halls is owing primarily to the efforts of Dr. Brown." In ensuing years, Brown advised (and worked as a spy for) the U.S. War Department, but continued to lead tours at AMNH, referring to the dinosaur fossils that he had collected as his "children." AMNH divided its dinosaurs into Saurischian and Ornithischian Halls. Although this approach was cladistically correct, critics complained that it incorrectly suggested that *Stegosaurus* lived with *Tyrannosaurus*, when, in fact, they were separated by more than 70 million years (*Stegosaurus* died off in the Late Jurassic, and *T. rex* appeared in the Cretaceous). The Saurischian and Ornithischian Halls at AMNH display about 100 specimens, 85 percent of which are made of fossils (i.e., not casts).

1942 Willis O'Brien begins work on a movie featuring a giant allosaur living in Mexico's Forbidden Valley. The project was soon abandoned, but in 1969 Ray

Harryhausen resurrected the project as the dinosaur-in-a-Western *The Valley of Gwangi*. Harryhausen often had little interest in accuracy, noting that he was not "making pictures for paleontologists" because he felt "it is far more important to create a dramatic illusion than to be bogged down with detailed accuracy."

1942 Albert Parr, the director of YPM, hires Yale senior Rudolph Zallinger (at a salary of $40 per week) to paint a mural for the museum. Soon thereafter, Zallinger began a six-week "crash course" in vertebrate paleontology and paleobotany. Zallinger's masterpiece, *The Age of Reptiles,* was completed in 1947, and was one of the largest mural frescoes in the world. The 110-foot-long, 16-foot-tall painting, which reads from right to left, spans more than 300 million years (i.e., from the Devonian to the Cretaceous) on the museum's east wall, with each period's expanse on the painting being proportional to the period's length. Trees in the foreground mark the beginning of the three Mesozoic periods. The end of the Mesozoic includes a smoldering volcano—a trademark of Zallinger's paintings of prehistoric life—that signaled dinosaurs' impending doom. The entire painting, which includes *Allosaurus, Brontosaurus, Stegosaurus, Ankylosaurus, Triceratops,* and *Tyrannosaurus,* dispelled the misconception that all dinosaurs lived together, but cannot be photographed because of obstructions such as the museum's mounted skeleton of *Apatosaurus*. *Tyrannosaurus* and *Brontosaurus* are "later" on the left half of the mural, but face to the right, while early figures on the right side of the mural face to the left. The transition from the Permian to the Mesozoic is depicted by *Dimetrodon, Edaphosaurus, Plateosaurus,* and *Allosaurus*, and *Stegosaurus* occupies the middle of the mural. The mouse-like mammal *Cimolestes*—just a few inches from Zallinger's signature—heralds the new world that followed dinosaurs' demise. Parts of Zallinger's fresco were reproduced in numerous books, magazines, and on more than 70 million postage stamps. Whereas Charles Knight had depicted *Dryptosaurus* as active, fighting animals, Zallinger's mural portrayed dinosaurs as slow, solid, and torpid; the only activity in the mural is *Allosaurus* scavenging a carcass. Nevertheless, the mural inspired many young paleontologists, including Robert Bakker, who noted in 1986 that he "had discovered an entire world, far, far away in time, that I could visit whenever I wanted via the creative labors of the paleontologists. And I made my mind up then and there that I would devote my life to the dinosaurs." Zallinger, whose father Franz was also an artist, has been blamed for fixing the classic images of "dinosaur orthodoxy" with dinosaurs as slow, doomed creatures.

1943 Allied bombings prompt officials at the Natural History Museum at Berlin's Humboldt University to dismantle its *Brachiosaurus* (the largest mounted dinosaur in the world) and store it in the museum's basement. Later that year, the museum's Reptile Hall was destroyed, and with it fossils of the Tanzanian stegosaur *Kentrosaurus*.

1943 Charles H. Sternberg, a towering figure in paleontology, dies in his sleep in Toronto at age 93. Sternberg had noted earlier "that whatever it might cost me in privation, danger, and solitude, I would make it my business to collect facts from the crust of the Earth."

1943 In ABC's *Lights Out,* a man intent on killing his wife takes her to the basement of a department store. After a dinosaur in the basement changes the man's opinion of his wife, the man sacrifices his life so his wife can live.

1943 Raymond Cowles (1896–1975) proposes that dinosaurs became extinct because of overheating.

1943 Sell Rite Giftware (SRG) begins producing its popular series of metal-cast ("Royal Bronze Finish") dinosaurs for gift shops in museums such as AMNH, BMNH, and CMNH. These toys, which were reissued in gray plastic in the 1960s, were some of the first successful mass-produced dinosaur toys, and today are prized by collectors. Many companies (e.g., Alva, H. S. Brumm, MPC, J. H. Miller, Ajax, Louis Marx) later mass-produced countless dinosaur toys; by the 1950s, toy dinosaurs were included in a variety of foods, including cereal boxes of Nabisco Rice and Wheat Honeys. In 2014, a small metal SRG *Mosasaurus* was selling for about $100.

1944 Carlton Morse's (1901–1993) *I Love a Mystery* radio show features "Stairway to the Sun," which includes dinosaurs atop a lost plateau. The episode was inspired by Arthur Conan Doyle's *The Lost World.* Other adaptations of *The Lost World* for radio followed in 1980 (New Zealand), 1997 (United States), and elsewhere.

1944 Charles Knight begins work on his final set of paintings, which were sold to the Natural History Museum of Los Angeles County.

1944 German merchant Waldemar Julsrud finds clay figurines at the foot of El Toro Mountain near Acámbaro, Mexico. Some of the dinosaur-shaped figurines were later used by young-Earth creationists to claim that dinosaurs lived with humans. Most archaeologists question the validity of the discovery.

1944 The Royal Air Force bombs Munich's Bavarian State Collections for Paleontology and Historical Geology, destroying the type specimens of *Aegyptosaurus, Carcharodontosaurus, Bahariasaurus,* and the only known specimen of *Spinosaurus* (which had been described in 1915 by Ernst Stromer von Reichenbach). That same year, during Operation Overlord (the code name for the Battle of Normandy), bombs also destroyed the University Museum in Caen, France (and its prized fossil of the theropod *Poekilopleuron*); Le Havre's Museum of Natural History and its *Megalosaurus* and stegosaur *Dacentrurus*, and Stuttgart's State Museum of Natural History (and its famed specimen of *Plateosaurus*). Fortunately, there were duplicates of some of the lost specimens. Although bombs destroyed large areas of each of the cities housing the three known fossils of *Archaeopteryx* (Berlin, Munich, and London), all three specimens emerged from World War II intact.

1945 Alfred Sherwood Romer (1894–1973) publishes *Vertebrate Paleontology*, which soon became the standard textbook for the field. Romer later recalled "all the old-timers I knew when young . . . liked Cope and almost to a man they hated Marsh's guts." In 1987, SVP began awarding the Romer-Simpson Medal to recognize lifetime achievement in vertebrate paleontology. Winners of this award, the society's highest honor, have included Edwin Colbert (1989), Loris Russell (1992), John Ostrom (1994), Malcolm McKenna (2001), and José Bonaparte (2008).

1945 American paleontologist Edwin "Ned" Colbert writes *The Dinosaur Book: The Ruling Reptiles and Their Relatives,* the first popular (i.e., nontechnical) book about dinosaurs. *The Dinosaur Book*—one of Colbert's many popular books about dinosaurs—was published by AMNH and introduced a new generation of scientists, amateur enthusiasts, and other people to paleontology. The lack of similar books for several years made Colbert an important influence on a generation of paleontologists and dinosaur enthusiasts. Jack Horner, who had problems reading, was especially influenced by artwork in Colbert's books.

1945 Charles Gilmore dies and is buried in Arlington National Cemetery. Gilmore named 25 dinosaurs, 13 (i.e., 52%) of which remain valid.

1945 Famed Hollywood director Cecil B. DeMille confesses to Charles Knight that "your drawings will give me many hours with one of my favorite subjects—the dawn of life on this planet, which is more fascinating than most of the activities of some of Earth's later inhabitants."

1945 In J. W. Pelkie's *King of the Dinosaurs,* cavemen encounter smart dinosaurs that use humans as balls in games.

1945 In the movie *Tarzan,* the hero finds a model that resembles a theropod, and then must fight the real thing.

1945 William Stokes describes Utah's Cleveland-Lloyd Dinosaur Quarry. Major excavations resumed at the quarry in 1960, after which the site became famous for its many dinosaurs.

1946 Charles Knight's second book, *Life through the Ages,* describes *Stegosaurus* as "no doubt, the stupidest member of a very moronic family," *T. rex* as "an enormous eating machine with an insatiable appetite and with practically no brains," and dinosaurs as having "been in existence too long, for they were stupid, unadaptable, and unprogressive." Knight's errors reflected the scientific thinking of his day.

1946 Charles Whitney Gilmore of the Smithsonian Institution posthumously names *Gorgosaurus lancensis* based on discoveries by the Cleveland Museum of Natural History in 1942. *G. lancensis,* a small tyrannosaurid, was classified as a dwarf *Albertosaurus* in 1970 by Canadian paleontologist Dale Russell, and later named *Nanotyrannus*.

1946 Edwin Colbert and his colleagues propose that dinosaurs were cold-blooded reptiles that grew to giant sizes that minimized daily variations in body temperature. This view of dinosaurs (i.e., in which body temperatures resulted from thermal inertia) was revived by J. R. Spotila in 1973.

1946 Soviet paleontologist Ivan Yefremov (1907–1972) leads a two-month expedition to the Nemegt Valley of the Gobi Desert. There, Yefremov found the remains of several dinosaurs and immediately began planning a second—and more extensive—expedition for 1948.

1946 The U.S. Postal Service commemorates the 100th anniversary of the Smithsonian Institution by issuing a 3¢ stamp. Fifty years later, the Smithsonian's 150th anniversary was commemorated with a 32¢ stamp.

1946 While working just outside Huxley in central Alberta, Charles M. Sternberg finds the first *T. rex* in Canada. Sternberg, the son of Charles H. Sternberg, was part of the greatest fossil-hunting family of all time. Sternberg's 40-foot-long *T. rex* was excavated in 1981 by ROM. Today, this specimen is displayed at ROM along with another *T. rex* found in 1981 in southern Alberta by three high school students.

1947 American-based artist Rudolph Franz Zallinger (1919–1995) completes his 110-foot-by-16-foot mural *The Age of Reptiles* on YPM's eastern inside wall. At one time, *The Age of Reptiles*—which spans nearly 350 million years of Earth's history—was the largest painting in the world. Zallinger's fresco depicted dinosaurs as slow, sluggish creatures.

1947 There are now 38 interpretations by 23 artists of *Stegosaurus*.

1947 *Time* magazine describes Edwin Colbert, a protégé of Henry Osborn, as "the Dick Tracy of the Mesozoic Age." Colbert's books, which often featured the art of Charles Knight, helped popularize dinosaurs.

1947 When Thomas Boylan dies, his widow Grayce continues to operate Fossil Cabin (Figure 26) near Medicine Bow, Wyoming, until the 1960s, when completion of Interstate 80 reduced traffic and visitors in the area. Grayce sold the property in 1974. In late 2013, the famous cabin remained closed and the property was for sale.

1947 While en route to Petrified Forest National Park in Arizona, American paleontologist Edwin Colbert and his assistant George Whitaker of AMNH find more than a dozen intertwined skeletons of the small Triassic theropod

Coelophysis (AMNH 7223) in the Late Triassic Chinle Formation at Ghost Ranch, New Mexico (Figure 31). (Colbert also named *Staurikosaurus,* one of the earliest known dinosaurs.) The site was less than a mile from the home of artist Georgia O'Keefe, who often visited the workers after excavations began. Colbert concluded that *Coelophysis* were swift runners and had bird-like postures. Colbert and his colleagues, who described the find as "a paleontological treasure beyond one's wildest dreams, abandoned [their] original plans and settled down for a summer of digging at Ghost Ranch . . . which involved much time, much profanity, and many scraped knuckles." The workers excavated several *Coelophysis*-laden blocks, including one that weighed eight tons. One block was sold to the Alberta's RTM, and another that was left at the site is now displayed in the Ruth Hall Museum of Paleontology (part of the Florence Ellis Hawley Museum) in Abiquiu, New Mexico.

Figure 31 Edwin H."Ned" Colbert (1905–2001) and his colleagues at Ghost Ranch, New Mexico, where they discovered thousands of bones of the Triassic theropod *Coelophysis*. Most of our knowledge of *Coelophysis,* which is the state fossil of New Mexico, comes from Colbert's work. Colbert, a protégé of Henry Fairfield Osborn (Figure 15), enjoyed a successful 40-year career as a paleontologist at the American Museum of Natural History. In 1969, Colbert discovered the therapsid *Lystrosaurus* in Antarctica. This dinosaur had earlier been found in South Africa, and Colbert's discovery confirmed that Antarctica and South Africa had once been attached. Colbert's popular books, most notably *The Dinosaurs: The Ruling Reptiles and Their Relatives* (1945), fed the public's growing interest in dinosaurs. (Bettmann/Corbis)

Knowing that the block would not fit through the museum's doors, the block was installed inside the building as the building was being constructed. Excavations at the 21,000-acre ranch later showed that several *Coelophysis* specimens described by Cope as different species were actually differently sized individuals of the same species. In the 1980s, the Carnegie Institution collected more specimens of *Coelophysis* at Ghost Ranch. *Coelophysis* is one of the best-known Triassic theropods.

1948 Ivan Yefremov (1908–1972) begins a second (and larger) expedition to the Gobi Desert. In addition to visiting Flaming Cliffs (where the group collected more *Protoceratops*), the expedition unearthed the huge *T. rex*-like carnivore *Tarbosaurus* (named by E. A. Maleev in 1955), duck-billed dinosaurs such as *Saurolophus* (which had been discovered earlier in the United States), and *Therizinosaurus* ("turtle-like scythe reptile," named by E. A. Maleev in 1954), which had three giant claws (each up to 28 inches long) on each digit of its forelimbs, and large, flat plates similar to those of a turtle shell. The finger bones of *Therizinosaurus* are larger than those of any other known animal. Most paleontologists suggested that the bizarre *Therizinosaurus* was a carnivore, but some suggested that it used its claws to tear open the nests of termites and other insects. (It is now suspected to be a herbivore.) The most bountiful collection of dinosaurs occurred in the Nemegt Valley at a quarry known as the "Tomb of the Dragons." The following year (1949), Yefremov's expedition to the Gobi focused on creatures that lived after dinosaurs. All in all, Yefremov's expeditions gathered more than 100 tons of specimens, many of which were displayed in Moscow's Paleontological Museum of the Academy of Sciences.

1948 The 76-minute movie *Unknown Island* ("See Man's Puny Attempts to Defeat Monstrous Beasts!") features a "lost world" where humans encounter dinosaurs on an uncharted island in the South Pacific. Humans arrive on the island after aerial photographs suggest that dinosaurs live there. Much of the film's footage was shot in the desert near Palmdale, California, and featured lizards dressed as dinosaurs, and human actors wearing rubber suits depicting ceratosaurs. Dinosaurs in *Unknown Island* were based on Charles Knight's paintings that appeared in the February 1942 issue of *National Geographic*. *Unknown Island* was the first "lost world" movie to present dinosaurs in full color.

1948 UFHNH opens in Vernal, Utah. Its first director was Ernest Untermann, Sr. (1864–1956), whose paintings hang in an upstairs gallery. The lobby of the remodeled Field House is dominated by a cast of Andrew Carnegie's *Diplodocus* from 1899.

1949 Civic leaders in Maidstone (a village near where Gideon and Mary Mantell discovered their fossils) incorporate the Mantells's *Iguanodon* into the town's coat of arms. The village's dinosaur also adorns jerseys of Maidstone's United Football Club.

1949 George Gaylord Simpson publishes his influential *The Meaning of Evolution.* After John Ostrom read Simpson's book, he decided to become a paleontologist. In the 1960s, Ostrom's discoveries began the dinosaur "renaissance."

1949 In *On the Town,* featuring Gene Kelly and Frank Sinatra, sauropods at a New York museum collapse because of the antics of the leading characters.

1949 Rudolph Zallinger receives a Pulitzer Fellowship in Art for his *The Age of Reptiles* in YPM.

1950 Barnum Brown's second wife Lilian "Pixie" Brown publishes *I Married a Dinosaur,* her first book describing her paleontological adventures with her famous husband. In the book, Roy Chapman Andrews recalled that Barnum Brown would "disappear from the museum, just fading out like the Vanishing American, and none of the staff knew where he had gone . . . but invariably his whereabouts were disclosed by a veritable avalanche of fossils descending in carload lots upon the museum. . . . He has discovered many of the most important and most spectacular specimens in the whole history of paleontology." Lilian Brown's subsequent books included *Cleopatra Slept Here* (1951) and *Bring 'em Back Petrified* (1956).

1950 Charles M. Sternberg names the bizarre *Pachyrhinosaurus canadensis* ("thick-nosed lizard"), a 20-foot-long dinosaur known only from its skull found near Fort MacLeod in southern Alberta, Canada. This skull resembles that of *Triceratops,* but without horns and with what resembles a broken-off tree trunk atop its snout (thus explaining its nickname, "old stump face"). *P. canadensis* was a star attraction in the Drumheller Town Museum before RTM opened in 1985.

1950 Forty years after he discovered the alleged "man tracks" near Glen Rose, Texas, Charlie Moss swears that he "personally saw and inspected from about 1908 to 1910, on the north bank of the Paluxy River . . . about 15 tracks or footprints of apparent giant human beings. The tracks measured 18 inches from heel to toe, the five toes being plainly marked. About 1910 a flood . . . carried away all of the footprints."

1950 German entomologist Willi Hennig's (1913–1976) *Grundzüge einer Theorie der phylogenetischen Systematik,* noting that biological similarities do not necessarily imply close phylogenetic affinity, proposes phylogenetic systematics as a way of assessing evolutionary relationships. When Hennig's book was translated into English in 1966 (*Phylogenetic Systematics*), it became a foundation for cladistics, a hierarchical classification of species based on evolutionary ancestry. This revolutionized systematics and taxonomy. Paleontologists were among the first zoologists to endorse Hennig's methods.

1950 In an article titled "When Giants Roamed the Earth" in the Seventh-Day Adventist magazine *Signs of the Times*, Clifford Burdick—a founder of the Deluge Society (one of the first creationist organizations in the United States)—becomes one of the first to claim that human tracks ("man tracks") occur alongside dinosaur tracks near Glen Rose, Texas. Burdick's paper, which included photographs of two large footprints on rock slabs, misrepresented statements by Roland Bird, who had examined the Paluxy River dinosaur tracks years earlier. Although Bird mentioned the carved tracks that prompted his trip to Glen Rose (as well as stories from local residents that "man tracks" could be found in the Paluxy Riverbed), Bird never reported human tracks in the Paluxy Riverbed. Burdick concluded that the Paluxy "man tracks" refuted evolution and supported a recent creation and the formation of fossils during Noah's Flood. Burdick's claims and photos of the Paluxy "man tracks" were later featured at several creation museums and in the influential *The Genesis Flood* (1961), the book that began the modern creationist movement.

1950 In Isaac Asimov's "Day of the Hunters," which was based on his unpublished "Big Game," reptiles having powerful weapons drive dinosaurs to extinction.

1950 Local dinosaur enthusiast Trygve "Tig" Seland builds a life-size *T. rex* outdoors in Drumheller, Alberta, Canada. Seland later became president of the local Museum Society.

1950 Ivan Yefremov (1908–1972) publishes *Taphonomy*.

1950 The low-budget movies, *Prehistoric Women* and *Two Lost Worlds,* feature footage from *One Million B.C.* (1940) from a decade earlier. In *Prehistoric Women,* women hate men, but keep some around for reproduction. The same approach was used two years later with *Untamed Women* (1952), *Robot Monster* (1953), and *Teenage Caveman* (1958). *Two Lost Worlds* promised fighting Godzillas, but these creatures never appeared. All of the movie's dinosaurs and prehistoric animals were pet-shop lizards, sometimes adorned with fins, and other times standing in front of a rear-screen projector to enhance their shadows. The movie, which starred James Arness in his first starring role, followed people shipwrecked on a desert island; just as the island appears destroyed by earthquakes and an erupting volcano, the survivors are met by a rescue ship.

1950 The movie *King Kong* is released for the fourth time.

1950 The National Science Foundation (NSF) is established as a federal agency because, in the words of President Harry Truman, "[G]overnment has a responsibility to see that our country maintains its position in the advance of science." NSF later funded many studies of dinosaurs.

1951 British artist Neave Parker's (1910–1961) monochromatic painting of *T. rex* is marketed on BMNH postcards. This image, as well as a subsequent one by Czech paleoartist Zdenêk Burian (in which *T. rex* attacks two duckbilled dinosaurs), was copied by makers of toys and other merchandise. Burian, whose work was influenced by Charles Knight, remains one of the most influential and recognizable artists (after Knight) of prehistoric animals. Burian's dinosaurs were used in *Journey to the Center of Time* (1966).

1951 Charles Knight paints his last painting, a mural for the Everhart Museum in Scranton, Pennsylvania. Knight's plan to build a display of life-size dinosaurs with Louis Paul Jonas never materialized.

1951 In *Jungle Manhunt*, Jungle Jim fights a dinosaur. The movie included scenes from *One Million B.C.*

1951 In Ray Bradbury's short-story "The Fog Horn" published in *Saturday Evening Post,* a ringing bell at a lighthouse lures a giant dinosaur from its underwater home by mimicking the dinosaur's mating call. In 1953, this story became part of the movie *The Beast from 20,000 Fathoms*, which, in turn, inspired the Japanese film *Gojira* (1954).[31] *Gojira*, which was directed by Eiji Tsuburaya, was released in the United States as the altered *Godzilla, King of the Monsters!* (1956; Figure 32), a successful movie that inspired many sequels, similar movies, and other products about monsters gone bad. (Among the scenes cut from American versions, a mother huddles with her child in Godzilla's shadow, and a fleeing woman sobs, "I barely survived the bombing of Nagasaki. And now this!"). Godzilla, the most popular stegosaur-like monster of all time, became so popular in the United States that it had its own television show, "The Godzilla Power Hour." Bradbury wrote four dinosaur-stories, of which "The Fog Horn"—the first popular story to bring a carnivorous dinosaur to the edge of civilization—was the most popular. Bradbury acknowledged he saw "*The Lost World* in 1925 and [I] fell in love with [O'Brien's] Knight-inspired dinosaurs . . . [that movie] changed my life."

1951 In the movie *The Lost Continent* ("Modern Man against Prehistoric Monster!") staring Cesar Romero and Hugh Beaumont (the "dad" character on *Leave It to Beaver*), a crew searching for a lost missile crash-lands on an uncharted Pacific island and encounters dinosaurs preserved by high amounts of radioactivity. The movie promised "Thrills of the atomic-powered future! Adventures of the prehistoric past!" and included a rare battle-to-the-death between dinosaurs of the same species (in this case, *Triceratops*).

1951 Kenneth A. Kermack concludes that sauropods could not have breathed through their nostrils if the rest of their bodies was submerged (as several paleontologists had suggested) because the water's pressure would have made it impossible to ventilate their lungs.

1951 On NBC's *Tom Corbett, Space Cadet,* a spaceship lands on a newly discovered planet that resembles Earth during the Mesozoic. Dinosaurs in the show were the Messmore and Damon models from "The World a Million Years Ago" (that were filmed while not moving). In another episode, the crew's ship was attacked by dinosaurs.

Figure 32 The movie *Godzilla, King of the Monsters!* (1956) was an American version of *Gojira,* which had debuted in Japan two years earlier. This movie, which began the flood of monster movies from Japan, featured a 400-foot-tall, bipedal, amphibious saurian with fiery, radioactive breath. Godzilla, who became a worldwide pop-culture icon, was inspired by "Rhedosaurus" created by animator Ray Harryhausen for *The Beast from 20,000 Fathoms* (1953). Godzilla was a dark metaphor for the nuclear weapons used on Hiroshima and Nagasaki. (Embassy Pictures Corporation/Photofest)

1951 Texas cartoonist and film historian George Turner debuts the comic strip *The Ancient Great Plains* in the Amarillo *Sunday-News Globe*. Turner, who often sought advice from West Texas State College geology professor Roy Reinhart for his drawings, sketched dinosaurs based on paintings by Charles Knight. Turner's comics were the first to feature anatomically correct dinosaurs.

1951 The "Eichstätt specimen" of *Archaeopteryx* (JM 2257) is found just north of Eichstätt, Germany. This specimen, the smallest known specimen of *Archaeopteryx*, is now housed at Eichstätt's Jura Museum. The specimen was initially described as a juvenile of *Compsognathus*, but in 1973 was classified as *Archaeopteryx* when it was noted that the fossil included impressions of feathers.

1951 To commemorate the centenary of Geological Survey of India, India issues a red 2-anna stamp featuring *Stegodon ganesa,* a prehistoric elephant. This was the first postage stamp to feature a prehistoric animal. The first postage stamp featuring a dinosaur, which was produced by China, would not appear for another seven years.

1952 In Arthur Clarke's "Time's Arrow," paleontologists discover dinosaur tracks near a physics lab. When the paleontologists excavate the last tracks, the physicists use their "Helium II" powered time-machine to go back in time. One of the travellers is chased and eaten by the theropod they are studying.

1952 In Ray Bradbury's "A Sound of Thunder," a company advertises "Safaris to any year in the past. You name the animal. We take you there. You shoot it." Time-travelers went back to the Late Cretaceous to hunt dinosaurs. The popular story, which appeared in *Collier's* magazine, was the most influential of the time-travel stories involving ancient life; it was later made into a television show, comic book, and movie.

1952 In *Untamed Women* ("They Feared No Monster . . . Yet Fell before the Touch of Man!"), World War II pilots discover a South Seas island populated with cavemen, beautiful women, man-eating plants, and dinosaurs.

1952 *King Kong* is re-released to theaters, prompting the release the following year of the similar *The Beast from 20,000 Fathoms*. In *The Beast from 20,000 Fathoms,* scientists tried to identify an unknown beast by studying paintings by Charles Knight.

1952 Paleontologists report the first discovery of dinosaur tracks from the Upper Triassic of Sweden.

1952 When AMNH begins remodeling what would become its Hall of Saurischian Dinosaurs, Ed Colbert (Barnum Brown's successor) convinces the ailing Roland Bird to interrupt his retirement to reassemble the dinosaur tracks he excavated from near Glen Rose, Texas, 14 years earlier. The tracks had been stored in the museum's courtyard, and most of the crates containing the tracks were rotted; labels and files regarding their reassembly had been lost. Nevertheless, in less than six months, Bird—assisted by Carl Sorensen and two stonemasons—reassembled more than 100 blocks to create a 29-foot-long, 8-foot-wide, and 18-inch-thick slab that shows six fore-footprints and six hind-footprints beneath and behind an *Apatosaurus*. This was the final act of Bird's famed career. (The bones of *Apatosaurus* are 145 million years old, while the sediments containing the tracks are 111 million years old.)

1953 After being dormant since Charles Gilmore and his colleagues left in 1924, DNM is again the site of excavations by the National Park Service. Over the next decades, DNM yielded more than 120 individuals and nearly 20 articulated skeletons, the largest collection of dinosaurs from a single locality ever found in the Morrison Formation. These specimens included *Camarasaurus, Barosaurus, Diplodocus, Allosaurus, Camptosaurus, Stegosaurus, Dryosaurus, Torvosaurus,* and *Ceratosaurus*. The most abundant fossil at DNM is *Unio,* a genus of clams.

1953 American artist Joe Kubert (1926–2012) premiers the comic-book series *1,000,000 Years Ago,* which is printed in 3-D and features Tor, a caveman who struggles to survive with dinosaurs. In 1974, Kubert illustrated *Tarzan* comics that featured prehistoric animals.

1953 Famed dinosaur-illustrator Charles Knight's last words to his daughter Lucy are, "Don't let anything happen to my drawings." She didn't: Knight's iconic work remains on display at AMNH, the Field Museum, the Bronx Zoo, and elsewhere. When the 79-year-old Knight died soon thereafter in New York, he was remembered as "the greatest genius in the line of prehistoric restoration and animal life that the science of paleontology has ever known. His work . . . will endure for all time." For decades, Knight's work had appeared in encyclopedias, comic books (e.g., Joe Kubert's *1,000,000 Years Ago, Tor*), movies (e.g., *The Beast from 20,000 Fathoms*), and other publications. Knight left behind more than 750 drawings and 150 oil paintings.

1953 Forrest J. Ackerman (1916–2008), one of the most avid promoters of science fiction, is awarded the inaugural Hugo Award as "#1 Fan Personality." Ackerman strongly influenced Ray Harryhausen, the stop-action animator of several dinosaur-related films. Ackerman, whose collection of movie memorabilia included models from numerous dinosaur-related films (e.g., *King Kong, The Lost World*), was a key figure in the wider acceptance by society of science fiction as a literary, artistic, and film genre.

1953 In Frederick Pohl's "The Ghost-Maker," a spirit is summoned into a *T. rex* at a museum. Luckily, the *T. rex* does not escape from the museum.

1953 *Life* magazine begins its "The World We Live In" by featuring on its cover Rudolph Zallinger's full-color painting of *Apatosaurus*. This article helped rekindle "dinomania" in the 1950s.

1953 On television's *Space Patrol,* the heroes meet a sauropod-like "Ice Demon." During one episode, which was broadcast live, the monster fell over.

1953 Richard Swann Lull publishes his third revision of Edward Hitchcock's study of dinosaur footprints. Ichnology, which had started with names to identify tracks, and then to names identifying the track makers, had now returned to names identifying the tracks. Lull interpreted the extinction of dinosaurs as a consequence of orthogenesis, or "racial senility" (i.e., driven to preordained extinction, independent of the environment).

1953 Roy Chapman Andrews's children's book *All about Dinosaurs* recounts his expedition's reaction to finding its first fossil in Mongolia: "We were so excited that we laughed and shouted and shook hands. We pounded each other on the back and did all the things men do when they are very happy." Andrews described *Brontosaurus* (i.e., *Apatosaurus*) this way: "The small brain weighed less than a pound. It shows that the creature was just about as stupid as an animal could be and still live."

1953 Rudolph Zallinger publishes a compositional study of *Tyrannosaurus* (produced in 1943 before he started his "The Age of Reptiles" mural for YPM) in the fifth installment of "The World We Live In" series in *Life* magazine. This portrayal of *Tyrannosaurus*, which was used in the movie *The Land Unknown,* inspired countless subsequent portrayals of *T. rex.* In 1967, Zallinger completed a 60-foot-long mural titled "The Age of Mammals" at YPM.

1953 The discovery of a deep underwater canyon called the Great Global Rift by American geophysicist Maurice Ewing (1906–1974) and Bruce Heezen (1924–1977) establishes plate tectonics and revolutionizes geology.

1953 The Hall of Dinosaurs reopens at AMNH after being renovated by Edwin Colbert. For the first time, dinosaur tracks excavated by Roland Bird near Glen Rose, Texas, were displayed in the hall.

1953 Trying to capitalize on another reissue of *King Kong,* the Warner Brothers movie *The Beast from 20,000 Fathoms* (based on Ray Bradbury's "The Fog Horn" from 1951) features a giant marine dinosaur that has been trapped in polar ice that is melted by a nuclear explosion. The monster then rampages New York City, its ancient breeding grounds. After eating a scientist, the "Rhedosaurus" eats a New York cop. When the army wounded the monster, people realized that everyone who came into contact with its blood died. The monster then started destroying Coney Island before being shot with a radioactive dart from atop the rollercoaster. *The Beast from 20,000 Fathoms* ("Raging Up from the Bottom of Time to Lose Its Crushing Fury on the City!") was the first movie about a prehistoric creature that awakened to menace modern civilization, as well as one of the first films to feature a marine dinosaur. *The Beast from 20,000 Fathoms* ("They couldn't escape the terror! And neither will you!"), which was Ray Harryhausen's first solo attempt at stop-action animation, cost $200,000 to make and was sold to Warner Brothers for $450,000. The film subsequently grossed more than $5,000,000 and helped trigger a decade of dinosaur-based movies, one of which was *King Dinosaur* (1955; "Terrifying! Fantastic! Startling!"), a movie made in less than two weeks that featured dinosaurs in outer space. (*T. rex* in this movie was an iguana held up by wires.) Like many other movies of this era, *King Dinosaur* included footage from *One Million B.C. The Beast from 20,000 Fathoms* was imitated with *The Giant Behemoth* (1959; "The Biggest Thing since Creation!") and *Gorgo* (1961), two British films directed by Eugene Lourie (the director of *The Beast from 20,000 Fathoms*).

1954 An episode titled "Operation Dinosaur" of the children's television-series *Rod Brown of the Rocket Rangers* is broadcast live on CBS. In the show, Rod used a time machine to return to the age of dinosaurs.

1954 Chung Chien "C.C." Young, "the Father of Chinese Paleontology," names *Mamenchisaurus* from bones unearthed two years earlier during construction of the Yitang Highway in Sichuan, China. Since the fossil was found near a construction site, Young named the type-specimen *Mamenchisaurus constructus. Mamenchisaurus* remains the largest dinosaur found in Asia; the largest specimen is more than 72 feet long (49 feet of which is the neck) and weighed 30 tons. Young described a variety of Chinese dinosaurs, including *Chialingosaurus,* China's first stegosaur.

1954 Rainer Zangerl (1912–2004) notes that although sauropods "have a great deal of public appeal, [our headless and truncated *Apatosaurus*] in its present condition does not convey a satisfactory idea of the animal. . . ." Two years later, Zangerl produced the popular *Dinosaurs, Predator and Prey*.

1954 Ed Delfs of the Cleveland Museum of Natural History begins excavations near Cañon City, Colorado. Delfs and his crew found a new species of the sauropod *Haplocanthosaurus* (*H. delfsi*), which Jack McIntosh and Mike Williams named in 1988. *Haplocanthosaurus* weighed about seven tons.

1954 Fossils of *Xenoceratops* are collected, but left unidentified, by Ottawa's Canadian Museum of Nature. The two-ton *Xenoceratops*, which was described in 2012 by Michael Ryan and David Evans, was a horned dinosaur that resembled *Triceratops* but lived 15 million years earlier (i.e., about 80 million years ago). *Xenoceratops* is one of the oldest known large-bodied horned dinosaur ever found in Canada.

1954 In "Through the Time Barrier," Superman, his friends from *The Daily Planet,* and a gangster go back in time to 50,000 BCE, where they meet dinosaurs.

1954 Roland Bird's article in *National Geographic* titled "We Captured a 'Live' Brontosaur" tells the world about the Paluxy River dinosaur tracks near Glen Rose, Texas. Bird was one of the first to interpret (rather than merely report the presence of) dinosaur tracks; he used the tracks to infer behaviors of dinosaurs. Although Bird knew that dinosaurs did not drag their tails, scientists disagreed. Bird was right. That same year (i.e., 1954), O. W. Willett reported finding six human footprints among dinosaur tracks in what is now DVSP.

1954 Russian paleontologist Evgeny Maleev (1915–1966) describes *Therizinosaurus,* a giant theropod discovered in the 1940s in Mongolia by a Soviet–Mongolian expedition. Maleev, who named 10 dinosaurs during his career (five of which are still considered viable), believed that the fossils could be from a turtle. Some claws of *Therizinosaurus* were up to a meter long.

1954 The Chialu Company begins selling dinosaur toys in Italy. These finely painted toys, made of plaster covering a metal frame, are among the rarest dinosaur-toys.

1954 The movie *Gojira* is released in Japan. The star of the movie, a green, 400-foot-tall, amphibious, two-legged *Stegosaurus*-like monster (which also had traits of *Tyrannosaurus* and *Iguanodon*) awakened from hibernation (since the Mesozoic) in the Pacific Ocean by atomic testing, terrorizes Tokyo. (*Gojira* was the nickname of an employee at Toho International Studios where the movie was made.) The movie—the first Japanese film to feature giant monsters—was meant to highlight problems unleashed by the atomic bomb, but its serious ideas were lost in the American version of the movie. When it was released in the United States by Embassy Pictures in 1956, *Gojira* was named *Godzilla, King of the Monsters!* Godzilla represented a living form of radiation, and the theme of the movie was the terror of nuclear weapons. The American version of the movie included footage featuring actor Raymond Burr (as reporter Steve Martin, one of the first people to see the monster rise from the ocean). Godzilla, the standard monster-gone-amuck, was an "incredible, unstoppable titan of terror" that was killed by an "oxygen destroyer" that eliminated free oxygen from the water in which Godzilla lived. *Godzilla, King of the Monsters!* included no stop-action animation; instead, the monster was played by actor (and World War II pilot) Haruo Nakajima (b. 1929) in a 200-pound Godzilla suit (Nakajima later wore suits as Rodan and other monsters). In America, *Godzilla, King of the Monsters!* was the most popular Japanese movie ever. In the more than 30 remakes and spin-offs of *Godzilla, King of the Monsters!,* Godzilla battled monsters that included giant lobsters (Ebirah), spiders (Spiga), snakes (Manda), pterosaurs (e.g., Rodan), King Kong, moths (Mothra), Space Godzilla, Mecha-Godzilla, sea monsters, extraterrestrial creatures (e.g., Ghidorah), 330-foot-tall bugs (Megalon), and shapeless blobs created by pollution (as in *Godzilla versus Hedorah* [1971]). However, none of the sequels had the impact of the original version. Japanese writer Shigeru Kayama later novelized Gojira from the movie script, but Kayama's novels were never published in English. American moviemaker Steven Spielberg described *Gojira* as "the most masterful of all the dinosaur movies because it made you believe it was really happening."

1954 There have been nearly 40 major paleontological expeditions to the area that now includes DPP during the past six decades. This area, which exposes sediments 78–74 million years old, houses the taxonomically richest American dinosaur fauna; 14 families and more than 40 genera of dinosaurs have been excavated from the park. Ornithischians outnumber theropods by 4:1; half of the ornithischians are hadrosaurs, and another 25 percent are ceratopsians.

1954 *Turok, Son of Stone,* a comic-book series published by Four Color Comics (which also produced classics such as *Woody Woodpecker*), debuts and features the fictional American character Turok, a Native American Kiowa warrior and his young friend Andar. Turok and Andar are trapped in Lost Valley inhabited by cavemen and dinosaurs they called "honkers." Every issue of *Turok, Son of Stone* featured a dinosaur or other prehistoric creature on its cover. Images of dinosaurs popularized by Rudolph Zallinger and Charles Knight appeared repeatedly in the comics. When *Turok, Son of Stone* was produced by Valiant Comics, the Lost Valley was moved from Dakota badlands to an alien dimension, where the dinosaurs became mean.

1955 After the Drumheller and District Museum Society is formed, a small museum to feature dinosaurs is developed in Drumheller, Canada. The first large exhibit was a 29-foot-tall *Edmontosaurus* loaned by the National Museum and installed by Charles M. Sternberg. Thereafter, Drumheller began promoting itself as the "Dinosaur Capital of Alberta."

1955 Aided by the Geological Survey and Chevron Oil, Loris Russell and Wann Langston find two skulls of *Pachyrhinosaurus* in southern Alberta.

1955 Although Godzilla was killed in its movie debut, it reappears in *Gigantis, the Fire Monster* (titled *Godzilla Raids Again* in the United States), the first of more than 20 sequels to *Godzilla* (1954). Copyright issues forced the movie's producers to change the name of Godzilla to Gigantis (viewers are told that Gigantis is Godzilla's "scientific name"). In the movie, two pilots find an isolated island at which Gigantis is battling an ankylosaur-like monster. (Paleontologists in the movie claimed that the monster had three brains; this was inspired by Othniel Marsh's suggestion in 1877 that *Stegosaurus* had two brains.) After the fighting animals fall into the ocean, Gigantis appears at Osaka Bay. Pilots use flares to draw Gigantis away from shore, but criminals start a fire in the bay and Gigantis returns. Osaka is almost destroyed, Gigantis sinks a Japanese fishing boat, and Gigantis is finally buried beneath tons of ice.

1955 Arthur Pack, who had earlier transformed Ghost Ranch (New Mexico) into a dude ranch, donates Ghost Ranch to the Presbyterian Church for use as an educational retreat. Four years later, Pack, with his wife Catherine and William "Bill" Carr, founded the Ghost Ranch Museum. In 1976, Ghost Ranch was designated a National Natural Landmark.

1955 As part of Alberta's 50th Jubilee Year, government officials in Alberta, Canada vote to protect part of the Red Deer River and ask Charles M.Sternberg for advice about a museum to display the area's dinosaurs. Sternberg recommended the now-disappeared town of Steveville (southeast of Drumheller), where almost half of the world's known Upper Cretaceous dinosaurs had been discovered. In 1955, Steveville Dinosaur Provincial Park was established. In 1962, the park's name was shortened to Dinosaur Provincial Park, which in 1979 became the first World Heritage Site ever established because of its paleontological importance. The first warden of Steveville Dinosaur Provincial Park (Alberta naturalist Roy Fowler) soon began preserving several of the local fossils. Today, the badlands along Red Deer River are within the 22,000-acre Dinosaur Provincial Park, and the area's museum showcases one of the best Late Cretaceous collections of dinosaurs in the world.

1955 British paleontologist Gavin de Beer (1899–1972) produces a 70-page monograph describing the London specimen of *Archaeopteryx*. This was the most comprehensive study of the specimen since Thomas Huxley's study 88 years earlier.

1955 Edwin Colbert proposes that the thick domes of skulls of pachycephalosaurs were an adaptation used as a battering ram. Later studies questioned this claim, citing the fact that injuries to the domes of the skulls are rare. (One of these rare injuries was reported in 2012 by Joseph Peterson and Christopher Vittore.) The thickened domes may have been social signals for other pachycephalosaurs.

1955 Ernie Nelson opens Prehistoric Gardens in southwestern Oregon. The attraction, which featured numerous life-size dinosaurs, grossed $18,000 during its first year.

1955 In A. Porges's "The Ruum," an alien spacecraft leaves an artifact (i.e., a *ruum*) on Earth during the Mesozoic that contains prehistoric animals in suspended animation. A modern-day miner finds the ruum and gets rich selling dinosaurs.

1955 In the Czechoslovakian movie *Journey to the Beginning of Time* (directed by Karel Zeman), four awestruck boys "living the excitement every boy dreams about" travel back in time on a river to "an age when gigantic monsters roamed the earth." The film opens with the boys' visit to AMNH, after which a rowboat rental on a lake in Central Park leads them to the river of time. *Journey to the Beginning of Time* was the first movie featuring stop-action dinosaurs projected in color.

These dinosaurs were modeled after Czech artist Zdenêk Burian's paintings from 1950 (which, in turn, had been inspired by Charles Knight). Burian produced 381 dinosaur-related oil paintings and 114 watercolors of prehistoric life. Zeman's movie was released in the United States in 1960, after which it became a favorite in children's shows.

1955 Louis Marx is declared "The Toy King" by *Time* magazine as his company begins producing toy dinosaurs depicting *Allosaurus, Ankylosaurus, Brontosaurus, Stegosaurus, Cynognathus, Trachodon,* and *Triceratops*. Louis Marx was the first inductee in the Toy Hall of Fame; in 1955, his company spent $312 on advertisements and had revenues exceeding $50 million.

1955 Sinclair Oil and its famous *Brontosaurus* logo, which appears at the company's 2,500 gas stations, is ranked #21 in the Fortune 500. By 1969, Sinclair ranked #58 in the same listing.

1955 The cheesy movie *King Dinosaur*—which promised "Daring Love!" and the "Mightiest Monster of Them All!"—features astronauts on the planet Nova fighting several monsters. The astronauts use an atomic bomb to destroy the planet. Posters promoting *King Dinosaur* promised *T. rex,* but the dinosaur was an iguana. This movie, which claimed that dinosaurs became extinct "as the result of fighting one another," was later satirized by *Mystery Science Theater 3000.*

1955 The dapper Barnum Brown returns to Montana for his last official field-season.

1955 The first U.S. postage stamp to feature prehistoric life is a 3¢ stamp showcasing the mastodon *Mammut americanus*. U.S. postage stamps featuring dinosaurs would not appear for another 15 years.

1955 While visiting his grandfather in New Jersey, fourth-grader Robert Bakker sees the September 7, 1953, issue of *Life* magazine. That issue, with *Brontosaurus* on its cover, included a full-color, six-page foldout of Rudolph Zallinger's *The Age of Reptiles* mural and an article titled "The Pageant of Life" by Lincoln Barnett describing life through the Mesozoic. According to Bakker, "I made up my mind then and there that I would devote my life to the dinosaurs."

1956 A. S. Romer's influential *Osteology of the Reptiles* depicts *Brontosaurus* with a *Camarasaurus* skull.

1956 Workers near Langenaltheim, Germany, find the "Maxberg specimen" of *Archaeopteryx* in the same quarry that produced the "London specimen." The specimen was described in 1959 by Florian Heller, a paleontologist from the Bavarian Erlangen University. Eduard Opitsch, the owner of the specimen, displayed the specimen at the Maxberg Museum in Solnhofen, but in 1982 he moved the fossil to his home. After Opitsch's death in 1991, the specimen—which consists of only a torso—was missing. Many residents of Solnhofen suspected that Opitsch buried the fossil to keep it from his heirs. The "Maxberg specimen" was the fourth known specimen of *Archaeopteryx*. Today, the location of this specimen is unknown; many suspect that it was sold secretly.

1956 Alfred Romer renames *Helopus* (which already referred to a bird) *Euhelopus*. (Although *Euhelopus* is also a genus of grass, the name was permitted because a genus name can refer to organisms in different kingdoms.) *Euhelopus* was a 50-foot-long, 15-ton sauropod that lived during the Early Cretaceous in what is now Shandong Province in China. Unlike most other sauropods, *Euhelopus* had longer forelimbs than hind limbs.

1956 AMNH's renovated Cretaceous Dinosaur Hall opens to the public. The hall includes *Tyrannosaurus* and *Triceratops*.

1956 Claire Patterson's article "Age of Meteorites and the Earth" is published in *Geochimica et Cosmochimica Acta*. Using three different radiometric dating methods, Patterson documented that ancient ocean-sediments and several meteorites were all 4.55 billion years old. Subsequent studies of other meteorites, lunar rocks, and other samples have consistently shown Earth to be 4.5–4.7 billion years old. If this period were condensed into a single year, the oldest known rocks would date from about mid-March, living organisms would appear in April, land plants and animals would appear near Thanksgiving, dinosaurs would dominate in mid-December (but disappear on Christmas Day, humanlike organisms would appear sometime in the evening of New Year's Eve, Rome would rule the Western world for five seconds (from 11:59:45 to 11:59:50), and Columbus would arrive in America at 11:59:57.

1956 Harry Sinclair, whose company helped popularize *Brontosaurus* by incorporating it as part of its logo, dies in Pasadena, California, and is interred in Roman Catholic Calvary Cemetery in eastern Los Angeles.

1956 In American writer Lyon Sprague de Camp's (1907–2000) "A Gun for Dinosaur," which appeared in *Galaxy* magazine, hunters use a time machine at the "big University" to go back in history and hunt dinosaurs. One of the hunters is killed by a lumbering theropod. In 1970, several British paleontologists admitted that their interpretations of *Pachycephalosaurus* ("thick-headed lizard")—that is, the animals' clashing of heads, as can be observed today in animals such as sheep—were inspired by de Camp's essay. De Camp wrote his essay after becoming annoyed with inaccuracies of other dinosaur-themed stories. "A Gun for Dinosaur" was adapted to radio, comic books, and other media. De Camp's *The Day of the Dinosaur* (1968) was coauthored by de Camp's wife, Catherine.

1956 Just before becoming director of the National Museum of Canada, geologist and Mayflower descendant Loris Russell (1903–1998) suggests that dinosaurs might have been warm-blooded; as he later noted, "I have long held that the great success of the dinosaurs over other reptiles implies a fundamental difference, and that this difference was the possession by the dinosaurs of a better circulatory system and an independent body temperature." Russell's suggestion was ignored.

1956 Paleontologist Max W. de Laubenfels (1894–1960) suggests that dinosaurs became extinct because of a meteoritic impact. Laubenfels's proposal was not taken seriously, but was revived in 1980 by Walter and Luis Alvarez.

1956 *The Beast of Hollow Mountain,* featuring dinosaurs designed by Marcel Delgado, is the first dinosaur-movie presented in the wide-screen "CinemaScope" format. *Allosaurus* starred in *The Beast of Hollow Mountain,* as well as in the 1969 movie *The Valley of Gwangi*, both of which combined dinosaurs with cowboys and Western themes (e.g., in one scene, Gwangi is in a wooden cage pulled by cowboys). *The Beast of Hollow Mountain* was advertised with ". . . It Came Out of Hiding to Kill! Kill! KILL!"

1956 The Chicago Natural History Museum (later the Field Museum) unveils mounted skeletons of *Gorgosaurus* looming over the duckbill *Lambeosaurus. Gorgosaurus* was the first dinosaur skeleton ever mounted without external supports.

1956 Toy companies such as Chicago's J. H. Miller Manufacturing Corporation and Louis Marx & Co. begin mass-producing plastic dinosaur-toys. The waxy toys made by J.H. Miller often melted if left in sunlight. Toys in the Louis Marx "Prehistoric Times" toy-sets were modeled after paintings by Charles Knight and Rudolph Zallinger's "Age of Reptiles" mural. Other companies that later mass-produced dinosaur toys included Orsenigo, Timpo, Alva, Abbeon, MPC, Cherilea, and Linde, whose toys—also based on Zallinger's mural—were often included as premiums in packages of coffee. In 2014, a Linde plastic dinosaur from a package of coffee was selling for about $15, and a Marx *Brontosaurus* was selling for about $40.

1956 Willis O'Brien and Ray Harryhausen (1920–2013; Figure 33) collaborate for the last time with their movie *The Animal World* ("Two billion years in the making!"), which included a 20-minute sequence featuring *Stegosaurus, Triceratops*, *Tyrannosaurus*, and an egg-laying *Apatosaurus*. The all-color movie was graphic, and some of its bloodiest scenes were deleted before the movie's release (especially the battle in which *Ceratosaurus* kills *Stegosaurus*[32]). Harryhausen used his stop-action "Dynamation" to create dinosaurs in movies such as *One Million Years* B.C. (1966) and *The Valley of Gwangi* (1969).

Figure 33 Raymond Frederick "Ray" Harryhausen (1920–2013) was an American writer, producer, and visual-effects specialist who brought dinosaurs to life in movies such as *The Beast from 20,000 Fathoms* (1953) and *One Million Years B.C.* (1966). Harryhausen, a protégé of Willis O'Brien (*The Lost World, King Kong*), was a master of stop-action animation. This photo from 1956 shows Harryhausen with one of the models used in his dinosaur movies. In 1992, Harryhausen received an Academy Award for Lifetime Achievement. (Warner Bros. Pictures/Photofest)

1956 Zdenêk Burian, working with Czech paleontologist and geologist Josef Augustana (1903–1968), publishes *Prehistoric Animals,* his most successful book. During his lifetime, Burian produced more than 1,000 oil paintings and illustrated more than 450 books and 500 short stories.

1957 A glass-walled museum is erected over DNM's quarry of bones. Although Earl Douglass had shipped tons of the bones to CMNH, he wanted the bone bed to be a museum at which visitors could see and appreciate paleontology. Douglass's dreams of fancy restaurants and an airport at the site remain unfulfilled.

1957 Chinese paleontologist Chow Minchin (formerly Zhou Min-zhen; 1918–1996) and C. C. Young found *Vertebrata PalAsiatica*, the world's first journal devoted exclusively to vertebrate paleontology. That same year, the Cenozoic Research

Laboratory changed its name to the Institute of Vertebrate Paleontology, and in 1960 to the Institute of Vertebrate Paleontology and Paleoanthropology (IVPP), which remains its current name.

1957 CMNH paleontology curator Leroy Kay (a protégé of Earl Douglass at the CMNH quarry that became DNM) persuades trustees of the newly established UFHNH in Vernal, Utah, to accept a cast of *Diplodocus carnegii.* Thanks to funds provided by the Vernal Lion's Club, the cast was shipped to Vernal, where—under Kay's supervision—a cement cast was mounted in front of the museum.

1957 In *Space Ship Sappy*, the Three Stooges are threatened by a dinosaur. That dinosaur was one of the giant lizards from the movie *One Million B.C.*

1957 In *Twenty Million Miles to Earth*, an expedition returning from Venus crashes in the Mediterranean. The only two survivors are the spaceship's captain and a small Venusian packed in a metal cylinder. When the cylinder washes ashore in Italy, boys open it and a tiny, two-legged, dinosaur-like creature called Ymir emerges. Ymir, who was brought to life by Ray Harryhausen's stop-action animation of an 18-inch-tall model, then escapes and grows to be 20 feet tall. It is captured and taken to Rome, where it again escapes and fights a circus elephant in the streets of Rome. In a scene reminiscent of King Kong atop the Empire State Building, Ymir is killed in the Colosseum.

1957 In *The Land Unknown*, four people survive a helicopter crash (caused by a collision with a pterosaur) to endure "a paradise of hidden terrors . . . deep in the icy Antarctic." There, more than 3,000 feet below the edge of the ice, they battle man-eating plants and dinosaurs such as *Tyrannosaurus* (portrayed by actors in dinosaur costumes based on paintings by Rudolph Zallinger). In 1965, the head of the *Tyrannosaurus* costume appeared as part of flame-breathing "Spot" in reruns of *The Munsters*.

1957 James A. ("Dinosaur Jim") Jensen (1918–1998), a high-school dropout who became a famous paleontologist, begins mounting dinosaurs with supports placed inside bones, thereby producing free-standing skeletons with no or few obvious supports.

1957 Sidney Enlow and Donald Brown discover that dinosaur bones have a dense Haversian system of vessels that can supply the tissue with large amounts of blood. The presence of such vessels typifies animals having high metabolic rates.

1957 Thanks to a gift from Caroline McGill (1879–1959), the Museum of the Rockies (MOR) opens on campus at Montana State University in Bozeman, Montana. In 1982, MOR had three specimens of dinosaurs, but today it houses the largest collection (more than 30,000 specimens) of North American dinosaur fossils in United States. It is also a national repository for dinosaur fossils collected on public lands, and houses one of only three histology labs in the world (and the only one in the United States) dedicated to dinosaurs. MOR also houses the world's largest collect of *T. rex,* one of which (named "Big Mike") greets visitors to the museum. MOR's curator of paleontology is John R. "Jack" Horner (b. 1946).

1957 The Sears Christmas Catalog advertises a 47-piece Marx "Prehistoric Life" toy set that includes *Brontosaurus, Hadrosaurus*, and *Tyrannosaurus* for $4.67. Five years later, the catalog included the last Flintstones set and the last Marx set of dinosaurs for many years.

1957 Wealthy engineer Joseph Tyrrell dies at age 99. Tyrrell discovered only one dinosaur (*Albertosaurus*), but that was the first dinosaur discovered in Canada. Tyrrell is memorialized by Alberta's RTM.

1958 A Visitor Center opens at DNM showing the original north "dinosaur wall" of the quarry to the public. Structural problems later forced officials to close the building, but renovations were completed for the building's reopening in 2011. The sandstone wall is 30 feet high and 190 feet long and exposes more than 1,500 bones (including those of a baby *Stegosaurus* and a partial skeleton of *Camarasaurus*). DNM is visited annually by thousands of visitors.

1958 Arthur Coggeshall, who had helped assemble casts of *Diplodocus carnegii* throughout the world, dies while directing the Santa Barbara Museum of Natural History.

1958 Cartoonist Johnny Hart (1931–2007) launches *B.C.,* a comic strip featuring cavemen and anthropomorphic talking animals. The cast of animals included a dinosaur called "Gronk," which had spinal plates like those of *Stegosaurus*. Like *Alley Oop*, B.C. was featured in several other media, including television specials, animated commercials, and video games. Until 2005, Johnny Hart's hometown (Broome County, New York) hosted a PGA tour event called the *B.C.* Open. The Broome County Parks Department continues to feature Gronk as its mascot.

1958 China issues the first postage stamp (costing 8 fen) anywhere in the world to feature a dinosaur (*Lufengosaurus huenei).* Since then, hundreds of different sets of dinosaur stamps have been issued by countries throughout the world.

1958 Following a visit to DNM, Utah artist Elbert Porter begins designing the 14 life-size dinosaurs that were displayed for six years at West Yellowstone, Montana, seven years at Orderville, Utah, and since 1977 at UFHNH.

1958 From Vernal, Utah, CMNH molds of *Diplodocus carnegii* are shipped to Rocky Mount Children's Museum in Rocky Mount, North Carolina. Because of age-related damage to the molds, a cast was never prepared. The molds then went to the Houston Museum of Science, where they were used to add the missing pieces of a *Diplodocus hays*. The specimen at the Houston Museum of Science was completed in 1975.

1958 In Brian Aldiss's short story "Poor Little Warrior," dinosaur hunter Claude Ford kills a brontosaurus, after which its lobster-sized parasites attack Ford.

1958 In Isaac Asimov's "A Statue for Father," a physicist uses a "Chrono-funnel" to retrieve dinosaur eggs from the Mesozoic. The scientist incubates and hatches the eggs. When one of the dinosaurs is killed, the scientist likes the taste of its meat, and later gets rich by marketing the dinosaur meat as "dinachicken." The scientist inadvertently ended world hunger, and a statue was erected in his honor.

1958 In Syd Hoff's children's story *Danny and the Dinosaur,* a toothless talking dinosaur enjoys a day with Danny, but they part ways when Danny notes, "We don't have room for a pet that size."

1958 The Field Museum unveils a reconstructed skeleton of *Apatosaurus,* fitted with a skull of *Camarasaurus* from CMNH. On the accompanying plaque, the dinosaur's name was changed to the incorrect (but more popular) *Brontosaurus*. The skeleton would not possess a valid skull (again provided by CMNH) for another two decades. *Apatosaurus* remains one of the most familiar and abundant dinosaurs in the Morrison Formation; it has been found in every state that exposes the Morrison Formation.

1958 Jim Jensen visits the Smithsonian Institution in Washington, DC, and sees a sauropod bone found by Vivian and Ed Jones of Delta, Colorado. In 1964, after moving to Utah, Jensen visited the Jones and learned where they had found the bone. This led in 1971 to the founding of Dry Mesa Quarry, the site of several discoveries of famous dinosaurs, including *Supersaurus* and *Brachiosaurus*.

1959 Elmer Riggs, now 90 years old, visits the Field Museum, where he views the *Apatosaurus* and says, "By, Joe, it was a pretty good job."

1959 Florian Heller describes the "Maxberg specimen" of *Archaeopteryx*. This specimen was later lost.

1959 In an article in *Systematic Zoology* commemorating the 200th anniversary of the 10th edition of *Systema Naturae*, W. T. Stearn declares Linnaeus the type specimen of *Homo sapiens*. (Linnaeus, apparently not hesitant about promoting himself, wrote five autobiographies.) Paleontologist Robert Bakker later suggested that Edward Drinker Cope be the type specimen for *Homo sapiens,* but his suggestion was rejected. Others who have been suggested as type specimen for *Homo sapiens* include Bob Hope, Arnold Schwarzenegger, and Raquel Welch.

1959 In Argentina, paleontologist Osvaldo Reig (1929–1992) and a goat herder named Victorino Herrera discover one of the oldest dinosaurs, *Herrerasaurus*.

1959 In *Natural History,* paleontologist George Gaylord Simpson (1902–1984) argues, "A negative cannot be proved in the full sense of the word. There is *some* probability that there are little men on the far side of the moon, but the probability is infinitesimally small. For numerous reasons . . . the probability that there are living dinosaurs is only a little larger."

1959 In *The Giant Behemoth*, London is terrorized—for the first time since *The Lost World* (1925)—by a giant *Brachiosaurus*-like "Paleosaurus" awakened by atomic testing. The wounded dinosaur rampages the banks of the River Thames, and is finally killed when it is hit by a radioactivity-laced torpedo shot from a submarine. *The Giant Behemoth* ("The Biggest Thing since Time Began!"), which was directed by Eugene Lourie (1903–1991; the director of *The Beast from 20,000 Fathoms*), was released in Britain under the title *Behemoth, the Sea Monster. The Giant Behemoth,* which repeated a car crushing scene three times, was Willis O'Brien's last dinosaur-film.

1959 Sinclair Oil registers its current trademark, featuring a green sauropod and the company name bordered by a pentagon.

1959 The cover of *Space Journal* features Harry Lange's painting "Interstellar Space Ship Astra-Alpha Lands on Planet 'X'", which includes space ships and dinosaurs modeled on Rudolph Zallinger's *Age of Reptiles* on an alien planet.

1959 The movie version of Jules Verne's *A Journey to the Center of the Earth* (1864) is released by 20th Century Fox. Although the public's perception of Verne's novel was associated with dinosaurs, there were no dinosaurs in the movie (in the tradition of *One Million B.C.*, "dinosaurs" were played by decorated lizards and baby alligators). *Dimetrodon,* a pelycosaur (i.e., not a dinosaur), was not in Verne's novel, but appeared in the film as an iguana fitted with a sail. Similarly, Verne's novel describes a fight between two prehistoric marine reptiles, but the movie includes a slow-motion fight between two decorated iguanas. The movie described how a group of people (portrayed by Pat Boone, James Mason, and others) go down an extinct volcano into the Earth's hollow core. In the movie's closing scene, the rubber-horned iguanas made up to resemble *Dimetrodon* resemble one of Charles Knight's classic murals at Chicago's Field Museum. The 2008 remake of *Journey to the Center of the Earth* (starring Brendan Fraser) included a chase scene with a computer-generated *T. rex*.

1960 For the first time in 19 years, excavation resumes at Utah's Cleveland-Lloyd Dinosaur Quarry when University of Utah paleontologists begin a five-year project with several cooperating schools. The excavations were directed by William Lee

Stokes, with help from James H. Madsen, Jr. That same year, Princeton University mounted "Malcolm" the *Allosaurus* from the Cleveland-Lloyd Quarry; this was the first publically displayed dinosaur from the quarry. The specimen was named for Malcolm Lloyd Jr. a Princeton law graduate who helped fund Princeton's original excavations at the site in 1939. The Cleveland-Lloyd Quarry has yielded more than 10,000 bones from more than 70 animals; more than 60 casts and original skeletons from these bones are displayed throughout the world.

1960 French priest and paleontologist Albert de Lapparent (1905–1975) notes, "As the geologic knowledge of all parts of the world increases, the discovery of [dinosaur] prints is being more and more frequently reported." That same year, de Lapparent discovered theropod tracks north of Spitsbergen, Norway. These tracks were at latitude 78° (until that time no tracks had been reported north of latitude 56°). Replicas of seven of those tracks are preserved in Oslo's Paleontological Museum.

1960 Inspired by the earlier work by Maurice Ewing and Bruce Heezen, geologist and former U.S. Navy officer Harry Hess (1906–1969) proposes that seafloor forms at mid-ocean ridges, after which it spreads to oceanic trenches, where it is subsumed. The next year, R. S. Dietz (1914–1995) independently developed the same idea and called it *sea-floor spreading*. Hess's idea, which was published in 1962, explained how continents move, thereby supporting Alfred Wegener's earlier claims about continental drift.

1960 Hanna-Barbera produces *The Flintstones* (Figure 34), television's first prime-time cartoon series, which debuts on ABC and brings dinosaurs into people's homes. The show, which portrayed a "modern stone-age family" of humans who lived with dinosaurs, saber-toothed tigers, wooly mammoths, and other extinct animals, lasted 166 episodes (until 1966) and was a combination of *Alley Oop* comedy with Jackie Gleason's "The Honeymooners." In *The Flintstones*, Fred Flintstone uses a brontosaur as a power shovel, characters eat "brontoburgers," and a pet dinosaur named Dino barks like a dog. *The Flintstones* was popular and prompted Hanna-Barbera to create other prehistoric animations; the most famous of these was 1974's *Valley of the Dinosaurs*, which described humans and dinosaurs living in an unknown valley in South America. Despite including more accurate portrayals of dinosaurs, *Valley of the Dinosaurs* never achieved the popularity of *The Flintstones. The Flintstones* was later revived in movies, television ads, and countless products such as toys, comic books, clothing, and vitamins. *The Flintstones,* which was Hanna-Barbera Productions' most successful cartoon series, was ranked by *TV Guide* in 2013 as the second-greatest television cartoon of all time. *The Flintstones* indoctrinated millions of children and adults with the notion that humans lived with dinosaurs. This claim has been repeatedly rejected by biologists, geologists, and other scientists, but remains a foundation of young-Earth creationism and virtually all creation museums. For example, AiG's Creation Museum includes dioramas

Figure 34 *The Flintstones* was an animated, American television show about a working-class Stone Age family and their neighbors. In the show's 166 original episodes that were broadcast from 1960 to 1966, humans used dinosaurs and other extinct animals as vacuum cleaners, showers, elevators, horns, clocks, and other machines. *The Flintstones*' portrayal of humans living with dinosaurs, a belief endorsed by almost half of U.S. citizens, is a foundation of creation museums such as Ken Ham's Creation Museum in Kentucky and Carl Baugh's Creation Evidence Museum in Texas. *The Flintstones* was television's first prime-time animated series. (ABC/Photofest)

depicting humans living with, playing with, and riding on dinosaurs. In 2002, NSF reported that 48 percent of Americans believed that humans and dinosaurs lived at the same time.

1960 In the comic book series *The War That Time Forgot*, American soldiers are stranded on an uncharted Pacific island inhabited by dinosaurs. The series, which featured "Moon-Boy, the first human," ended in 1968, and was revived as a limited series in 2008.

1960 In *The How and Why Wonder Book of Dinosaurs* by Darlene Geis, illustrator Kenyon Shannon includes fat sauropods in poses originally depicted by Charles Knight.

1960 Joseph Augustana's *Prehistoric Animals* includes a painting by Zdenêk Burian showing a *T. rex* attacking duckbilled dinosaurs.

1960 Neave Parker's (1910–1961) "The Last Days of the Dinosaurs" is published in *Illustrated London News*.[33] The illustration shows dinosaurs and smoldering volcanoes, with the foreground populated by mammals destined to survive the impending mass extinction. Parker's art was featured on 19 different postcards sold by BMNH; these cards are now valuable collectibles.

1960 Released 35 years after the debut of the original movie-version of *The Lost World,* the full-color remake of Sir Arthur Conan Doyle's *The Lost World* promises viewers that "Your Mind Won't Believe What Your Eyes Tell You!" They were right; most viewers were disappointed by dinosaurs that were lizards and alligators adorned with fins and horns (and which looked nothing like real dinosaurs). Producer Irwin Allen (1916–1991) hired Willis O'Brien (1886–1962) to oversee the film's special effects, but there is none of O'Brien's famed stop-action animation—decorated lizards and alligators were cheaper and faster to film. Professor Challenger, played by actor Claude Rains (1889–1967), points out a "brontosaur"—by now an iconic dinosaur well known to most viewers—that is a monitor lizard with a decoration on its short neck. Similarly, the popular *T. rex* was portrayed in the film by another lizard having a horn on its nose and a plastic fin on its back; viewers knew that this dinosaur did not have either. In the remake, one scientist—when asked if the dinosaurs he has seen are big—angrily responds that he has never seen a small dinosaur. For many critics, *The Lost World* (1960), which did not include the brontosaur-gone-amuck-in-London final scene, was a disappointing movie. However, clips from the movie later appeared in other productions, including *Voyage to the Bottom of the Sea.*

1960 Roy Chapman Andrews, the model for the movie character Indiana Jones, dies of a heart attack in California. Andrews spent his entire professional career at AMNH and became famous as a fossil hunter, despite the fact that he found few fossils himself. Although Andrews was not an influential scientist, he was a spectacular promoter of science and scientific expeditions. Andrews was buried in a family plot in Beloit, Wisconsin.

1960 Sixteen of Neave Parker's illustrations are displayed in "The Prehistoric World of Neave Parker."

1960 The first *T. rex* (AMNH 5866) found by Barnum Brown in Wyoming in 1900 is transferred to London's BMNH. Today, this specimen (BMNH R7994) and a second specimen (BMNH R7995) are displayed there. The specimen was originally named *Dynamosaurus imperiosus* ("powerful imperial lizard"), but Osborn abandoned this name in favor of *T. rex*. When young John Ostrom sought advice from Brown about a research topic, Brown suggested that Ostrom examine some fossils that Brown had collected three decades earlier on the Crow Indian Reservation in Montana. Ostrom didn't study these fossils for his doctoral dissertation, but studied them soon thereafter. Indeed, Ostrom took Brown's "Daptosaurus" out on loan from AMNH (along with Brown's notes on how to get to the Beauvais Creek sites where he had collected the specimens). This eventually led Ostrom to Montana, where he found *Deinonychus*. Ostrom's subsequent papers described the 100-pound theropod as "an animal so unusual in its adaptation that it will undoubtedly be a subject of great interest and debate for many years among students of organic evolution." Ostrom was right; *Deinonychus* became a bridge that transformed dinosaurs' images from slow, lumbering giants to fast, smart, and agile animals.

1960 The movie *Dinosaurus!* ("Alive after 70 Million Years! Roaring! Walking! Destroying!") includes stop-motion dinosaurs, one of which (*T. rex*) fights a steam shovel. Two frozen dinosaurs (*Brontosaurus* and *Tyrannosaurus*) and a caveman are found on an isolated island where a tourist resort is being built. The dinosaurs are struck by lightning and come back to life, along with a caveman who comes back to life after thawing. In *Dinosaurus!*, *Apatosaurus* (i.e., *Brontosaurus*) gets stuck in mud, much like one did 35 years earlier in the silent film *The Lost World.*

1960 The nodosaurid *Silvisaurus* ("forest reptile") is described by Theodore Eaton, Jr. (1842–1910). A decade later, Yale University's John Ostrom described *Sauropelta,* another early Cretaceous nodosaurid.

1960 The Piltdown Men's *Brontosaurus Stomp* is a top 10 hit in Australia. A decade later, The Move's *Brontosaurus* became a top 10 hit in the United Kingdom.

1960 The U.S. Postal Service issues a 4¢ commemorative stamp honoring industrialist and philanthropist Andrew Carnegie. Carnegie funded several expeditions to the American West to excavate dinosaurs, many of which are displayed in CMNH. In the early 1900s, Carnegie's *Diplodocus carnegii* was distributed to museums worldwide.

1960 Utah's Cleveland-Lloyd Quarry reopens under the auspices of the University of Utah, at which time L. Grant Stokes (the brother of William Stokes) and graduate student James H. Madsen Jr. organize the University of Utah Cooperative Dinosaur Project. The organizers told potential partnering-organizations that their goal was to obtain for them a mounted dinosaur skeleton "for the price of 'a good car'" (then about $2,000). Participating institutions included Yale University, Brigham Young University, Buffalo Museum, California Academy of Sciences, and the University of Michigan. The first year's work extracted 1,500 bones. The University of Utah Cooperative Dinosaur Project ended in 1966.

1960 William E. Swinton proposes that the name *Archaeopteryx lithographica* be transferred from a fossilized feather to the "London specimen" of *Archaeopteryx*. In 1961, Swinton joined the staff of ROM and became its director two years later. Swinton argued that dinosaurs went into trees for protection and learned to fly as they returned to earth (i.e., the so-called arboreal "from the trees down" theory).

1961 Jim Jensen pioneers the use of plastic foam for casting dinosaurs' bones to reduce weight. Jensen's results were published in 12-page article titled "A New Casting Medium for Use in Flexible and Rigid Molds" in *CURATOR The American Museum of Natural History*. Jensen described many dinosaurs (e.g., *Palaeopteryx, Ultrasaurus*), but his publications were sometimes questioned (Jensen had little formal training). The Brigham Young University Museum of Paleontology was built around Jensen's discoveries.

1961 *Allosaurus,* excavated in 1939–1942 from Cleveland-Lloyd Quarry, is assembled by preparators at Yale University and mounted at Princeton University. *Allosaurus* has longer arms than *T. rex,* a small, bony crest above and forward of each eye, and three-fingered hands (*T. rex* had two-fingered hands).

1961 American company Nu-Card Sales releases some of the first dinosaur-based trading cards. Only after the release of the movie *Jurassic Park* in 1993 did such cards become popular.

1961 An exhibit titled "The Career of Charles R. Knight" opens at AMNH. This was the first major exhibition of Knight's work.

1961 Edwin Colbert describes hadrosaurs as "alert and lively."

1961 In "The Odyssey of Flight 33" episode of *The Twilight Zone*, an airplane breaks the time barrier and flies into the Mesozoic. People aboard the plane see *Apatosaurus* (from the movie *Dinosaurus!,* which was released the previous year).

1961 In *Valley of the Dragons*, a comet transforms Earth into a prehistoric world of cavemen and pet-shop lizards dressed as dinosaurs.

1961 John Whitcomb Jr. and Henry Morris ask Moody Press publish *The Genesis Flood: The Biblical Record and Its Scientific Implications*, a 518-page, footnote-laden book that became the most important book (aside from the Bible) in the history of modern creationism. *The Genesis Flood* claimed that geologic strata have been arranged by deceitful, atheistic geologists, despite the fact that the patterns of fossils in geological strata were documented (long before the publication of *On the Origin of Species*) by geologists indifferent or hostile to the concept of evolution. *The Genesis Flood* also included photos of the Paluxy riverbed footprints to claim that humans lived with dinosaurs, adding that scientific claims contrary to young-Earth creationism "need to be challenged in the name of the Holy Scripture." Modern young-Earth creationism, with its popular claim that humans lived with dinosaurs, can be traced to this book.

1961 Renowned wildlife-sculptor Louis Paul Jonas, Jr. (1894–1971), who had studied with famed taxidermist Carl Akeley (1864–1926), is paid $250,000 by Sinclair Oil to build nine dinosaurs for the Sinclair Oil Dinoland Exhibit at the 1964–1965 New York World's Fair (Figure 35). (The exhibit's final cost exceeded $400,000.) The exhibit included brontosaur hatchlings sculpted by Jonas, who was advised by Barnum Brown; this was one of Brown's last projects. (Brown died in 1963 and never got to see the dinosaurs at the Fair; he was replaced by YPM's John Ostrom.) Over the next three years, Jonas and his team of 18 engineers, paleontologists, and robotics experts used more than 45,000 pounds of clay to build nine life-size dinosaurs that were floated 125 miles down the Hudson River to New York for display at the Fair. After completing the project, Jonas proclaimed, "If I never do another job, I shall be fulfilled." After the Fair ended, the dinosaurs were sent to places such as DNM in Utah (which got *Stegosaurus*), DVSP in Glen Rose, Texas (which got *T. rex* and *Brontosaurus*; Figure 37), Disneyland, Independence, Kansas (whose Riverside Park got *Corythosaurus*), and the National Zoological Park (which got the *Triceratops* named "Uncle Beazley").

Figure 35 The biggest attraction at the 1964–1965 New York World's Fair was the Sinclair Oil Dinoland Exhibit, which featured nine life-size dinosaurs. This photo shows the assembly of those dinosaurs, some of which had moving parts. After being seen by more than 10 million people, the dinosaurs were given to Dinosaur National Monument, Dinosaur Valley State Park, and other tourist sites in the United States. (Arthur Schatz/Time Life Pictures/Getty Images)

Some of Jonas's casts are displayed at the Boston Museum of Science and elsewhere.

1961 The Carbon County Prehistoric Museum (later renamed the College of Eastern Utah Prehistoric Museum) opens on the upper floor of the Price (Utah) City Hall. The museum displayed a variety of artifacts, including a complete *Allosaurus* from the Cleveland-Lloyd Dinosaur Quarry. In its first year, more than 13,000 people visited the museum.

1961 The movie *Gorgo* ("Towering over the Cities of the World, as Millions Flee in Awesome Terror!") is a British version of the actor-in-a-dinosaur-suit inspired by *Godzilla. Gorgo* was produced and written (with David Hyatt) by Eugene Lourie, who had directed *The Beast from 20,000 Fathoms* and *The Giant Behemoth*. In the movie, when a monster captured off the coast of Ireland is taken to London as a circus attraction, its 1,200-foot-tall mother follows her baby to London. The movie includes some of the best "actor in a rubber dinosaur suit" scenes. After the mother destroys London while searching for her baby, both the mother and baby return to the sea.[34] When *Gorgo* was being made, London residents were shocked when a full-size model of Gorgo was driven through the city's streets. *Gorgo* inspired a television series and several comic books.

1961 The second year of renewed excavations at Cleveland-Lloyd Quarry unearths the theropod *Ceratosaurus,* the only known carnivorous dinosaur having a horn on its nose and smaller horns over its eyes.

1962 Alfred Crompton and Alan Charig (1927–1997) name *Heterodontosaurus* ("different toothed lizard"), a small herbivorous dinosaur that lived in the early Jurassic. *Heterodontosaurus,* which was discovered in South Africa, had five fingers and, unlike most dinosaurs and reptiles, three different types of teeth. *Heterodontosaurus* was the first well-preserved heterodontosaurid.

1962 Bones of *Barosaurus* collected at DNM by Earl Douglass are traded to ROM in Toronto, Canada. Floor-space limitations prevented the reconstruction of *Barosaurus* in ROM's renovated dinosaur exhibit in 1970. In 2007, ROM's *Barosaurus* fossils were rediscovered, and today, "Gordo," the *Barosaurus,* greets visitors to the museum's dinosaur hall.

1962 Cleveland-Lloyd Quarry ships two skeletons of *Allosaurus* for mounting to Tokyo's National Science Museum of Japan, and another skeleton to the Los Angeles County Museum. In subsequent years, the quarry exported skeletons throughout the world.[35] The *Allosaurus* from Cleveland-Lloyd Quarry mounted at the National Science Museum in Tokyo was the first dinosaur ever exhibited in Japan.

1962 Edwin Colbert publishes the first systematic estimations and comparisons of the masses of dinosaurs. Colbert's model, which was based on displacement by scale models, estimated *Brontosaurus* to weigh 28–32,000 kg, *Diplodocus* 11,000 kg, and *Brachiosaurus* 78,000 kg. Since 2000, estimates (based on computer simulations) of the mass of *Brachiosaurus* have averaged 25,000 kg.

1962 Entomologist Stanley Flanders suggests that insects consumed the world's forests 65 million years ago, and the resulting lack of food led to the extinction of dinosaurs.

1962 In an article titled "Onward and Upward with Science," *The New Yorker* recounts the feud between Othniel Marsh and Edward Drinker Cope from the late 1800s.

1962 In *The Planet of the Storms*, Soviet cosmonauts—who learn that Venus is experiencing a geologic age similar to Earth's Mesozoic—land on Venus and fight dinosaurs.

1962 Moshe Avnimelech announces the first discovery of dinosaur tracks in Israel. The tracks, which were described more fully four years later, were found in a Jerusalem garden.

1962 *Reptilicus,* Denmark's first movie about a prehistoric monster, depicts a dinosaur-like creature being brought back to life, after which it attacks Copenhagen. The movie's star—a puppet with a hinged neck that wiggled from side to side—was advertised as "Invincible! Indestructible!" because it could clone itself. Reptilicus did not resemble any real dinosaur.

1962 Sidney Dillon Ripley (1913–2001), head curator at YPM, notes that the museum continues to open boxes of fossils collected by Othniel Marsh.

1962 The National Natural Landmarks Program is established by the U.S. Secretary of the Interior (under authority of the Historic Sites Act of 1935) "to identify and encourage the preservation for the full range of geological and biological features that are determined to represent nationally significant examples of the Nation's national heritage." Numerous dinosaur-related sites were later made National Natural Landmarks, including Como Bluff, Hell Creek Fossil Area, Bridger Fossil Area, Cleveland-Lloyd Dinosaur Quarry, Morrison Fossil Area, DVSP, Ghost Ranch, and Dinosaur Trackway.

1963 In the final year of his life, Barnum Brown renews his work as a consultant for Sinclair's Dinoland Pavilion at the upcoming World's Fair that will include nine life-size fiberglass models of dinosaurs, including a 70-foot-long *Brontosaurus* with a moving head, a *Triceratops* with a moving head, a 20-foot-tall *Tyrannosaurus rex* with a moving jaw, and a 4-foot-long *Ornitholestes*. These dinosaurs, which were the most lifelike full-size restorations of dinosaurs, were floated on a barge down the Hudson River to the 1964–1965 New York World's Fair. As famed dinosaur-hunter Roland Bird (Figure 29) noted, "The dinosaurs were towed around Manhattan Island for all who wished to come to the waterfronts to watch, as people did in hundreds of thousands. Quite a stunt." The Dinoland Pavilion was a highlight of the Fair; more than 10 million visitors viewed Sinclair's dinosaurs. (Today, the site of the fair is marked by the 12-story-tall Unisphere in Flushing Meadows-Corona Park in Queens, New York.) When the Fair closed, the dinosaurs were taken on a 37-city tour.

1963 Edwin Colbert, Barnum Brown's successor at AMNH, tells Brown, "There are, in [AMNH's] Tyrannosaur Hall, 36 North American dinosaurs on display. . . . You collected 27, an unsurpassed achievement." Soon thereafter, Brown died in New York (one week shy of his 90th birthday) and was buried in the River View Cemetery in Oxford, New York. Brown was interred atop a bluff overlooking the Chenango River beside his first wife, Marion (1877–1910). When his second wife, Lilian (1887–1971), died eight years later, she was buried on the opposite side of Brown, but slightly closer to him than Marion. Famed fossil-hunter Roland Bird had earlier described Brown as one of the "three kings" of the dinosaurs (the other two "kings" being Othniel Marsh and Edward Drinker Cope). From 1896 to 1942, Brown gathered more than 1,200 crates containing fossils that have been seen by more than 100 million visitors to AMNH. Sinclair Oil hired John Ostrom to complete Brown's work for the New York World's Fair the next year. During his career, Brown named 21 dinosaurs, 10 (i.e., 48%) of which remain valid. Many of the crates shipped by Barnum Brown to AMNH remain unopened.

1963 *King Kong vs. Godzilla* is the color comeback of two famous prehistoric monsters. In the movie, a reporter describes Godzilla as a hybrid of *T. rex* and *Stegosaurus*. (Special-effects animator Ray Harryhausen had described the title dinosaur as a combination of *T. rex* and *Allosaurus*.) In *King Kong vs. Godzilla,* Godzilla is turned into a hero. Unlike in the original *King Kong* (1933), Kong is not depicted by Willis O'Brien's stop-action animation, but instead by an actor in an ape suit. Viewers were not told how Kong escaped his earlier fall from the Empire State Building.

1963 Lutheran preacher Walter Lang (1913–2004) establishes the Bible-Science Association (BSA) "to stimulate an exchange of ideas on Bible-Science relationships" and begins publishing the mimeographed *Bible-Science Newsletter*. Despite its name, most members of the Association were not scientists. BSA publications endorsed biblical inerrancy and noted that "talking about dinosaurs, Noah's Ark . . . and UFOs can lead to . . . salvation through Jesus Christ." The association claimed that dinosaurs became extinct because of sin (*Dinosaurs and Sin,* 1973), that radiometric dating is unreliable (*What Is the Age of This Lava Flow?,* 1974), and that Nobel Prizes are given to older scientists because younger ones are evolutionists and not as productive (*Creationists: The Better Scientists,* 1978).

1963 Norman D. Newell (1909–2005) describes a major extinction (of more than one-third of all animal families) at the end of the Triassic (i.e., at the Triassic–Jurassic boundary), noting that dinosaurs replaced many groups of amphibians and reptiles.

1963 Sinclair Oil begins marketing dinosaur-shaped "beach toys made from petroleum chemicals."

1963 The British Museum Act makes BMNH an independent museum. In 1992, the Museums and Galleries Act formally changed BMNH's name from British Museum (Natural History) to Natural History Museum.

1963 The eight-acre "Prehistoric Forest" opens in Irish Hills, Michigan, and features dinosaurs sculpted by James Sidwell. The now-abandoned park, which closed in 1999, also included a smoking volcano.

1963 The Sinclair dinosaurs featured at the recent World's Fair ride through New York City aboard a float in the 37th Macy's Day Parade, accompanied by a 60-foot-long balloon of *Brontosaurus* (guided by 22 handlers), the Sinclair mascot. (The Sinclair balloon, which was made of 350 square yards of green fabric and held afloat by 7,000 cubic feet of helium, remained a hallmark of the parade until 1976, when it joined other famous balloons such as "Popeye" in retirement.) When the Smithsonian Institution declined the fiberglass dinosaurs, Sinclair shipped the *Tyrannosaurus* and *Brontosaurus* to what would in 1972 become DVSP near Glen Rose (Figure 37), Texas; *Stegosaurus* to DNM; a duplicate of *Stegosaurus* ("Steggie II") to the Cleveland Museum of Natural History; *Corythosaurus* to Independence, Kansas; *Struthiomimus* to the Milwaukee Museum; and "*Trachodon*" to the Brookville, Illinois, Zoo. *Ornitholestes* was stolen and never recovered, but duplicates were sent to New Jersey and Calgary. *Ankylosaurus* was sent to the Cleveland Zoo, but today its whereabouts are unknown.

1964 Along Interstate 10 at Cabazon, California, theme-park artist and sculptor Claude Bell (1897–1988) begins building "Ms. Dinny," a 150-foot-long, 45-foot-high, 150-ton brontosaur (Figure 36). Bell used more than 1,200 bags of concrete and steel bridge parts to create models of *T. rex, Apatosaurus,* and other dinosaurs. Bell, who at one point lived inside Ms. Dinny, finished the $300,000 project 11 years later when he paid a friend $1 and a case of Dr. Pepper to paint the giant dinosaur. Ms. Dinny, along with 60 other prehistoric creatures (including a 65-foot-tall *T. rex* started in 1981 and completed seven years later), were originally called "Claude Bell's Dinosaurs" and were designed to attract people to Bell's Wheel Inn Café, which had opened in 1958. The giant dinosaurs are among the most famous of all roadside dinosaurs; they have appeared in numerous movies and music videos (e.g., 1985's *Pee Wee's Big Adventure*) and are seen by 12 million passing motorists per year. In 2005, the park was transformed into a creationist museum.

Figure 36 In the 1960s, theme-park artist and sculptor Claude Bell (1897–1988) began building what came to be known as the "Cabazon Dinosaurs" in Cabazon, California, to lure customers to his Wheel Inn Café, which had opened in 1958. Bell's dinosaurs, which are along Interstate 10 just west of Palm Springs, California, are seen by more than 10 million passing motorists each year. In 2005, the dinosaurs were purchased by developer Gary Kanter, who transformed the park "from a tourist stop to a place of worship" that promotes young-Earth creationism. (Courtesy Randy Moore)

1964 French paleontologist Philippe Taquet (b. 1940) begins explorations in Africa that later unearth *Ouranosaurus* (an iguanodont) in the Sahara.

1964 In Spring Hill, Florida, Sinclair Oil opens Harold's Auto Center, one of its most unusual dinosaur-based gas stations. The station is covered by a 110-foot-long, 48-foot-tall *Brontosaurus*, with three arched service-bays in the dinosaur's side.

1964 On the last day of the field season, while walking along a hill near Bridger, Montana, Yale University's John Ostrom (1928–2005) and his assistant Grant Meyer find *Deinonychus* ("terrible claw"), a small, bird-like theropod from the Lower Cretaceous. (Thirty-three years earlier, the species had been discovered and informally named "Daptosaurus" by Barnum Brown, but never formally described.) The skeleton was light and had serrated teeth and big, sharp claws. This discovery, which Ostrom referred to as "startling," resurrected Thomas Huxley's claim that dinosaurs are closely related to birds. Ostrom's conclusions clashed with assumptions that dinosaurs were slow and stupid: "*Deinonychus* must have been anything but 'reptilian' in its behavior, responses and way of life. It must have been a fleet-footed, highly predaceous, extremely agile and very active animal . . . [and had] an unusually high metabolic rate." Ostrom documented derived traits that birds share with theropods, thereby pioneering the work that changed the perception of dinosaurs from big, stupid reptiles to agile, smart animals such as those featured in *Jurassic Park. Deinonychus* was the most common theropod in the Early Cretaceous Cloverly Formation. *Deinonychus* unveiled in the AMNH Hall of Saurischian Dinosaurs in 1995 is the only such specimen displayed anywhere that includes actual bones.

1964 In television's *Voyage to the Bottom of the Sea*, the crew becomes trapped in a prehistoric world depicted with sets and footage of decorated lizards from *The Lost World* (1960). In another episode of *Voyage to the Bottom of the Sea* in 1967, the crew time-travels to the Mesozoic, where dinosaurs are presented via clips from the movies *Animal World* (1956) and *The Lost World* (1960).

1964 The movie *Prehistoric Sound* includes an invisible dinosaur.

1964 The New York World's Fair continues to feature Walt Disney's "Magic Stairway" that carries visitors through scenes resembling those included in *Fantasia*. As had happened at the Chicago World's Fair three decades earlier, an exhibit of life-size models of dinosaurs exhibited by Sinclair Oil at the New York World's Fair triggered a wave of dinosaur-mania in the United States. Indeed, more than 10 million people visited the "Dinoland" attraction at the Sinclair Pavilion, which featured *Ankylosaurus, Apatosaurus, Corythosaurus, Ornitholestes, Stegosaurus, Struthiomimus, "Trachodon," Triceratops, Tyrannosaurus,* and *Archaeopteryx*. Today, the *Apatosaurus* and *T. rex* from that exhibit are part of DVSP just outside of Glen Rose, Texas (Figure 37).

1964 The Rebel Korn'r Gift Shop in White Post, Virginia, is converted to Dinosaur Land, which features tens of dinosaurs. Visitors also see totem poles, a 20-foot-tall King Kong, a 70-foot-long octopus, and a 60-foot-long shark created by James Sidwell.

1965 Alley Oop Fantasy Land opens in Iraan, Texas, near where the first major oil-strike west of the Pecos occurred. The boomtown attracted V. T. Hamlin, who created the *Alley Oop* comic strip in 1932 while working in oil fields. The park includes Dinny, Alley Oop's 65-foot-long, 16-foot-tall pet stegosaur.

1965 Cleveland-Lloyd Dinosaur Quarry (UT), which had been discovered in 1927, is designated a National Natural Landmark, and is thereby protected by the Historic Sites Act of 1935. Unlike at most other sites, at which herbivores comprise 90 percent of dinosaurs, herbivores at Cleveland-Lloyd comprise only 25 percent of the dinosaurs.

Figure 37 Visitors to Dinosaur Valley State Park (DVSP) near Glen Rose, Texas, are greeted by these life-size models of *Tyrannosaurus* and *Apatosaurus* that were originally part of the Sinclair Oil Dinoland Exhibit at the 1964–1965 New York World's Fair (Figure 35). While at DVSP, visitors can see dinosaur tracks near where Roland Bird (Figures 29, 30) excavated the tracks now displayed at the American Museum of Natural History and elsewhere. On the way to DVSP, visitors pass young-Earth creationist Carl Baugh's Creation Evidence Museum, which includes exhibits claiming that humans lived with dinosaurs. (Courtesy Randy Moore)

1965 Following a visit to Glen Rose, Texas (at the invitation of Clifford Burdick [1919–2005]), British chemist A. E. Wilder-Smith (1915–1995) publishes the profusely illustrated *Man's Origin, Man's Destiny,* which links Wilder-Smith's religious beliefs to claims about "man tracks" based primarily on misrepresentations of the work of Roland Bird. Wilder-Smith, a young-Earth creationist, tried to discredit Bird's statement that "no man lived back in the Age of Reptiles." When Bird subsequently learned that a book by astrologer and psychic Jeanne Dixon (1904–1997) had claimed that Bird "reported finding giant petrified human footprints along those of dinosaurs in a river bed in Texas," Bird "fired off a scathing letter to this female 'seer' demanding an immediate withdrawal of my name from all future reprinting. In turn I received an immediate apology from her ghost writer . . . saying my name would never appear again in future editions."

1965 In an article titled "Body Temperatures of Dinosaurs and Its Relationship to Their Extinction" in the *Journal of Paleontology*, L. S. Russell suggests that dinosaurs were warm-blooded. This idea was popularized in the 1970s and 1980s by Robert Bakker, who argued that mammals "were competitively inferior to dinosaurs during the day."

1965 Poland issues the first postage-stamps to feature *Tyrannosaurus rex, Apatosaurus, Stegosaurus, Corythosaurus,* and *Styracosaurus*. Today, *T. rex* appears on postage-stamps (128 stamps) more often than any other dinosaur; *Stegosaurus* (120 stamps) and *Triceratops* (94 stamps) are the other most-featured dinosaurs on stamps. However, more countries (64) have featured *Stegosaurus* than any other dinosaur on their stamps (*T. rex* and *Triceratops* are featured by 63 and 52 countries, respectively).

1965 Polish and Mongolian paleontologists begin expeditions to find dinosaurs in Mongolia. The expedition's leader, Zofia Kielan-Jaworowska (b. 1925), and its principal scientists were women. During the first year of the expedition, the paleontologists uncovered *Opisthocoelicaudia*, a 38-foot-long sauropod. This genus, which was not described until 1977, later appeared on postage stamps in Mongolia and the Republic of Guinea.

1965 Rudolph Zallinger's art is featured in Jane Werner Watson's *Dinosaurs and Other Prehistoric Reptiles*. In 1978, Zallinger's son, Peter, produced the children's book *Prehistoric Animals*.

1965 San Marino issues the first postage stamps to feature *Triceratops*.

1965 The Glen Rose (Texas) Chamber of Commerce and the Somervell County Historical Society map a "Dinosaur Trail" to lure tourists to Glen Rose and its famous dinosaur-tracks.

1965 Time-Life Books commissions Rudolph Zallinger, who had produced the famous *The Age of Reptiles* mural almost 20 years earlier for YPM, to illustrate human evolution for anthropologist F. Clark Howell's (1925–2007) *Early Man* (1965). Zallinger produced a fold-out section titled "The Road to *Homo sapiens*," a panorama depicting 25 million years of human evolution. Zallinger's iconic, controversial, and much-copied illustration showed 15 human ancestors lined up in a left-to-right march that became known as "the March of Progress."

1965 When the New York World's Fair closes, the robotic dinosaurs created by Louis Paul Jonas embark on a publicity tour that covers more than 10,000 miles through 25 states and 38 major cities. One of the most popular stops along the $500,000 tour was the Seminary South "Big Town" Mall in Ft. Worth, Texas (the 32nd stop of the 37-city tour), where the dinosaurs were visited by more than 350,000 people in five days. (*Dallas Morning News* reported that the dinosaurs were "disturbing the sleep of kids and the waking hours of drunks.") Residents of nearby Glen Rose, Texas, distributed more than 40,000 "Dinosaur Hunting Licenses"—valid for *Brontosaurus, Trachodon,* and *Allosaurus*—to promote their fledgling dinosaur-based tourism industry. After the tour, when the Smithsonian Institution declined the dinosaurs, Jonas's creations were dispersed throughout the country. In 1970, the Atlantic Richfield Company (which had purchased Sinclair Oil the previous year) donated two of the dinosaurs—*Brontosaurus* (now *Apatosaurus*) and *Tyrannosaurus*—to DVSP just outside of Glen Rose, Texas, where they remain on display today (Figure 37). In 1967, Disneyland installed dinosaurs from the 1964 World's Fair in its "Primeval World."

1966 Artesia, Colorado—"The Gateway to Dinosaur National Monument"—changes its name to Dinosaur, Colorado. The town, located just south of DNM, had a population of 338 in 2009. Many of the town's streets are named after dinosaurs (e.g., Tyrannosaurus Trail, Brontosaurus Boulevard, Triceratops Terrace).

1966 Como Bluff (WY) and Hell Creek Fossil Area (MT) are designated National Natural Landmarks, and are thereby protected by the Historic Sites Act of 1935.

1966 Excavation for a new state-building in Rocky Hill, Connecticut, uncovers 2,000 dinosaur tracks in the East Berlin Formation. A temporary building constructed to protect the tracks was destroyed by a storm. Two years later, Dinosaur State Park opened at this site, and in 1996, a permanent museum opened to display the tracks, many of which were left by large theropods.

1966 Brian Aldiss's "Heresies of the Huge God" includes a dinosaur-like, continent-sized creature so big that its mass changes Earth's orbit.

1966 In the cartoon *Space Ghost and Dino Boy*, Dino Boy (voiced by Johnny Carson) is stranded in an ancient world, where he makes friends with and rides "Bronty," an *Apatosaurus*. That same year, the low-budget situation comedy *It's about Time* featured astronauts who had broken the time barrier and encountered cave people and dinosaurs. Most of the dinosaur footage came from the movie *The Beast of Hollow Mountain*.

1966 Mongolia issues its first stamps (ranging from 5 to 100 mung each) featuring dinosaurs such as *Tarbosaurus, Talarurus*, and *Protoceratops*.

1966 *One Million Years B.C.* (A Time When There Was No Law and Man, Woman . . . and Beast Roamed the Erath—Untamed!"), a remake of the caveman classic *One Million B.C.*, includes several remarkable dinosaurs, not all of which lived at the same time during the Mesozoic. *Triceratops* uses its horns to fight off predators, and Tumak—a primitive human who is the hero of the story—tricks *Allosaurus* into impaling itself on a stake. Ray Harryhausen (Figure 33), a one-time protégé of Willis O'Brien, created the movie's stop-action dinosaurs. The movie—shot in winter on the desolate Canary Islands—was produced by British-based Hammer Films, which remade several old American films (e.g., *Dracula*). *One Million Years B.C.* featured a face-off between *Ceratosaurus* and *Triceratops,* despite the fact that these dinosaurs' lived millions of years apart. Rather than include a simplified language for the cave people, *One Million Years* B.C. has virtually no dialogue (except for a few simple words and grunts). The movie's dinosaurs were produced by Ray Harryhausen's stop-action animation. *One Million Years* B.C. marked the film debut of American actress Raquel Welch (b. 1940), who reprised the role of "Loana" of the Shell Tribe and was featured prominently in the movie's advertising as a scantily clad cavewoman in a rabbit-skin bikini. In the movie, Welch battled another such cavewoman played by Martine Beswick (subsequent sexy cavewomen were portrayed by actresses such as Caroline Munro, Victoria Vetri, and Julie Ege, "A Creature You'll Never Forget!"). *One Million Years* B.C. was a hit that, as in the past, inspired several new dinosaur-related movies. As in virtually all such dinosaur-related movies, cave people had unblemished skin, straight teeth, groomed hair, and perfectly fitting bearskin outfits and fur bikinis.

1966 Poland issues the first postage-stamp to feature *Archaeopteryx*. Today, more than 38 countries have featured *Archaeopteryx* on their stamps.

1966 Texas legislators propose that "Dinosaur Trail National Monument" near Glen Rose, Texas, be made a national monument. The area, which was tucked in a bend in the Paluxy River, later became DVSP. The publicity lured waves of tourists to tiny Glen Rose, where they paid $1 to see the dinosaur footprints. By this time, Glen Rose's courthouse square was adorned with a giant steel silhouette of a duckbill dinosaur (many believed that a duckbill dinosaur had made some of the area's tracks). The tracks in DVSP were made by *Paluxysaurus* (originally named *Pleorocoelus*) and *Acrocanthosaurus*, both of which lived millions of years earlier than *T. rex* and *Apatosaurus*. Many of the tracks that were visible when the park opened in 1972 have been destroyed by erosion.

1966 The Wistar Institute of Anatomy and Biology gives Edward Drinker Cope's skeleton (Specimen 4989) to the Museum of Archaeology and Anthropology at the University of Pennsylvania. There, anthropologist Loren Eiseley (1907–1977) stored Cope's skeleton in his office for "safekeeping."

1966 To celebrate its 100th anniversary, YPM publishes *Marsh's Dinosaurs,* which includes 150 lithographs (mostly unpublished) of Como Bluff dinosaurs. These lithographs were prepared by F. Berger and E. Crisand in the 1880s at a cost (funded by USGS) of more than $45,000.

1966 When dinosaur tracks are uncovered in New Jersey, reporters learn that local farmers—like those in Edward Hitchcock's day—had been describing them as "turkey tracks."

1966 While working in Montana's Garfield County, plumber and fossil hunter Harley Garbani (1923–2011) discovers a *T. rex* whose skull was later displayed in the Los Angeles County Museum of Natural History (LACM 23844). At the time, this specimen (60% complete, including a 5-foot-long skull) was the world's fourth and most complete specimen of *T. rex*. In the next 24 years, seven more *T. rex* were found in South Dakota, Alberta, and Montana. Paleontologist Jack Horner described Garbani, who found several specimens of *T. rex,* as "the world's greatest discoverer of *T. rex*." Specimens discovered by Garbani are now displayed at the University of California Museum of Paleontology (UCMP) in Berkeley, California.

1967 A partial skeleton of the large-brained *Stenonychosaurus* is found in Alberta, Canada. This skeleton was the basis for the first restoration of this species in 1982.

1967 Grigore Antipa (1867–1944), the first Romanian to reach the North Pole, is the first paleontologist to be featured on a postage stamp. The stamp, which was worth 40 bani, also featured the mammal *Deinotherium*. From 1892 to 1944, Antipa directed the Bucharest Natural History Museum, which now bears his name. Dinosaurs did not appear on Romanian stamps until 1993.

1967 In an episode titled "How to Hatch a Dinosaur" of ABC's *Batman*, Egghead tries to hatch a dinosaur egg, but the hatchling is Batman in a dinosaur costume.

1967 The 65-acre "Farwell's Dinosaur Park" opens in Beaver, Arkansas, as the largest dinosaur-park in the world. The park featured more than 100 dinosaurs, cavemen, and other prehistoric creatures, many of which were sculpted by Emmet Sullivan (who designed the dinosaurs at Rapid City and Wall Drug; Figures 27 and P.1). In the late 1970s, the park added a 40-foot-tall King Kong and was renamed "The Land of Kong." One of the park's new owners was American actor John Agar (1921–2002), who appeared in the 1976 version of *King Kong*. The park, which was featured in *It's Alive* (1968) and *Elizabethtown* (2005), closed in 2005.

1967 The City of Rapid City takes control of its Dinosaur Park.

1967 The influential glam-rock band "Tyrannosaurus rex" (later shortened to "T. rex") is formed by Marc Bolan. The band broke up in 1977 when Bolan died.

1967 With the proposal to create the Glen Rose, Texas, Dinosaur Trail National Monument mired in a committee in the U.S. Congress, the Texas legislature passes a bill creating DVSP. However, Governor John Connally line-item vetoed the appropriations for the park; this meant that Glen Rose had the state's approval for a park, but no money with which to operate it. Meanwhile, more than 100 carloads of tourists visited Glen Rose each weekend to see the area's dinosaur tracks.

1967 Writer and sculptor Sterling E. Lanier (1927–2007) produces a set of 48 bronze prehistoric animals that are sold at NMNH. The toys, which were based on images by Charles Knight and produced with the consultation of George Gaylord Simpson, were expensive and virtually indestructible. Lanier threw some of them into rivers, lakes, and canyons, hoping that they will one day be rediscovered.

1968 Dinosaur Trackway (CT) is declared a National Natural Landmark, and is thereby protected by the Historic Sites Act of 1935.

1968 French paleontologist Armand de Ricqlès (b. 1938) argues for endothermy in dinosaurs by claiming that the dense Haversian system of dinosaur bones indicated "high levels of metabolism, and hence probably endothermy, among dinosaurs." Similar claims were repeated by de Ricqlès in 1974 and by Robert Bakker in 1972.

1968 Rev. Stanley E. Taylor begins filming *Footprints in Stone* (released in 1973) to document "giant man tracks" near Glen Rose, Texas. Taylor's initial findings were first published by the Bible-Science Association, an early creationist organization. The next year, Taylor excavated what is now known as the "Taylor Site," a few hundred yards west of DVSP. The Taylor Site was later used to support young-Earth creationists' claim that humans and dinosaurs lived at the same time. Experts who studied the tracks at the Taylor Site concluded that the alleged "man tracks" there are infilled metatarsal dinosaur tracks.

1968 Robert T. Bakker (b. 1945), a student of John Ostrom who had attended Ostrom's 1964 excavation of *Deinonychus,* publishes his first paper about endothermy in dinosaurs. Bakker, an Ecumenical Christian minister, later became an outspoken proponent of warm-blooded dinosaurs.

1968 Texas uses the State Parks Bonds Program to purchase land just outside Glen Rose, Texas, that includes dinosaur tracks along the Paluxy River (Figure 38). The following year, DVSP, covering 1,523 acres just outside of Glen Rose, Texas, was designated a National Natural Landmark by the U.S. National Park Service. The park, which includes a large, meandering bend of the Paluxy River, is one of the best places in the world to see tracks of dinosaurs such as *Acrocanthosaurus* and *Paluxysaurus*. (Skeletons of these dinosaurs are displayed in the Fort Worth Museum of Science and History.) Trackways along the Paluxy River were made by at least 12 sauropods followed later by at least three theropods. Today, DVSP hosts more than 300,000 visitors per year.

1968 The Japanese film *King Kong Escapes* features a theropod called "Gorosaurus" portrayed by a human in a rubber suit. In *King Kong Escapes,* as well as all others that featured humans in dinosaur suits, the dinosaurs were either misshapen or distorted to accommodate the people in the suits. These films were seldom concerned with scientific accuracy, instead portraying dinosaurs as monsters rather than real animals.

Figure 38 A park ranger sweeps sediment out of theropod tracks in the Paluxy River at Dinosaur Valley State Park (DVSP) near Glen Rose, Texas. Dinosaur tracks in and along the Paluxy River are some of the most famous tracks in the world. The tracks shown here, which were originally visible from "The Overlook" (Site #3) at Dinosaur Valley State Park, were washed away in a 2007 flood of the Paluxy River. (Somervell County Heritage Center)

1968 The visitors' center at Cleveland-Lloyd Dinosaur Quarry is dedicated, along with a plaque designating the site a National Natural Landmark. For the next several years, more than 2,000 people per year visited the quarry each year.

1968 The Wall Drug Dinosaur (Figure P.1), a 37-foot-tall, 80-foot-long, 50-ton *Apatosaurus* in Wall, South Dakota, is moved (at a cost of $6,000) to the edge of town to ease construction of highway I-90 (Figure P.1).

1968 There are now more than 50 suggested causes of dinosaurs' extinction, including starvation, overeating, constipation, sexual problems, global warming, global cooling, cataracts, thin-shelled eggs, mammalian egg-eaters, magnetic reversals, supernova explosions, reduced amounts of carbon dioxide, predation, cancer, viruses, slipped discs, hurricanes, and stupidity.

1968 Two years after thousands of dinosaur tracks were discovered during excavations for a proposed state building, Dinosaur State Park opens near Rocky Hill, Connecticut. More than 500 of the tracks (probably made by *Dilophosaurus*) are now preserved and displayed in a museum; others were reburied for preservation.

1968 Yale University undergraduate Robert Bakker's article titled "The Superiority of Dinosaurs" in *Discovery* magazine calls for a reevaluation of dinosaurs from dumb, plodding hulks to quicker, smarter, active animals. The cover featured *Barosaurus* and two running, three-horned *Chasmosaurus*. Bakker, who restored an active and agile *Deinonychus*, noted that mammals had evolved at about the same time as dinosaurs, yet mammals were restricted to a few ecological niches while dinosaurs remained the largest and most successful terrestrial animals for the next 150 million years. For the next several decades, Bakker tried to alter dinosaurs' images from slow, lumbering dimwits to energetic, lively animals.

1969 135 years after Buckland's first description of a dinosaur, there are 170 valid genera of dinosaurs. Within 25 years, that number exceeded 300.

1969 At the North American Paleontological Convention in Chicago, Illinois, John Ostrom rejects the claim that dinosaurs were sluggish, cold-blooded animals.

1969 Dennis Dean claims that Edward Hitchcock—famed for his discovery of fossils that what were eventually identified as dinosaur tracks—was "the last American geologist to leave a personal mark upon our creative literature."

1969 Digby McLaren (1919–2004), the director of the GSC and the outgoing president of the Paleontological Society, proposes a mass extinction 400 million years ago caused by environmental changes triggered by the collision of Earth and a meteor. McLaren's work, which was published the next year in *Journal of Paleontology,* was the most convincing argument yet published for this cause of mass extinction. However, McLaren had no proof of an impact, and his paper generated little attention.

1969 DVSP, covering over 1,500 acres just outside of Glen Rose, Texas, is designated a National Natural Landmark by the U.S. National Park Service. The park, which includes a large, meandering bend of the Paluxy River, is one of the best places in the world to see tracks of dinosaurs such as *Acrocanthosaurus* and *Paluxysaurus*.

1969 Edwin "Ned" Colbert, a protégé of Henry Osborn, and Jim Jensen discover the skeleton of a 220-million-year-old *Lystrosaurus* (a bulky, herbivorous, therapsid from the Early Triassic) in Antarctica. Colbert's discovery, in combination with earlier discoveries of *Lystrosaurus* in South Africa, showed that Antarctica and

South Africa had once been joined (as a single continent, Gondwana), thereby providing strong evidence for continental drift. When Pangaea began to disintegrate 250 million years ago, Antarctica was the only continent that moved south. That same year (1969), Colbert—a founder and charter member of SVP—retired at AMNH.

1969 George Millecam builds Dinah, a 30-foot-tall pink Brontosaurus (with long eyelashes) to attract visitors to his Dine-a-Ville Motel in Vernal, Utah. When the motel closed, Dinah was moved to the east entrance to town.

1969 In *Bambi Meets Godzilla,* the most unusual movie to feature Godzilla, a happy Bambi is crushed by a giant reptilian foot. The 30-second animated movie, which was made by Marvin Newland, included only one scene.

1969 In the Warner Brothers movie *The Valley of Gwangi,* set in early 1912, a dinosaur (that's a combination of *Tyrannosaurus* and *Allosaurus*) named Gwangi is captured by cowboys in Forbidden Valley and used in a Mexican circus. The tail-dragging Gwangi (which does not appear until near the end of the movie) is described as *Allosaurus,* but legendary special-effects specialist Ray Harryhausen based his model on Charles Knight's vision of *Tyrannosaurus*. After Gwangi escapes and destroys a mad scientist (Professor Bromley), cowboys lasso Gwangi and put him on display. When Gwangi escapes, he is finally trapped in a cathedral, where he overturns a heater containing hot coals. In a spectacular final scene, Gwangi is burned to death in the flaming cathedral. The movie, in which "Cowboys Battle Monsters in the Strangest Roundup of All!", was filmed near Cuenca, Spain, where a famous Cretaceous fossil bed named Las Hoyas would be found in 1984 by amateur paleontologist Armando Diaz-Romeral. *The Valley of Gwangi* was a commercial failure; it was often paired with an adult-oriented sex film, and therefore missed the otherwise reliable audience of young people. *The Valley of Gwangi* inspired Virginia's Mark Cline in 2005 to open Dinosaur Kingdom, in which a velociraptor stalks Abraham Lincoln, while other dinosaurs help Confederates fight Union soldiers.

1969 José F. Bonaparte (b. 1928), a protégé of Alfred Romer, discovers the 33-foot-long prosauropod *Riojasaurus* ("Rioja lizard") in Argentina's La Rioja Province. Bonaparte, who later worked for the Museo Argentino de Ciencias Naturales "Bernardino Rivadavia" in Buenos Aires, described numerous South American dinosaurs and inspired a new generation of Argentine paleontologists.

1969 Robert Bakker's restoration of an active and agile *Deinonychus* appears in *Discovery* magazine. This image became one of the most recognizable restorations of a dinosaur. Bakker's restorations of *Deinonychus* later appeared in Adrian Desmond's *The Hot-Blooded Dinosaurs* (1975), Donald Glut's *The New Dinosaur Dictionary* (1982), and John Noble Wilford's *The Riddle of the Dinosaur* (1985).

Bakker later claimed that "fossils in general are the jumper cables for the human mind. They teach you to think sideways."

1969 Roland Bird (Figure 29), in a letter to creationist Mike Turnage, writes that "man tracks" near Glen Rose, Texas, "are definitely, repeat, definitely *not* human. . . . They were made by carnivorous dinosaurs wading through deep mud. . . . Anything else 'human' exhibited or reported 'found' in the area is the product of a very clever prankster with hammer and chisel."

1969 When Manchester, Connecticut's Hop Brook Bridge is demolished, Yale's John Ostrom collects the missing half of *Ammosaurus* that had been discovered (as part of the bridge) 85 years earlier. Marsh claimed that *Ammosaurus* had made some of the footprints collected by Edward Hitchcock in the mid-1800s in Connecticut.

1969 John Ostrom describes *Deinonychus* in *Osteology of Deinonychus antirrhopus, an Unusual Theropod from the Lower Cretaceous of Montana*. The cover of Ostrom's monograph was adorned with Robert Bakker's restoration of a sprinting, unfeathered *Deinonychus*. That same year, in an article titled "Terrible Claw" in *Discovery,* Ostrom argued that *Deinonychus* had a posture with its trunk nearly horizontal, its tail straight out behind (as a counterbalance), and its neck curved up like an ostrich. Ostrom claimed that "this is a much more natural-looking posture than the 'kangaroo' pose that is commonly illustrated for other carnivorous dinosaurs such as *Allosaurus* or *Tyrannosaurus*." Ostrom noted in a subsequent *Bulletin* of the YPM that the presence of a sickle-like claw on the second toe of each hind foot of *Deinonychus* meant that the dinosaur "must have been a fleet-footed, highly predaceous, extremely agile and very active animal . . . [having] an unusually high metabolic rate." Most paleontologists have since abandoned the tail-dragging posture of dinosaurs.

1970 Berney Neufeld and his coworkers from Loma Linda University study the dinosaur tracks and "man tracks" near Glen Rose, Texas. Five years later, Neufeld became the first creationist to claim that none of the tracks were made by humans. Also skeptical was Wilbert Rusch, president of the Creation Research Society.

1970 BMNH paleontologist Barney H. Newman estimates that the tail of the famous *T. rex* mounted in AMNH was about 10 feet too long.

1970 British paleontologist Peter Galton (b. 1942) revives the idea that ornithischian dinosaurs are the closest relatives of birds. This proposal was later abandoned.

1970 Charles M. Sternberg's final publication discusses what happens to an animal after it dies (i.e., taphonomy).

1970 Claude Bell, the owner of "Claude Bell's Dinosaurs" in Cabazon, California, notes that Ms. Dinny—whose stomach houses a store—is "the first dinosaur in history, so far as I know, to be used as a building."

1970 Dale Russell claims that Henry Osborn's *Albertosaurus sarcophagus* is the same as *Gorgosaurus*.

1970 Five hundred people attend the dedication of DVSP just outside of Glen Rose, Texas. Texas governor Preston Smith (1912–2003) officiated the ceremony, which included the dedication of a plaque from the Department of the Interior recognizing the park as a National Natural Landmark. The ceremony was attended by a nostalgic Roland Bird, a famed dinosaur-hunter who noted that "unlike many naturalists of the present day, I came up the hard way." Smith had asked Atlantic Richfield (at this point the parent company of Sinclair Oil) to donate the now-retired Sinclair dinosaurs to Texas so they could be displayed in Glen Rose to advertise the area's dinosaur tracks. Atlantic Richfield promised that Texas could have all of the dinosaurs, but later had to scale back their offer when "word got out" and "everybody in the free world wanted a dinosaur." Glen Rose was given *Brontosaurus* and *Tyrannosaurus* (both of which today greet visitors to DVSP; Figure 37), two baby brontosaurs "with simulated eggs," and *Archaeopteryx*. One of the baby brontosaurs (a Sinclair promotional model) went to Glen Rose's Oakdale Park, and now stands in front of the Glen Rose Visitors Center.

1970 Glen Rose resident Wayland Adams discloses how his uncle (George Adams) and others carved "man tracks" in slabs of Cretaceous rocks from the Paluxy riverbed. Nevertheless, several organizations continue to claim that such "man tracks" prove that humans lived dinosaurs.

1970 *Gorgosaurus libratus* is renamed *Albertosaurus libratus*.

1970 Much as Richard Lull (1867–1957) had blamed the demise of dinosaurs on "racial senility" in 1924, W. E. Swinton argues that dinosaurs' extinction resulted from "racial old age (phylogeronty) [and the] lack of plasticity, both physical and mental." Collaborations of Swinton with artist Neave Parker in the 1950s had not depicted dinosaurs as agile.

1970 The Italian company Panini issues a set of 360 cards featuring dinosaurs and other prehistoric life. This set dwarfed the 60-card set issued several years earlier by Nu Card.

1970 The U.S. Postal Service issues its first stamps that feature dinosaurs. The 6¢ stamps featured *Diplodocus, Apatosaurus, Stegosaurus, Camptosaurus,* and *Compsognathus*. The four-stamp set, titled "Natural History," also featured birds, mammals, and an ancient canoe. The set was issued to celebrate the 100th anniversary of the opening of AMNH, but it included art from Rudolf Zallinger's *The Age of Reptiles* of YPM. The U.S. Postal Service again issued stamps featuring dinosaurs in 1989, 1995, 1997, and 2000.

1970 *When Dinosaurs Ruled the Earth,* a follow-up by Hammer Films to *One Million Years B.C.*, features some of the most lifelike prehistoric animals ever included in a movie. Stop-action animator Jim Danforth (b. 1940) was nominated for an Academy Award for his work on this movie. The movie was similar to *One Million Years B.C.* The ads also featured scantily clad 1968 Playboy Playmate of the Year Victoria Vetri (b. 1944; using the name Angela Dorian)[36] as the cavewoman Sanna, "who could tame any man . . . even man's deadliest enemies." (Other cavewomen in the movie included Magda Konopka and Imogen Hassall.)[37] In the movie, Vetri was saved from hostile cavemen by a dinosaur she helped when the dinosaur was young. Despite its title, *When Dinosaurs Ruled the Earth* ("Enter an Age of Unknown Terrors, Pagan Worship and Virgin Sacrifice!") included only one dinosaur (the horned *Chasmosaurus*); *T. rex* was advertised, but was not in the film. Danforth later created dinosaurs for the comedy *Caveman* (1981).

1970 While working in southern Mongolia, paleontologists unearth *Deinocheirus,* a 15-foot-long theropod from the Late Cretaceous. Its distinguishing trait is its 8-foot-long forelimbs that were used "in tearing dead or weakly agile prey asunder."

1970 John Ostrom borrows a fossil named *Pterodactylus crassipes* by von Meyer in 1855; Ostrom took the specimen (insured for $1 million) out of the Netherlands' Teylers Museum in a shoebox. Ostrom later renamed the specimen *Archaeopteryx*. As Ostrom noted, "My find is a classic example of why a paleontologist or museum should not throw things away that can't be absolutely classified as worthless." Ostrom's discovery prompted him to reexamine *Archaeopteryx*, which he later noted as being similar to the theropod *Deinonychus*. The curator of Teylers Museum told Ostrom, "You have made our museum famous."

1971 After an eight-year absence, the Sears Christmas Catalog again advertises a "57-piece Prehistoric Set" of cavemen, dinosaurs, and other prehistoric life. The set, which cost $5.99, was made by Louis Marx.

1971 As part of a study for the U.S. National Parks Service, paleontologist John Ostrom notes, "The Haddonfield site [in New Jersey, where *Hadrosaurus foulkii* was discovered] is one of great scientific and historical importance." The site was made a national landmark in 1994 "as a significant representative of our Nation's heritage."

1971 Dale Russell (b. 1937) and Wallace Tucker argue that a supernova's explosion, and the resulting increased levels of radiation, caused the extinction of dinosaurs. (In 1974, artist Eleanor Kish painted a dying ankylosaur scorched by a supernova.) Russell also claimed that non-dinosaurian traits in *Deinonychus* "were reminiscent of human attributes."

1971 *Dinosaurs . . . The Terrible Lizards* includes another face-off of *T. rex* and *Triceratops*.

1971 Harvard paleontologist Alfred Sherwood Romer (1894–1973) names *Lagosuchus,* a small bipedal archosaur. This genus, now *Marasuchus,* is considered by many to be the closest relative of dinosaurs.

1971 In an article titled "Dinosaur Physiology and the Origin of Mammals" in the journal *Evolution*, Robert Bakker presents his most famous "heresy"—namely, that dinosaurs were warm-blooded. Bakker's evidence included dinosaurs' adaptations for an energetic lifestyle, the cavities and pores in their vertebrae (e.g., as evidence of an air-sac system as in birds for dispelling excess heat), the correlation of physiology to posture, and that only erect, warm-blooded dinosaurs could have kept mammals in narrow (e.g., small-body and probably nocturnal) niches.

1971 In the Gobi Desert, Mongolian-born vertebrate paleontologist Rinchen Barsbold (b. 1935) and his colleagues with the Polish–Mongolian Expedition find a fossilized *Velociraptor* locked in combat with a pig-sized *Protoceratops*. Barsbold's later work helped bring Mongolian paleontology into world prominence. Near the end of his career, Czech artist Zdenêk Burian painted *Velociraptor* as an agile, fast predator.

1971 Peter Dodson (b. 1946) of the University of Pennsylvania is the first to apply taphonomy to dinosaur fossils when he studies how dinosaur bones in DPP moved in ancient stream beds. Dodson was later an editor of *The Dinosauria* and author of *The Horned Dinosaurs* (1998).

1971 Vivian Jones unearths a bone of a large theropod southwest of Delta, Colorado. When Jim Jensen of Brigham Young University learned about the discovery, he and Kenneth Stadtman began a decades-long excavation at the scenic site that became known as Dry Mesa Quarry. The first major discovery there occurred in 1972 when workers unearthed a 30-foot-long theropod that would eight years later be named *Torvosaurus* ("savage lizard"), a two-ton theropod and owner of the bone found by Jones. (*Torvosaurus* was also the first "megalosaurid" found in the Morrison Formation.) That same year, Jensen found at Dry Mesa Quarry the largest dinosaur known at the time, which he informally named "Supersaurus." If the proportions of "Supersaurus" were similar to those of *Brachiosaurus*, then "Supersaurus" would have weighed more than 50 tons and been approximately 90 feet long. In 1985, "Supersaurus"—a sauropod—was formally described, thereby becoming *Supersaurus* ("super lizard"). Dry Mesa Quarry, which became famous for its large sauropods, yielded more than 20 genera of dinosaurs; the diversity of dinosaurs there exceeded that of any other quarry in the Morrison Formation.

1972 British paleontologist Alick Walker (1925–1999) proposes that modern birds share several features with *Sphenosuchus,* a crocodile-like Triassic archosaur.

1972 Donald Glut (b. 1944) publishes *The Dinosaur Dictionary,* the first "A to Z" compilation of dinosaurs. In subsequent years, this format was copied by several other publishers. Glut is one of the world's leading authorities on dinosaurs in movies.

1972 *Extraterrestrials in History* claims that the extinction of dinosaurs resulted from a giant explosion in outer space that was triggered by extraterrestrial intelligences so that humans could evolve.

1972 *Footprints in Stone*, Stanley Taylor's film about the evidence that humans and dinosaurs lived contemporaneously, is distributed throughout the United States. Young-Earth creationist Carl Baugh later described Taylor's film as "the greatest tool for the creationist explanation" of dinosaur tracks and "man tracks" near Glen Rose, Texas. To counter the impact of *Footprints in Stone,* the American Anthropological Association released the videotape *The Case of the Texas Footprints,* which explained the "man tracks" as remnants of dinosaur tracks (i.e., as Roland Bird had interpreted the tracks).

1972 Geologist Peter Vogt of the Naval Research Laboratory notes that the K-T extinction coincided with increased volcanic activity throughout the world (e.g., by India's Deccan Traps, whose eruptions spanned the K-T boundary). Three years later, this idea was developed by Dewey McLean, who claimed that the Deccan Traps' eruption had caused a greenhouse-effect at the end of the Mesozoic. The Deccan Traps began erupting 68.6 million years ago (± 80,000 years), and the eruption stopped 65.0 million years ago (± 110,000 years), suggesting that these eruptions may have been an accomplice in the K-T extinction. (The end-Permian mass-extinction, which occurred 250 million years ago, coincided with massive volcanic eruptions across Siberia). The overlap of the K-T impact and the eruptions of the Deccan Traps makes their effects hard to separate.

1972 In a paper in *Nature* titled "Anatomical and Ecological Evidence of Endothermy in Dinosaurs," Robert Bakker argues that bones of dinosaurs, birds, and mammals are all highly vascularized, and that this similarity is evidence for endothermy in dinosaurs.

1972 Jack Horner ends his formal education at the University of Montana in Missoula without a degree and returns to his hometown (Shelby, Montana) to work in his father's sand-and-gravel business. Horner later became a renowned dinosaur paleontologist who was awarded honorary doctorate degrees.

1972 Marvel Comics' *Shanna the She-Devil* features a femme fatale who battles dinosaurs in a jungle.

1972 Mike Raath announces the first discovery of dinosaur tracks in Rhodesia (now Zimbabwe). The tracks were made by *Vulcanodon* ("Vulcan tooth"), one of the earliest sauropods.

1972 The Republic of Maldives issues postage stamps featuring dinosaurs, including *Stegosaurus, Diplodocus, Triceratops*, and *Tyrannosaurus*. The *Tyrannosaurus* was based on Rudolph Zallinger's mural at YPM; other images were from Bertha Parker's *Life through the Ages* (1947) and *Animals of Yesterday* (1952).

1973 Albert Bennett and Bonnie Dalzell argue that conclusions in Robert Bakker's *Evolution* article in are "neither logically nor biologically sound." Others, such as Alan Feduccia, leveled other criticisms at Bakker, with some linking Bakker's ideas with his mentor John Ostrom.

1973 Bridger Fossil Area (MT), Garden Park Fossil Area (CO), and Morrison Fossil Area (CO) are designated National Natural Landmarks, and are thereby protected by the Historic Sites Act of 1935.

1973 C. C. Young names the duckbill *Shantungosaurus giganteus* following a four-year excavation on China's Shandong peninsula. The 51-foot-long hadrosaur had hooved front feet, suggesting it walked on all fours.

1973 In *Dinosaurs and Sin,* the Bible-Science Association attributes the extinction of dinosaurs to human sin. The next year, in *Human Footprints and Dinosaur Tracks,* the Bible-Science Association cited the Paluxy "man tracks" to support its claim that humans and dinosaurs lived at the same time.

1973 In *Fossil Footprints Found,* Carol Taylor uses the Paluxy River "man tracks" as evidence that humans lived with dinosaurs: "The proof is here before us, Man lived with *Allosaurus*, (Or one like him.)"

1973 John Ostrom, responding to British anatomist Alick Walker's (1925–1999) rejection of birds being an descendant of dinosaurs, writes in *Nature,* "The skeletal anatomy of *Archaeopteryx* is almost entirely that of a coelurosaurian dinosaur—not thecodont, not crocodilian, and not avian." Ostrom, unlike Gerhard Heilmann in the 1920s, knew that several newly discovered dinosaurs (e.g., *Segisaurus, Velociraptor*) had clavicles. Ostrom's subsequent suggestion that the feathered forearms of *Archaeopteryx* were used to swat insects gained little support. The claim that birds are the descendants of dinosaurs was also challenged by the nascent organization BAND (Birds Are Not Dinosaurs).

1973 Nobel laureate Harold Urey (1893–1981) suggests that comet impacts caused several mass extinctions, including the one that closed the Mesozoic. Most scientists rejected Urey's idea. Paleontologist David Raup later suggested that some of the resistance to Urey's idea was a commitment by geologists to the uniformitarianism proposed by James Hutton and Charles Lyell.

1973 Peter Galton's suggestion that ornithischians had cheeks leads to significant changes in scientists' views of these animals.

1973 Stanley Taylor's 16-mm film *Footprints in Stone* continues to be used by many creationists to support their claims that humans lived with dinosaurs. The film describes Taylor's interpretation of four shallow sets of tracks (i.e., Giant Run, Turnage, Taylor, and Ryals) along the Paluxy River west of Glen Rose, Texas, and concludes that the prints were "unquestionable" evidence of humans living with dinosaurs. Although some creationists and creationist organizations later rejected Taylor's claims, the tracks continue to be cited as evidence of dinosaurs living contemporaneously with humans.

1973 Walter Alvarez begins studying paleomagnetism and the foraminifera of the pink Gubbio limestone in deep-sea sediments near the medieval city of Gubbio in northern Italy. Although the K-T extinction did not correlate with a magnetic reversal, Alvarez noted that the boundary was marked by a layer of gray clay 1–2 centimeters thick. These studies culminated in Alvarez and his father Luis (a Nobel Prize–winning physicist) proposing in 1980 that the K-T extinction was triggered by a meteor impact. The Alvarezes were not the first to propose dinosaurs' extinction from an extraterrestrial impact, but they were the first to have physical evidence to support their claim.

1974 A. J. de Ricqlés argues from studies of bone histology that dinosaurs had a metabolism intermediate between warm-blooded and cold-blooded organisms. This "intermediate metabolism" that gave dinosaurs "extraordinary abilities to be active and grow rapidly on a limited food supply" was first fully described in 1980 by Philip Regal and Carl Gans.

1974 Citing research that included John Ostrom's claim that birds are descendants of dinosaurs, Robert Bakker and Peter Galton argue in an article in *Nature* titled "Dinosaur Monophyly and a New Class of Vertebrates" that dinosaurs were endotherms having metabolisms more like birds and active mammals than to any living reptiles. Bakker and Galton also proposed the controversial new class Dinosauria that would include birds and dinosaurs (i.e., Saurischia, Ornithischia, and Aves), with snakes, turtles, crocodiles and their allies classified as "reptiles." This was the first time that anyone had proposed that dinosaurs be regarded as a separate, monophyletic class (rather than a polyphyletic group consisting of two different orders). Bakker and Galton claimed that flight alone was not an adequate criterion for animals to be categorized in an independent class.

1974 Fred Beierle's *Giant Man Tracks* includes photos of dinosaur tracks and humanlike tracks from the Paluxy River near Glen Rose, Texas.

1974 In a set of short-stories titled *A Little Girl to Whom Nothing Ever Happens* by Soviet writer Kirill Bulychev (1934–2003), a little girl named Alice from the future becomes friends with a brontosaur hatched from a frozen egg found on the banks of the River Yenisei.

1974 In Alberta's DPP, Irene Vanderloh discovers what in 1978 will be named *Saurornitholestes* ("lizard-bird-stealer"). This carnivorous dinosaur was about the size of a wolf.

1974 In CBS's *Valley of the Dinosaurs*, a science teacher and his family on a rafting adventure capsize and end up in an underground cavern. There, they encounter cave people and dinosaurs as they try to survive and get home. In the same year as the debut of *Valley of the Dinosaurs,* NBC's dinosaur-based series *Land of the Lost* described the adventures of a family rafting the Colorado River that is taken by a "dimensional vortex" to a world of dinosaurs and cavemen. There, viewers met Grumpy, the tyrannosaur, Dopy, the cute baby brontosaur, Big Alice, the *Allosaurus*, and Spike, the triceratops (all animated with stop-action video and hand puppets). The ambitious *Land of the Lost,* which was produced by Marty and Sid Krofft (who had begun their careers as puppeteers), was a live-action series, not a cartoon.

1974 In *United States v. Diaz,* the Ninth Circuit Court of Appeals rules that the phrase "object of antiquity, ruins, and monuments" in the Antiquities Act of 1906 is unconstitutionally vague, prompting federal officials to be hesitant when using the law to prosecute looters. The problem was remedied with passage of the Archaeological Resources Protection Act in 1979.

1974 In *Dinosaurs Before Adam?,* gap creationist Robert Gentet argues that dinosaurs were extinct before the six-day creation week in the Bible.

1974 In the novel *Our Children's Children*, people from the 25th-century fleeing extraterrestrial attackers travel back in time. During the Upper Cretaceous, they kill all of the dinosaurs. A similar extinction of dinosaurs by extraterrestrials (in this instance, while searching for sources of energy) was described earlier in Ivan Yefremov's *Star Ships* (1948).

1974 Invicta Plastics of Leicestershire, England, in conjunction with BMNH, begins producing a line of 23 dinosaur toys. These hard, monochromatic, rubber toys, which were the first museum-quality dinosaur toys, were produced for two decades. In 1988, Safari Ltd. began producing colorful toys, soon after which Invicta colorized its toys.

1974 John Ostrom's paper titled "*Archaeopteryx* and the Origin of Flight" revives Thomas Huxley's century-old claims that birds are the descendants of dinosaurs. Ostrom had been the first person to clearly emphasize that *Archaeopteryx* resembled certain carnivorous dinosaurs (coelurosaurs). The following year, Ostrom detailed his ideas in a monograph about *Archaeopteryx*. Even artist Charles Knight, who had no training in science, had noted the resemblance of birds and dinosaurs, commenting that *Archaeopteryx* was "a sort of dinosaur with wings."

1974 Self-trained paleontologist Joan Wiffen (1922–2009), "the Dragon Lady of New Zealand," finds a tail vertebra of a theropod; this was New Zealand's first recognized dinosaur fossil.

1974 The "Invasion of the Dinosaurs" episode of the British science-fiction television-series *Doctor Who* depicts dinosaurs as political capital used by time-travelers trying to change history's mistakes. The time-travelers' plans are foiled by Doctor Who.

1974 China's Sichuan Province yields *Tuojiangosaurus*, a 23-foot-long stegosaur. The fossil was named three years later.

1975 "Valley of the Dinosaurs" opens on the grounds of Silesian Zoological Garden in Chorzow, Poland. The outdoor park included 16 life-size dinosaurs, several of which were unearthed by the 1963–1971 Polish–Mongolian expeditions to the Gobi Desert. The park's first visitor was 12-year-old Karol Sabath, who became a paleontologist.

1975 Ann Coates's children's book titled *Dinosaurs Don't Die* describes how the Sydenham dinosaurs built by Benjamin Waterhouse Hawkins come to life for a child named Daniel, who lives nearby. Daniel becomes friends with "Rock," the *Iguanodon,* and they visit BMNH.

1975 Antievolution crusader Kelly Segraves's *The Great Dinosaur Mistake* argues that humans lived with dinosaurs, that dinosaurs did not become extinct millions of years before humans, and that dinosaurs are mentioned in the Bible. Five years later, Segraves expanded his arguments in *Everything You Always Wanted to Know about Dinosaurs, Flying Saucers, the Beginning and the End.*

1975 Berney Neufeld becomes the first creationist to claim that "man tracks" alongside dinosaur tracks near Glen Rose, Texas, are poor evidence for the coexistence of dinosaurs and humans. However, the following year, Clifford Burdick claimed that "man tracks" near Glen Rose are "so well documented" that researchers should move on to other projects. Other creationists—including John Morris of ICR and Walter Lang, director of the Bible-Science Association—continued to promote the tracks as evidence that humans lived with dinosaurs. This interpretation contradicted much evidence but was consistent with their religious belief that humans and dinosaurs were created on day six of creation week. The "Burdick Track"—a 14-inch-by-6.5-inch "man track" found near Glen Rose, Texas, and now displayed at Carl Baugh's Creation Evidence Museum of Texas—honors Burdick.

1975 Edgar Rice Burroughs's book *The Land That Time Forgot* (1918) is made into a movie in which the "lost world" resides in the South Pacific. The movie featured erupting volcanoes and actors in dinosaur suits, as well as dinosaurs portrayed by hand-puppets. The hero (actor Doug McClure) and Lisa (actress Susan Penhaligon) battle cavemen and dinosaurs, and get married near the end of the movie. Unlike most "lost world" movies, the animals "evolved" according to their location on the island rather than time. Three years later, a movie was made of *The People That Time Forgot*.

1975 Harvard historian Adrian Desmond (1947) adds support for endothermy in dinosaurs with his 352-page book *The Hot-Blooded Dinosaurs: A Revolution in Paleontology*. As Desmond noted, "*Deinonychus* embodies all that is distinctly non-reptilian in dinosaurs."

1975 In an article titled "Dinosaur Renaissance" in *Scientific American,* Robert Bakker—a former student of John Ostrom—brings Ostrom's argument for dinosaur endothermy to the public; as Bakker later noted, "We need endothermic dinosaurs; evolutionary theory demands them; the empirical data confirm that they existed." Bakker's article, which included an argument about how ratios of predator–prey biomass relate to endothermy, included the first restoration (by paleoartist Sarah Landry) of a feathered Triassic dinosaur (*Syntarsus*). (Critics such as James Farlow argued that ratios of predator–prey biomass as a basis for dinosaur endothermy were inconclusive.) Bakker pointed out "a particularly happy implication" for dinosaur enthusiasts: "The dinosaurs are not extinct; the colorful and successful diversity of the living birds is a continuing expression of basic dinosaur biology."

1975 Jack Horner begins working as a preparator at Princeton University. Horner later discovered many dinosaurs, was an advisor for movies (e.g., *Jurassic Park*), and revolutionized ideas about the social life of dinosaurs.

1975 Jack McIntosh and David Berman reopen the controversy about the skull atop CMNH's *Apatosaurus* by noting that *Apatosaurus* had a long-snouted skull with narrow teeth, while that of *Camarasaurus* was short-snouted, crowned, and had large teeth. Four years later, the *Camarasaurus* skull atop *Apatosaurus* was restored with a replica of a *Diplodocus*-like skull found at DNM. Soon thereafter, similar changes were made at other museums, including the Field Museum, AMNH, and YPM, the home of the first *Apatosaurus*.

1975 Paintings by Eleanor (Ely) Kish at the Canadian Museum of Nature are among the most accurate illustrations of dinosaurs and their environments. Kish's work included accurate plants and above-ground tails on some dinosaurs. Kish later illustrated several other books about dinosaurs, including Dale Russell's *A Vanished World: The Dinosaurs of Western Canada* (1977) and *An Odyssey in Time: The Dinosaurs of North America* (1989).

1975 Paleontologist Peter Dodson emphasizes the difficulties in distinguishing male and female dinosaurs, and suggests that several different species are merely variations among individuals of the same species.

1975 Teresa Maryanska and Halszka Osmólska (1930–2008) describe the 2-foot-long *Bagaceratops,* which was found in Omnogov, Mongolia. The previous year, Maryanska and Osmólska had become the first women to describe several new kinds of dinosaurs.

1975 The Utah Division of State History begins five years of work at the Cleveland-Lloyd Quarry. The light-gray bone-layer there, which ranges from 1.5 to 4.3 feet thick, has produced prolific numbers of *Allosaurus,* as well as *Ceratosaurus, Torvosaurus*, and *Stokesosaurus* (named in honor of W. Lee Stokes). In 2008, additional discoveries showed *Stokesosaurus* to be an early cousin of *Tyrannosaurus* that lived nearly 100 million years earlier than its famous relative. Much of the early lineage of tyrannosaurs occurred on other continents.

1975 William Dankenbring's (b. 1941) *The First Genesis* endorses gap creationism and claims that the K-T extinction was caused by a cosmic battle between God and Satan. Three years later, in *Beyond Star Wars*, Dankenbring claimed that the Great Pyramid is a memorial to the Flood and that UFOs are meant to convince people to reject Jesus.

1975 The home of Edward Drinker Cope at 2102 Pine Street in Philadelphia, Pennsylvania, is made a National Historic Landmark by the National Parks Service (Figure 39).

1976 A Canadian team led by paleontologist James Kitching discovers 190-million-year-old dinosaur eggs in South Africa's Golden State Highlands National Park. In 2010, two of these eggs were shown to contain the oldest known embryos of terrestrial vertebrates. The babies had large heads and long forelimbs, but the 5-meter-long adults had small heads and long necks, indicating that their necks and hind limbs grew faster than their forelimbs and head. When paleontologists returned to the site, they discovered that *Massospondylus* returned to the same nursery year after year, and infants stayed at their nests until they had doubled in size.

1976 A remake of the movie *King Kong* (1933) includes no dinosaurs.

Figure 39 The bitter feud between paleontologists Othniel Marsh (Figure 9) and Edward Drinker Cope (Figure 10) fascinated the public, and was covered by newspapers throughout the United States. This monument on Philadelphia's Pine Street marks the site of Cope's residence, which was filled with fossils when he died. Cope's fossils were bought by the American Museum of Natural History (AMNH), and Cope's desk is housed in AMNH's library. In his will, Cope asked that his remains be made the holotype of *Homo sapiens*. (Courtesy Randy Moore)

1976 American fantasy artist William Stout (b. 1949) illustrates dinosaurs in *Prehistoric Portfolio*, and later in *The New Dinosaur Dictionary* (1978), *The Dinosaur Scrapbook* (1980), and *The Dinosaurs: A Fantastic New View of a Lost Era* (1981), which included influential images of raptors (especially *Deinonychus*).

1976 Atlantic Richfield divests its right to the Sinclair logo and brand, thus reestablishing Sinclair as an independent company.

1976 Charles McGowen's *In Six Days* uses the Paluxy "man tracks" to claim that humans lived with dinosaurs.

1976 French paleontologist Philippe Taquet (b. 1940) of the Natural History Museum in Paris names and describes *Ouranosaurus* ("brave lizard"), a large Early Cretaceous ornithopod discovered in modern-day Niger. *Ouranosaurus* was 22-foot-long and weighed up to four tons. That same year, Taquet received a box of fossils that included what he and Hans-Dieter Sues (of NMNH) later named *Majungatholus atopus,* a pachycephalosaur.

1976 Ghost Ranch, a canyonland north of Santa Fe, New Mexico, made famous by artist Georgia O'Keeffe, is designated a National Natural Landmark, and is thereby protected by the Historic Sites Act of 1935.

1976 In a letter to paleontologist Wann Langston Jr., Roland Bird interprets the dinosaur tracks he excavated at Glen Rose, Texas, as belonging to a carnivore attaching a sauropod, both of whom walked with their legs directly beneath their bodies.

1976 In his college textbook *History of Life*, Richard Cowan invokes Hutton's and Lyell's uniformitarianism to claim that the K-T mass extinction resulted from "climatic change, induced in the first place by continental movements."

1976 In John Eric Holmes's *Mahars of Pellucidar,* the last of Edgar Rice Burroughs's seven Pellucidar stories, psychologist Chris West is transported to an ancient world, where he witnesses a battle between *Tyrannosaurus* and *Triceratops*.

1976 In *Nature*, John Ostrom claims that were it not for its feather imprints, *Archaeopteryx* "would be identified unquestionably as coelurosaurian theropods."

1976 In the movie *Allegro Non Troppo,* dinosaur-like creatures evolve to strains of Ravel's *Boléro*.

1976 James Madsen publishes *Allosaurus fragilis: A Revised Osteology,* the authoritative monograph on the genus.

1976 Mary Wade and Richard Thulborn from Queensland, Australia, find more than 4,000 tracks left by carnivorous dinosaurs. The site was made a public park in 1982. In 2013, the Queensland Museum in Brisbane opened an exhibit of dinosaurs and their tracks collected from throughout Queensland.

1976 Paleontologists find eggs of the prosauropod *Massospondylus* from the Early Jurassic in South Africa.

1976 Sinclair Oil begins distributing cute brontosaurs made of fiberglass at its gas stations.

1976 The Federal Land Policy and Management Act (Public Law 94-579) strengthens protections of paleontological resources (including fossils) on public lands. That same year, the Fruita Paleontological Area became the first area set aside by the U.S. Bureau of Land Management to protect its fossils.

1977 A parade celebrates the arrival of 14 life-size fiberglass replicas of prehistoric animals to UFHNH in Vernal, Utah. The animals had been built in the early 1960s by Utah sculptor Elbert Porter, and were later displayed in West Yellowstone, Montana, and Orderville, Utah. Today, these creatures are part of the Field House's "Dinosaur Gardens."

1977 American geologist Stephen Jay Gould dedicates his first book, *Ever Since Darwin*, to "my father, who took me to see the *Tyrannosaurus* when I was five."

1977 Dale Russell, a former student of Edwin Colbert, publishes *A Vanished World: The Dinosaurs of Western Canada*. This book helped popularize the discoveries of dinosaurs from Alberta and nearby regions.

1977 Frederick Beierle's *Man, Dinosaurs, and History* uses the alleged "man tracks" at Glen Rose, Texas, to advocate Christianity.

1977 Gregory S. Paul (b. 1954) restores *Archaeopteryx*.

1977 In ABC's made-for-television movie *The Last Dinosaur*, workers for an oil company find a lost world beneath the polar ice cap. A similar plot—that is, the discovery of an underground world—had been used by Edgar Rice Burroughs in *At the Earth's Core*. In *The Last Dinosaur*, the oil company's owner is a big-game hunter who drills into underground chambers searching for prehistoric monsters. There, he encounters several creatures, including "the last dinosaur," a tyrannosaur played by an actor in a costume.

1977 In *Dinosaurs: Those Terrible Lizards*, famed debater Duane Gish of ICR promotes the Paluxy prints near Glen Rose, Texas, as proof that humans lived at the same time as dinosaurs. According to Gish, dinosaurs were herbivores until Adam and Eve sinned, and became extinct because of climate change after the collapse of the pre-Flood water canopy. Gish, a young-Earth creationist, included a discussion of bombardier beetles to suggest that dinosaurs breathed fire.

1977 In *Ontogeny and Phylogeny*, Stephen Jay Gould describes Edward Drinker Cope as "Americas' first great evolutionary theorist."

1977 In *The Dragons of Eden*, Carl Sagan (1934–1996) speculates about intelligent dinosaurs, wondering what would have happened if they had not gone extinct.

1977 In the movie *The People That Time Forgot* (a sequel to *The Land That Time Forgot*), a rescue plane is attacked by a pterosaur and forced to land. The rescuers finally leave the island after using a stegosaur to move their plane.

1977 Jack Oehlert completes building more than 30 life-size replicas of prehistoric animals for Lion Country Safari in Irvine, California. When the Safari never opened, the dinosaurs endured a fate common to many roadside dinosaurs: they began years of deterioration in a parking lot.

1977 James Farlow (b. 1951) and his colleagues suggest that plates of *Stegosaurus* were used primarily to regulate body temperature, not for protection.

1977 James Kitching (1922–2003) finds seven more 190-million-year-old eggs of *Massospondylus* in South Africa's Golden Gate Highlands National Park. It would be almost three decades before paleontologists would try to extract the eggs' embryos. These embryos, which were announced in *Science* in 2005, were the oldest dinosaur embryos ever found.

1977 Officials at DPP discover what became known as Quarry 143, where Philip Currie noted. "It was difficult to walk anywhere without stepping on bone." Excavations at Quarry 143 began two years later and showed that more than 80 percent of the 20–60 bones per square meter in the quarry were the ceratopsian *Centrosaurus* (now sometimes called *Eucentrosaurus*). The fossils documented a herd of young, juvenile, and middle-aged animals. This was in sharp contrast to other areas of the park, where duckbills account for 40 percent of the fossils.

1977 Peter Galton names the dinosaur *Othnielia* as a tribute to Othniel Marsh. During his career, Galton named 20 dinosaurs, 15 (i.e., 75%) of which remain valid.

1977 Posters for the movie *The Crater Lake Monster* ("A Beast More Frightening than Your Most Terrifying Nightmare!") promise a theropod, but the movie features a plesiosaur.

1977 The movie *Legend of the Dinosaurs* features several Mesozoic reptiles, but no dinosaurs.

1977 The science fiction comic *2000 A.D.* is published. The popularity of the comic (weekly sales exceeded 200,000 copies) helped readers visualize dinosaurs as active, dynamic creatures. Almost 20 years later, Steven Spielberg brought these dinosaurs to life in *Jurassic Park*.

1977 The U.S. Department of the Interior makes the Ghost Ranch Quarry a National Natural Landmark. The *Coelophysis* quarry at Ghost Ranch is one of the most famous Triassic dinosaur locales in the world.

1978 AAAS hosts a symposium to debate endothermy in dinosaurs. Some of the papers presented at the symposium—for example, the one by John Ostrom and Robert Bakker—supported dinosaurian endothermy; other papers rejected dinosaurian endothermy outright. Proceedings of the symposium—*A Cold Look at the Warm-Blooded Dinosaurs* (1980)[38]—later concluded that no "resolution of the controversy over whether dinosaurs were scaled-up cold-blooded reptiles or warm-blooded surrogate 'mammals' is reached here, although the weight of current opinion lies between the extremes." In *Dinosaurs Rediscovered* (1992), Don Lessem (b. 1951) described the symposium as a "dead end" regarding dinosaurs being warm-blooded. Lessem later helped reconstruct the skeleton of the 70-ton, 100-foot-long *Argentinosaurus* and the world's largest meat-eater, *Giganotosaurus* (also from Argentina). In 2013, a computer simulation revealed that *Argentinosaurus,* which had ligaments and tendons similar to those of horses, could walk about five miles per hour.

1978 Benjamin Waterhouse Hawkins's models at Sydenham are repaired at a cost of £3,375. Near this time, two more of Hawkins's models were found in a storeroom owned by the Greater London Council.

1978 Bernard Heuvelman's (1916–2001) *The Last Dragons of Africa* discusses African folklore and myths that could have been attributed to dinosaurs. The most famous of these (Mokèlé-mbèmbé) involved an elephant-sized, long-necked sauropod-like creature from Congo. Some young-Earth creationists used these stories to support their belief that dinosaurs lived with humans.

1978 Controversy about the skull atop the CMNH *Apatosaurus* skull is resolved by *Skull and Relationships of the Upper Jurassic Sauropod Apatosaurus (Reptilia/Saurischia),* a monograph by David S. Berman (an assistant curator at CMNH) and John S. McIntosh (a protégé of Barnum Brown and an expert regarding sauropods). After studying the field-notes of Earl Douglass, Berman and McIntosh concluded that a *Diplodocus*-like skull (CM 11162) rather than the *Camarasaurus* skull belonged atop *Apatosaurus louisae*. In October, 47 years after the most recent (and incorrect) skull had been placed on *Apatosaurus,* a cast of CM 11162 skull was finally mounted atop *Apatosaurus* (the original CM 11162 skull, which was too fragile to mount, was displayed in a case beside the skeleton.) Paleontologist Robert Bakker described the controversy this way: "Not only did *Brontosaurus* have the head of the wrong species, it had the head of the wrong genus, and not only the wrong genus but the wrong family. And not only the wrong family, but the wrong family of families. In fact, the head given to *Brontosaurus* for a hundred years was just about as wrong as you could get and still be within the grand order of brontosaurians. To make a mistake of the same magnitude, you'd have to put the head of a giraffe on the body of a goat."

1978 David Gerrold's *Deathbeast* describes a safari to the Early Cretaceous to kill *Tyrannosaurus*, "the Deathbeast, god of prehistoric fear." Deathbeast was depicted as warm-blooded. *Deathbeast* marked the fictional debut of *Deinonychus*.

1978 Dong Zhiming (b. 1937) and colleagues describe *Yangchuanosaurus shangyouensis,* one of China's best-preserved theropods.

1978 Glen Penfield, a geologist working for Pemex (Mexico's national oil company), conducts airborne magnetic surveys of the Yucatán. Anomalies around the northern Yucatán prompted Penfield to suggest that the so-called Chicxulub Crater is the K-T impact site. In his report titled *Preliminary: Geophysical Interpretation Report—Progreso area, written for Petroleos Mexicanos*, Penfield identified the Chicxulub crater in the Yucatán Peninsula as being "at or near the K-T boundary."

1978 In Bob Buckley's "The Runners," time-travelers return to the end of the Cretaceous and see human-like dinosaurs that can use primitive tools, and a supernova explosion threatens dinosaurs with extinction. Eight years earlier, Isaac Asimov's "Those Dying Lizards" implicated supernova explosions as the cause of dinosaurs' extinction.

1978 In Clifford Simak's (1904–1988) *Mastodonia,* a couple meets Catface, an alien whose spaceship has crashed on Earth. Catface transports the couple to the Late Cretaceous, where they see *Triceratops, Ankylosaurus*, and *Tyrannosaurus*, and establish a base called Mastodonia. Catface was later called home.

1978 In Marvel Comics' *Devil Dinosaur,* a red *Tyrannosaurus*-like dinosaur and its ape-like companion Moon-Boy live on the planet Dinosaur World where dinosaurs coexist with human-like creatures. The series lasted less than a year.

1978 In the movie *Planet of Dinosaurs* (which resembles *The Lost World* set in outer space), astronauts aboard spaceship "Odyssey" crash on a life-supporting planet in its own Mesozoic, where they confront dinosaurs (depicted by puppets designed by Stephen Czerkas). The astronauts saved themselves by luring the dinosaurs to a poison stake. The poorly received movie, which promised "A Nightmare World of Prehistoric Monsters," was not released in theatres until 1981.

1978 Ismael Ferrusquia-Villafranca and his colleagues announce the first discovery of dinosaur tracks in Mexico.

1978 Louis Marx & Company's dinosaur toys continue to be popular. In Sears catalogues, the toys promised to "recreate the world as it was millions of years ago when dinosaurs ruled the earth."

1978 Paleontologist Walter P. Coombs Jr. (b. 1942) determines many of the basic aspects of dinosaur locomotion.

1978 Roland Bird dies in his home in Homestead, Florida, and is buried in Grahamsville, New York, beneath the inscription "Discoverer of Sauropod Dinosaur Footprints."[39] Bird was buried with a dinosaur gastrolith in his pocket.

1978 The Black Hills Institute of Geological Research (BHI, formerly Black Hills Minerals, which was founded in 1974) is incorporated in Hill City, South Dakota. BHI was later involved in discovering and excavating several dinosaurs, including "Sue," the *T. rex* now featured at Chicago's Field Museum. *T. rex* "Stan" stands at BHI; its skull is one of the best-preserved and most-complete dinosaur-skulls yet discovered. "Stan" is named for its discoverer, Stan Sacrison, who found the specimen in 1987.

1978 To correct historical mistakes, a group of British time-travelers in *Doctor Who and the Dinosaur Invasion* bring dinosaurs to London, where they are destroyed by the army.

1978 "Uncle Beazley," a *Triceratops* from the 1964–1965 New York World's Fair, stars in an adaptation of *The Enormous Egg* (1956), a children's book by Oliver Butterworth about a *Triceratops* that hatches from a mutant chicken-egg.

1978 Virginia Tech's Dewey McLean proposes that the accumulation of carbon dioxide at the end of the Cretaceous changed oceanic currents and global climate, thereby leading to the K-T mass extinction.

1978 While prospecting for dinosaurs near Rudyard, Montana, "with a bag of plaster of Paris and a case of beer," paleontologists Jack Horner and Robert "Bob" Makela (1940–1987) visit a rock shop in tiny Bynum, Montana, owned by Marion Brandvold, David Trexler, and Laurie Trexler. There, Brandvold showed Horner and Makela a small fossil that Horner believed was part of a baby duck-billed dinosaur. Horner later said that the bones "weren't much to look at, just two dusty pieces of grey bone, but it was immediately obvious to me that they were the hip end of a duckbill thighbone and a bit of a rib—except that they were the wrong size. The femur, or thighbone, of a typical duckbill might be four feet high and as thick as a fencepost. The femur that Mrs. Brandvold handed me, if the bone had been whole, would have been the size of my thumb. . . . What I had in my hand was a bone from a baby dinosaur, a duckbill—exactly what I wanted, in a place I never expected it." Horner and Makela later went to James and John Peebles's Ranch outside Choteau, where they unearthed the small meat-eating *Troodon formosus* (which Joseph Leidy had originally named) and a nest of 15 fossilized baby dinosaurs in mudstone of the Two Medicine Formation. These were the first baby dinosaurs ever found. (In the 1920s, Roy Chapman Andrews found dinosaur eggs, but not babies, in Mongolia.) Their subsequent discovery in Late Cretaceous sediments of several colonial nesting grounds—including unhatched grapefruit-sized eggs, nests six feet across, foot-long babies with worn teeth, and still other three-foot-long babies—indicated that hadrosaurs nested in colonies and that their babies stayed in their nests for some time, were fed like young birds are fed, were cared for by adults, and had poorly developed joints similar to those of altricial birds (which remain in their nests). Horner and his coworkers eventually found more than 60 whole or partial skeletons, and 300 whole or partial eggs, each of which was approximately 8 inches long (i.e., the approximate size of the eggs found by Roy Chapman Andrews in Mongolia in the 1920s). Horner named the new duckbill *Maiasaura peeblesorum* ("Good Mother Lizard"). The species name, *peeblesorum,* honored the Peebles family, who owned the land where the specimens were collected.

1979 All species of "Anatosaurus" are assimilated into *Edmontosaurus* except for "*Anatosaurus*" *copei* (which in 1990 was named *Anatotitan*). Hundreds of fossils of *Edmontosaurus* have been found at the Ruth Mason Dinosaur Quarry in Faith, South Dakota. Reconstructions of these skeletons have been sold to museums throughout the world; for example, the *Edmontosaurus* "Mr. Ed" (80% of which is made of fossils, not molds) displayed in 1991 at the University of Wisconsin's Geological Museum was Wisconsin's first mounted dinosaur.

1979 Altangerel Perle (b. 1945) and Rinchen Barsbold (b. 1935) describe *Segnosaurus,* the first segnosaurid.

1979 At a symposium in Copenhagen, Swiss geologist Kenneth Hsu (b. 1929) speculates that a comet hitting Earth caused cyanide poisoning that killed marine organisms and the dinosaurs.

1979 Como Bluff, Wyoming, is declared a United Nations Educational, Scientific, and Cultural Organization (UNESCO) World Heritage Site. That same year, DPP (founded in 1955) became the first such site chosen for its paleontological significance. More than 900 people attended the dedication of DPP, which included a tribute to Charles M. Sternberg and tours of an area where bones of a large dinosaur-bone-bed were exposed for viewing. The park has yielded more than 500 partial to complete dinosaur skeletons from 35 species, and the nearby badlands of Drumheller have yielded another 25 species. More than 50,000 people visit the park per year.

1979 Construction of an industrial plant leads to the development of the mid-Jurassic Dashanpu Quarry near Dashanpu, China. The quarry later produced numerous dinosaurs, including two primitive sauropods (*Datousaurus* and the club-tailed *Shunosaurus*) and the stegosaur, *Huayangosaurus*.

1979 George Olshevsky completes the first restoration of *Hylaeosaurus* in more than a century. In 2000, Olshevsky admitted that the restoration was "dated" and a "synthesis" of *Polacanthus* and *Hylaeosaurus*.

1979 In Roger Zelazny's (1937–1995) novel *Roadmarks*, Marquis de Sade schemes to use *Tyrannosaurus* to eliminate a powerful industrialist.

1979 In *Valley of the Giants,* chiropractor Cecil Dougherty claims that humans left footprints alongside those of dinosaurs in Lower Cretaceous limestone of the Glen Rose Formation at Glen Rose, Texas.

1979 Princeton Univeristy undergraduate Fran Tannenbaum finds a fossilized egg sitting atop a rock ledge atop a small hill just south of Choteau, Montana. Jack Horner and Bob Makela subsequently found 10 nests (about 20 feet apart, with as many as 24 eggs per nest) of dinosaurs (e.g., the hypsilophodontid *Orodromeus makelai*) at this 76.7-million-year-old site, which became known as Egg Mountain. The nests were at different levels, prompting Horner to conclude that young dinosaurs "either remained in the colony or returned to the site frequently, either of which may have been a result of parental care." As Horner later noted, "The fact that numerous baby hadrosaurs had been eating and staying together suggests the presence of extended parental care comparable to that practiced by warm-blooded mammals. If food was being brought to the nest, someone had to do it. And if the babies ranged out of the nest, it's unlikely they would find their way back without parental supervision." This was the first evidence of colonial nesting in dinosaurs. An interpretive sign between mile-posts 57 and 58 on U.S. Highway 287 describes Egg Mountain, the first site in the Western Hemisphere to produce clutches of whole eggs attributable to dinosaurs.

1979 Jim Jensen discovers what he would later name *Ultrasaurus* (BYU 9044), a creature more than 90 feet long and weighing more than 130 tons (i.e., even larger than *Supersaurus*). Jensen formally described *Ultrasaurus* in 1985, at which time it was discovered that the name had already been used by Haang Mook Kim two years earlier to describe a dinosaur found in South Korea. Consequently in 1991, the name *Ultrasaurus* was changed to *Ultrasauros.* Several paleontologists claimed that *Ultrasauros*—which had a 9-foot-long shoulder blade—was merely a large specimen of *Supersaurus*, and *Ultrasauros* became a junior synonym of *Supersaurus* (which had been named first). Today, the name *Ultrasauros* has been abandoned.

1979 John Ostrom suggests his "insect-net hypothesis"—that is, ancestors of the first birds were small coelurosaurs covered by small feathers used to gather insects as food. Ostrom's hypothesis was soon abandoned, but it triggered much research.

1979 Kenneth Ham (b. 1951) founds, with John Mackay, The Creation Science Foundation in Australia, which later merges with the Creation Science Association. Ham then moved to the United States to work with ICR, and in 1994 opened a U.S. branch of his Australian organization now called Answers in Genesis (AiG). By 2005, AiG-U.S. had separated from other branches of AiG. Ham, a young-Earth creationist, wrote several books and produced a radio program that claimed that dinosaurs lived with humans. Ham argued "if something disagrees with the Bible, it is wrong, regardless of the evidence." In 2007, AiG opened the sprawling $27,000,000 Creation Museum near Cincinnati, Ohio, that uses dinosaurs to promote its fundamentalist message. Within three years, the museum had attracted more than one million visitors.

1979 Martin Vince of the University of Tucumán, a preparator for José Bonaparte, finds *Mussaurus patagonicus* ("mouse lizard") in Argentina's Santa Cruz Province. These were the first prosauropod eggs and hatchlings to be discovered.

1979 On CMNH's "Dinosaur Day" (October 20), the *Camarasaurus* skull atop *Apatosaurus* is replaced with an accurate skull. YPM followed suit in 1981, as did AMNH in the mid-1990s.

1979 Peter Galton and Jim Jensen name *Tarbosaurus,* which had been discovered eight years earlier near Dry Mesa Quarry in Colorado. *Tarbosaurus* was one of the largest theropods of the Jurassic. Dry Mesa Quarry also produced *Diplodocus, Camarasaurus, Brachiosaurus*, and *Apatosaurus*. Tons of materials excavated from Dry Mesa Quarry are stored under the football stadium at Brigham Young University, where they await examination.

1979 Philippe Taquet and Hans-Dieter Sues name *Majungatholus atopus,* a theropod from the Late Cretaceous found in modern-day Madagascar. This was the first report of a pachycephalosaur from the Southern Hemisphere.

1979 Teenager India Wood, while spending the summer at a friend's ranch in northern Colorado, discovers a bone bed just south of DNM. An *Allosaurus* from this site now greets visitors to the Denver Museum of Nature and Science.

1979 The Archeological Resources Protection Act (Public Law 95-341) protects archeological resources on public lands, including fossils "found in an archeological context."

1979 The film *The Great Dinosaur Mystery* claims that *Stegosaurus* lived in medieval Ireland.

1979 Walter Alvarez reports his findings about the Gubbio clay at a meeting of the America Geophysical Union. Clay at the K-T boundary lacked fossils but contained more than 30-times more iridium than adjacent sediments (i.e., 10 versus 0.3 parts per billion, respectively), suggesting that the K-T mass extinction coincided with an extraterrestrial impact. At this point, there were more than 80 published hypotheses for the extinction of dinosaurs; some of the theories were plausible but lacked evidence (e.g., thin eggshells, disease, poisoning by an accumulation of selenium in plants), whereas others (e.g., overhunting by aliens) seemed absurd. The K-T extinction eliminated 85 percent of species, 75 percent of genera, and 50 percent of taxonomic families. At the time of the K-T extinction, dinosaurs comprised 1–2 percent of the worldwide fauna.

1979 While hiking in New Mexico, musician Arthur Loy finds tail vertebrae of a dinosaur that will in 1991 be named *Seismosaurus.* This 171-foot-long dinosaur was estimated to be the longest dinosaur ever found. *Seismosaurus* is now a junior synonym of *Diplodocus.*

1980 After noting that the controversy about dinosaur endothermy "has no more significance than that relating to the sex of angels," bone histologist Armand J. de Ricqlés offers partial support for Robert Bakker's claims about warm-blooded dinosaurs by arguing that dinosaurian physiology was intermediate between that of conventional reptiles and avian-type endotherms.

1980 At a conference of Orthodox Jewish scientists, physicist Lee Spetner declares *Archaeopteryx* a deliberate fraud. Spetner's 1996 book, *Not By Chance! Shattering the Modern Theory of Evolution,* continued his attack on evolutionary biology.

1980 Canada features Bigfoot/Sasquatch on a 39¢ stamp.

1980 Fred Beierle's *Man, Dinosaurs, and History* claims that tracks in the Glen Rose Taylor Site are human tracks 19 inches long. According to Beierle, these tracks beside those of dinosaurs prove young-Earth creationists' claims about Earth's age, Noah's Flood, and the coexistence of humans and dinosaurs.

1980 Glen Kuban begins studying the dinosaur and "man tracks" near Glen Rose, Texas, and concludes that the "man tracks" were not made by humans. That same year, ICR's John Morris released *Tracking Those Incredible Dinosaurs . . . and the People Who Knew Them,* a book that defended the authenticity of some of the "man tracks" with dinosaur tracks near Glen Rose. Morris described "15 tracks or footprints of apparent giant human beings" to support his claim that "man and dinosaur walked together at the same time and place." Although many subsequent studies refuted the claims that the "man tracks" were made by humans, some creationists later charged (without supporting evidence) that Kuban had altered some "human footprints" along the Paluxy River.

1980 Improved methods of radiometric-dating change the Permian–Triassic boundary from 237 to 250 million years ago.

1980 In a bold, long (i.e., 14 pages), multidisciplinary paper titled "Extraterrestrial Cause for the Cretaceous-Tertiary Extinction" published in *Science,* American physicist and Nobel laureate Luis Alvarez (1911–1988), his geologist son Walter (b. 1940), and nuclear chemists Frank Asaro and Helen Michel (all of the University of California at Berkeley) suggest that the K-T extinction was caused by a giant asteroid (10 kilometers in diameter) that hit Earth 65 million years ago and released the energy equivalent of 100 million megatons of TNT (i.e., an amount of energy several orders of magnitude greater than that of all current nuclear weapons). The Alvarezes argued that the increased levels of iridium at the K-T boundary (documented in Italy, Denmark, and New Zealand) was "direct physical evidence for an unusual event at exactly the time of the extinctions . . . that appears to offer a satisfactory explanation for nearly all the available paleontological and physical evidence." The Alvarezes' claim was initially ridiculed, but subsequent evidence (including the presence of global iridium anomalies and the discovery in 1991 of the impact crater) supported their argument. Before the Alvarezes' claim, hypotheses for dinosaurs' extinction centered on slow declines, changes in plants' metabolism, destruction of Earth's ozone layer, the "greenhouse effect," thin eggshells, magnetic reversals, extreme temperatures, altered amounts of precipitation, egg-eating mammals, excessive ultraviolet radiation, stress, explosions of supernovas, constipation, and competition from mammals. Many of these explanations had been likened to Rudyard Kipling's *Just So Stories* because they were documented only by wishful thinking. Headlines in the *Science* tell the story about the Alvarezes' claim of a meteor impact triggering the K-T mass extinction: "Asteroid Theory of Extinctions Strengthened" (10/31/1980), "Impact Looks Real, The Catastrophe Smaller" (11/20/1981), "Isotopes Add Support for Asteroid Impact" (11/11/1983), "Asteroid Impact Gets More Support" (5/8/1987), "Was There a Prelude to the Dinosaurs' Demise?" (2/12/1988), "Huge Impact Is Favored K-T Boundary Killer" (11/11/1988), "Huge Impact Tied to Mass Extinction" (8/14/1992), and "How Lethal Was the K-T Impact?" (9/17/1993). Soon after the appearance of the Alvarezes' paper, Harvard paleontologist Stephen Jay Gould noted, "Many geologists, including myself, have long found themselves in the uncomfortable position of viewing extraterrestrial catastrophes as inherently plausible but rooting strongly against them because we know no way of obtaining evidence for such catastrophes even if they had occurred. The asteroidal theory has changed all that." The paper by Alvarez and others changed discussions of, research about, and perceptions of mass extinctions.

1980 José Bonaparte (b. 1928) and Jaime E. Powell describe *Saltasaurus* ("lizard from Salta"), a 39-foot-long sauropod from Late Cretaceous sediments in what is now northwestern Argentina. This was the first genus of dinosaur known to have bony plates embedded in its skin.

1980 *Journal of Vertebrate Paleontology* is founded at the University of Oklahoma, and the following year its premier volume comprises 442 published pages. In 2013, it published 1,487 pages.

1980 Phil Bjork of the South Dakota School of Mines finds a 40 percent–complete specimen of *T. rex*. The skull of this specimen is displayed in the school's museum.

1980 Poland issues a stamp featuring a skeleton of *Tarbosaurus* beside a map of where it was collected.

1980 Ralph Molnar describes *Minmi paravertebra,* the first ankylosaur described from the Southern Hemisphere.

1980 Readers of *Science-80* read "Living Dinosaurs," which claims that, "as fantastic as it seems," an elephant-sized dinosaur lives in the swamps of western Africa.

1980 Robert Kofahl, of the Creation Science Research Center, claims in *Handy Dandy Evolution Refuter* that the Paluxy "man tracks" support young-Earth creationism.

1980 Rudolph Zallinger receives the Addison Emery Verrill Medal for *The Age of Reptiles*. The medal was accompanied by this citation: "Artist and teacher. Your great natural history murals at the Peabody Museum are a fusion of scientific accuracy and artistic genius. Guided by your own diligent research and painstaking collaboration with scientists, your imagination has allowed us a glimpse into past worlds no human eye ever witnessed." Zallinger's work was commemorated by toys issued by Marx and other companies.

1980 Walter Coombs Jr. discovers additional fossils of *Psittacosaurus* in the materials collected in the Asiatic expeditions in 1922–1925. The hatchlings' tiny skulls (1.1 to 1.6 inches long) remain some of the smallest dinosaur skeletons so far discovered.

1980 Wilbur Fields's *Paluxy Field Explorations* describes the Taylor Site tracks as "very large probably man tracks . . . 15–16 inches long."

1981 A presentation at the Vancouver Planetarium titled "Death of the Dinosaurs" begins a decade of "dinomania" in Canada. In 1983, Calgary Zoo opened its renovated Prehistoric Park that included more than 25 life-size dinosaurs, and that same year, the National Museum of Natural Sciences opened a giant dinosaur-exhibit in Ottawa. By 1986, life-size models of *Styracosaurus* were exhibited at Vancouver's Expo '86. In 1989, the Dinosaur Project found a specimen of *Hesperornis,* an extinct genus of flightless bird from the Cretaccous. And in 1990, the Hitachi Dinoventure-90—exhibited in Japan and sponsored by the Alberta Culture and Multiculturism Office—attracted more than two million visitors.

1981 *Coelophysis bauri,* which is featured in the logo of the New Mexico Museum of Natural History, becomes the state fossil of New Mexico.

1981 *Dinosaur ABC's* introduces children to the Paluxy "man tracks" as evidence of humans living with dinosaurs.

1981 Geologists Carl Orth and Bob Tschudy publish strong support for an asteroid impact at the end of the Cretaceous, and Glen Penfield and Antonio Camargo (who had been born in northern Yucatán) conclude a presentation at the Society of Exploration Geophysics by noting, "We would like to note the proximity of the [Yucatán] feature in time to the hypothetical Cretaceous-Tertiary boundary event responsible for the emplacement of iridium-enriched clays on a global scale and invite investigation of this feature in light of the meteoric climatic alteration hypothesis of the late Cretaceous Extinction." The enormous crater, near the town of Chicxulub, was the crater that the Alvarez group was looking for. Pennfield and Camargo's announcement was featured on the front-page of newspapers and magazines (e.g., *Sky and Telescope*), but overlooked by the Alvarez group and most other scientists. In the 1990s, the Chicxulub crater was identified as the K-T impact site by Alan Hildebrand, Kevin Pope, and others.

1981 Herman and Kia Regusters go to the Congo searching for a living *Apatosaurus*-like sauropod referred to as *Mokèlé-mbèmbé* by natives of the village of Boa. A spokesman later claimed that they saw the creature in Lake Tele, but no one produced any physical evidence to support these claims. Despite this lack of evidence, Mokèlé-mbèmbé continues to be cited by several young-Earth museums claiming that non-avian dinosaurs lived with humans, and may live today.

1981 In Jack Dann's and Gardner Dozois's "A Change in the Weather," dinosaurs metaphorically rain from the skies and become a nuisance.

1981 In Los Angeles, California, Griffith Park Observatory opens its "Death of the Dinosaurs" exhibit to illustrate the Alvarezes' theory that dinosaurs became extinct because of a meteoric impact.

1981 In *The Ape Men*, when students complain that cavemen did not live with dinosaurs, a Christian disagrees; his evidence includes the Bible and the Paluxy River "man tracks."

1981 In *The Mismeasure of Man,* paleontologist Stephen Jay Gould admires Edward Drinker Cope's skills as a theoretician and paleontologist, but condemns his racist views.

1981 In the slapstick comedy movie *Caveman*, Atouk the Caveman (played by former Beatles' drummer Ringo Starr) subdues a chubby tyrannosaur (created by special-effects expert Jim Danforth) by feeding it sleep-inducing fruit. Dinosaurs crow like a rooster at dawn and howl at the moon at night, and Ringo staggers while carrying a dinosaur egg bigger than he is.

1981 James Farlow, a former student of John Ostrom, reports that carnivorous dinosaurs in Texas could run faster than 40 kilometers per hour.

1981 Le Monde Prehistorique (Prehistoric World) opens in Morrisburg, Ontario, Canada. The park features more than 40 prehistoric animals sculpted by brothers Serge and Paul Dupuis, who were inspired by Charles Knight and Ely Kish's paintings. Although a 100-foot-long *Apatosaurus* dominates the park, other displays along the winding trail showcase *Huayangosaurus* (a plated dinosaur from Asia), *Kentrosaurus* (a stegosaur from South Africa), *Cetiosaurus* (a sauropod from Europe), and *Minmi* (an armored dinosaur from Australia).

1981 NASA convenes the first of three workshops at Ames Research Center to discuss the evolution of life, advanced life in the universe, and extraterrestrial influences on evolution. That same year, a conference titled "Geological Implications of Impacts of Large Asteroids and Comets on Earth" convened at Utah's Snowbird ski resort to discuss the Alvarez hypothesis for the K-T extinction. At that symposium, David Raup (b. 1933) told attendees of his earlier computer simulations suggesting that a typical mass extinction in a region could not occur unless the affected region had an area that approached half of Earth's surface. In 1986, Raup wrote in *The Nemesis Affair*, "*I made no attempt to publish this research because I knew it would be laughed at, or worse*." (emphasis by Raup)

1981 Paleontologists from RTM in Drumheller, Alberta, collect a 30 percent–complete specimen of *T. rex* in central Alberta; this specimen had been found in 1946 by Charles M. Sternberg. That same year, a second, more-complete specimen of *T. rex* was found by high school students in southern Alberta, Canada. This specimen, named "Black Beauty," was mounted in RTM in 1992. It took a crew of five people three years to prepare and mount the *T. rex*.

1981 Paul Taylor, son of *Films for Christ* founder Stan Taylor (who produced *Footprints in Stone*), argues in *The Great Dinosaur Mystery and the Bible* that "evidence from the Bible and from scientists" proves that Earth and dinosaurs are only a few thousand years old, that fossilized dinosaurs were produced by Noah's Flood, and that human hunters killed the dinosaurs."

1981 *Quest for Fire* is another movie showing humans living with dinosaurs.

1981 Researchers from CMNH, the New Mexico Museum of Natural History, YPM, and the Museum of Northern Arizona reopen the *Coelophysis* quarry at Ghost Ranch, New Mexico. A large block from that excavation is now displayed at Ghost Ranch Museum.

1981 Russian paleontologist Sergei Kurzanov (b. 1947) describes *Avimimus* ("bird mimic'), a 5-foot-long slender theropod having quill knobs on its forearms.

The fossil was discovered in late Cretaceous deposits in Mongolia. In 2008, paleontologists from Mongolia, Canada, and the United States led by Phil Currie reported a large bone-bed of *Avimimus* in the Gobi Desert. Although *Avimimus* was originally claimed to be a closer ancestor of dinosaur to birds than *Archaeopteryx*, more recent analyses suggest that *Avimimus* is a bird-like oviraptorosaurus.

1981 The first dinosaur-track is found in Alberta's DPP. The print was described eight years later by Phil Currie.

1981 The Sternberg family dynasty of dinosaur hunters ends with the death of Charles M. Sternberg. Charles M. Sternberg helped establish DPP and, in his later years, worked at the National Museum of Canada while helping produce paleontology exhibits at the Calgary Zoo. His sibling George spent his later years at what became Fort Hays State University, and Charles's other sibling Levin worked at Toronto's ROM. Charles M. Sternberg named 19 dinosaurs, eight (i.e., 42%) of which remain valid.

1981 While hiking in a remote area called Rabbit Valley west of Grand Junction, Colorado, J. D. Moore and Pete Mygatt (and their wives) find bones in a shallow drainage area. This site became the Mygatt-Moore Quarry that ultimately produced thousands of bones from several genera of dinosaurs. Among these dinosaurs were *Allosaurus, Camarasaurus, Diplodocus*, *Apatosaurus,* and *Mymoorapelta*, the oldest armored ankylosaur (and first Jurassic ankylosaur) found in North America.

1981 While working in northern Malawi, Louis Jacobs of Southern Methodist University and Elizabeth Gomani of Malawi unearth the titanosaurs *Malawisaurus* and *Karongasaurus*.

1982 At a presentation to the NAS in Washington, DC, Luis Alvarez tells a large audience that "the asteroid hit, and that the impact triggered the extinction of much of the life in the sea—are no longer debatable points," but that the impact's effect on the extinction of dinosaurs "is very much open to debate, although I believe that it very definitely did." Others, such as University of Colorado scientist Erle G. Kauffman, concluded, "I don't rule out a meteorite or an ecological collapse as the final straw, but it wasn't the cause."

1982 Baptist preacher Carl Edward Baugh (b. 1936) pays Emmitt McFall of Glen Rose, Texas, $20 for the rights to excavate tracks from McFall's property along the Paluxy River. Two days later, Baugh announced that he had discovered human and dinosaur tracks side-by-side and that his discoveries would have "unparalleled historic significance"—the Dallas Ft. Worth *Star-Telegram* described the claim with the headline "Tracks Step on Evolution." Baugh claimed to have found more than 50 "man tracks" on the south bank of the Paluxy River at the "McFall Ledge," which is southwest of DVSP. Two years later, Baugh opened the Creation

Evidence Museum of Texas just outside of DVSP in Glen Rose. The popular museum consisted of a small group of trailers and a larger building identified as a "scientifically chartered museum." Baugh used the museum to discredit evolution by arguing that people lived contemporaneously with dinosaurs. Baugh later hosted a weekly show on Trinity Broadcasting Network for 11 years titled "Creation in the 21st Century," where he was referred to as the "foremost doctor on creation science." Baugh claimed that Earth at creation was 12 percent smaller with twice the atmospheric pressure (and three-times the amount of oxygen), that early Earth was surrounded by a several-inches-thick layer of crystalline hydrogen that reflected the sun's rays so humans could live for 900 years (this layer of hydrogen shattered when God shouted), that creation water can heal people, that people can use their minds to alter the rates of radioactive decay, that a tomato plant grown in conditions simulating pre-Flood conditions grew to 30 feet tall and produced 5,000 tomatoes, and that the worldwide flood pushed the hydrogen canopy into space (where it now forms a belt of asteroids that will fall to Earth during the Great Tribulation). Baugh's work has been widely criticized, including by young-Earth creationists; for example, AiG's Andrew Snelling noted that Baugh's work "is not the mainstream of the creation movement," and the Creation Science Foundation announced, "We do not regard [Carl] Baugh as representative in any way of the mainstream creation movement." Baugh responded "I don't mind being controversial. . . . Controversy was introduced in the Garden of Eden, and it . . . will follow me after I'm gone."

1982 Canadian paleontologist Philip Currie (b. 1949) and his colleagues note the many similarities of dinosaurs that lived in Canada and China.

1982 David Thomas' first life-size bronze model of a dinosaur ("Spike" the *Pentaceratops sternbergii*) debuts at Albuquerque's New Mexico Museum of Natural History. Thomas's later work was featured at MOR (*Maiasaura peeblesorum*), Ghost Ranch Museum (*Coelophysis bauri*), UFHNH (*Utahraptor ostrommaysi*), and Taiwan's National Museum of Natural History (*Tyrannosaurus rex*). The New Mexico Museum of Natural History features an impressive mount of *Allosaurus maximus* attacking *Diplodocus hallorum* (these giant dinosaurs were previously called *Saurophaganax* and *Seismosaurus*, but were found to be larger versions of already known genera).

1982 Extending the thinking of Carl Sagan, Dale Russell (b. 1937) and model-maker Ron Séguin reconstruct a hypothetical "evolved" troodontid that had an encephalization quotient similar to that of modern humans. These upright creatures, called dinosauroids, have been featured in numerous books and articles.

1982 In *Ancestors and Archetypes,* historian Adrian Desmond (b. 1947) analyzes the sociological aspects of the debate between Richard Owen and Thomas Huxley regarding the connection between birds and dinosaurs. Desmond concluded that the anti-progressivism arguments of Owen, as well as the pro-progressivism claims of Huxley, were both ideologically driven.

1982 Kenneth Carpenter suggests that taphonomic bias influences why most dinosaur eggs and eggshells have been found in Upper Cretaceous strata, especially in dry, arid regions.

1982 Paleontologist Jack Horner becomes curator of vertebrate paleontology at MOR in Bozeman, Montana.

1982 Sylvia Czerkas organizes the exhibit *Dinosaurs, Mammoths, and Cavemen: The Art of Charles R. Knight* at the Natural History Museum of Los Angeles County. The exhibit was accompanied by a book by Czerkas having the same title, as well as by articles in *Natural History* and *Terra*. Knight, who produced approximately 1,000 paintings, sketches, and sculptures of ancient animals, was the most influential of all artists who have restored extinct animals.

1982 Thanks to a two-year campaign by elementary school students, *Stegosaurus* is chosen as the state dinosaur of Colorado.

1982 The world's oldest known *Iguanodon* is discovered at Rabbit Valley, about 30 miles west of Grand Junction, Colorado.

1982 When Martin Lockley and his colleagues renew studies of the Purgatoire River dinosaur tracks, they discover about 1,000 tracks in more than 50 trackways (Figure 40). Some of the trackways extend for more than 600 feet.

1983 After creationists claim that "man tracks" in the Paluxy River near Glen Rose, Texas, prove that humans lived contemporaneously with dinosaurs, Steven Schafersman of the Texas Council for Science Education reports in *Geotimes* that "some [of the tracks] are poorly preserved dinosaur tracks, some are the rear-heel slide mark of single dinosaur footprints, some are dinosaur tracks made in extremely fluid mud in which the mud flowed back after the dinosaur extracted its foot, some are random depressions on the limestone surface with trace-fossil burrow casts misinterpreted as ridges between 'toes,' and some are artfully improved 'human footprints' showing superficial 'toes' carved in the softer lower limestone bed. . . . The molds of casts and photographs which the creationists have shown me are of decades-old, obviously carved footprints." Numerous other studies have discredited the claims by creationists that they have discovered "man tracks" in the same sediments as those of dinosaurs.

1983 Eruptions of Hawaiian volcanoes are shown to release iridium. This finding was used by several scientists—most notably geologists Charles Drake and Charles

Figure 40 These tracks along the Purgatoire River in southeastern Colorado comprise part of the largest mapped trackway of dinosaur prints in North America. The area's more than 1,300 tracks—left 150 million years ago by more than 100 dinosaurs—were discovered in the 1930s. This photo shows paleontologist Martin Lockley examining tracks left by sauropods. (Louie Psihoyos/Corbis)

Officer—to argue that the K-T extinction could have been produced by volcanism rather than a meteor impact. Iridium anomalies have been reported from more than 50 K-T boundary sites from around the world.

1983 In Montana, Jack Horner finds what will later be named *Cerasinops* ("lesser horned face"). This 40-pound, 3-foot-tall biped may link primitive horned dinosaurs in Asia with those in America. Horner argued that hadrosaurs survived in part because of a "survival strategy" similar to that of some modern birds that involves food sharing and parental supervision (including parents bringing food to their young).

1983 Jan Weaver concludes that *Brachiosaurus* and other large sauropods could not gather food fast enough to maintain endothermy.

1983 John Ostrom notes, "If [Robert] Bakker had moved more cautiously, we'd have had less heat and a better reception. I was never able to convince him that understatement was more powerful than overstatement."

1983 London plumber William J. Walker finds a 12-inch-long broken claw of a dromaeosaur in a clay pit just south of Dorking in Surrey, the United Kingdom. The find was officially described three years later and became known as *Baryonyx walkeri* ("Walker's heavy claw"). On the thumb of each hand of *Baryonyx* was a claw that was up to 12 inches long. The carnivorous saurischian, which was 30 feet long, 10 feet tall, and weighed more than two tons, was an early relative of spine-backed *Spinosaurus. Baryonyx* is one of the few known fish-eating dinosaurs.

1983 *Ripley's Believe It or Not!* notes that Brigham Young University paleontologist "Dinosaur Jim" Jensen has collected over 100 tons of fossil bones.

1983 *Science* reports "now that, for many at least, asteroid impact has been accepted as a causative agent in mass extinction, attention turns to the wider view." Later that same year, *Science* reported that isotopic analyses support an "asteroid impact 65 million years ago but cannot exclude a huge volcanic eruption."

1983 The "Dinosaur Committee" of the Seventh-Day Adventists' Office of Education releases the two-volume *A Creationist View of Dinosaurs*, which claims that dinosaurs "once lived on our earth" and were produced by "the combining, or mixing, of living things to produce other living things that God did not originally create." The committee also concluded that dinosaurs were not aboard Noah's Ark because they were not originally created by God.

1983 The Soviet News Agency reports humanlike tracks among the 1,500 dinosaur tracks in Turkmenia.

1983 While working at Egg Mountain south of Choteau, Montana, Jack Horner and his coworkers become the first in the world to find dinosaur embryos. The site has since yielded numerous eggs, embryos, and baby skeletons of *Maiasaura*. In 2013, Liberty University's Marcus Ross claimed that these *Maiasaura* were "trying to care for their young in the chaos of Noah's Flood."

1983 While working at the Philmont Scout Ranch in northeastern New Mexico, American geologist Charles Pillmore finds the first fossilized footprint of *T. rex*. The print, which was 33 inches long and 28 inches wide, was described by Martin Lockley and Adrian Hunt in 1994.

1983 Zhao Xijin and Philip Currie excavate *Monolophosaurus* ("single-crested dragon"), which was a 15-foot-long theropod having an unusual bony crest down the middle of its head. *Monolophosaurus,* which lived in central Asia 150 million years ago, was not closely related to *Dilophosaurus* ("double-crested reptile," from the early Jurassic) or the Permian reptile *Trilophosaurus* ("triple-crested reptile").

1984 Amateur fossil-collector Armando Diaz-Romeral of Spain discovers Las Hoyas, a famous 115-million-year-old bed of Cretaceous fossils, many of which are dinosaurs.

1984 An article titled "Dinosaur Experts Resist Meteor Extinction Idea" in the *New York Times* reports a survey of 500 geophysicists, paleontologists, and geologists in North America and Europe showing that 61 percent of respondents accept the basic evidence for a major impact at the end of the Cretaceous. Only 12 percent of respondents did not think that there was an impact or a mass extinction at the end of the Cretaceous. The following year, a survey of 118 paleontologists at SVP's annual meeting in Rapid City, South Dakota, showed that 90 percent of respondents accepted the evidence for an impact, but only 4 percent attributed the impact as a major cause of dinosaurs' extinction. In a 1985 poll, 27 percent believed that there was no mass extinction to be explained. The main argument for this reasoning was the claim that populations of dinosaurs had been declining long before the meteorite struck Earth.

1984 As RTM is being constructed in Drumheller, Alberta, Canada, paleontologists at the site find a lower jaw of the small, birdlike *Troodon* from the Late Cretaceous. This was one of the first such fossils ever discovered. *Troodon* had a large brain-to-body-mass ratio and large, retractable, sickle-shaped claws on their second toes.

1984 Artist Mark Hallett's (b. 1947) *Dawn of a New Day* imagines the transition from the Age of Dinosaurs to the Age of Mammals by showing several small mammals eating the last remains of a dinosaur.

1984 At a ravine used as a dumping ground, 13-year-old Boy Scout Christopher Brees locates where *Hadrosaurus foulkii* had been found in 1858 by William Parker Foulke in Haddonfield, New Jersey. Brees's work was supported by grants from Philadelphia's Academy of Natural Sciences and Haddonfield's Junior Women's Club. Today, the famous site—at the end of Maple Avenue off Grove Street—is visited by more than 100 visitors per month, many of whom leave small dinosaurs in tribute to the site's historic importance. The site is marked by a 700-pound stone adorned with a plaque showing a profile of *Hadrosaurus foulkii* and this text: "*Hadrosaurus foulkii*. In a marl pit on the John E. Hopkins farm in October, 1858 the world's first nearly-complete dinosaur skeleton was unearthed by William Parker Foulke. This find was adjacent to this point. This was also the first dinosaur skeleton ever to be mounted. The bones represented a 25-foot, seven- to eight-ton, herbivorous hadrosaurus (reptile). Its height probably ranged from six to ten feet at the hip. Some 55 of an estimated 80 bones were discovered. This creature lived 70 to 80 million years ago during the Cretaceous period at the end of the dinosaur age. This site was developed in 1984 as an Eagle Scout project by

Christopher Brees, Troop 65. Major project funding by the Academy of Natural Sciences, Philadelphia, Pennsylvania" (Figure 41). Brees's efforts rekindled local and national interest in *Hadrosaurus* and the famous site.

1984 Australian young-Earth creationist Ken Ham (b. 1951), who will become CEO of Answers in Genesis, writes in "What happened to the dinosaurs?" that "human footprints have also been found in dinosaur footprint strata in Russia." Another article that same year noted that reports of Russian "man tracks" were "without factual foundation."

1984 Fossils of what in 1991 will be named *Amargasaurus cazaui* are found near La Amarga, Argentina. This 30-foot-long, Early Cretaceous sauropod had two parallel rows of spikes extending from its vertebrae. Such spines had traditionally been associated with the presence of a sail.

Figure 41 In 1858, William Parker Foulke and a crew from Philadelphia's Academy of Natural Sciences excavated *Hadrosaurus foulkii* from a ravine in Haddonfield, New Jersey. The skeleton of *H. foulkii*—the first dinosaur named in North America—was displayed at the academy a decade later, and from 1868 to 1883 was the only articulated dinosaur mounted anywhere in the world. The monument shown here was erected in 1994 near where *H. foulkii* was found; appreciative visitors often leave plastic dinosaurs to show their gratitude. (Courtesy Randy Moore)

1984 High concentrations of iridium have been detected at more than 50 K-T boundary sites throughout the world.

1984 In *Search for Dinosaurs* (the second book in the *Time Machine* series), readers go to the Mesozoic to photograph *Archaeopteryx*.

1984 Invertebrate paleontologist David Raup (b. 1933) and Jack Sepkoski (1948–1999) use a statistical analysis of extinctions during the past 250 million years to conclude that mass extinctions have occurred every 26 million years. (A similar idea had been suggested seven years earlier by Princeton University scientists A. G. Fisher and Michael Arthur.) In their five-page paper in *PNAS*, Raup and Sepkoski did not identify a cause for the extinctions, but suggested that it was probably extraterrestrial (e.g., the result of the passage of the Solar System through the Milky Way galaxy).

1984 Jack Horner describes baby *Maiasaura*. Their poorly developed joints suggested that these babies did not leave their nests, and therefore, that adults brought food to them.

1984 R. E. H. Reid uses reptilian "growth rings" in dinosaur bones as evidence that dinosaurs had a physiology intermediate between that of "cold-blooded" and "warm-blooded" organisms.

1984 The 10-minute film *Prehistoric Beast* features *T. rex* having proper posture and anatomy. *Prehistoric Beast* was made by Phil Tippett with go-motion animation.

1984 The alternate rock band Dinosaur forms in Massachusetts. When legal issues arose about the band's name, it was changed to Dinosaur Jr.

1984 The first dinosaur-only art show occurs in Boston.

1984 UCMP Robert Long discovers what is believed to be the earliest known dinosaur. This fossil, which was about the size of a small ostrich, was 225 million years old.

1984 When geologist Bruce Bohor and his colleagues at USGS examine mineral grains at the K-T Boundary near Hell Creek, Montana, they see "shock lamellae" on the grains' surfaces. Nearly 25 percent of the quartz grains at the boundary were crisscrossed by the wavelike features produced by an impact. These lamellae form only in craters produced by nuclear bombs and meteors. Within three years, shocked quartz had been reported at nine other K-T sites from throughout the world.

1984 While examining a site called "Lot's Wife" in Petrified Forest National Park, prospectors find the first North American staurikosaur. They gave it the informal name "Gertie" to honor the first cartoon that featured a dinosaur.

1985 After four years of planning and construction, the Tyrrell Museum of Paleontology opens in Drumheller, Alberta, Canada. Two years later, a satellite display opened in DPP. The Tyrrell Museum of Paleontology, one of the top dinosaur museums in the world, displays *Albertosaurus, Coelophysis, Hadrosaurus, Lambeosaurus, Tyrannosaurus, Dromaeosaurus,* and other dinosaurs. Near the museum, Jack Horner and Philip Currie excavated a small jaw spotted two years earlier by Horner. Currie described the discovery like this: "For the first time we had a *Stenonychosaurus* with *Troodon* teeth in it. So we were able to show *Stenonychosaurus* and *Troodon* were one and the same dinosaur." Currie's subsequent studies of *Troodon* described a passage connecting the inner ears—a trait common in birds but not yet described in dinosaurs. Currie was convinced: "What's interesting about the whole concept of birds coming from dinosaurs is suddenly you're dealing with an animal that isn't extinct any more. The dinosaurs are still alive."

1985 Although Walter Coombs Jr. had suggested *Hylaeosaurus* and *Polacanthus* were synonyms, they are listed as separate genera in *Illustrated Encyclopedia of Dinosaurs*.

1985 An editorial titled "Miscasting the Dinosaur's Horoscope" in the *New York Times* discredits a meteoric impact (and periodicity) as being the cause of dinosaurs' extinction. The editorial concluded "terrestrial events, like volcanic activity or changes in climate or sea level, are the most immediate possible causes of mass extinctions. Astronomers should leave to astrologers the task of seeking the cause of earthly events in the stars."

1985 As Walt Disney preferred, *T. rex* in the Walt Disney movie *My Science Project* has three fingers. In reality, *T. rex* had two.

1985 Bill Watterson's (b. 1958) comic strip *Calvin and Hobbes* debuts in newspapers. In years ahead, dinosaurs were a favorite subject of the popular strip.

1985 Caleb Curriculum releases *Dinosaurs and the People Who Knew Them,* a coloring book claiming that Noah took baby dinosaurs aboard the Ark.

1985 Charles Pellegrino's speculative "Dinosaur Capsule" in *Omni* magazine explores the use of DNA of insects preserved in amber. This article inspired Michael Crichton when Crichton was writing *Jurassic Park*.

1985 Chicago hairstylist John J. Lanzendorf (b. 1946) buys a sculpture of *Tyrannosaurus rex*. In subsequent years, Lanzendorf added hundreds of additional pieces, producing one of the largest collections of dinosaur art in the world. Lanzendorf's dinosaur-related art was featured in *Dinosaur Imagery: The Lanzendorf Collection* (2000).

1985 Daniel Whitmire and John Matese suggest in *Nature* that the 26-million-year periodicity of mass extinctions is due to meteoric bombardments caused by movements of an unidentified "Planet X" near Neptune.

1985 Dartmouth geologists Charles B. Officer and Charles L. Drake suggest in an article in *Science* titled "The Cretaceous-Tertiary Boundary" that "iridium and other associated elements were not deposited instantaneously at the [K-T] transition, but rather that there was an intense and variable influx of these constituents during a relatively short geologic time interval on the order of 10,000 to 100,000 years . . . the available evidence favors a mantle origin rather than a meteoric origin for these constituents."

1985 David Norman publishes the classic *Illustrated Encyclopedia of Dinosaurs,* which includes more than 700 illustrations of the 68 best-known dinosaurs.

1985 Following an on-site inspection of the Paluxy "man tracks" near Glen Rose, ICR's John Morris, Paul Taylor, and Marian Taylor (the son and widow of Stanley Taylor) admit that they cannot identify any fossilized markings anywhere in or near the Paluxy River as human or probably human. *Films for Christ* withdrew its *Footprints in Stone* from circulation, but continued to sell its *The Great Dinosaur Mystery and the Bible* and a film having the same title that promoted the "man tracks" as evidence that humans lived with dinosaurs.

1985 Fossilized teeth and jaws found by amateur geologist Marj Averett near Grand Junction, Colorado, are identified as part of a primitive iguanodont, the first of these dinosaurs to be discovered in the Americas.

1985 In *Science* magazine, chemists Wendy Wolbach, Roy Lewis, and Edward Anders report large amounts of soot in clays at the K-T boundary, which they attribute to massive wildfires triggered by a comet's impact. The 0.4-inch-thick layer was assumed to have formed in less than a year, whereas the same thickness of limestone above and below the boundary formed over a span of 100 years. In 1988, Wolbach and her colleagues concluded that more than half of Earth's vegetation burned in the aftermath of the impact.

1985 In the comedy movie *My Science Project*, a misguided science project produces *T. rex* in the school gymnasium. The dinosaurs were depicted with rod puppets. Dinosaurs, instead of scientists, travel through time.

1985 In the Disney movie *Baby—Secret of the Lost Legend* ("The Greatest Adventure Ever Born!")—a movie about an orphaned, cute-faced brontosaur—dinosaurs have human traits (e.g., one cries when it sees its dead father). A young couple in Africa unearth a vertebra of *Brontosaurus*, and are then led to a village of sick people, where a dying man describes the creature that the village's residents just ate (and were sickened by). The couple then finds a family of *Brontosaurus*. Dinosaurs in the movie were played by actors in rubber suits. One of these dinosaurs was a sauropod, which required the actor to walk on all fours. In the movie's final scene, the mother *Brontosaurus* and her baby swam off into the sunset. *Baby—Secret of the Lost Legend* is a take-off on the legend of Mokèlé-mbèmbé, the dinosaurs allegedly living in Africa. Like the legends of Yeti, Bigfoot, and Loch Ness Monster, there is no physical evidence for Mokèlé-mbèmbé.

1985 In the documentary *Dinosaur,* paleontologist Dale Russell speculates that a dinosaur may live in Congo.

1985 Jack Horner and James Gorman's *Maia: A Dinosaur Grows Up* traces the life of a baby *Maiasaura* from hatching until it protects its own offspring from predators. That same year, Horner discovered the bizarre *Einiosaurus* ("buffalo lizard") in Montana's Two Medicine Formation. *Einiosaurus*, which probably lived in groups, is characterized by a forward-curving nose-horn that resembles a can opener. *Einiosaurus*, which was described and named in 1995 by Scott Sampson, was featured on a 32¢ stamp issued by the U.S. Postal Service in 1997.

1985 Jim Jensen names *Dystylosaurus* ("two beam lizard") after finding a vertebra more than 3 feet tall. That same year, Jensen also named the gigantic *Ultrasaurus* (and estimated its mass to be 180 tons), but soon learned that the name was used two years earlier by paleontologist Haang Mook Kim to describe a sauropod discovered in Korea.

1985 *Maiasaura* becomes the first dinosaur to go into outer space when a *Maiasaura* bone fragment and eggshell are launched aboard Space Shuttle "Challenger" (STS-51-F). That same year, *Maiasaura peeblesorum* became the state fossil of Montana.

1985 Mike Perry (director of the Museum of Western Colorado) and James Madsen, Jr. (Utah's first state paleontologist) invent "The Dinosaur Triangle," whose points and primary attractions consist of Grand Junction, Colorado (including Riggs Hill, Dinosaur Hill, Rabbit Valley), Price, Utah (including Cleveland-Lloyd Dinosaur Quarry, College of Eastern Utah Prehistoric Museum), and Vernal, Utah (including DNM, UFHNH). Most dinosaur fossils in and around this area are found in the Morrison Formation.

1985 More than 4,000 people attend the opening of RTM, on the banks of the Red Deer River, in Drumheller, Alberta, Canada. In its first year, the museum attracted more than 500,000 visitors and hired 18 staff and three associates (as well as more than 50 volunteers). The $27-million museum, which continues to be visited by more than 350,000 people per year, includes more than 40 mounted dinosaur skeletons and 11,000 fossil specimens.

1985 José Bonaparte and Fernando Novas describe the Late Cretaceous theropod *Abelisaurus comahuensis*. More abelisaurs, including *Carnotaurus* (1990) and *Xenotarsosaurus* (1986), were soon discovered.

1985 Phil Tippett (b. 1951), who was inspired to become a special-effects artist by the work of Ray Harryhausen, depicts realistic dinosaurs in CBS's animated *Dinosaur!* This documentary won Tippett his first Emmy for Special Visual Effects. The hour-long program was narrated by Christopher Reeve (1952–2004, the actor who portrayed Superman) who, six minutes into the show, asks, "What terror came out of the skies to end the dinosaurs?" Unlike similar shows that had relied on skeletons in museums, *Dinosaur!* presented recent discoveries with dinosaurs brought to life with stop-action animation.

1985 Princeton gives its collection of 15,000 vertebrate fossils to YPM.

1985 The cover of *Time* magazine depicts *T. rex* with the headline "Did Comets Kill the Dinosaurs?" The accompanying article described the Alvarezes' "bold new theory about mass extinction," which was based on a meteorite more than 10 kilometers in diameter hitting Earth at a speed exceeding 50 kilometers per second.

1985 The Museum of Western Colorado opens the 1.5-mile, 14-stop Rabbit Valley Trail through Time and, the following year, interpretive trails at Riggs Hill and Dinosaur Hill. Four of the stops along the Trail through Time show exposed dinosaur fossils, including *Camarasaurus* and *Iguanodon*.

1985 The *New York Times* reports that only 4 percent of paleontologists attending the annual meeting of SVP accept a meteoric impact as the cause of dinosaurs' extinction at the end of the Cretaceous. Paleontologist Robert Bakker, who helped popularize ideas about dinosaurs being warm-blooded, claimed, "The arrogance of [the advocates of dinosaurs' extinction being triggered by a meteoric impact] is simply unbelievable. They know next to nothing about how real animals evolve, live and become extinct. But despite their ignorance, the geochemists feel that all you have to do is crank up some fancy machine and you've revolutionized science. The real reasons for the dinosaur extinctions have to do with temperature and sea level changes, the spread of diseases by migration and other complex events. But the catastrophe people don't seem to think such things matter. In effect, they're saying this: 'We high-tech people have all the answers, and you paleontologists are just primitive rock hounds.'"

1985 The *New York Times* reports that some paleontologists fear that research not consistent with the Alvarezes' impact hypothesis is being rejected by top-tier journals.

1985 To promote the movie *Godzilla 1985,* Godzilla—a human in a rubber suit—walks along Chicago's Magnificent Mile and gives interviews to reporters and passersby.

1985 While working in northern New Mexico, Utah paleontologist David Gillette excavates *Seismosaurus hallorum* ("Earth-shaker lizard"), a sauropod that was up to 150 feet long. This name was later changed to *Diplodocus hallorum*.

1985 While working near Buffalo, Wyoming, Patrick McSherry finds a new species of stegosaur that in 2001 is named *Hesperosaurus. Hesperosaurus* is among the most primitive stegosaurs known from the Morrison Formation.

1986 "Discovering Dinosaurs" opens at ANS in Philadelphia. The exhibit featured dinosaurs remounted by paleontologist Kenneth Carpenter to reflect modern ideas about their behaviors, as well as Edward Drinker Cope's infamous *Elasmosaurus* (which had originally been mounted with its skull at the wrong end).

1986 After again visiting Glen Rose, Texas, to examine the alleged "man tracks" along the Paluxy River, John Morris of ICR cautions against using the tracks as evidence against evolution. However, Morris later suggested that the tracks could be authentic.

1986 Albuquerque, New Mexico, hosts the First International Symposium on Dinosaur Tracks and Traces. The meeting was supplemented by a week-long trip through Texas, New Mexico, Oklahoma, Arizona, Utah, Colorado, and Arizona to examine tracks. Presentations at the symposium were published three years later in *Dinosaur Tracks and Traces,* the first book devoted entirely to dinosaur ichnology.

1986 Argentinian scientists from the Museum of La Plata discover fragments of an ankylosaur (*Antarctopelta oliveroi*) in Late Cretaceous rocks on James Ross Island off the Antarctic Peninsula. This was the first dinosaur found in Antarctica. Three years later, fossils of a hypsilophodontid were found on nearby Vega Island.

1986 Bob Morales sculpts several miniature *Stegosaurus*, with one defending its baby from *Ceratosaurus*.

1986 Bryan Small unearths *Chindesaurus* ("Chinde Lizard") from Chinde Point in Petrified Forest National Park. The 4-foot-long herrerasaurid lived in the Late Triassic (approximately 216 million years ago) and was formally described in 1995. Few other dinosaurs have been found at Petrified Forest National Park.

1986 Humanlike footprints near dinosaur tracks at Glen Rose, Texas, are shown to be either "inept carvings" from the 1930s or dinosaur tracks that have not weathered like other tracks. These findings prompted ICR to question the validity of the Glen Rose "man tracks," claiming in a letter to its followers that the tracks are "illustrative, not definitive" and do not affect the "over-all case against evolution." Some creationists stopped promoting the film *Footprints in Stone,* but others rejected the evidence refuting their belief that dinosaurs lived with humans.

1986 In *Archaeopteryx, the Primordial Bird: A Case of Fossil Forgery,* astronomer and science-fiction writer Sir Fred Hoyle (1915–2001) and Chandra Wickramasinghe (b. 1939) claim that feathers of the "London specimen" and "Berlin specimen" of *Archaeopteryx* are forgeries. Hoyle and Wickramasinghe likened the fossil to the famous Piltdown Man hoax while claiming that Richard Owen knew of the forgery, probably had authorized the forgery, and may have arranged the death of Andreas Wagner, who had questioned the authenticity of the fossil shortly before his death. According to Hoyle and Wickramasinghe, Owen wanted to use the fraud to discredit Darwinian evolution. Critics quickly pointed out that there was no evidence supporting Hoyle's and Wickramasinghe's allegations; for example, in *Science,* Alan Charig (1927–1997)—curator of BMNH and custodian of the "London specimen" of *Archaeopteryx*—summarized Hoyle and Wickramasinghe's charges as being "based on a plethora of faulty observations, incorrect data, wrong interpretations, untrue statements and misleading arguments; which, in turn, are due to sheer carelessness, lack of knowledge of the relevant subjects, false logic and a fertile imagination." (In 1981, Wickramasinghe had testified in an Arkansas trial on behalf of "creation science.") Although Owen was not well liked, there is no evidence that he knew about, or committed any, forgery of the *Archaeopteryx* specimen.

1986 In *Dinosaurs Past and Present*, artist Mark Hallett (b. 1947) coins the term *paleoartist* to describe his work in a "neat and tidy way." In 2001, SVP defined paleoart as "the scientific or naturalistic rendering of paleontological subject matter pertaining to vertebrate fossils."

1986 In Lawrence Yep's *Monster Makers, Inc.*, a scientist produces bioengineered dinosaurs and a miniature flame-breathing Godzilla as pets used to settle other planets.

1986 In *The Dinosaur Heresies,* Robert Bakker develops earlier claims that dinosaurs were warm-blooded. Bakker's popular book attracted much attention to this idea. Bakker opened his book by recollecting how, late one night in YPM, "the thought struck me! 'There's something very wrong with our dinosaurs.' I was standing in the Great Hall . . . at the foot of the *Brontosaurus* skeleton, It was 3:00 A.M., the hall was dark, no one else was in the building. 'There's something very wrong with our dinosaurs.' The entire Great Hall seemed to say that." Bakker also revived Othniel Marsh's century-old claim that sauropods were characterized by live-birth; the following year (1987), artist Doug Henderson painted *Apatosaurus* giving live-birth to a baby. (This claim was refuted with the discovery of sauropod embryos and egg shells.) Bakker, who noted that *T. rex* is the only dinosaur known to the general public by its full scientific name, added, "*Tyrannosaurus rex* is just irresistible to the tongue." Bakker was one of the advisors for the movie *Jurassic Park* and for the 1992 PBS series, *The Dinosaurs*. In *Jurassic Park*, paleontologist Robert Burke, who was eaten by a *T. rex,* was loosely based on Bakker.

1986 Jack Horner becomes the first paleontologist to receive a MacArthur Foundation Award. Over the next five years, Horner got $204,000.

1986 John Morris of ICR abandons his earlier claims about the Paluxy prints when he writes in *Impact* that "none of the four trails at the Taylor site [in Glen Rose, Texas] can today be regarded as unquestionably of human origin. . . . The various controversial prints labeled as human by Carl Baugh in recent years are of uncertain origin, and at best are not comparable in quality to prints at the sites above, thereby providing no support for the original position" that humans coexisted with dinosaurs. *Films for Christ* responded in *Origins Research* that "there is no question that the 'human-like' tracks excavated by Mr. Taylor are real impressions—not carved hoaxes or erosion marks, as numerous people have long charged without first-hand examination . . . it would be hasty to discredit all earlier conclusions of able researchers and Glen Rose residents regarding the identification of certain tracks as 'human' and 'human-like.' However, the new evidence at hand certainly requires a re-evaluation and possible reinterpretation of all the data gathered. We highly recommend that no one represent any of the Paluxy tracks as proven evidence of human existence during the Cretaceous until final, reliable conclusions can be reached regarding new and old data. . . . In light of these developments, none of the four trails at the Taylor site can today be regarded as unquestionably of human origin." Although Morris urged other creationists to stop using the tracks as evidence against evolution, he emphasized several alleged "mysteries" about the tracks. Many young-Earth creationists and creationist museums continue to use the alleged "man tracks" at Glen Rose to claim that humans lived with dinosaurs.

1986 Louis Marx & Company produces its last sets of dinosaur toys.

1986 Paleontologist Jack Horner, who attended the University of Montana for seven years (and flunked out seven times) but did not graduate, is awarded an honorary doctorate of science from the University of Montana. By 2010, Horner had published more than 100 papers and several popular books. In 2006, Horner received another honorary doctorate, this one from Pennsylvania State University.

1986 Philip Currie, Dale Russell, and other paleontologists from Canada, and Beijing's IVPP collaborate in the Canada–China Dinosaur Project, the first cooperative paleontological project between China and the West since the Central Asiatic Expeditions in the 1920s. The expeditions produced eight new genera and 11 new species of dinosaurs and reptiles. Currie later helped describe some of the first feathered dinosaurs (e.g., *Protarchaeopteryx, Caudipteryx*).

1986 Polish geologist Andrzej Pszczólkowski finds thick, coarse debris in Cuba near the K-T boundary, and suggests it is evidence of an impact. Pszczólkowski's paper, which was written in Spanish and published in a Polish journal, attracted relatively little attention.

1986 Robert Sloan and his colleagues argue in *Science* that "dinosaur extinction in Montana, Alberta, and Wyoming was a gradual process that began 7 million years before the end of the Cretaceous and accelerated rapidly in the final 0.3 million years of the Cretaceous, during the interval of apparent competition from rapidly evolving immigrating ungulates."

1986 Sculptor Stephen Czerkas calls a press conference to announce that the back plates of *Stegosaurus* were in one row, with small plates near the neck tilted at alternating directions. Soon thereafter, Charles Gilmore showed that *Stegosaurus* had two rows of alternating plates.

1986 The Bible-Science Association forms a task force to investigate tracks in the Paluxy Riverbed near Glen Rose, Texas.

1986 The Bureau of Land Management creates the self-guided Mill Canyon Dinosaur Trail about 15 miles north of Moab, Utah, between Arches National Park and Canyonlands National Park. The trail includes 15 interpreted areas where visitors can see the remains of dinosaurs such as *Allosaurus, Stegosaurus, Camptosaurus,* and *Camarasaurus*.

1986 The cover of Kenneth Hsu's (b. 1929) book *The Great Dying: Cosmic Catastrophe, Dinosaurs, and the Theory of Evolution* shows dinosaurs turning away from a fireball.

1986 The first age-estimates (based on potassium-argon methods) of Iowa's Manson Crater yield ages of approximately 70 million years, but the variability of the data makes it impossible to eliminate the crater as a K-T impact site.

1986 The first symposium devoted exclusively to dinosaur systematics occurs at the Tyrrell Museum of Paleontology in Alberta. The results of the symposium were published in 1990 and edited by Kenneth Carpenter and Philip Currie.

1986 The Museum of Western Colorado begins a three-year project to acquire and protect the 32 acres comprising Riggs Hill in Grand Junction, Colorado. In

1900, Elmer Riggs found the remains of the first known *Brachiosaurus* at Riggs Hill. *Brachiosaurus* was once believed to be the biggest dinosaur, but has been superseded by *Supersaurus* and *Argentinosaurus*.

1986 Vertebrate paleontologist Jacques A. Gauthier produces the first cladistic study to include birds and dinosaurs; the results supported John Ostrom's claim that theropods such as *Deinonychus* are the closest extinct relative of birds. After examining more than 100 traits of birds and dinosaurs, Gauthier concluded that birds belong to a clade of coelurosaurian dinosaurs.

1987 "Dinosaurs Past and Present" becomes the first international art show about dinosaurs to tour the world.

1987 A Canadian/Chinese dinosaur expedition is described by *Life* magazine as "the most expensive dinosaur field trip ever." The $7-million expedition discovered new species of the large carnivore *Jiangjunosaurus* (first described in 1983) and the 90-foot-long, *Mamenchisaurus*-like *Titanosaurus*, the largest sauropod ever found in Asia. (*Titanosaurus* described by Marsh in 1877 is now known as *Atlantosaurus*).

1987 A chapter of James A. Michener's (1907–1997) book *Centennial* describes the life and death of *Diplodocus*.

1987 A tooth claimed by Carl Baugh to be from a human ("Glen Rose Man") in Cretaceous limestone from the McFall II site near Glen Rose, Texas, (and therefore suggesting that humans lived with dinosaurs in the Cretaceous) is shown to be from a fish. Several years later, Baugh's museum claimed to have found the fossilized finger of a girl's left hand in Cretaceous sediments nearby. Replicas of the alleged finger were soon for sale in the museum.

1987 Paleontologist Stan Sacrison discovers remains of *T. rex* in the Hell Creek Formation outside Buffalo, South Dakota. This site was excavated in 1992 by Peter Larson of BHI of Hill City, South Dakota; the 30,000 hours of work required for Stan's excavation and preparation produced the 65 percent complete, 40-foot-long, 12-foot-high (at the hips) *T. rex* "Stan" that is now displayed at BHI (BHI 3033); replicas of "Stan" sold by BHI (for $100,000 each) are found in other museums throughout the world. Studies of "Stan" helped document that *T. rex* had keen senses of smell and eyesight, the largest brain, and some of the most powerful jaws known for any dinosaur.

1987 American paleontologists Joel Hammond and Peter Dodson, while noting that dinosaur-related art was increasingly based on science, suggest that artists' views of dinosaurs have illustrated, defined, and answered questions about dinosaurs.

1987 Art Chadwick of Loma Linda University concludes that the Glen Rose "man tracks" were made by dinosaurs, not humans. Answers in Genesis later included the Glen Rose "man tracks" in an article titled "Arguments we think creationists should NOT use." Nevertheless, AiG continues to claim that "there is much evidence that dinosaurs and humans co-existed."

1987 Brigham Young University begins three years of excavations at the Cleveland-Lloyd Quarry.

1987 Carl Baugh's *Dinosaur: Scientific Evidence That Dinosaurs and Men Walked Together* claims that tracks in the Paluxy riverbed show that humans and dinosaurs lived together. Baugh named the makers of the human tracks in the riverbed "Humanus bauanthropus." Baugh, who claims to have discovered more than 90 human footprints in the Paluxy's Cretaceous limestone, also described reports of brontosaurs still being alive in the Congo (called "Mokèlé-mbèmbé" by natives) and claims that "artifacts relating to the presence of man have been discovered through the [geologic column] from top to bottom."

1987 CMNH begins marketing anatomically accurate toys of prehistoric life, including dinosaurs. Germany's Bullyland Company made casts for the original 17 animals, which included *Stegosaurus, Tyrannosaurus, Apatosaurus, Brachiosaurus, Triceratops, Apatosaurus, Allosaurus, Diplodocus,* and *Maiasaura* (on a nest). Several of these models have now been retired or replaced with new figures.

1987 Dallas teacher Sheryl Leach (Figure 42) creates Barney, a friendly *T. rex* who stars in a series of home videos called *Barney and the Backyard Gang*. Barney became famous in 1992 when Leach's shows were revamped for the television series *Barney & Friends* on PBS. The show's theme-song described Barney as a "dinosaur sensation." In each episode, children imagined something and a plush Barney doll transformed in the "real" Barney that children were imagining. Episodes ended with "I Love You" and Barney winking to the audience as he reverted to being a doll. Leach left the show in 2002. Barney, who became the most famous *T. rex*, inspired countless licensed products and starred in several radio shows, movies, and specials (e.g., his television show was broadcast throughout the world). At one point, Barney was receiving more than 10,000 letters per week and was watched by 80 percent of 2- to 5-year-olds who were watching television.

1987 Dhananjay Mohabey discovers a clutch of dinosaur eggs in India. Twenty-three years later, Jeffrey Wilson and his colleagues reported that the discovery also included a predatory Cretaceous snake that apparently ate the sauropod hatchlings. Contrary to *The Land Before Time* and *Baby—Secret of the Lost Legend*, dinosaurs such as *Apatosaurus, Supersaurus,* and *Argentinosaurus* laid clutches of 10–20 eggs.

Figure 42 Barney, a purple, anthropomorphic *T. rex,* was created in 1987 by Sheryl Leach and subsequently starred in 268 episodes of the American children's television series *Barney & Friends*. The optimistic Barney used dances and songs to convey educational messages. Original episodes of *Barney & Friends* aired on Public Broadcasting Service (PBS) stations from 1992 to 2010 but Leach left the show in 2002. Although no new episodes of *Barney & Friends* have been produced since 2009, reruns continue on PBS. (Franziska Krug/Getty Images)

1987 Douglas Henderson illustrates the Deccan Traps volcanic eruptions in Sylvia and Stephen Czerkas's *Dinosaurs: A Global View*. (The word *trap* comes from the Swedish and Danish word *trappa,* meaning *stairs*; lava in the Deccan Traps is often eroded in terraced, step-like formations).

1987 In "A Dinosaur on a Bicycle," a civilized dinosaur returns to the Cretaceous, where it encounters other time-traveling animals.

1987 In *Dinosaurs, Dinosaurs, Dinosaurs,* Gary Owens becomes a dinosaur. He sends a friend to Crystal Palace to get "healing water," after which he gets a job as part of a museum exhibit.

1987 In his article titled "Return of the Dancing Dinosaurs" in *Dinosaurs Past and Present,* Robert Bakker tries to restore dinosaurs to active, energetic animals as originally conceived by early paleontologists such as Edward Drinker Cope and Thomas Huxley. That same year, Bakker said Charles Gilmore's work "had a plodding adequacy devoid of the depth of biological insight by Mantell, von Meyer, Huxley, and Riggs. Gilmore described bones as nearly totally inanimate creations and never displayed any first-hand experience with the muscular anatomy or joint structure of extant species. Gilmore's lead-footed reconstructions became the universal standards for textbooks and museums."

1987 In Joseph Wallace's *The Rise and Fall of the Dinosaur,* artwork by Douglas Henderson shows *Maiasaura* dying after a volcanic eruption. *Maiasaura* disappeared millions of years before the K-T extinction, but the figure's caption describes how volcanic activity condemned "the great dinosaurs to extinction."

1987 John Fischner sculpts a skull for the *Apatosaurus* at DVSP near Glen Rose, Texas. However, in 1995, the original (and inaccurate) *Camarasaurus*-like skull was reattached to the dinosaur because officials believed that (1) the accurate skull looked too small, and (2) *Apatosaurus* and other Sinclair dinosaurs from the 1964–1965 World's Fair were pieces of history in their own right (Figure 37). The tracks displayed near the entrance to DVSP are casts (arranged to create the appearance of a trackway) of those found in the 1930s by Roland Bird.

1987 Joseph Wallace suggests that *Maiasaura* in Montana might have been killed by the meteor that created the 22-mile (diameter) crater near Manson, Iowa.

1987 Mark Hallett, inspired by paintings by Charles Knight, illustrates *Dinosaurs Past and Present*. Hallett's subsequent work included *Dinosaurs: A Global View* (1991), *Puzzle of the Dinosaur Bird: The Story of Archaeopteryx* (1996), *Dinosaurs: The Encyclopedia* (1997), and numerous guides, magazines, and other books. Hallett's work also inspired Steven Spielberg during production of the movie *Jurassic Park*.

1987 Molecular biologist Roy Mackal (1925–2013) claims in *A Living Dinosaur: In Search of the Mokèlé-mbèmbé* that at least four types of dinosaurs are living in the Likouala Swamp of People's Republic of the Congo. However, Mackal provided no physical evidence for his claims. Mackal, whose expeditions were funded by the Society for Cryptozoologists, also claimed to have seen the Loch Ness monster.

1987 Namsarai B finds the turkey-size, flightless theropod *Mononychus* ("one claw") in Late Cretaceous deposits near Bogin Tsav, Mongolia. *Mononychus* was the dinosaurian equivalent of modern anteaters, for it had a long, tubular snout and small, conical teeth. After *Mononychus* was named (and C. F. Payne's restoration appeared on the cover of *Time* magazine) in 1993, it was discovered that the name *Mononychus* had already been assigned to a beetle in the early 1800s. To correct the problem, the dinosaur's name was changed from *Mononychus* to *Mononykus*. *Mononykus*, which had short arms bearing a single large claw, has many primitive traits (e.g., teeth, a long tail) lost later in birds. Soon after its discovery, researchers at AMNH found a skeleton labeled "birdlike dinosaur" collected during the 1923 expedition to Flaming Cliffs in Mongolia that strongly resembled *Mononykus*.

1987 Peter Dodson notes that 40 percent of the 265 genera of currently recognized dinosaurs have been described since 1969.

1987 *Science* reports "the global distribution of shocked quartz at Cretaceous–Tertiary boundary argues for an asteroid or comet impact and against a volcano as a cause of the mass extinction."

1987 ANS in Philadelphia installs two 10-foot-tall bronze *Deinonychus* statues outside its main entrance. The statues were created by Kent Ullberg (b. 1945).

1987 The English edition of Dong Zhiming (b. 1937) and Angela Milner's *Dinosaurs from China* is the first popular book to tell Westerners about Chinese dinosaurs. Zhiming is one of China's most prominent dinosaur paleontologists.

1987 The television series *Dinosaucers* features smart dinosaurs as heroic protagonists from the planet Reptilon that come to Earth to save the planet from other animals.

1987 The Zigong Dinosaur Museum (ZDM) opens outside the city of Zigong in the Sichuan Province of southern China. This museum was the first museum in Asia that specialized in dinosaurs. Like DNM in the United States, ZDM was built atop an ongoing excavation so visitors could view fossils in place. Today, ZDM is one of the most important Middle-Jurassic dinosaur localities.

1987 While at Devil's Coulee in Alberta, Canada, teenager Wendy Sloboda finds pieces of eggshells. These eggshells eventually made their way to RTM. When paleontologists from the museum returned to the site, they found near-intact 75-million-year-old eggs, alone and in clutches, containing embryos of duckbill dinosaurs. Phil Currie and Jack Horner later named the dinosaur *Hypacrosaurus stebingeri*. Eggs of *Hypacrosaurus* had a volume of nearly four liters and a diameter of about 20 centimeters. For comparison, eggs of the hadrosaur *Maiasaura* had a volume of approximately 0.9 liters and a diameter of 12 centimeters. The nests at Devil's Coulee, which are now a protected national resource, were the first dinosaur nesting sites found in Canada, and the second in North America.

1987 While working near Delta, Colorado, geologist Bob Young reports the first fragments of dinosaur eggs in the Morrison Formation.

1987 Young-Earth creationists Carl Baugh and Don Patton (of the Metroplex Institute of Origin Science) claim that the Taylor Site near Glen Rose, Texas, includes dinosaur tracks having human tracks inside them. Baugh and Patton claimed to find more "man tracks" in 1999. These claims, like earlier ones about human tracks alongside dinosaur tracks, were soon refuted by scientists and others who examined the site. David Tyler noted, "We have passed the stage where the activities of Baugh and his friends can be tolerated or ignored. Their activities are a cause of dishonor coming upon the name of Christ."

1988 "The Last Thunder Horse West of the Mississippi" in Isaac Asimov's *Science Fiction Magazine* uses a loose portrayal of the Cope–Marsh feud to show how ruthless competition leads to destruction. To ensure that the dinosaur (the fictional "Johaurus") remained in Marsh's possession, workers strapped dynamite to its body. The dinosaur is later blown up in a gunfight.

1988 A skeleton of *T. rex* confirms that the "tyrant lizard" had only two fingers on its forelimbs. Most dinosaur-themed movies since the 1920s (e.g., *One Million* B.C.*, The Land Unknown, Dinosaurus!*) had depicted *T. rex* with three fingers on its forelimbs.

1988 A study documents the poor scientific literacy in the United States; for example, only 12 percent of the respondents believe that astrology is "not at all scientific." Meanwhile, a survey of 400 biology teachers shows that only 25 percent have biology degrees. Half of the teachers believe (or were not sure) that "some races of people are more intelligent than others," and one-third believe (or were not sure) that humans and dinosaurs lived at the same time.

1988 *Allosaurus* becomes the state fossil of Utah, and *Astrodon* becomes the state dinosaur of Maryland.

1988 An indoor ride called "Kingdom of the Dinosaurs" opens at Knott's Berry Farm in Buena Park, California. Visitors aboard the ride travelled back in time to see dinosaurs and other prehistoric life. A similar ride named "Jurassic Park—The Ride" later opened at Universal Studios Hollywood.

1988 Douglas Dixon's "The New Dinosaurs: An Alternative Evolution" shows dinosaurs in a parallel world as if they had not become extinct.

1988 George Holmes, the CEO of Crown Manufacturing, commissions Chicago sculptor John Kearney (b. 1924) to make five "chromosaurs" from chrome bumpers of junked cars. When Holmes's business failed, the chromosaurs were sold; today, you can see them at *Ripley's Believe It or Not* in Grand Junction, Texas.

1988 Gregory S. Paul, inspired by Charles Knight, publishes the highly regarded *Predatory Dinosaurs of the World*. Paul, who was among the first (and often imitated) artists to portray dinosaurs as active in the 1970s, later was an advisor for the movie *Jurassic Park*. Paul worked with paleontologists, named several dinosaurs (e.g., *Giraffatitan, Dollodon bampingi*), and wrote several more books, including *The Princeton Field Guide to Dinosaurs* (2010).

1988 In *Acts and Facts*, ICR's John Morris claims that he had returned to Paluxy to examine "new evidence" gathered by Carl Baugh and Don Patton for a human footprint inside a print of a dinosaur. After Morris acknowledged that the claim by Baugh and Patton "may sound bizarre" and that ICR did not endorse the claim, Morris endorsed the claim by noting that it was "supported by the existence of somewhat human-like impressions, each rather consistent in length" and "in several cases, toe-like impressions are seen in the proper location. Some are best denoted by accentuated discoloration." Although Morris's new claims were quickly refuted by a variety of studies, some creationists began asking for the re-release of the film *Footprints in Stone*.

1988 In its "Readasaurus" project, the U.S. Postal Service uses an egg of the imaginary dinosaur "Readasaurus" to urge kids to read books.

1988 In the *Dino-Riders* television series, people from another planet fleeing oppressors arrive on Earth during the Mesozoic. They then use dinosaurs to fight their enemies. In 1988, Tyco Industries began producing its popular Dino-Riders action figures, the first set of which included *Diplodocus, Deinonychus, Styracosaurus,* and a pterodactyl. Paleontologist Robert Bakker, who advised Tyco about the figures, quit the project when Tyco refused to make the figures scientifically accurate.

1988 In the family cartoon-movie *The Land Before Time,* a baby orphaned *Apatosaurus* named "Littlefoot" searches for the legendary "Great Valley," where it will be safe from carnivores. Along the way, Littlefoot meets other orphaned dinosaurs (Ducky the Duckbill, Cera the Triceratops). Carnivorous dinosaurs such as "Sharp Tooth" the *Tyrannosaurus* are portrayed as violent and evil, and herbivorous dinosaurs such as "Spike" the *Stegosaurus* are portrayed as amicable. Littlefoot also endures earthquakes and volcanic eruptions. By 2008, there were 12 remakes of *The Land before Time*. Evangelist Phil Phillips later denounced *The Land before Time* for its "decidedly ungodly message."

1988 In the popular *What Really Happened to the Dinosaurs,* young-Earth creationist John Morris (of ICR) and Ken Ham (now of AiG) report "some of the many things that ICR scientists have learned about one of God's mightiest creation—the dinosaurs." These discoveries included that Adam and Eve—the first two humans—"must have lived at the same time" as dinosaurs, that all dinosaurs (as well as all other animals) were originally herbivorous, that Adam and Eve's sin caused dinosaurs to start eating meat, that dinosaurs were aboard Noah's Ark, that fossils of dinosaurs came from Noah's Flood, and that many dinosaurs were killed by humans. By 1997, *What Really Happened to the Dinosaurs* was in its seventh printing.

1988 Jim Jensen names the nearly complete *Cathetosaurus lewisi* that he collected from the Dominguez-Jones Quarry south of Grand Junction, Colorado. This specimen, which was later shown to be *Camarasaurus,* is displayed at the Earth Sciences Museum of Brigham Young University.

1988 Jody Bourgeois, Thor Hansen, and Patricia Wibers announce in *Science* that beds of Texas's Brazos River—which expose rocks from the Upper Cretaceous and lower Tertiary—were made by a tsunami initially at least 1,000 feet high. This meant that the K-T impact site was probably less than 3,000 miles away in the Caribbean, western Atlantic, or Gulf of Mexico.

1988 Joyce Berry, a fourth-grade teacher in Haddon Township, New Jersey, launches a campaign to make *Hadrosaurus foulkii* the state dinosaur of New Jersey. Although *H. foulkii* was the first dinosaur to be exhibited anywhere in the world, nine other states had already chosen their official dinosaurs and fossils. From 1988 to 1991, 97 students from Berry's school lobbied legislators, and in 1991, Governor Jim Florio signed "Chapter 161 of the Laws of New Jersey of 1991: An Act Designating the *Hadrosaurus foulkii* as the New Jersey State Dinosaur."

1988 New argon-argon studies of shocked feldspar samples from Iowa's Manson Crater make the crater a primary suspect as the K-T impact site. The discovery the next year of Russia's Kara Crater—initially described as 66 million years old—provided another candidate.

1988 Nicolas Steno, who established the foundation of modern geology, is beatified by Pope John Paul II on October 23. Steno is the only geologist to ever be beatified.

1988 Officials discover that the dinosaur tracks excavated by Roland Bird that are on display at the Texas Memorial Museum in Austin, Texas, are deteriorating because the building housing them is poorly ventilated. As dinosaur-track expert Mike O'Brien noted, "If there is a lesson in this, it is that dinosaur tracks should be preserved in the ground where they are found. If they are in pristine condition like the Glen Rose tracks, they should be molded and displayed as casts." In 2013, the Texas Memorial Museum was in the process of restoring and moving the tracks into the climate-controlled museum.

1988 One year after receiving his doctorate degree in geology from Columbia University, paleontologist Paul Sereno (b. 1957) begins six weeks of fieldwork in Argentina's Andes, where he and Alfredo Monetta later excavated the first complete skeleton of the 228-million-year-old *Herrerasaurus,* a primitive theropod. Sereno described the discovery as "the most complete early dinosaur skeleton, and the first herrerasaur skull ever discovered." When Sereno found the fossil in the Valley of the Moon National Park, he started crying, later noting, "It was one of the greatest moments of my life." A model of *Herrerasaurus* is now displayed at Chicago's Field Museum, as well as at the entrance to the dinosaur exhibits at CMNH. Other dinosaurs described by Sereno included *Afrovenator, Deltadromeus, Eoraptor* (a relative of *Herrerasaurus*), *Gracilceratops, Rajasaurus,* and *Raptorex*.

1988 Tom DeRosa establishes the Creation Studies Institute, which argues that humans lived with dinosaurs. The institute, which acknowledges that the "man tracks" near Glen Rose, Texas, are "disputed," also sponsors the Creation Discovery Museum, which also claims that humans lived with dinosaurs.

1988 Paleontologist Harley Armstrong of the Museum of Western Colorado identifies the site of Quarry 12 between Fruita and Grand Junction, Colorado, where Elmer Riggs found *Camarasaurus* in 1900–1901. Today, downtown Fruita features Dinosaur Journey Museum and a 20-foot-tall dinosaur that was named "Grrrreta" by local school children in 2000. Several dinosaurs at Dinosaur Journey Museum came from the Dinosaur Valley Museum (Dinosaur Discovery), which closed in 2000.

1988 Paleontologist Walter P. Coombs Jr. suggests that *Trachodon mirabilis* is a *nomen dubium* (questionable name).

1988 Paleontologists find more than 200 dinosaur tracks along a stream in Africa's Koum Basin. Soon thereafter, some local residents claimed that the tracks were interspersed with impressions of the knees, elbows, and foreheads of Muslim worshippers. Follow-up investigations found no such impressions.

1988 Peter Wellnhofer (b. 1936) of Munich's BSP describes the "Solnhofen specimen" of *Archaeopteryx* (BSP 1999), which was found the previous year by Jura Museum curator Gunter Viohl in the private collection of Friedrich Müller (a former mayor of Solnhofen). This specimen was originally classified as *Compsognathus*, and is now housed at the Bürgermeister-Müller-Museum in Solnhofen, Germany. The "Solnhofen specimen" is the largest known specimen of *Archaeopteryx*.

1988 Playskool begins producing Definitely Dinosaurs toys. These popular toys, each branded with a stamp of authenticity, were often included with meals at restaurants such as Wendy's.

1988 *Science* notes, "A large impact rather than a volcano is widely taken to be the primary agent of destruction at the end of the dinosaur age."

1988 The Safari Ltd. Company begins working with paleontologists such as Lanzendorf Award winner (2008) Luis V. Rey (b. 1955) to produce "the most accurate scale-model toy dinosaurs in the world." The following year, Safari released the first 17 dinosaurs of the Carnegie Collection. These hand-painted dinosaurs, most of which were sculpted by Forest Rogers, were based on fossils at CMNH and remain popular today. Since 1997, the collection has grown 1–3 creatures per year. Most of the models are 1:40 scale; the largest is *Diplodocus* (24 inches long) and the shortest is *Dimetrodon* (3 inches long).

1988 Thomas M. Lehman describes "a gigantic skull" of the horned dinosaur *Pentaceratops sternbergii* from New Mexico that is 3 meters long. This is the largest known dinosaur skull. The 6.8-meter-long body of *P. sternbergii* was not exceptionally large.

1988 TOPPS releases *Dinosaurs Attack!,* a card game involving dinosaurs moving to the present day. Images on many of the 55 cards, 11 stickers, and poster were reminiscent of the *Mars Attack!* cards released in 1962. Also in 1988, TOPPS produced 3-inch-tall, soft plastic dinosaur toys that were included in bags of candy. TOPPS released similar toys in 1993 to coincide with the release of the movie *Jurassic Park.*

1988 *Triceratops* becomes the state fossil of South Dakota.

1988 Wendy Wolbach and her colleagues note that most of the soot at the K-T boundary is restricted to the lowermost 3 millimeters of the K-T boundary. They argued that most of Earth's vegetation burned in the days and months after the K-T impact event.

1988 While fishing with her family in Montana's Fort Peck Reservoir, Kathy Wankel finds the first shoulder and arm of *T. rex* ever discovered. The next May, the site was visited by MOR's Pat Leiggi, who began excavating the bones. Thanks to help from the Army Corps of Engineers, the 80 tons of bone-laden sandstone were excavated for less than $5,000. The so-called Wankel *T. rex* (Figure 43), which was 85 percent complete and 11.5 meters long, is set to be the centerpiece of Smithsonian's new dinosaur-hall, scheduled to open in 2019.

1989 A bone of a 10-foot-long tyrannosaur is found in Dinosaur Cove in Victoria, Australia.

1989 A mold of *Diplodocus carnegii* made of fiberglass and polyester is erected inside UFHNH in Vernal, Utah, only about 300 miles from where *Diplodocus carnegii* was discovered in 1899. The economy of Vernal depends heavily on dinosaur-based tourism (Figure 44).

1989 Australian paleontologists Patricia Vickers-Rich and her husband Tom describe *Leaellynasaura,* a small, ornithischian dinosaur from the Early Cretaceous.

Figure 43 Famed paleontologist John R. "Jack" Horner (b. 1946) standing beside "Wankel" *T. rex* near Fort Peck Reservoir in McCone County, Montana, in 1990. At the time, this specimen was the largest and most complete *T. rex*, and the first to have been found with an arm. After being displayed in the Museum of the Rockies for many years, this specimen was loaned in 2014 for 75 years to the Smithsonian Institution for display in its new dinosaur hall. In 2013, Horner received the Romer-Simpson Prize, the highest honor a paleontologist can receive from the Society of Vertebrate Paleontology. (AP Photo)

Leaellynasaura was named for Leaellyn Rich, the couple's 10-year-old daughter who discovered the fossils. Patricia (b. 1944) was the inspiration for "Paleontologist Barbie," a doll released in 1997 by toy manufacturer Mattel. In 2011, the Smithsonian Institution began marketing the doll.

1989 Canada's first postage stamp (38¢) to feature a prehistoric animal showcases *Albertosaurus. Albertosaurus* was again featured on Canadian stamps in 1993 (43¢) and 1998 (45¢).

1989 Dale Russell's *An Odyssey in Time: The Dinosaurs of North America* includes Ely Kish's painting (from 1985) of the K-T impact. Russell commented that the meteor "seemed to hang in the sky like a second moon, or the eye of God, but no dinosaur looked at it with understanding. It suddenly swelled in the sky, and then a dark mantle spread across the firmament. Why did it have to be this way?"

1989 Dinosaur tracks are found in Colorado sediments only 0.37 meters below the K-T boundary. Later, tracks were found even closer to the boundary in the Pyrenees of northeast Spain.

Figure 44 The economies of "dinosaur towns" such as Vernal, Utah (near Dinosaur National Monument; population 9,089) and Glen Rose, Texas (near Dinosaur Valley State Park; population 2,444) rely on dinosaur-based tourism. Shown here is Vernal's "Dinah," a pink, 30-foot-tall sauropod that welcomes motorists to town. (Ron Chapple/ Dreamstime.com)

1989 *Discover* magazine dedicates its October issue to extinction. The issue included an article by American paleontologist Stephen Jay Gould titled "An Asteroid to Die For."

1989 During its fourth year of operation, RTM attracts its two-millionth visitor.

1989 In Barry Longyear's *The Homecoming*, a rocket filled with five-fingered dinosaurs returns to Earth after being away for 79 million years.

1989 Creation Truth Foundation is established in Alabama to counteract "the godless philosophy of evolution." The foundation, which moved to Oklahoma in 1994, promotes young-earth creationism and its claim that humans lived with dinosaurs.

1989 In comic-strip artist Gary Larson's (b. 1950) *Wildlife Preserves,* a scientist (with a time machine in the background) approaches the rear end of a *Diplodocus* with a giant thermometer. The caption reads, "An instant later, both Professor Waxman and his time machine are obliterated, leaving the cold-blooded/warm-blooded dinosaur debate still unresolved."

1989 The U.S. Postal Service issues a set of 25¢ stamps featuring winged reptile *Pteranodon*[40] and the dinosaurs *Tyrannosaurus*, *Stegosaurus*, and *Brontosaurus*. In the margin of the sheets of stamps was "Stamposaurus." The stamps, which were launched at Orlando, Florida's Disney World, were accompanied by promotions including T-shirts, videocassettes of *The Land before Time*, and a dinosaur "discovery kit" that could be purchased for $3.95. However, there was soon a controversy about using the name *Brontosaurus* instead of the more-accurate (but less popular) *Apatosaurus*. As noted on the editorial page of the *New York Times*, "The Postal Service has taken heavy flak for mislabeling its new 25-cent dinosaur stamp, a drawing of a pair of dinosaurs captioned '*Brontosaurus*.' Furious purists pointed out that the "brontosaurus" is now properly called *Apatosaurus*. They accused the stamp's authors of fostering scientific illiteracy, and wanted the stamps recalled." The U.S. Postal Office, in its Postal Bulletin Number 21744, responded by noting that "although now recognized by the scientific community as *Apatosaurus*, the name *Brontosaurus* was used for the stamp because it is more familiar to the general population. Similarly, the term 'dinosaur' has been used generically to describe all the animals, even though the Pteranodon was a flying reptile." The Post Office's decision was defended by Stephen Jay Gould, who noted the popularity of *Brontosaurus* (e.g., in Sinclair advertisements, in the movie *Fantasia*). Restorations by self-taught illustrator John Gurche (b. 1951)[41] used to create these stamps were only 1.5-times larger than the stamps. The Postal Service produced more than 400 million of the stamps. Gurche, whose works are displayed at AMNH and NMNH, won a 2000 Lanzendorf Paleoart Prize from SVP for his mural of "Sue" at the Field Museum. Gurche's *Deinonychus* attacking *Tenontosaurus* was a popular poster sold by NMNH.

1989 Students at Minnesota's Zion Lutheran School claim that because dinosaurs on postage stamps promote evolution, the U.S. Postal Service should give equal time to the belief that God created the world in six days. As the students' leader noted, "It's fine if they want to do dinosaur stamps, but Jesus was more important. He died for us. Dinosaurs didn't."

1989 James Farlow names the dinosaur that created the sauropod-tracks at Glen Rose, Texas, *Brontopodus birdi*. Most tracks at Glen Rose are 55–70 centimeters long and 25–40 centimeters wide.

1989 Paul Taylor's *The Great Dinosaur Mystery and the Bible* argues that dinosaurs were aboard Noah's Ark and that dinosaurs lived for "a few centuries" after the Ark landed.

1989 Phil Currie describes the first dinosaur-print found in DPP. The print (TMP 81.34.1), which had been found eight years earlier, is exhibited at the DPP Field Station.

1989 SVP announces, "The barter, sale, or purchase of scientifically significant vertebrate fossils is not condoned, unless it brings them into or keeps them within a public trust."

1989 The Associated Press designates the understandings of mass extinctions as one of the "10 Top Scientific Advances of the Past Decade."

1989 The television broadcast of *A Claymation Christmas Celebration* is hosted by a smart *Tyrannosaurus* and a dimwitted *Styracosaurus*.

1989 The Valley of Prehistory, one of the largest (32 hectares) theme parks in the world, opens in the Santiago Province of Cuba. The park includes more than 200 life-size prehistoric animals (including an 89-foot-long *Diplodocus*), many of which were based on artwork by Zdenêk Burian. In 1987, the park was featured on Cuban stamps.

1990 A Yale paleontologist notes that the term *Dinosauria* "was abandoned as an official grouping many years ago, but the general public never gave up on it."

1990 After an absence of 60 years, AMNH resumes fieldwork in Mongolia. This work, led by Mark Norell, Malcolm McKenna, and Michael Novacek, uncovered numerous mammals and dinosaurs, as well as an embryonic *Oviraptor* (which showed that some of the eggs found earlier in Mongolia's Bayn Dzak [i.e., Flaming Cliffs] and attributed to *Protoceratops* had been laid by *Oviraptor*). Unlike earlier AMNH expeditions to China, all fossils of the 1991 expedition remained the property of Mongolia.

1990 Alan Hildebrand and others find beds of glass spheres several feet thick (instead of the usual few inches) at the K-T boundary along the southern coast of Haiti. Unlike at other K-T sites, where the spheres were about the size of sand grains, spheres in Haiti were as big as pea gravel. Hildebrand and his coworkers concluded that a large crater on the northern Yucatán peninsula is the K-T impact site.

1990 American actor Doug McClure (1935–1995) hosts *The Hollywood Dinosaur Chronicles,* a documentary about dinosaur-themed movies. McClure had starred in several dinosaur-themed movies, including *The Land That Time Forgot* (1974), *At the Earth's Core* (1976), and *The People That Time Forgot* (1977). A similar documentary—*Dinosaur Movies*—was produced in 1993.

1990 Brian Franczak (b. 1955) illustrates *Encyclopedia of Dinosaurs*. Franczak later became one of the premier illustrators of prehistoric life.

1990 Carl Baugh announces that he has found Noah's Ark in Turkey, but the ark has been broken into four pieces.

1990 David Elliott of The Ohio State University finds the first dinosaur bones on the Antarctic mainland. The fossils were found at Mt. Kilpatrick, about 400 miles from the South Pole.

1990 In "What Caused the Mass Extinction?" published in *Scientific American,* Walter Alvarez and Frank Asaro (b. 1927) argue that an impact was responsible for the K-T extinction, while Vincent Courtillot (b. 1948) argues that the extinction was due to volcanism.

1990 In a chapter titled "*Triceratops*: An Example of Flawed Systematics" in *Dinosaur Systematics: Perspectives and Approaches* (edited by Kenneth Carpenter and Philip Currie), paleontologists John Ostrom and Peter Wellnhofer examine type-specimens of the 16 species of *Triceratops* and conclude "that the known sample of *Triceratops* specimens cited here represents a single species—*Triceratops horridus*." Ostrom and Wellnhofer also claimed that *Torosaurus* "might also be the male form of *Triceratops*," and "on stratigraphic and geographic distribution grounds, *and* current understanding of anatomic variation in living populations, is that only a single species of *Triceratops* can be substantiated. . . . The only valid species is *Triceratops horridus*."

1990 In *A Nymphoid Barbarian in Dinosaur Hell* ("The Pre-Historic and the Pre-Pubescent, Together at Last!"), a sexy girl (played by Linda Corwin) is besieged by dinosaurs (and other threats) created by mutations of friendly mammals. Advertisements for the movie featured a dinosaur not in the movie.

1990 In the decade following the pioneering paper by Luis Alvarez and his colleagues claiming that the K-T mass extinction was triggered by a meteoric impact, more than 2,000 articles in scientific journals have been published about the K-T extinction.

1990 In the movie *Edward Scissorhands,* Edward creates a topiary of *T. rex*.

1990 James Kelly's "Mr. Big" describes a human society in which "twanking" can transform humans into any morphological shape. The boy "Stennie" resembles a raptor-like dinosaur.

1990 James Farlow responds to Robert Bakker's attacks on (and disdain for) many other paleontologists and their ideas: "Unfortunately, the strongest impression gained from reading the literature of the dinosaur physiology controversy is that some of the participants have behaved more like politicians or attorneys than scientists, passionately coming to dogmatic conclusions via arguments based on questionable assumptions and/or data subject to other interpretations. Many of the arguments have been published only in popular or at best semi-technical works, accompanied by rather disdainful comments about the stodgy 'orthodoxy' of those holding contrary views; what began as a fresh way of considering paleontological problems has degenerated into an exercise of name-calling." Other paleontologists criticized Bakker's claims as unoriginal popularizations of other people's ideas.

1990 *Jurassic Park,* a science fiction novel by Michael Crichton, describes a "biological preserve" off the coast of Costa Rica featuring 292 genetically engineered dinosaurs. The dinosaurs, created from DNA recovered from mosquitoes entombed in fossilized tree resin, are exceptionally active. (*Velociraptor* was the first baby dinosaur seen by Alan Grant in John Hammond's lab.) Crichton's novel, which was more accurate than the movie (e.g., the novel acknowledges that there is no evidence that *Dilophosaurus* could spit), included 15 kinds of dinosaurs, but the movie included only seven. In 1993, Steven Spielberg adapted Crichton's book into the film *Jurassic Park,* which won numerous awards (including three Oscars).

1990 Marilyn Houck of the University of Arizona and her coworkers analyze all of the known specimens of *Archaeopteryx* and conclude that the specimens represent one species.

1990 MOR excavates from Fort Peck Reservoir (Montana) one of the most complete skeletons of *T. rex,* and the first such specimen having preserved arms with two fingers. These short limbs were not vestigial; they had large areas for the attachment of muscles. Another skeleton of *T. rex* discovered the same year, "Sue" from South Dakota, overshadowed this discovery. Elsewhere in Montana in the early 1990s, Jack Horner and his colleagues discovered mass graveyards of juvenile *Diplodocus* in Jurassic rocks of the Morrison Formation. The Morrison Formation has yielded more than 90 types of vertebrates, barely one-third of which are dinosaurs. The most abundant dinosaurs in the Morrison Formation are *Apatosaurus, Camarasaurus, Diplodocus, Stegosaurus,* and *Allosaurus.*

1990 NMNH exhibits dinosaurs sculpted by Jim Gary, a paleoartist whose work has been featured in encyclopedias, textbooks, videos, and magazines for decades. Gray's exhibit attracted more visitors than any of the museum's previous exhibits. Four years earlier, in the movie *Howard the Duck,* Howard walked under a *Stegosaurus* sculpted by Gray.

1990 Peter Dodson notes that 540 genera (285 of which are valid[42]) and 800 species (336 of which are valid) of dinosaurs have been named since 1825. Nearly half of the genera were based on one specimen, and only 20 percent were based on skulls and complete skeletons. Nearly half of the genera were from the Late Cretaceous, and genera had an average lifespan of 7.7 million years. An average of six species had been reported per year since 1970, and each genus included an average of 1.2 species. The most abundant fossils in museums included *Maiasaura* (more than 200 species), *Psittacosaurus* (120 specimens), and *Coelophysis* (100 specimens). Dodson estimated that 900–1200 genera of dinosaurs existed, and that the fossil record was 25 percent complete.

1990 Robert Bakker and his colleagues name the small, 145-million-year-old biped *Drinker nisti* in honor of famed paleontologist Edward Drinker Cope. The species name refers to the **N**ational **I**nstitute of **S**tandards and **T**echnology. A closely related animal is the small ornithopod *Othnielia* (now *Othnielosaurus*), which was named after Othniel Marsh, Cope's rival.

1990 Russia's Kara Crater is shown to be 73–76 million years old, thus eliminating it as the K-T impact site. Three years later, Iowa's 22-mile-wide Manson Crater was also eliminated when it was shown to be 74 $\pm$ 0.5 million years old (i.e., too old to have been responsible for the K-T extinction).

1990 Scientists report that the 120-mile-wide Chicxulub crater off the Yucatán Peninsula was formed precisely at the K-T extinction. Geologists hypothesized that the original impact triggered an earthquake one-million-times greater than any recorded earthquake in history.

1990 The new College of Eastern Utah Prehistoric Museum greets more than 34,000 visitors. Today, attendance exceeds 40,000 people per year, many of whom come to see bones of *Utahraptor*. Other dinosaurs, most from the nearby Cleveland-Lloyd Quarry, include *Allosaurus*, *Camarasaurus*, *Camptosaurus,* and *Stegosaurus*.

1990 While prospecting near Faith, South Dakota, and in her fourth season with BHI, self-taught paleontologist Sue Hendrickson (b. 1949) discovers the largest and most (more than 85%) complete *Tyrannosaurus rex* fossil protruding from the Hell Creek Formation along a hillside (Figure 45). The tyrannosaur's femur was 54 inches long, suggesting that she was 13 feet tall at the hips and 41 feet long. Peter Larson's BHI (a commercial fossil-hunting company) in nearby Hill City excavated the bones over 17 days, for which he paid $5,000 to Maurice Williams, the Sioux rancher who owned the land on which Sue was found. (At the time, land there was selling for approximately $60 per acre). The check's memo-line read, "For theropod skeleton Sue/8–14–90-MW," and Williams cashed the check the following day. Soon thereafter, Williams wrote to Larson, "I didn't sell the fossil to you. I only allowed you to remove it and clean it and prepare it for sale." Two months after the discovery, the fossil was claimed by the Cheyenne River Sioux, a tribe to which Williams belonged. More than 2,000 people came to BHI to see Sue being prepared for display. However, federal agents later seized Sue and

Figure 45 Sue Hendrickson (b. 1949) stands beside "Sue," a 67-million-year-old *Tyrannosaurus rex* that she discovered outside Faith, South Dakota, in 1990. It took six workers from the Black Hills Institute 17 days to excavate the 42-foot-long, 13-foot-high (at the hips) fossil, and 10 museum preparators more than two years (30,000 hours of work) to prepare "Sue's" more than 250 teeth and bones for display. "Sue" (FMNH PR2081)—whose unveiling in 2000 generated publicity worldwide—was bought by the Field Museum for $8.4 million, the largest amount ever paid for a fossil. "Sue" is the most famous real dinosaur in the world. (Black Hills Institute of Geological Research, Photographer: Peter L. Larson)

removed the fossils from the site. After a protracted legal battle, the fossil was purchased in 1997 at a Sotheby's auction. Bidding began at $500,000, but less than 10 minutes later it sold for $7.6 million ($8,362,500 including the commission) to the Field Museum (Chicago, Illinois), where it was unveiled in 2000. McDonald's Corporation, which has a restaurant in the museum, helped fund Sue's purchase in exchange for the rights to use Sue's image to market its products. The discovery of "Sue" and her auction set off another wave of dinosaur mania that included movies (e.g., *Jurassic Park*) and the popular children's show *Barney & Friends.* Today, "Sue" remains a featured attraction at the Field Museum, and in 2013, the museum's Field Bistro began selling "Tooth and Claw," a lager inspired by "Sue."

1990 Young-Earth creationist and future dinosaur-promoter Kent Hovind offers $10,000 "to anyone who can give any empirical evidence (scientific proof) for evolution." Hovind, who later increased the "standing offer" to $250,000, declared bankruptcy in 1996 and has never given away the money (an example of the "scientific proof" for evolution would be to reproduce the Big Bang and create a new universe). Hovind's offer was denounced by Answers in Genesis (another young-Earth organization) as a "gimmick." Hovind also promoted the alleged "man tracks" at Glen Rose, Texas, as evidence against evolution and for a young Earth, noting that "This fossil is one of many that makes the evolution theory of an old earth look silly . . . Dinosaurs lived with man!"

1991 After featuring Charles Darwin on its 29-pence stamps in 1982, Great Britain issues its first stamps featuring dinosaurs. These dinosaurs were *Iguanodon* (22 pence), *Stegosaurus* (26 pence), *Tyrannosaurus* (31 pence), *Triceratops* (37 pence), and *Protoceratops* (33 pence). The stamp-sets also included a plesiosaur depicting the Loch Ness monster.

1991 Alan Hildebrand and his coworkers report in *Geology* that one-third of the quartz and feldspar grains from the Chicxulub crater are shocked. Three independent analyses of cores from the crater dated the crater at 64.94 ± 0.11, 64.97 ± 0.07, and 65.00 ± 0.08 million years old, providing an average age of 64.98 ± 0.05 million years. This was the first scientific paper that described a crater formed at the K-T boundary. This crater, named for the nearby fishing village of Puerto Chicxulub, is 120 miles in diameter. In 1980, Luis Alvarez and his colleagues had predicted that a crater from the meteor associated with the K-T mass extinction would be 90–125 miles in diameter.

1991 AMNH begins a four-year, $48 million renovation of its two dinosaur halls. That same year, AMNH unveiled *Barosaurus* in its 8,000-square-foot Theodore Roosevelt Rotunda, which is the main entrance to the museum. Before this time, the rotunda had been kept empty (except for temporary exhibits) to honor the importance of Roosevelt to the museum (a statue of Roosevelt greets visitors just outside the museum's entrance). The 100-foot-high *Barosaurus* (one of only two such specimens displayed anywhere in the world) is the world's tallest free-standing dinosaur.

1991 Disney Studios' comedy *Dinosaurs* describes a civilization of smart dinosaurs that includes a family called "The Sinclairs" and B. D. Richfield, Earl Sinclair's boss. The shows used dinosaurs to address social issues such as workplace problems, women's rights, teenagers' angst, and environmental problems. After *Dinosaurs* was canceled, it became popular in reruns.

1991 Footprints of *Eubrontes giganteus* become the state fossil of Connecticut.

1991 Fossil Cabin (Figure 26) on U.S. Highway 30 near Como Bluff (and 50 miles northwest of Laramie) is sold to the Fultz family. In 2011, the cabin was closed to the public and put up for sale.

1991 Glen Penfield writes in *Natural History* that he had identified the Yucatán crater at the K-T impact site more than a decade earlier.

1991 *Godzilla v. King Ghidorah* describes how a dinosaur in 1944 saves the Japanese army. Nuclear tests in the Bikini Atoll later converted this dinosaur into Godzilla.

1991 *Hadrosaurus foulkii* becomes the state dinosaur of New Jersey.

1991 In *Dinosaurs of Eden,* evangelist Ken Ham tells readers that "Adam and Eve, and the dinosaurs, were also created about 6,000 years ago," that Adam and Eve and all animals were originally vegetarians (but that God allowed humans to eat meat about 4,300 years ago), that "there was no death, disease, suffering, or bloodshed in the world" until Adam and Eve sinned, that Adam died 930 years after he sinned, that "sin affected dinosaurs" (as well as "plants, dirt, the stars, [and] each human being"), and that scientists "are sinners." Ham concluded, "We can say, 100 percent, absolutely for sure, that people lived with dinosaurs! The reason we can be so sure is that God, who doesn't tell a lie, told us in His Word that land Animals and Adam and Eve were made on the sixth day of Creation!"

1991 In the $2.95 comic book *Death Dreams of Dracula*, "Dinosaurs Seek Blood for Dracula."

1991 In the four-part miniseries *Dinosaur!* produced by the Arts & Entertainment Network, host Walter Cronkite notes that dinosaurs "were intelligent enough to survive for 150 million years. You can't be too dumb and make it that long."

1991 Mongolia issues postage stamps honoring *The Flintstones*.

1991 Paleobotanist Jack Wolfe suggests that an "impact winter" occurred at the K-T boundary, but subsequent research questioned this conclusion.

1991 Paul Sereno and his colleagues, working in Argentina's Valley of the Moon, discover the 3-foot-long, bipedal *Eoraptor* ("dawn stealer"), a slender, primitive carnivorous dinosaur and cousin of *Herrerasaurus. Eoraptor* was discovered in upper Triassic sediments corresponding to the dawn of dinosaurs, approximately 228 million years old. *Eoraptor*, like other earlier dinosaurs such as *Herrerasaurus* and *Staurikosaurus,* was slender and upright, suggesting that dinosaurs' upright posture evolved before dinosaurs' giant sizes. Sereno, Catherine Forster, and their colleagues named *Eoraptor* in 1993. *Eoraptor,* which was found in the same valley as *Herrerasaurus,* was the earliest and most archaic dinosaur ever found.

1991 Peter Sheehan and his colleagues study the ecological diversity of Hell Creek, Montana and North Dakota, and find "no evidence . . . of a gradual decline of dinosaurs at the end of the Cretaceous."

1991 Phil Tippett (b. 1951) is hired by Steven Spielberg to use his "Go Motion" techniques to coordinate the special effects for the movie *Jurassic Park*. When

Spielberg subsequently decided to instead use computer-generated images for the movie, Tippett exclaimed, "I've just become extinct" (a line that Spielberg included in the movie), but remained a consultant for dinosaur-related scenes in *Jurassic Park*. The declining costs of computer-generated images increased their usage in subsequent productions such as *Walking with Dinosaurs* (1999), *When Dinosaurs Roamed America* (2001), *Animal Armageddon* (2009), and *Planet Dinosaur* (2011).

1991 Researchers find a giant blade-like claw in Early Cretaceous sediments of Utah's Gaston Quarry. Two years later, the 20-foot-long, one-ton dinosaur was named *Utahraptor,* and in 1994 it appeared in the movie *Jurassic Park. Utahraptor*, which had a 1.5-foot-long skull and 15-inch claws, is the oldest (125 million years old) and largest of all known raptor dinosaurs. *Utahraptor* was similar to *Deinonychus*, but was twice its size.

1991 Sankar Chatterjee describes fossils (found in 1984) of *Protoavis,* which he claims is a bird 210–225 million years old (i.e., far older than *Archaeopteryx*). Because the fossils were in poor condition, not especially bird-like, and collected from different locations, many paleontologists continued to view *Archaeopteryx* as the earliest known bird, and many considered *Protoavis* to be an unusual small theropod. Subsequent studies of birds such as *Sinornis* (from the Lower Cretaceous of China) and *Iberomesornis* (from the Lower Cretaceous of Spain) have been consistent with *Archaeopteryx* being close to the primary line of bird evolution. A study in 2013 claimed that *Archaeopteryx* was in the midst of losing its ability to fly (i.e., becoming "secondarily flightless").

1991 Skulls and other bones of the Early Jurassic theropod *Cryolophosaurus* ("cold crested lizard") are found by paleontologists William Hammer and William Hickerson on James Ross Island adjacent to the Antarctic Peninsula. The 25-foot-long *Cryolophosaurus,* the first carnivorous dinosaur found in Antarctica and the first dinosaur of any kind to be named from that continent, had an unusual comb-like crest on its head; this was why it was informally referred to as "Elvisaurus."

1991 Spherules from the K-T boundary in Haiti are dated to 65.01 ± 0.08 million years, an age statistically indistinguishable from those taken from the center of the crater.

1991 The 95 percent–complete *Allosaurus* "Big Al" (MOR 693) is uncovered near Wyoming's Howe Quarry. The eight-meter-long specimen, which had been found by Kirby Siber and crews from the Sauriermuseum in Aathal, Switzerland, was on land managed by the Bureau of Land Management, and was excavated by MOR. Siber's crew later found several other specimens at nearby Howe Quarry, including *Allosaurus, Stegosaurus, Camarasaurus, Dryosaurus,* and *Othnielosaurus*. In 1996, Siber and his colleagues found "Big Al Two," the best preserved skeleton of its kind.

1991 The Denver Museum of Natural History starts its "parapaleontologist" program to train amateurs to find and collect fossils.

1991 Two years after founding Creation Science Evangelism in Pensacola, Florida, evangelist Kent Hovind opens Dinosaur Adventure Land, a "theme park and science museum that gives God the glory for His creation." Hovind, who was sentenced to 10 years in prison in 2006, offered a "$250,000 reward for proof of evolution!" (As of 2013, Hovind had not paid the reward.) At Dinosaur Adventure Park, Hovind told visitors that (1) no dinosaurs died before sin, (2) God commanded the first animals to be vegetarians, (3) people started to die after Adam's sin, (4) dinosaurs lived with humans, and (5) the Bible told Job about dinosaurs (i.e., "behemoth" described in Job 40:15–24).

1991 While working in Montana's Hell Creek Formation, amateur fossil-hunter Kenneth Olson discovers a *Triceratops* bone having bite marks inflicted by *T. rex*. These specimens were later used to estimate that the bite of *T. rex* exerted 1,440–3,011 pounds of force, which is greater than the crushing force exerted by any known creature. For comparison, humans exert a maximum force of approximately 170 pounds with their rear teeth, an African lion 937 pounds, and an alligator just less than 3,000 pounds. Olson also excavated the largest known skull of *Torosaurus*.

1991 In *Bully for Brontosaurus*, Stephen Jay Gould defends the U.S. Postal Service's decision to issue stamps featuring the name *Brontosaurus* instead of *Apatosaurus*. Gould claimed that humans' interest in dinosaurs may be "intrinsic and endemic, a necessary and permanent fact of life . . . perhaps dinosaurs act as the trigger for a deep Jungian archetype of the soul; perhaps they rank as incarnations of primal fears and fascinations, programmed into our brains as the dragons of Eden." Gould refused to refer to *Brontosaurus* as *Apatosaurus* partly because Charles Knight had always referred to the dinosaur as *Brontosaurus*.

1992 A near-complete *Stegosaurus stenops* found by Bryan Small near Cañon City, Colorado, shows that Charles Gilmore was correct in 1915 when he claimed that *Stegosaurus* had staggered rows of spinal plates.

1992 AMNH starts planning to move its famous *T. rex* to what is now its Hall of Saurischian Dinosaurs. The *T. rex*, which had been on display since 1915, was fragile, and the move was time-consuming. More than a year later, the repositioned fossil—this time, with its head down, jaws shut, and back horizontal—was unveiled to the public.

1992 AMNH's *Apatosaurus* is disassembled and later remounted to reflect current ideas about how it looked and lived. Today, a tiny model of *Apatosaurus* in its original pose (e.g., with its tail dragging) can be found near the specimen's front left foot.

1992 Another remake of *The Lost World* promises viewers that "They Were Searching for the Unknown. . . . What They Found Was Fantastic!"

1992 Author and illustrator James Gurney's (b. 1958) fictional *Dinotopia: A Land Apart from Time* shows how Arthur Dennison and his son are shipwrecked in 1862 on an uncharted tropical island (Dinotopia) populated by peaceful dinosaurs "moving with terrific speed despite their size." Dinosaurs had lived on Dinotopia for millions of years, and had sought shelter in caverns ("The World Beneath") as other dinosaurs became extinct. On Dinotopia, humans coexist with dinosaurs that work as teachers, baby-sitters, and librarians. *Dinotopia: A Land Apart from Time*, a best-seller that sold over two million copies in 18 languages, was followed by the sequel *Dinotopia: The World Beneath* (1999), the prequel *Dinotopia: First Flight* (which included a board game; 1999), and *Dinotopia: Journey to Chandara* (2007). As had previous books in the series, *Dinotopia: Journey to Chandara* included new discoveries, such as *Einiosaurus* and dinosaurs found in China's Liaoning Province. *Dinotopia* spawned other books (written by other authors), television series, movies, and video games.

1992 Carl Baugh claims that people can change rates of radiometric decay with their minds, eggs do not hatch outside of Earth's magnetic fields, *T. rex* was a herbivore, and that Baugh possesses teeth of a *T. rex* "impacted" with chlorophyll.

1992 Famed dinosaur special-effects animator Ray Harryhausen (Figure 33) receives an Academy Award for Lifetime Achievement.

1992 Headlines of the tabloid *Weekly World News* announce, "Space Probe Finds Dinosaurs on Mars." According to the accompanying article, a "mind-boggling NASA photograph" taken 16 years earlier by *Viking I* showed live dinosaurs on Mars, but the government did not release the information. The "dinosaurs" in the photo were iguanas.

1992 ICR's Duane Gish (1921–2013) argues in *Dinosaurs by Design* that, before the Noachian Flood, humans lived with dinosaurs. Gish's book includes a picture of *Baryonyx* being killed by St. George, and tells readers, "All dinosaurs not on Noah's Ark were drowned in the Flood," "God created man and dinosaurs on the sixth day of creation," "the Bible does describe an animal that must have been a dinosaur," "humans and dinosaurs really did live at the same time," that God "put most or all of [the animals aboard Noah's Ark] into a form of hibernation," and that Adam and Eve's sins caused dinosaurs and all other animals to become violent and start killing.

1992 In "The Virgin and the Dinosaur," a sexy paleontologist and a time-traveling guide go through a Hell Creek wormhole to become the first time-travelers to visit the Mesozoic. They see the Mesozoic aboard *Challenger*, a fusion-powered balloon, but then crash. When the pair consummate their relationship, they find *Alamosaurus* foraging in the tree.

1992 In a PBS documentary, Robert Bakker cites discoveries at Como Bluff in 1877 as making dinosaurs famous worldwide.

1992 In *Dr. Mordrid,* a "good" mastodon and an "evil" tyrannosaur from outer space battle in a museum.

1992 In Gregory Benford's "Shakers of the Earth," scientists use DNA from a fossilized bone to create a herd of *Seismosaurus* for a "Kansas Sauropod Park."

1992 In Harry Turtledove's "The Green Buffalo," a commentary on endangered species, a shop owner in Lusk, Wyoming, is hired by John Bell Hatcher (a collector for Othniel Marsh) to hunt for buffalo. After being transported back in time to the Mesozoic, some hunters returned to 1891 and resumed their search for buffalo. They saw *Torosaurus* amidst a herd, but assumed that it was a diseased green buffalo. After killing the dinosaur, the hunters brought Hatcher some of its meat at the dig site (it tasted like chicken). Neither Hatcher nor the hunters realized that they'd eaten a dinosaur.

1992 In J. F. Rivkin's *Age of Dinosaurs: Tyrannosaurus rex,* Christine Fawcett investigates the disappearance of her relatives in the 1920s in a South American rainforest, during which she discovers a fresh piece of *T. rex* bone in a box shipped by her ancestors. Christine is taken to a Mayan pyramid, which is a time machine. She returns to the Late Cretaceous, where she finds her uncle.

1992 In Michael Bishop's "Herding with the Hadrosaurs," two boys who have time-travelled to the Late Cretaceous are orphaned when their parents are eaten by a tyrannosaur. The boys are saved when they are befriended by a herd of hadrosaurs.

1992 In Stephen Leigh's (b. 1951) *Dinosaur World*, time travelers enter a parallel universe and find civilized dinosaurs.

1992 In the 88-minute *Adventures in Dinosaur City,* three teenagers playing with their parents' time-machine are transported to a real-life version of *Dinosaur City,* their favorite television cartoon.

1992 Louis Jacobs helps direct the excavation of a near-complete *Acrocanthosaurus* north of Glen Rose, Texas. *Acrocanthosaurus* probably left many of the famous tracks in the Paluxy River at Glen Rose.

1992 ICR opens its Museum of Creation and Earth History in Santee, California (Figure 46). The museum and its educational programs tell visitors that dinosaurs were aboard Noah's Ark and that humans lived with dinosaurs. In 2013, when the musem was denied membership in the San Diego Museum Council, Tom Cantor (the president of the museum) blamed the vote on "the old prejudice against God" and compared the museum's fight for respect with the American civil rights movement. The museum, which is now owned and operated by Cantor's Light and Life Foundation, hosts more than 30,000 people per year.

1992 In Carenque, a suburb of Lisbon, Portugal, a 141-meter-long trackway of dinosaur tracks is saved by diverting a road through a tunnel under the tracks. The trackway, the longest known trackway at the time, was made by large, bipedal dinosaurs in the Cretaceous.

1992 Munchehagen's Dinosaur Park opens in Hanover, Germany, and features more than 200 life-size, prehistoric animals (including a 148-foot-long *Seismosaurus*)

Figure 46 The Creation and Earth History Museum—the oldest creationism museum in the world—was founded in 1992 by Henry Morris's Institute of Creation Research to promote young-Earth creationism. Today, the museum is operated by Light and Life Foundation. Like other creation museums (such as Ken Ham's Creation Museum in Kentucky and Carl Baugh's Creation Evidence Museum in Texas), the Creation and Earth History Museum uses dinosaurs to lure visitors, and includes exhibits depicting how humans lived with dinosaurs. (Courtesy Randy Moore)

along a 2.5-mile, circular "life through time" trail. The park, which also includes more than 200 139-million-year-old tracks, features *Velociraptor* and *Protoceratops* fighting, as well as a *Seismosaurus* whose neck directs visitors to bathrooms. Many of the dinosaurs were sculpted by Krzysztof Kuchnio.

1992 Paleontologist Jack Horner (Figure 43) is hired by Globe Toys to help produce realistic dinosaur toys.

1992 The British Museum (Natural History) is renamed the Natural History Museum.

1992 Paleontologists find a 125-million-year-old neck bone of a sauropod on the Isle of Wight. The 75-cm-long bone came from a 50-ton dinosaur more than 20 meters long, which was the biggest dinosaur ever found in Britain. The fossil is now displayed in Dinosaur Isle Museum on the Isle of Wight.

1992 Robert Bakker and his colleagues describe *Edmarka rex* from the Como Bluff area. *Edmarka* is similar to *Torvosaurus,* and may be an individual variant of that genus.

1992 *Science* reports "radioisotopic dating now has forged the final link" between the K-T impact and dinosaurs' extinction.

1992 Stephen Leigh promotes Ray Bradbury's classic short-stories about dinosaurs with *Ray Bradbury Presents Dinosaur World* and, in the following year, *Ray Bradbury Presents Dinosaur Planet.*

1992 The Dinocardz Company issues the 80-card "Dinocardz." The following year, Australia's Dynamic Marketing issued the 60-card "Escape of the Dinosaurs," First Glance Productions (of the Canadian Museum of Nature) issued its small 24-card "Dinosaur Art" (based on art by Ely Kish), Kitchen Sink Press issued its 36-card "Dinosaur Nation," and Redstone Marketing issued its 50-card "Dinosaurs of the Mesozoic" based on the art of Brian Franczak.

1992 The Dinosaur Museum opens in Blanding, Utah, as a nonprofit public charity. Today, this museum is one of the most entertaining and educational museums in the region; it includes posters for dinosaur movies, the model of *Brontosaurus* used in *King Kong,* a tribute to Charles Knight, and a replica of one of the *Edmontosaurus* mummies discovered by Charles H. Sternberg.

1992 The first major paleontological excavations begin at Dinosaur Ridge since the work by Yale University in the 1870s. Dinosaur Ridge has yielded more than 70 different dinosaurs found in 10 different strata. The area includes numerous dinosaur tracks, as well as *Apatosaurus* bones that Arthur Lakes left behind when he quit excavating the area in 1879. Tracks on the eastern side of Dinosaur Ridge are approximately 100 million years old, whereas the bones on the western side of the ridge are 150 million years old. Nearby "Triceratops Trail" exposes sediments that are 68 million years old.

1992 The novel *The Ugly Little Boy* depicts how a company retrieving objects from the past collects a small green dinosaur that is described as a long-tailed chicken.

1992 The popular children's show *Barney & Friends* debuts on television, starring an optimistic, anthropomorphic, purple and green *Tyrannosaurus rex* who sings and dances while conveying positive, friendly messages. His theme song was "Barney Is a Dinosaur," which was sung to the tune of "Yankee Doodle." Although *T. rex* was a carnivore, Barney liked fruit, vegetables, milk, and peanut-butter-and-jelly sandwiches. Barney's friends included Baby Bop (a green *Triceratops* who wears a pink bow and pink ballet slippers, and carries a yellow blanket), B.J. (a yellow *Protoceratops*—and Baby Bop's older brother—who wears red sneakers and a red baseball cap), and Riff (an orange hadrosaur who wears green sneakers and who is Baby Bop and B.J.'s cousin). In *Cracked* magazine, in which a parody of Barney ate a chicken, Barney asked, "What do ya expect? I am a Tyrannosaurus, you know." New episodes of *Barney & Friends* ended in 2009, but the show remains popular as reruns on many PBS stations.

1992 There are now more than 800 books written in English about dinosaurs.

1992 Thirty FBI agents, assisted by the National Guard, raid BHI and confiscate files, fossils, and 10 tons of still-plastered "Sue," the *T. rex* discovered two years earlier by Sue Hendrickson. Authorities claimed that Sue's excavation violated the Federal Antiquities Act, which "prohibits outsiders from making contracts with people of the Indian community without Federal permission." For the next five years, legal battles waged over Sue's ownership. Finally, a U.S. district court ruled that the famous fossil belonged to the government, which had held the land (in trust) for the rancher on whose property Sue was discovered. In 1997, that rancher (Maurice Williams) sold the mostly unprepared parts of Sue at an auction to Chicago's Field Museum for more than $8 million, the most ever paid for a fossil. Sue, who was 28 years old when she died, continues to be a primary attraction in Chicago's Field Museum.

1992 Walter Alvarez (b. 1940) and his colleagues find sandstone with shocked quartz at the K-T boundary extending hundreds of miles into northern Mexico. The lowest layers of the sandstone (ejected from the impact and containing the shocked quartz) overlay thick layers of sandstone containing coastal debris. Atop these layers is a thinner layer of fine clay containing large amounts of iridium.

1993 The cover of *Time* magazine features *Mononychus* ("a new link between dinosaurs and birds"), promises readers "the truth about dinosaurs," and proclaims "Surprise: Just about everything you believe is wrong." When it was discovered that *Mononychus* was the name of a beetle, the Late Cretaceous dinosaur's name was changed to *Mononychus* to *Mononykus*.

1993 A 33-page, 39-count indictment of BHI results in *T. rex* "Sue" being confiscated by the federal government. Although SVP supported the seizure, paleontologists such as Phil Currie, Robert Bakker, and Jack Horner denounced the government's actions; as Horner noted, "Who's going to walk off with a tyrannosaurus?" Two years later, following a two-and-a-half-week trial, the sale of "Sue" to BHI was declared "null and void." In 1995, Peter Larson was convicted of two felonies and, in 1996, was fined $5,000 and sentenced to two years in prison. Elementary school children from Hill City, South Dakota (population approximately 650) pleaded for FBI agents not to take "Sue," and Larson got $47.50 raised by seventh-graders in "Sue's Freedom Fund."

1993 A cover of *Nature* shows a zircon crystal scarred with fractures produced by extreme shock. The accompanying article described how studies of these crystals corroborated the claim in 1980 by Luis and Walter Alvarez that the K-T extinction was associated with a meteoric impact.

1993 AMNH's Mark Norell (b. 1957) finds the emu-sized *Citipati osmolskae* ("funeral pyre lord") over a clutch of 6-inch-long eggs. The fossils came from Late Cretaceous sediments in what is now Mongolia. This was the first direct evidence of nesting behaviors in theropods.

1993 An Israeli dairy—hoping to capitalize on the dinosaur craze triggered by *Jurassic Park*—sells milk with free dinosaur stickers, but is threatened with loss of its kosher status because "this is like seeping sacrilege. . . . Dinosaurs symbolize a heresy of the creation of the world because they reflect Darwinistic theories."

1993 Companies ranging from Tyco ("Dino-Riders") and Hasbro ("Baby Sinclair") to Toys 'R' Us and McDonald's begin marketing dinosaurs keyed to the movie *Jurassic Park* to boost sales. Tiger Electronics packaged dinosaur-themed characters with a "genuine fossil" from the Jurassic. That same year, a plastic "Godzilla the Dinosaur" toy was named one of the "10 Worst Toys" because of its sharp spinal plates. Before the release of *Jurassic Park*, MCA sold more than 100 licenses for more than 1,000 products that were expected to gross more than $100 million.

1993 Dale Russell and Z. Zheng publish the first phylogenetic analysis of sauropods.

1993 During his third season of fieldwork in the Gobi Desert with the Mongolian Academy of Science-AMNH expedition, Michael Novacek and his coworkers examine Ukhaa Tolgod ("small brown hills"). During the next seven years, they excavated or recorded more than 200 dinosaurs (and almost 1,000 mammals and 1,400 lizards) from an area about one-third the size of New York's Central Park. Their discoveries included *Oviraptor philoceratops* (which they named "Big Mamma") crouched in a brooding position atop a nest of 20 8-inch-long elliptical eggs (that were previously identified as *Protoceratops* eggs). These fossils preserved an adult in a position identical to that of modern birds when they brood their young. Two years later, a team from AMNH made a similar discovery nearby, indicating that the adults were not eating the eggs, but instead were brooding them.

1993 Geologist Chuck Pillmore reports a track of a Colorado hadrosaur only 37 centimeters below the K-T boundary.

1993 German veterinarian and fossil enthusiast Burkhard Pohl and his colleagues begin excavating the red sediments of the Morrison Formation near Thermopolis, Wyoming. These excavations later helped stock the Wyoming Dinosaur Center. The center features *Triceratops* (the state dinosaur of Wyoming), the "Thermopolis specimen" of *Archaeopteryx* (the only such specimen in North America), and a reconstructed *Seismosaurus* that is 135 feet long. In 2011, this specimen was reported to contain remnants of soft tissue.

1993 Guinea issues a 16-stamp set of postage stamps (ranging from 50 to 250 francs each) showing dinosaurs from the Triassic, Jurassic, and Cretaceous (i.e., all three periods of the Mesozoic). The stamps also showed how Pangaea split apart to produce today's continents.

1993 Guyana issues its first set of postage stamps featuring dinosaurs. Today, Guyana has issued more postage stamps featuring dinosaurs than any other country.

1993 In "Last Rights," dinosaur-rights activists free trapped dinosaurs ("prisoners of zoos") created by genetic engineering of the DNA from chickens and other birds. After being set free, one activist is killed by *Deinonychus,* who was assumed to be herbivorous.

1993 In Barbara Delaplace's "Fellow Passengers," *Deinonychus* escapes from Zoological Gardens and begins killing cattle. When the dinosaur is recaptured, animal-rights activists free the dinosaur, prompting a reporter for the *Blatant Enquirer* to reflect that dinosaurs, like the last passenger-pigeon, will soon be extinct.

1993 In David Gerrold's "Rex," a tyrannosaur living in a family's basement escapes and begins eating other dino-pets in the neighborhood. After Rex attacks burglars, the family realizes its value as a guard dinosaur.

1993 In London, a clutch of dinosaur eggs is sold to an American collector for £50,600.

1993 In Michelle Sagara's "Shadow of Change," outcast April Stephens feels ill, eats raw meat, and notices her skin is becoming scaly and her teeth sharper. After eating her pet cat and a visitor who came to her house to check on her, April becomes a carnosaur.

1993 In the children's film *We're Back: A Dinosaur's Story,* dinosaurs appear as benevolent friends in suburban America. When Professor New Eyes feeds "brain grain" to four dinosaurs, they become smart and hear children in the future wish they could see dinosaurs; the four dinosaurs go there so the children can see them. *T. rex* plays golf.

1993 In *The Complete T. Rex: How Stunning New Discoveries Are Changing Our Understanding of the World's Most Famous Dinosaur,* Jack Horner and Don Lessem (b. 1951) argue that *Tyrannosaurus* was slow and probably could not run because, like most large theropods and modern elephants, the ratio of their femur to their tibia exceeded 1.0. In 1991, Lessem, David Weishampel (b. 1952), and Peter Dodson founded the Dinosaur Society, which was the first nonprofit society devoted to dinosaurs. According to a statement released by the society, ". . . billions are made on dinosaur products. Few of these products succeed in accurately communicating the discoveries and the excitement of dinosaur science." The society disbanded in 1997.

1993 In the movie *Mrs. Doubtfire,* Robin Williams and others on a television show teach about dinosaurs.

1993 In the movie *Super Mario Brothers,* dinosaurs living in a different universe evolve into smart, self-aware, humanlike animals. That universe was one of two parallel universes created 65 million years ago when a meteorite hit Earth.

1993 Jim Ian Kirkland (b. 1954) and his colleagues describe *Utahraptor* ("Utah's predator"), a large dromaeosaur from the Lower Cretaceous that he and his colleagues had discovered in Grand County, Utah, two years earlier. *Utahraptor* had single claws (each up to 9 inches long) on the second toe of each of its hind feet. The following year, this giant velociraptor became famous when it appeared in the blockbuster movie *Jurassic Park. Utahraptor*, which also appeared in the 1999 BBC series *Walking with Dinosaurs,* was more than 20 feet long from head to tail, and weighed at least 1,000 pounds; for comparison, *Deinonychus*—the stand-in for its relative *Velociraptor* in *Jurassic Park*—weighed about 200 pounds. *Utahraptor ostrommaysorum,*[43] the type specimen, was named to honor Yale paleontologist John Ostrom and Dinamation animator Chris Mays.

1993 Mike Fredericks premiers *Prehistoric Times,* a magazine that covers all things prehistoric.

1993 Paleontologist Louis Jacobs (b. 1948) dismisses claims about human footprints with dinosaur footprints as "so outrageously false that the whole story would be comical were it not for the implications of public education of the fundamentalist agenda. Yet this nonsense is amazingly resilient. . . . Believers in the association of dinosaur and human footprints reject virtually all intelligent explanations." Jacobs added, "There are no living dinosaurs in Africa."

1993 Peter Wellnhofer describes the "Munich specimen," which is "the seventh specimen of *Archaeopteryx*" (S6, formerly known as the "Solnhofen-Aktien-Verein specimen"). The "Munich specimen," which was found in 1991, is the only specimen of *Archaeopteryx* with an intact sternum (breastbone). In 1999, the "Munich specimen" was sold to the Bavarian State Collection of Paleontology and Historical Geology (now BSP in Munich).

1993 Researchers from SUNY-Stony Brook and Madagascar's University of Antananarivo uncover the titanosaur *Rapetosaurus* on Madagascar. Subsequent discoveries there included the carnivore *Masiakasaurus* and *Majungatholus*.

1993 Robert Bakker names Edward Drinker Cope as the type specimen for *Homo sapiens*. Bakker's claim was rejected, and a subsequent tour by National Geographic photographer Louie Psihoyos (b. 1957) of famous dinosaur sites with Cope's skull (specimen 4989 from the University of Pennsylvania's Museum of Archaeology and Anthropology) produced controversy and criticisms of grandstanding, boorish behavior.

1993 Rubén Dario Carolini discovers a thigh bone of *Giganotosaurus* ("giant lizard"), "the largest carnivore that ever walked the earth," in Late Cretaceous sediments of Patagonian Argentina. Today, a 42-foot-cast of *Giganotosaurus* hovers above the reception desk of the Argentina Academy of Natural Sciences. *Giganotosaurus* was described earlier by Eberhard Fraas (1862–1915), but the name was already in use, so it was renamed *Torniera* by R. Sternfeld in 1911. *Giganotosaurus,* which lived about 30 million years before *T. rex*, had one of the longest skulls of any carnivore yet discovered.

1993 *Science* reports, "Scientists have new doubts about the [asteroid impact's] ability to kill the dinosaurs."

1993 Sculptor and welder Joe Barrington finishes his 65-foot-long, 35-foot-tall *Apatosaurus*. The dinosaur, which took more than $12,000 and a year to make, is displayed at the Cimarron County Historical Society Museum in Boise City, Oklahoma.

1993 Steven Spielberg's blockbuster movie *Jurassic Park* presents the most realistic portrayals of dinosaurs ever seen in movies (Figure 47). Instead of using stop-action animation, Spielberg used full-scale robotic models and computer-generated graphics to depict dinosaurs. *Jurassic Park* ("An Adventure 65 Million Years in the Making!") was an adaptation of Michael Crichton's earlier novel (1990) of the same name. Spielberg's movie portrayed dinosaurs not as lumbering, stupid animals, but instead as the active, clever animals advocated by John Ostrom and his successors. In the movie, a rich philanthropist created a theme park (Isla Nublar) that included genetically engineered dinosaurs such as *Dilophosaurus, Tyrannosaurus, Velociraptor* (referred to by the movie's characters as "raptors"), *Triceratops, Stegosaurus, Apatosaurus,* and *Maiasaura*. The movie, a milestone in the use of computer technology, grossed more than $910 million (mostly overseas), making it the top-grossing film at that time (it was surpassed four years later by *Titanic*). Alan Grant, played by actor Sam Neill, is a paleontologist (first seen excavating *Velociraptor* in Montana) modeled after paleontologist Jack Horner. The movie stars *T. rex*, which was represented by a 20-foot-tall, 13,000-pound animated model

Figure 47 The movie *Jurassic Park* (1993), an adaptation of Michael Crichton's book of the same name from three years earlier, brought dinosaurs to life with full-scale robotic models (as shown in this photo) and computer-generated images. *Jurassic Park,* made for $63 million, grossed more than $900 million (mostly overseas), making it the top-grossing movie at the time. The success of *Jurassic Park* spawned numerous dinosaur-themed books, comics, snacks, toys, video games, advertising campaigns, and other products. (Murray Close/Sygma/Corbis)

(Horner commented that it "was the closest I've ever been to a live dinosaur"), and its antagonist was *Velociraptor* (which was modeled after the larger *Deinonychus,* a close relative of *Velociraptor*). Cooperative hunting by *Deinonychus* was subtly acknowledged in the movie; Muldoon's (the game warden) last words, after being mauled by the predators, were "clever girl" (this quote doesn't appear in Crichton's novel). Spielberg also acknowledged the business of dinosaurs: Upon seeing *Jurassic Park* for the first time, a lawyer for the park's owner says, "We're gonna make a fortune off this place . . . a fortune." (*Jurassic Park* was accompanied by a massive, $60 million marketing campaign that included thousands of products such as toys and video games). In *Jurassic Park* and its sequel, *Velociraptor* lacks feathers; subsequent discoveries showed that *Velociraptor* was covered with feathers. (In *Jurassic Park III, Velociraptor* has quill-like structures on the back of its head and neck.) In the movie, *Dilophosaurus* was depicted at only half its actual size, and *Velociraptor* was made larger to appear more threatening. Despite the film's name, most of the dinosaurs in *Jurassic Park*—which won three Oscars (Visual Effects, Sound Effects Editing, and Sound)—were not alive until the Cretaceous.[44] Spielberg followed *Jurassic Park* with *The Lost World* (a homage to Sir Arthur Conan Doyle's classic book and silent movie). Spielberg's movies were commercially successful and stimulated the public's interest in paleontology.

1993 The Cheyenne River Sioux Tribe invokes business licensing laws to sue in tribal court for the forfeiture of *T. rex* "Sue." The next year, the forfeiture was denied.

1993 The friendly purple dinosaur *Barney* is featured on the cover of *TV Guide.*

1993 The movie *Father Hood,* about a deadbeat dad on the run from cops with his kidnapped children, includes a confrontation with *T. rex*. Today, the model for that dinosaur greets at visitors to Cascade Caverns in Fair Oaks Ranch, Texas.

1993 The movie *Prehysteria!* portrays a home overtaken by five dinosaurs hatched from newly discovered eggs ("They're the World's Oldest Party Animals!"). Each pet-sized dinosaur was named for a famous musician (e.g., "Jagger," "Elvis," "Madonna"). When *Prehysteria!* generated more than $100 million in revenues, there were two sequels.

1993 The relatively low-budget movie *Carnosaur* ("Driven to extinction. Back for revenge!"), unlike many earlier dinosaur-based movies, features only carnivorous dinosaurs (e.g., *Velociraptor, Tyrannosaurus*) as villains. These genetically engineered dinosaurs were animated with puppets and robots, some of which were created by John Buechler. In *Carnosaur*, as well as in the sequel *Carnosaur II* (1994; "Back for Another Bite!"), the dinosaurs are killed by humans. After breeding a *T. rex* by combining DNA from dinosaur and chickens, the mad scientist infects women with a virus that causes them to give birth to baby carnosaurs. The mechanical dinosaurs in *Carnosaur* reappeared in *Carnosaur*'s two sequels and in *Dinosaur Island* (1994).

1993 The Republic of China issues a set of four postage stamps featuring the Chinese creation story. On one stamp, Pan Gu (in Chinese mythology, the first living being and creator of all) is shown holding dinosaurs.

1993 There are now 949 known species of dinosaurs; a new species is being discovered about every 40 days.

1993 Virgil "Buck" Sharpton describes the K-T impact site as the Chicxulub crater, which is along the northern Yucatán. This crater, which is nearly the size of New Hampshire, is one of the largest craters on Earth.

1993 When red blood cells are found in a *Tyrannosaurus* bone, headlines claim "Tyrannosaurus DNA May Be Extracted." This announcement nearly coincided with the release of the *Jurassic Park* movie.

1993 While looking for fossils at a construction site in northwestern North Korea, high-school students discover the remains of what the magazine *Korean Pictorial* dubs "Archaeopteryx." Soon thereafter, North Korea's president Kim Il-Sung (1912–1994) ordered the fossil be named "Proornis coreae" ("founding bird of Korea"). Since the fossil has not been formally described, "Proornis coreae" is a *nomen nudum*. Subsequent analyses of the fossil have suggested it is a closer relative of the crow-sized bird *Confuciusornis* than *Archaeopteryx*.

1993 While working in Niger, Paul Sereno of the University of Chicago discovers the 135-million-year-old, 27-foot-long theropod *Afrovenator* ("African hunter"), a relative of William Buckland's *Megalosaurus*. The next year, Sereno and others described *Afrovenator abakensis* in *Science* (*abakensis* refers to Abaka, the Tuareg name for the area of Niger where the fossils were found). In Morocco, Sereno excavated *Carcharodontosaurus* ("shark toothed reptile"), a giant theropod having a skull about 6 feet long. This skull was thought to have the longest skull of any theropod until the even-larger *Giganotosaurus* was discovered in 1993. In 1996, skulls of *Carcharodontosaurus* and *Homo sapiens* adorned the cover of *Science*.

1994 A cast of *Brachiosaurus* collected by Elmer Riggs is displayed in the Field Museum. In 1999, with the approaching unveiling of *T. rex* "Sue," this cast was moved to Concourse B at Chicago's O'Hare airport. A second cast of *Brachiosaurus* was displayed outside the Field Museum.

1994 AiG opens for business in Kentucky and begins denouncing evolution and other sciences while promoting young-Earth creationism. AiG soon became the most dominant of the many antievolution organizations. AiG later opened a $27 million Creation Museum, which, like all such museums, relied heavily on dinosaurs to promote its religious, anti-science message. In 2014, AiG was developing an amusement park centered on a life-size replica of Noah's Ark.

1994 Battat, Inc., in conjunction with the Boston Museum of Science and paleo-sculptors Dan LoRusso and Gregory Wenzel, begins producing anatomically accurate, 1:40-scale dinosaur toys.

1994 Bernardino Pérez-Moreno and his colleagues describe *Pelecanimimus polyodon,* the first-known European ornithomimosaur.

1994 *Dinosaur Eggs and Babies* is edited by Karl Hirsch, Kenneth Carpenter, and Jack Horner.

1994 *Dinosaur Impressions,* published in French, chronicles the long career of Philippe Taquet (b. 1940), one of the world's most noted paleontologists. The book was translated into English in 1998 by paleontologist Kevin Padian (b. 1951).

1994 During a joint expedition of AMNH and the Mongolian Academy of Sciences, researchers find hatchlings of *Protoceratops* having skulls 1 inch long.

1994 Embryos from Flaming Cliffs originally believed to be *Protoceratops* are shown to be those of the theropod *Oviraptor*.

1994 In "The Bite of the Bronto," Robert Bakker invokes the active brontosaurs in *King Kong* (1933) and *The Lost World* (1925).

1994 In *Dinosaurs, the Bible, Barney and Beyond*, Phil Phillips condemns *We're Back? A Dinosaur's Story* for its occult symbols and promotion of evolution, *Dinosaurs!* for being "straight out of evolutionary science books," *The Flintstones* for being a "spoof of the story of Christ's birth," and *Barney & Friends* because Barney "does what Jesus Christ *alone* can truly do in a person's life." Phillips urged believers to "arm your children with Bible-related facts about dinosaurs . . . not the distorted evolutionary theories" because "evolutionary theory is in more disrepute and under greater threat of 'extinction' now than in just about any time during the last one hundred years." Earlier, Phillips's *Turmoil in the Toybox* (1990) condemned the evil influences of Smurfs, Care Bears, Mighty Mouse, and other toys.

1994 In *Hunting Dinosaurs*, Louie Psihoyos purports the type-specimen of *Homo sapiens* to be Edward Drinker Cope. In 1959, W. T. Stearn (1911–2001) had designated Carolus Linnaeus as the type specimen for *H. sapiens*.

1994 In Robert J. Sawyer's *End of an Era*, paleontologists aboard the spaceship *Sternberger* go back to the Late Cretaceous, where they encounter dinosaurs in

Alberta's (Canada) badlands. However, Earth was dominated by Martians, who had suppressed gravity and bioengineered giant dinosaurs. When the paleontologists used radio waves to destroy the Martians' gravity suppressors, Earth was hit by a meteor that caused the K-T extinction.

1994 In the movie *Dinosaur Island*, airmen end up in a "lost world" island inhabited by beautiful women and dinosaurs. The *T. rex* from the movie *Carnosaur* (1993) was worshipped as "The Great One" as bikini-clad women waited for a man who could kill their god.

1994 In *Wes Craven's New Nightmare*, a child threatened by villain Freddy Krueger looks to a toy *Tyrannosaurus* for protection.

1994 Robert Bakker notes, "*Jurassic Park* needed an intelligent villain, one with formidable powers, one that could almost catch and kill and eat the human heroes. . . . It is the raptors that have come to symbolize all the revised scientific thinking about dinosaurs."

1994 The BSA's "Paluxy Task Force" self-publishes *Texas Tracks and Artifacts: Do Texas Fossils Indicate Coexistence of Men and Dinosaurs*, a 109-page booklet promoting the often-discredited claims by Carl Baugh that humans and dinosaurs lived at the same time.

1994 The Field Museum opens its "DNA to Dinosaurs" (later renamed "Life through Time") exhibit. Some dinosaurs (e.g., *Apatosaurus excelsus*) were remounted in more dynamic poses, and Charles Knight's paintings were moved to new positions. Some dinosaurs (e.g., *Parasaurolophus*) were displayed for the first time.

1994 In *Footprints and the Stones of Time,* Carl Baugh and Clifford Wilson claim that the "man tracks" in Glen Rose, Texas, are "academically documented extensive trails of human footprints," and that "the only reason for rejecting (the 'man tracks') as a human print is the 'Establishment' argument that dinosaurs had died out long before man was around." Baugh and Wilson also claimed that "Evolution did not take place at all; it is a religious concept."

1994 The first Antarctic dinosaur species, the theropod *Cryolophosaurus ellioti,* is formally named and described.

1994 The first Dino Fest (later renamed Dinofest) International opens in Indianapolis. About 30 people attend.

1994 The first *Dinosaur Talks* premiers, and the first paleontologist to be interviewed is Edwin H. Colbert.

1994 *The Flintstones* is made into a feature movie, and is followed six years later by the less successful sequel, *The Flintstones in Viva Rock Vegas*. In both movies, everything in the past parallels the present, with rocks and dinosaurs replacing electricity and pets.

1994 The fragmented comet Shoemaker-Levy 9 smashes into Jupiter, showing that giant impacts (such as the one proposed as a cause of the K-T extinction) still occur in the solar system.

1994 The site near Haddonfield, New Jersey, where *Hadrosaurus* was discovered is made a National Historic Landmark. The following year, the site was dedicated at a ceremony whose attendees included Christopher Brees, the former Boy Scout who, a decade earlier, had reestablished the site. Also present was Joyce Berry, a local teacher whose students had lobbied for several years to have *Hadrosaurus* be named New Jersey's official state dinosaur. The actual dig-site is about 75 yards beyond the marker toward the Cooper River just to the left of the ravine (as you face the monument).

1994 The theropod *Saurophaganax maximus* becomes the state fossil of Oklahoma.

1994 To date, more than 2,500 publications address the extinction of dinosaurs.

1994 Tony Fiorillo of the Dallas Museum of Natural History argues that dinosaur fossils from DNM were probably deposited over only a few years, indicating that the animals found there had all lived at the same time.

1994 *Triceratops* becomes the state dinosaur of Wyoming.

1994 While searching for petrified wood, geology student Rich Thompson discovers the 20-ton brachiosaur *Sonorasaurus*. Thompson's discovery was later featured in a display at Tucson's Arizona-Sonora Desert Museum.

1994 Workers at a limestone quarry in central Bolivia discover what comes to be known as the Dinosaur Park of Cal Orck'o (the name of a nearby mountain), the largest known quarry of dinosaur footprints (Figure 48). The quarry, located four miles outside Sucre, Bolivia, displays more than 150 different kinds of dinosaur footprints in 462 distinct trails along a 325-foot-long slab of near-vertical limestone. (In 2010, a large section of the quarry collapsed, taking with it more than 200 footprints made by two titanosaurs.) The site also includes Parque Cretácio, a dinosaur museum that opened in 2006 that features 24 life-size dinosaur replicas and an area from which to view the tracks. The more than 4,000 65-million-year-old prints are part of a UNESCO World Heritage Site.

Figure 48 At El Parque Cretácico (established in 2006) near Sucre, Bolivia, more than 5,000 dinosaur tracks left 68 million years ago decorate a near-vertical wall of limestone more than 100 meters high. These tracks comprise the largest known quarry of dinosaur footprints in the world. (AP Photo/Dado Galdieri)

1995 A six-year, $7.7-million renovation of the Denver Museum of Natural History includes the new exhibit "Prehistoric Journey." The museum's *T. rex* was remounted upright on one foot.

1995 An imprint allegedly of a human hand is found in Cretaceous rock in Palo Pinto County, Texas. The imprint is now displayed at Carl Baugh's Creation Evidence Museum in Glen Rose, Texas.

1995 As increasing numbers of scientists and others begin to accept the Chicxulub crater as the site of the K-T extinction event, the August 7 issue of *Nature* features a gravity map of the crater on its cover.

1995 Dan Chure (b. 1951), a paleontologist at DNM, changes the name of *Saurophagus* to *Saurophaganax* ("king of the reptile eaters") after determining that the original name had been used to identify another animal. The 42-foot-long *Saurophaganax* was the largest carnivore known from the Morrison Formation; it was almost as big as *T. rex*, which appeared 85 million years later.

1995 Dinosaur Depot opens in Cañon City, Colorado, and features Jurassic fossils from the Garden Park area. The museum closed in 2013.

1995 Edwin Colbert's *The Little Dinosaurs of Ghost Ranch* describes his famous discoveries in New Mexico.

1995 Fantasy Art issues its "Dinotopia" set of 91 cards and a poster. The following year, Steve Jackson Games issued the 109-card "Dino Hunt."

1995 *Films for Christ* uses Paul Taylor's book *The Great Dinosaur Mystery and the Bible* as a basis for its *The Great Dinosaur Mystery,* which argues that humans lived with dinosaurs.

1995 In *Dinosaur in a Haystack*, Stephen Jay Gould attributes dinosaurs' ongoing appeal to their being "big, fierce, extinct—in other words, alluringly scary, but sufficiently safe." Gould also noted "creationist detractors charge that evolution is an unproved and unprovable charade—a secular religion masquerading as science. They claim, above all, that evolution generates no predictions, never exposes itself to test, and therefore stands as dogma rather than disprovable science. This claim is nonsense. We make and test risky predictions all the time; our success is not dogma, but a highly probable indication of evolution's basic truth."

1995 In *Dinosaurs, Diamonds, and Other Things from Outer Space*, David Carlisle revives the idea that supernova explosions led to the extinction of dinosaurs.

1995 In *First Frontier,* the crew of Star Trek finds dinosaurs in a parallel universe on Earth.

1995 Karl Shuker suggests that, because scientists disagree on how dinosaurs became extinct, dinosaurs may not be extinct after all.

1995 Louis Jacobs's *Lone Star Dinosaurs* includes one of the first depictions of the tsunami associated with the meteor impact associated with the K-T extinction.

1995 Mark Norell and his colleagues report *Oviraptor* atop its eggs from Late Cretaceous sediments of the Gobi Desert. This discovery provided more evidence of bird-like brooding by dinosaurs.

1995 Michael Crichton (1942–2008) publishes the 429-page techno-thriller *The Lost World*, a sequel to his earlier book, *Jurassic Park*. Crichton's novel, like Arthur Conan Doyle's 1912 novel of the same name, described an expedition to a Central American area where dinosaurs were still alive (in Crichton's book, this was because of genetic engineering). Various characters fended off attacks by *Velociraptor, Procompsognathus*, and *Tyrannosaurus* while discussing topics such as evolution and extinction.

1995 Nomenclature expert George Olshevsky publishes *Dinosaur Genera List* online. This was the closest thing to an "official" list of dinosaurs.

1995 Pei-ji Chen and his coworkers from China's Nanjing Institute of Geology and Palaeontology begin unearthing the chicken-sized, 125-million-year-old theropod *Sinosauropteryx* ("Chinese lizard wing") and other feathered dinosaurs from China's Liaoning province. *Sinosauropteryx*, an early relative of *T. rex*, was the first-known fossil of a non-avian dinosaur with feathers, and supported the claim that birds descended from dinosaurs. *Sinosauropteryx* later was the first dinosaur to have its colors revealed.

1995 Rodolfo Coria (b. 1959) and Leonardo Salgado (b. 1962) describe the 45-foot-long, 12-foot-tall (at the hips), 7-ton *Giganotosaurus carolinii*, South America's largest known theropod. The dinosaur, from the Late Cretaceous, was named for Rubén Carolini, an automobile mechanic and amateur fossil-hunter.

1995 Rudolph Zallinger, who painted the famous "Age of Reptiles" mural at Yale's Peabody Museum of Natural History, dies of cancer at the age of 75 in Branford, Connecticut.

1995 José Sanz and his colleagues announce the discovery of more than 20 seaside nesting-sites of dinosaurs in northeastern Spain. The Bastús site contains an estimated 30,000 eggs.

1995 Scott Sampson (b. 1961) names *Einiosaurus*, a ceratopsian with a forward-curving horn on its nose. Two years later, *Einiosaurus* appeared on one of "The World of Dinosaurs" U.S. postage stamps. In 2009, Sampson began hosting the children's show *Dinosaur Train* on PBS.

1995 Several members of SVP buy a tombstone for John Bell Hatcher's grave in Pittsburgh's Homewood Cemetery. Hatcher's grave (Section 9, Lot 653), which had been unmarked for 91 years, is now marked with a stone that lists Hatcher's name, his daughter Ruth's name (who is buried beside him), and a sandblasted image of *Torosaurus*.

1995 Soviet paleontologist Lev Alexandrovich Nessov's (1947–1995) *Dinosaurs of Northern Eurasia* names several genera of dinosaurs. Nessov, who had discovered 80 dinosaur localities in the USSR, died soon after publication of his monograph.

1995 SVP organizes Save America's Fossils for Everyone (SAFE) to lobby for the protection of fossils on public lands.

1995 The 65-million-year-old birdlike theropod *Rahonavis ostromi* is discovered in northwestern Madagascar and, three years later, is restored in the magazine *Discover*.

1995 The futuristic movie *Theodore Rex* features a genetically engineered, talking dinosaur that helps a cop find a mad scientist who is intent on destroying the world. At the time, this $33-million movie was the most expensive direct-to-video film ever made.

1995 The gray, windowless Wyoming Dinosaur Center opens in Thermopolis, Wyoming (population approximately 3,100). This was Wyoming's first facility dedicated to the excavation, preparation, and display of dinosaurs. Painted dinosaur-tracks lead visitors from Thermopolis's main street, across the Bighorn River, and to the Dinosaur Center.

1995 The movie *Toy Story* features Rex, a funny and shy *T. rex* who befriends the main character Woody.

1995 The television documentary series *PaleoWorld* premiers on The Learning Channel. The first episode featured Paul Sereno's work in the Sahara Desert. In 1998, comedian Jay Leno commented, "It is little wonder that a great big dinosaur would go extinct searching for fish in the middle of the Sahara Desert."

1995 The U.S. Postal Service issues the "Comic Strip Classics" set of twenty 32¢ stamps, one of which honors *Alley Oop*. The stamp shows Oop atop his *Stegosaurus*.

1995 With ads proclaiming "The Dinosaurs Return. A New Era Begins," AMNH opens its Halls of Saurischian and Ornithischian Dinosaurs. The new halls reflected modern cladistic methods of classifying dinosaurs, and several of the dinosaurs were repositioned to reflect advances in knowledge of their behaviors (e.g., the tail of *T. rex* was raised above the ground). An exhibit in the Saurischian Hall told visitors that "birds are dinosaurs."

1996 "Claude Bell's Dinosaurs" in Cabazon, California, is expanded with a gift shop and a 60-room hotel.

1996 Chen Pei Ji tells colleagues at the meeting of SVP of *Sinosauropteryx prima*, the first feathered dinosaur to be reported. The specimen was described in English two years later in *Nature*.

1996 Chinese paleontologists describe *Confuciusornis,* a primitive pigeon-sized bird from the early Cretaceous that bridges the morphological gap between *Archaeopteryx* and more modern birds. *Confuciusornis* had a short tail, a shoulder girdle similar to that of modern birds, no teeth, and several features of primitive theropods not present in more-advanced birds. *Confuciusornis*, the oldest known bird to have a beak, had long, modern feathers on its wings.

1996 David Martill and his colleagues describe *Irritator challengeri*, an Early Cretaceous dinosaur that had been found in modern-day Brazil. The name *Irritator* described the feelings of the paleontologists when they discovered that the skull's snout had been elongated with plaster to increase its commercial value, and *challengeri* honored Professor Challenger of Arthur Conan Doyle's *The Lost World*. Andrew Kellner later recognized *Irritator* as a spinosaur, a croc-snouted, often sail-backed predatory dinosaur (*Irritator* may be a synonym for *Angaturama*). Spinosaurs starred in *Jurassic Park III*.

1996 Fossil collector Peter Larson goes to prison; he is released almost 19 months later into home confinement for two months (and then two years of probation). Seven weeks after his release, "Sue"—the best preserved, most complete, and first *T. rex* to reveal detailed information about the dinosaur's sensory abilities—was sold to the Field Museum.

1996 *How Dinosaurs Learned to Fly*, an animated film for children produced by the National Film Board of Canada, shows how dinosaurs avoided extinction by learning to fly.

1996 In an article in *Journal of Chinese Geology* (written in Chinese), Ji Qiang and Ji Shuan name and describe *Sinosauropteryx* ("Chinese reptilian wing"), the first feathered dinosaur to be reported. (The specimen was described in English two years later in *Nature*.) *Sinosauropteryx*, a turkey-sized theropod found by farmer and fossil enthusiast Li Yingfang in Liaoning Province in northeastern China in Early Cretaceous (125-million-year-old) deposits, was the first genus of non-avian dinosaur found with fossilized impressions of feather-like structures (the impressions were less than 0.2 inches long). When paleontologist Phil Currie saw the fossil, he told the *New York Times*, "I was bowled over." One slab of the fossil went to Chen Pei Ji in Nanjing (who announced the fossil at the annual meeting of SVP), and the counter slab was sold to Ji Qiang at the National Geological Museum in Beijing for $750. *Sinosauropteryx,* one of the smallest known non-avian theropods, had short arms and an unusually long tail. Some people later claimed that the filamentous structures around *Sinosauropteryx* were collagen fibers, not primitive feathers. China's feathered dinosaurs provided some of the most spectacular fossils of the 20th century, and were featured on a variety of shows, stamps, and greeting cards (e.g., by Beijing's Geological Museum of China in 1997). Numerous dinosaur fossils having feathers or integumentary filaments have been found near Liaoning.

1996 In *Carnosaur III* ("Terror Will Never Be Extinct!"), terrorists hijack several genetically engineered dinosaurs. A group of Americans is then sent to eliminate the indestructible dinosaurs. At one point, when a *Tyrannosaurus* is blown up, viewers can see the metal "skeleton" of the puppet used for the scene. The *Carnosaur* sequels included stock footage from *Dinosaur Island* and other movies that featured dinosaurs.

1996 In *Dinosaurs and the Bible*, David Unfred claims that *T. rex* ate melons, despite the fact that melons reached North America only 400 years ago. In 2006, Andy McIntosh and Bodie Hodge revived this claim by showing *T. rex* holding a watermelon in its mouth, with a caption saying, "*T. rex* originally ate vegetables" (despite the fact that watermelon is a fruit, not a vegetable). Other books by young-Earth creationists have shown *Dilophosaurus* eating leaves, *Compsognathus* eating berries, *Ceratosaurus* eating a banana plant, and *Deinonychus* eating fruit. All of these dinosaurs were theropods.

1996 In its foyer, the Natural History Museum of Los Angeles County unveils *Tyrannosaurus* and *Triceratops* facing off. The actual skull of *Tyrannosaurus* was too heavy for mounting, so it was displayed in a separate case.

1996 In *The Great Dinosaur Extinction Controversy,* Charles Officer and Jake Page attack the Alvarezes' claim that the K-T mass extinction was caused by an extraterrestrial impact. Officer and Page argued that "the Alvarez hypothesis has collapsed under the weight of accumulated geologic and other evidence to the contrary, as well as from an increasingly obvious absence of scientific evidence proffered in its support," "the impact theory has not stood up at all to expert scrutiny," "most of the 'science' performed by the Alvarez camp has been so inexplicably weak," "some skeptics have wondered if the entire affair was not, on the impact side, some kind of scam," and that "dinosaurs disappeared outside North America well before the Cretaceous came to an end. . . . And in North America, the dinosaur demise began as a gradual process. . . . The two last known dinosaur species were the herbivore *Triceratops* and the carnivore *Tyrannosaurus*." Many of these claims were challenged by paleontologists. However, if dinosaurs had become extinct before the K-T event, then the K-T boundary and the corresponding cosmic evidence would be irrelevant to dinosaurs' fate.

1996 In *The Origin and Evolution of Birds*, ornithologist Alan Feduccia rejects birds' dinosaurian ancestry and argues for a trees-down origin of avian flight. That same year, Feduccia—who believed that dinosaurs evolved from early archosaurs such as *Longisquama*—was quoted in *Science* as saying, "The theropod origin of birds will be the greatest embarrassment of paleontology in the 20th century." Feduccia's claim continues to be used in young-Earth creationists' attacks on evolution.

1996 Japanese scientists, Carl Baugh, and Joe Taylor excavate a three-toed dinosaur track and an adjacent "man track" near Glen Rose, Texas, prompting the Japanese workers to later release a documentary "verifying" that humans lived with dinosaur. Also in 1996, NBC's hour-long *The Mysterious Origins of Man*, a special narrated by Charlton Heston, presented "controversial evidence that could rewrite man's history." The special promoted claims by Baugh and Don Patton that humans lived with dinosaurs, a claim based on alleged "man tracks" found beside dinosaur tracks at the "Taylor Site" near Glen Rose, Texas. No paleontologists were interviewed for the program. Although *The Mysterious Origins of Man* implied that the "Burdick Track" and other "man tracks" near Glen Rose prove that humans lived with dinosaurs millions of years ago, both Baugh and Patton rejected this conclusion, instead believing that humans lived with dinosaurs just a few thousand years ago. Claims about "man tracks" by Baugh, Patton, and others have been rejected by numerous studies, including those by several young-Earth creationists such as Ken Ham (AiG) and Don Batten (Creation Ministries International).

1996 Robert Bakker's self-illustrated *Raptor Red* describes a year in the life of a female *Utahraptor* and several of her kin in the Early Cretaceous of North America. Two years earlier, Bakker had written about a fictitious *Utahraptor* in "Cretaceous Park" in *Earth* magazine.

1996 Sculptor David A. Thomas delivers his 17-foot-long *Utahraptor ostrommaysorum* to Utah's Field House of Natural History.

1996 The *Dechronization of Sam Magruder*—a novel written by American paleontologist George Gaylord Simpson between 1970 and 1984, and discovered by his daughter Joan Simpson Burns—describes a 22nd-century time-traveler (Sam Magruder) who has been accidentally transported to the Cretaceous. During his adventure, Magruder confirms that dinosaurs are clumsy, lumbering animals that have no memory; they quickly forget about their prey when the prey hide. Stephen Jay Gould described Simpson as ". . . unquestionably, the greatest vertebrate paleontologist of the 20th century, perhaps the greatest of all time."

1996 The educational film *Before La Brea* prepared for "Dinosaur Theatre" explains why there are no dinosaurs at Rancho La Brea.

1996 The first stegosaur tracks are reported in the Morrison Formation from near Price, Utah. These tracks were hind footprints.

1996 The Institute for Creation Research (ICR) publishes *Noah's Ark: A Feasibility Study*, by young-Earth creationist John Woodmorappe ([b. 1954], the pen-name for Jan Peczkis). Woodmorappe argued that Noah and his family could have fed all 16,000 animals, including several dinosaurs, aboard the Ark if they had used modern production techniques. One reviewer evaluated Woodmorappe's claim that Noah trained all of the animals to defecate and urinate on demand into buckets this way: "This, of course, makes Noah the greatest animal trainer in history." In 1999, Woodmorappe's *The Mythology of Modern Dating Methods* rejected radiometric dating.

1996 The low-budget movie *Dinosaur Valley Girls* (which is similar to *Dinosaur Island* [1994]) depicts how a time-traveling, fading film-star meets dinosaurs and bikini-clad cavewomen who try to outrun dinosaurs in a prehistoric world. The anatomically correct dinosaurs included *Allosaurus, Stegosaurus*, the armored sauropod *Saltasaurus*, the horned meat-eater *Carnotaurus*, and—for the first time in a movie—the herbivore *Camptosaurus. Allosaurus* was the dinosaur that got the most screen time.

1996 The Museum of Ancient Life opens in Lehi, Utah, and features one of the largest collections of mounted dinosaurs in the world.

1996 The UK Dinosaur Society issues a Christmas card showing duckbilled dinosaurs eyeing a "Christmas Star" that is a meteor streaking toward Earth.

1996 There are now 787 named dinosaurs. During the next four years, this number increased to 895, and to 1,415 in 2013.

1996 Thieves excavate the only known track of a stegosaur in Australia. The 66-pound fossil was recovered a year later when the thieves tried to sell it in Asia.

1996 Thieves ransack Utah's Cleveland-Lloyd Quarry.

1996 YPM is featured in *The Simpsons* episode titled "Burns, Baby Burns."

1996 *Zuniceratops,* the earliest-known ceratopsian to have eyebrow horns, is discovered by eight-year-old Christopher Wolfe (son of paleontologist Douglas Wolfe) in Middle Cretaceous sediments of New Mexico.

1997 After seeing *The Lost World* (the sequel to *Jurassic Park*), paleontologist Jack Horner—the model for the movie's "Dr. Grant"—notes, "There is no mystery left. Now that we know exactly how they looked, sounded, and moved, the fascination [with dinosaurs] may be lost." (Dr. Grant's initial reaction to seeing the dinosaurs was to exclaim "We're out of a job.") The movie featured several dinosaurs-who-eat-humans-who-won't-listen-to-warnings, as well as dinosaurs who defended their young.

1997 An expedition of American and Argentinian paleontologists exploring an extinct volcano in Patagonia's Neuquén Province discovers a vast nesting ground for sauropods that includes several intact 6-inch eggs. Some of the 80-million-year-old eggs containing embryos were the first unequivocal sauropod embryos ever found, the first dinosaur embryos ever found in the Southern Hemisphere, and the first fossils of dinosaur embryos having skin ever found. Several giant dinosaurs (e.g., *Argentinosaurus, Futalognkosaurus*) have been found in Patagonia.

1997 An international team of paleontologists returns from China and announces that it found the first internal organs of dinosaurs, a dinosaur fossil containing a mammal that it had just eaten, and a theropod with a fiber-like crest.

1997 Chicago's Field Museum teams with McDonald's and Walt Disney World Resorts to place the winning bid of $8.4 million for *T. rex* "Sue," a tyrannosaur discovered in 1990 by Sue Hendrickson in South Dakota. This was the largest amount ever paid for a fossil. "Sue" (FMNH PR2081) is now a primary attraction at the Field Museum.

1997 Creationist Michael Oard claims that dinosaurs became extinct in a global flood.

1997 Dinosaur tracks have been found at more than 1,500 places around the world.

1997 Donald Glut publishes the first of three volumes (1997, 1999, 2002) of *Dinosaurs: The Encyclopedia.*

1997 Fossils of *Tyrannosaurus imperator* are unearthed at Fort Peck, Montana. Its 52.4-inch pubis bone was the largest ever, far exceeding that of *Giganotosaurus* (46.5 inches).

1997 Geologist Erik Kvale and his colleagues discover three-pronged, clawed dinosaur footprints in Wyoming. This site is now known as the Red Gulch Dinosaur Tracksite.

1997 In "The Land That God Forgot," Nicholas DiChario (b. 1960) and Jack Nimershein (b. 1949) recount how God tells Pope John Paul IV in 2212 to download his consciousness into a computer. The pope then tries to convert dinosaurs to Catholicism before an asteroid impact. In the impact's aftermath, small animals gather and listen attentively when the pope discusses religion and philosophy with (and reads the Bible to) a *T. rex* named "Rexanne."

1997 In a special edition of *Star Wars Episode IV: A New Hope*, George Lucas (b. 1944) adds *Brachiosaurus*-like animals based on models used in *Jurassic Park*.

1997 In Bud Sparhawk's "Fierce Embrace," two paleontologists falling in love find a pair of tyrannosaurs who died while copulating in a "fierce embrace."

1997 In Marc Cerasini's novel *Godzilla 2000,* Godzilla and Rodan (a mutated *Pteranodon*) battle monsters produced by a meteor shower.

1997 In Maureen McHugh's (b. 1959) "Down on the Farm," Grace Sabiston—who raises herds of *Oviraptor*—faces losing her farm. A cunning lawyer saves the day when he shows that the oviraptors were genetically enhanced, not genetically engineered.

1997 In Niger's Ténéré Desert, Paul Sereno and his colleagues discover *Suchomimus tenerensis* ("crocodile mimic from the Ténéré"), a sail-backed spinosaur from the Early Cretaceous. *Suchomimus tenerensis* was more than 30-foot-long and 12-foot-high at its hips.

1997 In Susan Schwartz's "Drawing Out Leviathan," the dinosaur-like Leviathan cannot get aboard Noah's Ark because she's the last of her kind.

1997 In *T. rex and the Crater of Doom*, Walter Alvarez notes that "impact as a geologic process . . . must be recognized as a rare but significant kind of event, and evidently the cause of at least the K-T mass extinction . . . can volcanism be dismissed from the list of catastrophic events with global effects? Not yet. . . . I would have dismissed the apparent age match between the Deccan Traps and the K-T impact-extinction event as a strange coincidence, if it were not that a second such coincidence has turned up. . . . Recently, Paul Renne . . . has obtained reliable dates on both the Siberian Traps and the Permian-Triassic boundary . . . they are indistinguishable. A good detective shouldn't ignore even a single coincidence like the K-T-Deccan match in timing, and when it is bolstered by a second coincidence . . . it just has to be significant. . . . Right now . . . there is an intriguing mystery, some obviously significant clues, and nobody has any idea what the explanation will be."

1997 In *The Complete Dinosaur,* while contemplating the scene in the movie *Jurassic Park* in which *T. rex* eats the lawyer, Michael Brett-Surman (b. 1950) and James Farlow ask, "How many lawyers would it take to properly feed a captive *T. rex*?" The answer? 73 lawyers per year if *T. rex* was cold-blooded, and 292 if *T. rex* was warm-blooded. Surman wrote several books (e.g., *The Complete Dinosaur, The World of Dinosaurs*, and *Dinosaur Field Guide*) and was a senior consultant for dinosaur stamps issued by the U.S. Postal Service.

1997 In *The Mistaken Extinction*, Lowell Dingus and Timothy Rowe argue that dinosaurs exist today as birds.

1997 In western Colorado, 14-year-old Stephanie Willen finds the first dinosaur eggs from the Mygatt-Moore Quarry. That same year in Montana, 15-year-old Sherri Flamand found the most complete maiasaur fossil known.

1997 Michael Crichton's novel *The Lost World* is made by Steven Spielberg into the movie *The Lost World: Jurassic Park*. There are numerous differences between the two works, with the biggest being that in the movie, *T. rex* attacks San Diego (this was added when Spielberg was inspired by the attack of *Brontosaurus* on London in the 1925 adaptation of Conan Doyle's *The Lost World*). Because of this attack, the existence of Isla Sorna is revealed, thereby setting up *Jurassic Park III*. After a record-setting opening weekend, *The Lost World: Jurassic Park* grossed $618 million worldwide, which was almost $300 million less than the original *Jurassic Park*.

1997 Michael Kohl and John McIntosh edit and release *Dinosaurs in the Old West: The Field Journals of Arthur Lakes*.

1997 MOR's exhibit "Dinosaur Families" opens at the Field Museum. The exhibit featured mechanical dinosaurs and casts of dinosaurs and their eggs.

1997 Paleontologist Larry Walker finds several fossils of what would later be named *Falcarius utahensis* in the Utah desert. Walker secretly excavated the public land without a permit, and in 2002 was indicted for selling claws for $15,000. The 4-meter-long, 126-million-year-old (Early Cretaceous) *Falcarius utahensis,* which was described in 2004, was an intermediate between predatory dinosaurs and herbivorous therizinosaurs; it had smaller meat-cutting teeth relative to leaf-shredding teeth, and a larger gut to digest the plants it ate. The discovery of *Falcarius utahensis* suggested that herbivorous dinosaurs evolved from velociraptor-like ancestors.

1997 Paleontologist Paul Sereno is named one of *People* magazine's 50 Most Beautiful People.

1997 *Pleurocoelus* is named the Lone Star State Dinosaur of Texas. In 2007, these fossils were reevaluated and assigned to *Paluxysaurus* (a new genus of sauropod); two years later, *Paluxysaurus* became the Lone Star State Dinosaur of Texas.

1997 R. E. H. Reid revives earlier claims (e.g., of A. J. de Ricqlés in 1974) that dinosaurs had an intermediate metabolism not found in modern organisms (i.e., dinosaurs have "no true modern physiological counterparts").

1997 The Field Museum's exhibit of "*Archaeopteryx*: The Bird that Rocked the World" is the first time *Archaeopteryx* is shown outside of Europe.

1997 The front page of the *New York Times* announces the discovery of feathered dinosaurs with the headline, "In China, a Spectacular Trove of Dinosaur Fossils Is Found." The article reported a press conference called by a team of American and European paleontologists who described feathered dinosaurs the year before in China's Liaoning Province. No Chinese scientists were invited to the press conference.

1997 The movie *The Lost World: Jurassic Park* debuts as a home-video.

1997 The U.S. Dinosaur Society is disbanded.

1997 Three years before displaying the entire skeleton of *T. rex* "Sue," the Field Museum displays one of her 12-inch-long teeth.

1997 To coincide with the release of Steven Spielberg's *The Lost World: Jurassic Park* (the sequel to 1993's *Jurassic Park*), the U.S. Postal Service issues 219 million "The World of Dinosaurs" stamps (Figure 49) featuring the artwork of James Gurney (b. 1958). Gurney, the author of the popular *Dinotopia* series of books, consulted Jack Horner and other paleontologists as he was designing the stamps, which included images of the carnivore *Ceratosaurus* ("horned reptile"), the herbivore *Camptosaurus* ("bent lizard"), *Camarasaurus* ("chambered lizard"), *Brachiosaurus* ("arm lizard"), *Stegosaurus* ("roof lizard"), *Allosaurus* ("different lizard"), the coelurosaur *Daspletosaurus* ("frightful lizard"), and the hadrosaur *Corythosaurus* ("Corinthian helmet lizard"). The sheets of fifteen

Figure 49 In 1997, the U.S. Postal Service issued "The World of Dinosaurs," a set of fifteen 32¢ stamps featuring dinosaurs from the Cretaceous of Montana (75 million years ago) and the Jurassic of Colorado (150 million years ago). The stamps, which were designed by artist James Gurney (b. 1958), were issued at Dinosaur Valley Museum in Grand Junction, Colorado. In 2013, Australia Post released a set of six postage stamps featuring Australia's "Age of Dinosaurs." Those stamps were also designed by Gurney. (AP Photo/USPS)

32¢ stamps depicted two scenes—one from Montana during the Cretaceous (approximately 75 million years ago) and another from Colorado during the Jurassic (approximately 150 million years ago). "First Day of Issue" ceremonies for the stamps were held at the Dinosaur Valley Museum in Grand Junction, Colorado. The stamps were popular; collectors saved more than 38 million of the stamps. For comparison, the most popular stamp of all time was a 29¢ stamp issued in 1993 featuring Elvis Presley; collectors have saved more than 124 million of those stamps.

1997 An exhibit of 59 paintings from James Gurney's (b. 1958) *Dinotopia* opens in Three Rivers, Michigan.

1997 *Turok, Dinosaur Hunter* is released for the *Nintendo 64* game console. Several *Turok* sequels followed, as did numerous books, DVDs, and movies.

1997 While working in the Sahara Desert, Paul Sereno and his coworkers discover a 95 percent–complete skeleton of *Jobaria*, a sauropod named after a local legendary creature named "Jobar." The long-necked, mid-Jurassic dinosaur was a primitive sauropod that may have been able to rear up on its hind legs. Two years later, Sereno included a skeletal reconstruction of *Jobaria* in his *Science* article titled "The Evolution of Dinosaurs."

1998 A headline in the October 30 issue of the London *Times* announces "Rock Solid Proof That Comet Killed Dinosaurs." The story described how two scientists from the Scripps Institute of Oceanography had found a chromium spike at the K-T boundary, thus suggesting an extraterrestrial impact.

1998 A *Nature* cover-story titled "A Dinosaur with Feathers" by Chen Pei Ji and colleagues describes a reconstructed *Caudipteryx* ("tail feather"), a peacock-sized dinosaur having a tiny head, a long striped tail plume, and four small teeth at the end of its beak. *Caudipteryx*, which was found in early Cretaceous deposits of China's Liaoning Province, was the first dinosaur to show unambiguous evidence for modern feathers on a non-avian dinosaur. These feathers, especially those along the backs of its arms, lacked specializations for flight. Berkeley paleontologist Kevin Padian noted, "The work of Ji *et al.* should lay to rest any remaining doubts that birds evolved from small coelurosaurian dinosaurs."

1998 Anthony and Christina Bush open Noah's Ark Zoo Farm, a hands-on zoo and working farm just outside of Bristol, United Kingdom. The Bushes argue that dinosaurs roamed Earth between the time that God created Earth (100,000 years ago) and Adam and Eve (21,000 years ago). More than 130,000 people visit Noah's Ark Zoo Farm each year.

1998 Bruce Rothschild and colleagues describe metastatic cancer in Jurassic dinosaurs found in Colorado.

1998 Canadian paleontologist Phil Currie (b. 1949) uses old photographs to locate the site (in what is now Alberta's Dry Island Buffalo Jump Provincial Park) at which Barnum Brown found the hind feet of *Albertosaurus*. Brown's fossils languished in AMNH for years before being prepared.

1998 *Capitalsaurus* becomes the "state" dinosaur of the District of Columbia (even though the District of Columbia is not a state).

1998 The Mt. Blanco Fossil Museum opens in Crosbyton, Texas, and claims that it is "digging up the facts of God's Creation: One fossil at a time." The museum, which began in 1986 as a smaller museum, exhibits "man tracks" from near Glen Rose, Texas, which it claims show that humans lived with dinosaurs.

1998 *Coelophysis* becomes the second dinosaur to go into outer space when a *Coelophysis* skull from CMNH is aboard the launch of Space Shuttle *Endeavour* (STS-89).

1998 Congress recognizes that agencies managing federal land must protect public resources, including "fossilized paleontological specimens, which provide valuable clues to the Earth's history."

1998 Donations totaling $260,000 help AMNH restore several of Charles Knight's paintings following 25 years in storage.

1998 Douglas Wolfe and James Kirkland describe *Zuniceratops christopheri*, one of the few North American dinosaurs from the Middle Cretaceous.

1998 Hal Hellman's *Great Feuds in Science* describes the Cope-Marsh feud as one of science's "ten liveliest debates ever."

1998 In *Darwin's Error*, German engineer Hans-Joachim Zillmer (b. 1960) repeats much-criticized (and long-rejected) claims of human footprints beside those of dinosaurs.

1998 In Greg Bear's lighthearted sequel to *The Lost World*, "Dinosaur Summer," scientists in the 1940s return circus dinosaurs to Venezuela's El Grande Plateau.

1998 In Saskatchewan, Canada, Royal Saskatchewan Museum paleontologists find the largest known coprolite from a theropod. The 65-million-year-old fossilized dung, which was 17 inches long and 6 inches high, probably came from *T. rex*. Today, the museum displays a replica of the fossil.

1998 In *Taking Wing*, Pat Shipman describes the "Berlin specimen" of *Archaeopteryx*—a primitive bird that is a mosaic of avian and reptilian traits—as "more than a stony record of an extinct species. It is an icon—a holy relic of the past that has become a powerful symbol of the evolutionary process itself. It is the First Bird."

1998 In the book *The Last Dinosaur Book*, William J. Thomas Mitchell (b. 1942) claims that Charles Knight's portrayal of the confrontation between *T. rex* and *Triceratops* (Figure 19) is an allegory of unbridled capitalism, that Osborn's depictions of dinosaurs were linked "to questions of male potency" and "the phallic overtones of 'bones,'" and that the green color of Knight's creations may be "an emblem of the white man's burden" while representing "both sides of the 'color line.'" Mitchell believed that "facts" about dinosaurs cannot be separated from the cultural icons that they have become.

1998 In the IMAX film *T-Rex: Back to the Cretaceous*, actor Tuck Milligan (b. 1985) plays Charles Knight, who travels back in time to paint dinosaurs. The movie, which featured a time-traveling *T. rex*, also included actor Laurie Murdoch portraying Barnum Brown.

1998 In what is described as "a unique approach to the topic of dinosaurs," Ken Ham's *The Great Dinosaur Mystery Solved!* claims that "dinosaurs were created on day six" of creation week (the same day that Adam and Eve were created) "around 6000 years ago." Ham also claimed that "dinosaurs are only a mystery if you accept the evolutionary story of their history," that "Christians can authoritatively state that dinosaurs did live with people," and that "the evolutionist teachings on dinosaurs" underlie social ills such as abortion, homosexual behavior, pornography, suicide, and lawlessness. Ham has received several honorary degrees, including from Temple Baptist College (1997) and Liberty University (2004).

1998 Luis Chiappe and colleagues report the first known embryos of sauropods, which were found in Late Cretaceous sediments in Patagonia, Argentina. Two years later, Chiappe and others reported the discovery of 200 intact eggs in 25 square meters of the area.

1998 *Nature* reports a 9-inch-long specimen—"little more than a hatchling"—of *Scipionyx samniticus,* a new genus of theropod. The early Cretaceous specimen, which had been collected in the 1980s near Naples, Italy, by Giovanni Todesco, included small patches of tissue along its tail and upper chest.

1998 Paul Taylor's book *The Great Dinosaur Mystery and the Bible* claims that dinosaurs lived with humans, and that "the mystery about dinosaurs is easy to solve with the help of God's word." *The Great Dinosaur Mystery and the Bible* was endorsed and promoted by Henry Morris, president and founder of ICR.

1998 Peter Larson and his colleagues install *T. rex* "Stan" at the Smithsonian. Larson was one of the first paleontologists to study bone pathology in *T. rex*.

1998 The first-named stegosaur tracks are reported near Utah's Arches National Park. These tracks were made by the stegosaur's front and back legs. The shapes of these tracks matched the predictions made about them in 1990.

1998 The most complete skeleton of *Haplocanthosaurus* is found northeast of Vernal, Utah. This fossil is now exhibited in the UFHNH.

1998 Thomas Holtz (b. 1965) argues that, based on foot and leg anatomy, tyrannosaurs and their relatives were the fastest large theropods. That same year, P. Christiansen proposed that *T. rex* could move at a maximum speed of 25 miles per hour (about the speed of a human sprinter). Tracks at the "Park North East" site in Glen Rose, Texas, were made by a dinosaur moving at about 28 miles per hour.

1998 University of Chicago paleontologist Paul Sereno and his coworkers report *Suchomimus tenerensis* ("crocodile mimic"), a large theropod with a crocodile-like mouth. *Suchomimus tenerensis* was found in the 112-million-years-old Elrhaz Formation in Niger's Ténéré Desert.

1998 The Jurassic Foundation is established to provide money for dinosaur paleontology. The foundation has distributed more than $2 million, making it one of the largest private sources of money for dinosaur-related research.

1999 Sixteen-year-old Tyler Lyson finds a mummified hadrosaur in the badlands of North Dakota. Subsequent analyses published in 2010 showed that "Dakota" had scaly, double-layered skin, that its tail contained more muscle than previously suspected, and that the 67-million-year dinosaur could have run more than 25 miles per hour (i.e., faster than *T. rex*).

1999 A symposium at Yale University honors John Ostrom for his revolutionary ideas about dinosaurs. Ostrom's work initiated the "dinosaur renaissance."

1999 At the urging of paleontologist Paul Sereno, SVP begins awarding its John J. Lanzendorf PaleoArt Prize to recognize achievement in several categories of paleoart (e.g., two-dimensional, three-dimensional, scientific illustration), which it describes as "one of the most important vehicles for communicating discoveries and data among paleontologists, and is critical to promulgating vertebrate paleontology across disciplines and to lay audiences." The Lanzendorf PaleoArt Prize is now supplemented by the National Geographic Digital Modeling and Animation Award; these are the only art-awards presented "by the scientific community that specifically recognize accomplishments in paleontological visualization." In its inaugural year (2000), there were more than 250 entries, and the winners included Michael Trcic for his bronze statue of *Jobaria*.

1999 Chinese paleontologist Xiao-Chun Wu and his colleagues discover a well-preserved feathered dinosaur that they nickname "Dave the fuzzy raptor," after a character in a Cheech and Chong routine. This birdlike, non-avian theropod was later assigned to the genus *Sinornithosaurus*. (The next "fuzzy" discovery would be nicknamed "Chong.") *S. millenii* ("Chinese bird lizard"), a meat-eating dinosaur found in volcanic ash deposited 120 million years ago, was the first dromaeosaur having a coat of downlike "proto-feathers" conceivably used for insulation, but not for flight. This discovery supported claims that birds evolved from dinosaurs.

1999 Construction workers building a road near Tibet's Bangda Airport unearth several 160-million-year-old sauropod tracks. Some local residents claimed the tracks had been left by the "Deity of the Mountain" that was scared away by the construction.

1999 Great Britain issues its "British Scientists' Achievements" postage stamps, one of which (26 pence) features Charles Darwin and *Archaeopteryx*. Darwin has been featured on stamps throughout the world, as has *Archaeopteryx* (38 countries as of 2010).

1999 In "Dinosaurs and the Bible," Ken Ham claims that dinosaurs, Adam, and Eve were made on the sixth day of creation, "there could not have been any animal fossils (and no dinosaur bones) before sin," "dinosaurs must have been on the Ark," "the dinosaur fossils which were formed as a result of [the] Flood were probably formed about 4,500 years ago, not millions of years ago," "dinosaurs—all the different kinds—must have lived alongside of people after the Flood," dinosaurs did not evolve, humans lived with dinosaurs 6,000 years ago, and no dinosaurs died until Adam sinned. In 2007, Ham added that dinosaurs are mentioned in the Bible (e.g., Genesis 1:21, Psalm 74:13, Job 41, Isaiah 30:6).

1999 In a letter to his followers, ICR's John Morris asks "Wouldn't it be wonderful to remove the stumbling block of radioisotope dating?" Soon thereafter, ICR announced a five-year plan promising to discredit radiometric dating.

1999 In Michael Swanwick's "Riding the Gigantosaurus," the mind of a business executive having terminal cancer is transferred into a theropod. The theropod celebrates his first kill and has sex with three females (Eve, Slut, and Scarface). He is then rescued by scientists and becomes more human than ever before.

1999 In *Science*, Paul Sereno argues that dinosaurs dominated mammals because of opportunistic replacement rather than competitive displacement. Sereno noted that there was a 15-million-year gap between the initial radiation of dinosaurs and their rise to dominance.

1999 In the most accurate skeletal reconstruction of *Stegosaurus* ever completed, paleoartist Tracey Ford orients the tail spikes outward and horizontal.

1999 José Bonaparte describes *Lessemsaurus,* a large sauropod from the Late Triassic whose name honors writer "Dino Don" Lessem.

1999 Mary Higby Schweitzer and her colleagues report that fossil feathers of *Shuvuuia deserti* test positively for beta-carotene, a protein found in birds' feathers.

1999 *National Geographic* announces "Archaeoraptor," a fossil of a feathered bird from China's Liaoning Province having a long dinosaur-like tail. The fossil was bought at an Arizona gem show for $80,000 by the Dinosaur Museum of Blanding, Utah, and was described as a "missing link between terrestrial dinosaurs and birds that could actually fly. . . . This fossil is perhaps the best evidence since *Archaeopteyx* that birds did, in fact, evolve from certain types of carnivorous dinosaurs. . . . We can now say that birds are theropods just as confidently as we say that humans are mammals." "Archaeoraptor" was said to be a feathered bird having a long dinosaur-like tail. Soon thereafter, in an article titled "Feathers for *T. rex*?" that was accompanied by a press conference and much fanfare, the fossil was announced by artist Christopher Sloan of *National Geographic.* However, scientists later showed the fossil was a fake consisting of 88 pieces mounted on shale and glued into place with builder's grout; the head and upper body were from a primitive fossil bird (*Yanornis martini*) and the tail from *Microraptor zhaoianus,* a small, gliding theropod, thus prompting *U.S. News & World Report* to dub "Archaeoraptor" "The Piltdown Chicken." Although *Science* and *Nature* had rejected papers describing "Archaeoraptor" because of suspicions it was a hoax and smuggled illegally out of China, Bill Allen (editor of *National Geographic*) claimed to be angry to learn that the fossil was a fraud. Scientists have found many examples of feathered dinosaurs, but creationists continue to use the "Archaeoraptor" scandal to try to question evolutionary theory and scientists' claim that birds evolved from dinosaurs. The scandal highlighted the often-sloppy work of popular magazines when pursuing a story, as well as illegal fossil-sales in China and elsewhere.

1999 *Nature* magazine announces the discovery of *Beipiaosaurus* ("Beipiao lizard"), a 7-foot-long dinosaur from the Early Cretaceous near the city of Beipiao in China's Liaoning Province. The type-specimen (*B. inexpectus*) was named for "the surprising features of this animal," which included features similar to those of *Sinosauropteryx* as well as longer, simpler, and unique feathers (found in a second specimen discovered in 2009). *Beipiaosaurus,* which had a toothless beak, was one of the largest feathered dinosaurs; its feathers, which could not be used for flight, were probably used for insulation or mating.

1999 One hundred years and six days after the discovery of *Diplodocus carnegii,* a life-size (84 feet long, 24 feet tall) model of "Dippy" the *Diplodocus carnegii* made of gel coat and fiberglass is erected outside the northwest corner of CMNH (Figure 50). This model, a fleshed-out replica of the 1907 model inside CMNH, had its tail posed in the air.

1999 Paul Sereno and his colleagues discover *Nigersaurus taqueti,* an elephant-sized sauropod that roamed in what is now Niger. Sereno and his colleagues later used 3-D imaging to show that the more than 50 columns of teeth were replaced faster than those of any other dinosaur, with up to 10 replacement teeth stacked up behind each tooth. As Sereno noted, "Among dinosaurs, *Nigersaurus* sets the Guinness record for tooth replacement." The head of *Nigersaurus* faced down, at a 90° angle to the neck.

1999 Researchers in North Dakota discover a mummified hadrosaur so well preserved that they will eventually be able to calculate its muscle mass.

Figure 50 In 1999, 100 years after workers from the Carnegie Museum of Natural History (CMNH) found *Diplodocus carnegii,* a life-size model of "Dippy" was unveiled outside of CMNH. "Dippy," which is a Pittsburgh landmark, is a 3,000-pound, fiberglass replica of the skeleton of *Diplodocus carnegii* inside the museum. Andrew Carnegie's distribution of casts of *D. carnegii* to museums throughout the world (e.g., Figure 18) led to many people referring to CMNH as "The House That Dippy Built." (Courtesy Randy Moore)

1999 Robert Nicholls's painting "Impending Doom" shows a group of *T. rex* with a comet hurtling to Earth in the background.

1999 The Alaska Museum of Natural History opens in Anchorage, and includes the hadrosaur "Lizzie," Alaska's oldest known dinosaur.

1999 The Discovery Channel's *Dinosaur Attack!* prominently features the famous dinosaur-tracks near Glen Rose, Texas.

1999 The falling cost of computer animation leads to several documentaries about dinosaurs, such as *Walking with Dinosaurs* (1999), *When Dinosaurs Roamed America* (2001), and *Animal Armageddon* (2009). Despite these lower costs, *Walking with Dinosaurs* was the most expensive documentary series (on a per-minute basis) ever made to that time. The six-part BBC series, for which artist Luis V. Rey (b. 1955) was the advisor, won three Emmy Awards and was watched by nearly one-third of the population of the United Kingdom when it aired. In 2013, *Walking with Dinosaurs* was made into an $80 million movie; its hero was Patchi the *Pachyrhinosaurus*.

1999 The website for the antievolution, pro-Christianity Genesis Park is unveiled "to showcase the evidence that dinosaurs and man were created together." The website promotes ideas of young-Earth creationists such as Carl Baugh, Kent Hovind, AiG Duane Gish, and ICR.

1999 While working in Carbon County, Montana, J. D. Harris and Peter Dodson discover the 45-foot-long basal sauropod *Suuwassea*. Harris and Dodson named *Suuwassea* in 2004.

1999 William Ables suggests that *T. rex* used infectious saliva to kill prey.

2000 "The World's Largest T-Rex" opens in Drumheller, Alberta, Canada. The 85-foot-tall dinosaur (about five-times taller than a real *T. rex*) included an internal stairway to an observation deck, from where visitors could view a 20-foot-tall *T. rex* (built in 1960 and relocated here in 1992) through the dinosaur's teeth.

2000 A partial skeleton of the "Bürgermeister-Müller specimen" of *Archaeopteryx* is found and loaned to the Bürgermeister-Müller Museum in Solnhofen. The specimen, which consists of a single wing, is popularly known as the "chicken wing" specimen of *Archaeopteryx*.

2000 A specimen of the small, herbivorous *Thescelosaurus* named "Willo" (unearthed by Michael Hammer in northwestern South Dakota in 1993) is reported by *Science* to include remnants of a fossilized four-chambered heart. Many scientists subsequently disputed this claim, and in 2011 another group of paleontologists reported that the "heart" was merely a piece of iron-rich rock. This specimen is now displayed at the North Carolina Museum of Natural Sciences.

2000 A team of NMNH scientists produces the first-ever digital reconstruction of an entire dinosaur. This reconstruction of *Triceratops* was accompanied by a short video and a remounted version of the museum's "real" *Triceratops,* which in 1905 had become the first mounted horned dinosaur.

2000 Academic Press releases *Dinosaur Imagery* featuring the work of John Lanzendorf (b. 1946). That same year, as part of the opening of the display of *T. rex* "Sue," Chicago's Field Museum of Natural History displayed 70 of Lanzendorf's sculptures and paintings; this was the first time that such art had been exhibited in a major museum. The following year, Lanzendorf sold his dinosaur collection to the Children's Museum of Indianapolis, whose "Dinosphere" (established in 2004) is one of the largest exhibits of dinosaurs in the world. The Dinosphere's exterior features an *Alamosaurus* and her two babies crashing through the museum's wall. In Evansville, a 20-foot-tall "Millie the Dinosaur"—a *T. rex*—points the way to Children's Museum of Evansville.

2000 Artist Gregory Paul suggests that Charles Knight often painted misty scenes "possibly because of his poor long-range vision."

2000 As part of its "Celebrate the Century" commemoration, the U.S. Postal Service issues a 33¢ stamp honoring *Jurassic Park*.

2000 Commercial fossil collector Nathan Murphy finds a mummy of a 77-million-year-old duckbill *Brachylophosaurus,* with more than 90 percent of its skin intact, in the Judith River Formation in Phillips County, Montana. The fossil, which was named "Leonardo" because of some graffiti on a nearby rock face, was later declared the world's best-preserved dinosaur by *Guinness Book of World Records*. The dinosaur, whose gut was infected by several parasites, is now featured in the Phillips County Museum in Malta, Montana, along with "Elvis" the *Brachylophosaurus*, one of the best articulated dinosaur skeletons every found. In 2009, Murphy served time in jail for stealing fossils.

2000 Hundreds of believers gather in northern Kentucky to dedicate land for the future site of AiG's $27-million Creation Museum and Family Discovery Center. According to Ken Ham, the CEO of AiG, this museum will undo "the damage done . . . when Clarence Darrow put William Jennings Bryan on the stand" at the Scopes Trial. John Whitcomb Jr., the coauthor of the influential *The Genesis Flood*, promised that if the museum ever dishonored God, God would destroy it. The Creation Museum relies heavily on dinosaurs to lure visitors and promote the anti-science religion of young-Earth creationism. During its first three years of operation, the Creation Museum attracted 1.2 million customers.

2000 In Eric Garcia's *Anonymous Rex* and its prequel *Casual Rex* (2001)—both of which are spoofs of *Jurassic Park*—dinosaurs survive the K-T extinction and blend with human society. Examples of dinosaurs masquerading as humans have included Elvis Presley, J. Edgar Hoover, and the first woman to burn her bra (because she "had dealt with one strap too many"). Both novels are told from the view of Vincent Rubio, a *Velociraptor* who is a private investigator who enjoys deception. A movie version of *Anonymous Rex* was released by SciFi Pictures in 2005.

2000 In *Radioisotopes and the Age of the Earth*, Larry Vardiman of ICR claims that "observation and theory should always be subservient to a proper understanding of the Word of God." ICR claims that humans lived with dinosaurs.

2000 In the Sahara, Paul Sereno and his colleagues discover *Rugops primus* ("first wrinkle face"). This dinosaur's similarity to South American fossils suggested that Africa separated from Gondwana more recently than previously thought.

2000 In *The Scientific American Book of Dinosaurs*, Thomas J. Holtz claims, ". . . the creatures that most immediately come to mind when the word 'dinosaur' is mentioned are the giant sauropods." However, that same year, Robert Bakker described *Tyrannosaurus* as "the most popular dinosaur among people of all ages, all cultures, and all nationalities," and Philip Currie (curator of dinosaurs at RTM) described *T. rex* as "the standard against which other dinosaurs are measured."

2000 In Thomas Hopp's *Dinosaur Wars*, dinosaurs on the moon avoid the K-T extinction, after which they return to Earth and fight humans. Two years later, Hopp's novel *Dinosaur Wars: Counterattack* reprised the eternal conflict between prehistoric life and modern man.

2000 Jack Chick releases *In the Beginning,* a small, popular booklet promoting young-Earth creationism and claiming that humans lived with dinosaurs. Chick argued, "There are photos of men's footprints walking next to dinosaur footprints. They were found in Glen Rose, Texas . . . God doesn't lie." When one of the characters is told that "that dinosaurs and man were both created on the same day [by God] roughly 6000 years ago" (because "that's what God's word tells us"), the character responds, "Man, that's too hard to swallow."

2000 John R. "Jack" Horner (Figure 43) discovers the fossilized *Tyrannosaurus* leg bone (MOR 1125, informally known as "B-rex") from which Mary Higby Schweitzer would later (in 2007) retrieve proteins.

2000 Mathew Wedel and colleagues describe *Sauroposeidon proteles* ("earthquake god lizard"), the largest known brachiosaur. The neck of this sauropod was 39 feet long. When the first fossils of this dinosaur were found in 1994, they were mistaken for petrified wood.

2000 McDonald's begins including one of eight talking dinosaur-toys in its "Happy Meals."

2000 Michael Benton claims that Robert Bakker's "vision of *Deinonychus* in life was so startling that the news spread . . . generating new excitement."

2000 *National Geographic* admits that "Archaeoraptor" is a chimera. Two years later, it described the "Archaeoraptor" affair as "a tale of misguided secrecy and misplaced confidence, of rampant egos clashing, self-aggrandizement, wishful thinking, naïve assumptions, human error, stubbornness, manipulation, backbiting, lying, corruption, and, most of all, abysmal communication." Despite the fact that scientists uncovered the fraud, creationists continued to use the incident to denounce the validity of evolution, the truthfulness of scientists, and the relatedness of birds and dinosaurs. As evangelist Charles Colson (of Watergate fame) noted, "What was supposed to be startling news has turned out to be yet one more example of the scientific community peddling fraud as scientific fact."

2000 North of Jordan, Montana, workers from MOR find a *T. rex* whose jacketed femur is too heavy to be airlifted by a helicopter to lab. Paleontologist Jack Horner (Figure 43) decided to break the bone, and sent his former student Mary Schweitzer some fragments of the broken bone. Schweitzer and her colleague Jennifer Wittmeyer determined that "B. rex" (named for MOR crew-chief, Bob Harmon) was pregnant because of the presence of medullary bone along the inside shafts of long bones in the hind limbs (medullary bone stores material for creating eggshells). These results indicated that non-avian dinosaurs had physiological responses to pregnancy.

2000 Paleontologists find the first eggs of *Deinonychus*.

2000 Pascal Godefroit and colleagues describe the giant *Charonosaurus jiayinensis*, the first tube-skulled hadrosaur from Asia.

2000 Phil Currie suggests that *T. rex* was a social animal that hunted in packs.

2000 Sheldon Johnson discovers dinosaur tracks in Early Jurassic sediments in St. George, Utah. These tracks later became part of the St. George Dinosaur Discovery Site at Johnson Farm, which is less than 100 yards from where Johnson made his initial discovery. The site includes the only known fossils that show clear impressions of dinosaurs' hands.

2000 Stephen Czerkas tells paleontologists that he and his wife made "an idiot, bone-stupid mistake" with the forged "Archaeraptor." Two years later, Czerkas's *Feathered Dinosaurs and The Origin of Flight* attacked other paleontologists' "blinding influences of preconceived ideas." The book named six species, five of which were disputed. In 2004, Czerkas's traveling exhibit having the book's name opened in South Dakota amidst claims that some of the fossils had been smuggled out of China illegally. Czerkas denied the claims.

2000 *T. rex* "Sue," discovered a decade earlier by Sue Hendrickson, is unveiled at the Field Museum in Chicago, and starts another wave of dinosaur-mania. "Sue" increased attendance at the museum by 50 percent; 1.6 million people visited the Field Museum in 1999, and 2.4 million visited in 2000. Peter Larson, who had excavated Sue, missed the unveiling.

2000 *T. rex* "Z-rex," which was discovered on private land in South Dakota in 1992, is put up for auction on *eBay* with an asking price exceeding $8 million. The fossil did not sell and was later bought by Graham Lacy, who renamed it Samson (in honor of the biblical figure of the same name). CMNH began preparing Samson in 2004, after which it was returned in 2006 to its owner.

2000 The National Science Foundation (NSF) reports that almost half of the general public in the United States believes that humans and dinosaurs coexisted.

2000 The prosauropod *Isanosaurus,* one of the earliest known dinosaurs and the earliest known sauropod, is discovered in northeastern Thailand.

2000 The two-pound, 2-foot-long "four-winged dinosaur" *Microraptor* is discovered in Early Cretaceous deposits of Liaoning, China. The discovery was later lauded as the "most exciting find on the question of the origin of flight." *Microraptor,* one of the first dinosaurs discovered having evidence of feathers and wings, was among the most abundant dinosaurs in the Liaoning ecosystem. *Microraptor* was described in 2003. In 2011, Jingmai O'Connor and her colleagues reported bones of birds in the abdomen of *Microraptor,* suggesting that the dinosaur ate birds. In 2013, another study found the remains of fish in the dinosaur's stomach, and concluded that *Microraptor* could spear fish like a kingfisher.

2000 The world learns of *Byronosaurus,* a small troodontid discovered in 1993 in the Gobi Desert.

2000 To commemorate Gideon Mantell's contributions to paleontology, the Mantell Monument is unveiled at Whiteman's Green, Cuckfield: "In a quarry on this site, Gideon and Mary Ann Mantell first discovered the 125-million-year-old fossilized remains of the dinosaur named Iguanodon by Mantell in 1825." The Mantell Monument is the only dinosaur monument in the United Kingdom.

2000 Walt Disney Pictures releases *Dinosaur*, which includes the first onscreen appearance of the two-horned *Carnotaurus* ("flesh-eating bull"). *Dinosaur* is about an orphaned *Iguanodon* named Aladar, who was raised by a family of lemurs. Like *The Land before Time*, *Dinosaur* featured talking dinosaurs.

2000 Working near DVSP outside Glen Rose, Texas, Alvis Delk and convicted murderer James Bishop of Stephenville, Texas, discover a theropod track in a pile of rocks along a creek flowing into the Paluxy River. The large track, in a 140-pound piece of limestone, later became the "Alvis Delk Cretaceous Footprint" when Delk found an 11-inch-long "man track" beside the dinosaur track (Figure 51). This track is displayed at (and is the symbol of) Carl Baugh's Creation Evidence Museum in Glen Rose, Texas, and is used by many young-Earth creationists to claim that humans lived with dinosaurs. In 2008, Kristi Delk—James Delk's daughter—expressed her doubts about the claims, noting, "I am skeptical, actually."

Figure 51 The "Alvis Delk Cretaceous Footprint" was discovered in 2000 by Alvis Delk and James Bishop near Glen Rose, Texas. Eight years later, Delk reexamined the track and found a human footprint beside the theropod track. Delk sold the print to young-Earth creationist Carl Baugh (b. 1936), who features it prominently in his Creation Evidence Museum's claims that humans lived with dinosaurs. (Courtesy Randy Moore)

2000 Godzilla reappears in the American-made movie *Godzilla 2000* as a lizard-like monster that attacks New York.

2000 Researchers in China's Liaoning Province uncover an intact feather-covered dromaeosaur, a dinosaur having many of the physical traits of a bird. Dromaeosaurs lived from the late Jurassic to the end of the Cretaceous.

2001 A juvenile *T. rex* named "Jane" is found in Montana's Hell Creek formation by a crew from the Burpee Museum of Natural History of Rockford, Illinois. Jane's healed skull wounds suggest that tyrannosaurs were bit in the face when fighting. Today, Jane (BMRP 2002.4.1, which is 50% complete) is exhibited in the Burpee Museum.

2001 After an article in *Nature* by Qiang Ji and his colleagues documents a theropod having proto-feathers, news outlets such as ABC News suggest that "maybe even mighty *T. rex* had feathers" and "maybe baby *Tyrannosaurus* looked something like a cute, fuzzy baby chick."

2001 Arkansas state representative Jim Holt (b. 1965) asks young-Earth creationist Kent "Dr. Dino" Hovind (b. 1953) to testify as an expert before the State Agencies and Governmental Affairs Committee to support Holt's bill requiring "that when public schools refer to evolution that it be identified as an unproven theory." In 1989, Hovind founded Creation Science Evangelism in Pensacola, Florida, and two years later opened his antievolution amusement park known as Dinosaur Adventure Land ("where dinosaurs and the Bible meet"). Hovind claimed that evolution is impossible because "whenever a farmer crossbreeds a cow he expects to get a cow, not a kitten." In 2006, after being convicted of 58 federal charges (including tax evasion), Hovind began serving a 10-year prison sentence.

2001 Dinosaur Isle opens at Sandown on the Isle of Wight, one of the most important sites of dinosaur remains in Europe. The museum, which is shaped like a pterosaur, was Britain's first purpose-built dinosaur museum. Dinosaurs discovered on the Isle of Wight include *Iguanodon, Hypsilophodon,* and *Baryonyx*.

2001 G. P. Burness and colleagues admit "the never-since-surpassed size of the largest dinosaurs remains unexplained."

2001 In *Man, Dinosaurs, and Mammals Together*, John Allen Watson uses evidence from phosphate beds of South Carolina to argue that humans lived with dinosaurs.

2001 In *Drawing Out Leviathan: Dinosaurs and the Science Wars*, Keith Parsons describes Mitchell's *The Last Dinosaur* as "doctrinaire, self-indulgent, and lacking in the most basic scholarly standards. I do not know how to say this nicely."

Parsons asks, "How can our understanding of dinosaurs be enhanced by joining paleontology to a field where the operative rule apparently is to accept whatever sounds plausible to Professor Mitchell?"

2001 In his final interview just before death, Edwin Colbert recalls that he studied mammals early in his career, but when Barnum Brown faced mandatory retirement in 1942, administrators at AMNH "just put me in his slot. I sorta inherited the dinosaurs." Colbert, who never worked with Brown, described famed AMNH paleontologist Roy Chapman Andrews as "not a dinosaur expert at all. He was an expedition leader."

2001 In the *Journal of Paleontology*, paleontologist Christopher Brochu summarizes the evidence linking birds and dinosaurs, noting "that birds are derived theropod dinosaurs is no longer the subject of scholarly debate."

2001 James Clark and his colleagues name *Khaan,* an oviraptorid from the Late Cretaceous of Mongolia. The 4-foot-long specimen was found with another specimen; the couple is informally referred to as "Romeo and Juliet."

2001 James Kirkland and Douglas Wolfe describe *Nothronychus mckinleyi,* a herbivorous theropod having a beak, bird-like hip, and four-toed feet. *Nothronychus* was found in Late Cretaceous sediments near the Arizona–New Mexico border.

2001 Ornithologist Richard Prum and colleagues from IVPP establish that several raptor dinosaurs were fully plumaged with feathers having a modern structure. Feathers evolved in dinosaurs before the origin of flight and prior to the origin of birds. The feathers on dinosaurs probably functioned in sexual selection and social communication.

2001 Paleontologists report a leg bone of a hadrosaur in a Paleocene formation from 64.5 million years ago, suggesting that some dinosaurs survived at least 500,000 years into the Cenozoic. Most scientists dismissed this claim, instead arguing that the bone had been washed from its original (and older) sediment.

2001 Robert Bakker reminisces, "I was the first to put feathers on dinosaurs. That was back in 1975 in a *Scientific American* article."

2001 Scott Sampson and his colleagues name the small theropod *Masiakasaurus knopfleri* ("vicious lizard of Knopfler") for guitarist Mark Knopfler of the band Dire Straits (the paleontologists found bones only when Knopfler's music was playing). *Masiakasaurus* lived around 70 million years ago, and its front teeth projected forward instead of down, suggesting that the dinosaur had a specialized diet.

2001 Stephen Hutt (b. 1949) and his colleagues describe *Eotyrannus lengi* ("dawn tyrant"), the first well-preserved basal tyrannosaurid from Europe.

2001 The American Film Institute names *Jurassic Park* the 35th-most thrilling film of all time. Three years later, *Empire* ranked *Jurassic Park* as the sixth most influential film of the magazine's lifetime.

2001 To date, the movie *Jurassic Park III* is the only movie in the *Jurassic Park* series neither based on a book by Michael Crichton nor directed by Steven Spielberg. *Spinosaurus* replaced *T. rex* as the lead dinosaur, and *T. rex* appeared only briefly in a fight (which it lost). The movie, which was set on a Pacific island named Isla Sorna, was based on a divorced couple's attempts to find their son. The movie received mixed reviews, but was a commercial success; it was made for $93 million and grossed $368,780,809, making it the 8th-highest grossing film of the year. *Jurassic Park III* was later nominated for a Golden Raspberry Award as the "Worst Remake of a Sequel."

2001 Vance Farrell's book *Evolution Cruncher* promotes the Glen Rose tracks as evidence that humans lived with dinosaurs.

2001 AMNH reopens its famous display of a menacing *Allosaurus* and a majestic, towering *Barosaurus* (AMNH 6341) protecting her offspring (Figure 52). The adult *Barosaurus* was collected at what is now DNM in Utah in 1912–1914 by crews from the Smithsonian Institution, the University of Utah, and the Carnegie Museum. All of the parts of the adult *Barosaurus* were gathered in trades between 1929 and 1933 by Barnum Brown for AMNH. The juvenile *Barosaurus* was reconstructed with casts derived from a rare juvenile of *Diplodocus* (AMNH 7530) collected at Howe Quarry in Wyoming.

2002 Attendees at a meeting of the SVP are greeted by a hoax poster titled "The Case for *Stegosaurus* as an Agile, Cursorial Biped." The alleged author of the poster was T. R. Karbek (a rearrangement of the name R. T. Bakker) of the fictional Steveville Academy of Palaeontological Studies.

2002 Colorado and Utah christen the Dinosaur Diamond National Scenic Byway, a 512-mile loop linking famous dinosaur-sites in Vernal, Price, Fruita, Grand Junction, Dinosaur, and other towns in the two states.

2002 *Cryptovolans,* the first-known dinosaur to have asymmetric primary flight feathers, is discovered in Early Cretaceous sediments in China. Asymmetric feathers, which create lift, are necessary for flight. Some paleontologists argued that *Cryptovolans* might be *Microraptor*.

Figure 52 Today's visitors to New York City's American Museum of Natural History are greeted by this spectacular display of *Barosaurus* rearing skyward to protect its baby from an attacking *Allosaurus*. This *Barosaurus* skeleton, which is made of casts of fossils collected at Dinosaur National Monument, is the tallest freestanding dinosaur mount in the world; it is also one of only two *Barosaurus* specimens on view anywhere in the world. (Louie Psihoyos/Corbis)

2002 Darren Tanke and Bruce Rothschild's *Dinosores: An Annotated Bibliography of Dinosaur Paleopathology and Related Topics—1838–2001* catalogs what went wrong with dinosaurs, including insect bites, bone damage, gout, and cancer.

2002 Fecheng Zhang and colleagues describe the 6-inch-long *Epidendrosaurus ningchengensis*, a tiny theropod that was perhaps one of the closest relatives of birds, from the Late Jurassic.

2002 Geologist Kevin Pope argues that an asteroid 10 kilometers in diameter (as hypothesized by Luis Alvarez and his colleagues) was not big enough to inject enough dust into the atmosphere to suppress photosynthesis. Instead, Pope argued that the culprits were sulfate aerosols that produced acidic rain.

2002 Gerald Mayr reports bristle-like structures on the tail of the early ceratopsian *Psittacosaurus*.

2002 In "Mesozoic Miasma," Allen and Diane Debus describe dinosaurs whose farts are destroying the atmosphere. A decade later, a research study produced the same conclusion.

2002 In Douglas Lebeck's *Memories of a Dinosaur Hunter*, a depressed paleontologist is sent to the Late Cretaceous, where he encounters *Velociraptor* and witnesses a confrontation between *Triceratops* and *T. rex*.

2002 NSF's *Science and Technology Indicators* reports that 49 percent of U.S. citizens believe that humans and dinosaurs lived at the same time.

2002 Paleontologists and others gather at 2102 Pine Street in Philadelphia, Pennsylvania, to dedicate a historical marker in front of the former home of Edward Drinker Cope: "Internationally renowned vertebrate paleontologist and zoologist, Cope lived and worked here in his later years. He wrote many scientific papers describing hundreds of fossils and living animals and is famous for his long-standing feud with O.C. Marsh of Yale" (Figure 9). Cope's former homes were divided into several apartments.

2002 The board game *Evo: The Last Gasp of the Dinosaurs* is named "Game of the Year" by *Games* magazine. In *Evo*, which went out-of-print in 2007, climatic changes often shifted the balance of power, and a meteor strike could end the game at any time.

2002 The finch-sized *Epidendrosaurus* is described from Middle Jurassic sediments in northeastern China. *Epidendrosaurus* was the first known non-avian dinosaur having long hands, well-developed forelimbs, and curved claws for climbing.

2002 The finch-sized feathered coelurosaur *Scansoriopteryx heilmanni* from China's Liaoning Province is named in honor of Gerhard Heilmann. This unusual dinosaur had a long third-finger, big eyes, and a short snout. Some paleontologists consider *Scansoriopteryx* to be a synonym of *Epidendrosaurus*.

2002 The first of James Gurney's *Dinotopia* stories becomes a television miniseries. Gurney's work has been featured in several exhibits at the Smithsonian Institution.

2002 The Jurassic Park Institute announces plans to release its 329-card "Dinopedia" set of cards, but the cards are never published.

2002 *Yanornis* is described in *Nature* as "Archaeoraptor's better half."

2003 After examining 10,000 vertebrae from several hundred specimens of dinosaurs such as *Stegosaurus, Tyrannosaurus,* and *Triceratops*, Bruce Rothschild and his colleagues conclude that cancer was common in hadrosaurs. *Edmontosaurus* was the most cancer-prone, and one had a malignant tumor.

2003 Alan Feduccia claims in *Discover* that "'Archaeoraptor' is just the tip of the iceberg. There are scores of fake fossils out there, and they have cast a dark shadow over the whole field."

2003 An 18-foot-long, 8-foot-tall bronze *Hadrosaurus foulkii* sculpted by John Giannotti is erected in downtown Haddonfield, New Jersey, to commemorate the discovery of the famous fossil near there in 1858 (Figure 53). The dedication of the $140,000 project was attended by more than 2,000 people, including ancestors of William Parker Foulke, who had helped excavate the original fossil. Haddonfield was the first U.S. "dinosaur town."

2003 Carl Baugh's Creation Evidence Museum announces the discovery of a 9.75-inch-long "Beverly" track allegedly left by a human beside dinosaur tracks in Cretaceous limestone near Glen Rose, Texas.

2003 The exhibit *Honoring the Life of Charles R. Knight* begins touring the United States.

2003 Two groups of researchers, working 2,000 miles apart, find two new Antarctic dinosaurs—one a Cretaceous theropod and the other a Jurassic sauropod—in the same week.

Figure 53 This 8-foot-tall bronze statue of *Hadrosaurus foulkii* was unveiled in 2003 in Haddonfield, New Jersey, not far from the former home of Edward Drinker Cope and the site where *H. foulkii* was found 145 years earlier (Figure 41). (Courtesy Randy Moore)

2003 Writing for *Answers in Creation* ("Bringing the Bible and Science Together Without Conflict"), Greg Neyman claims that there is "no credible evidence" for the many claims on young-Earth creationist Kent Hovind's *Creation Science Evangelism* website that dinosaurs lived with humans. As Neyman noted, "*Creation Science Evangelism* articles present no proof of recent dinosaurs (because there is none!) and they go to great stretches of the imagination in the hopes of fooling their followers into believing the young-Earth lie. In essence, they embarrass the young-Earth science community."

2003 Xu Xing (b. 1969) of China's IVPP describes the four-winged *Microraptor*. Xu Xing has discovered more dinosaur species than any living scientist.

2003 A partial skeleton of *Antetonitrus ingenipes* found in Upper Triassic sediments in South Africa is reported to be "the oldest known definitive sauropod."

2003 As China continues to crack down on the illegal smuggling of its fossils, two men are sentenced to 10 years in prison and fined $24,000 for smuggling fossils from Liaoning to Korea.

2003 The tooth sent in 1874 by Arthur Lakes to Othniel Marsh is rediscovered in YPM, and researchers note that it may be the first specimen of *Tyrannosaurus* ever collected.

2004 A nearly complete skull and four vertebrae of *Dracorex* from the Hell Creek Formation in South Dakota are donated to the Children's Museum of Indianapolis. The flattened skull of *Dracorex* had spiky horns, bumps, and a long muzzle. The herbivore, named *D. hogwartsia* ("dragon king of Hogwarts," from "Hogwarts Academy" of the Harry Potter series of books by J. K. Rowling), was described two years later by Robert Bakker and his colleagues, but was later suggested to be a juvenile of *Pachycephalosaurus*.

2004 A study of the 1,401 species of dinosaurs named between 1824 and 2004 shows that (1) the quality of the type (i.e., name-bearing) materials has steadily improved, (2) recently named dinosaurs are probably new species rather than redundancies, and (3) there is no evidence that paleontologists are rushing new dinosaurs' names into print at an excessive rate. The historical rate of error in dinosaur species, which exceeds 50 percent, has probably decreased because the quality of materials used to name them has improved. Of the 726 currently invalid species of dinosaurs in 2004, 582 (80.2%) were based on isolated teeth and bones, whereas only 247 (36.6%) of 675 valid species were based on such materials.

2004 BMNH celebrates the bicentenary of its founder by opening an exhibit titled "Richard Owen—The Man Who Invented Dinosaurs." The exhibit included the first public display of the 6-inch bone fragment that Owen received from Walter Mantell to identify the moa.

2004 Clear imprints of the tail, pelvis, heel, and feet of a dinosaur are unearthed at Johnson Farm near St. George, Utah. These fossils provided paleontologists with the first evidence of a squatting dinosaur. The Johnson family donated the initial tracks to the city of St. George, which began building the interpretive center that today is the St. George Dinosaur Discovery Site at Johnson Farm in St. George.

2004 Eddie, a mummified *Edmontosaurus*, is discovered in 65-million-year-old sediments in southwestern North Dakota.

2004 *Hypsibema missouriensis* becomes the state dinosaur of Maryland.

2004 In China's Liaoning Province, Mark Norell and Xu Xing unearth the duck-sized *Mei long* ("sleeping dragon"), a primitive troodontid that was preserved with its head tucked between its elbow and body, and with its tail encircling its body. This pose, which mimics that of sleeping or resting birds, supported the claim that birds evolved from dinosaurs. *Mei long* was later featured in ITV's *Prehistoric Park*, in which several of the dinosaurs were suffocated by volcanic gases in their "sleeping" position.

2004 In his autobiography *An Animated Life*, famed special-effects artist Ray Harryhausen (Figure 33) pays tribute to Charles Knight, noting, "Long before Willis O'Brien, myself, and Steven Spielberg, Charles Knight put flesh on creatures that no human had ever seen."

2004 Patricio Dominguez Alonso and colleagues at the University of Madrid use scanning X-ray computed tomography to report that the "London specimen" of *Archaeopteryx* had a brain similar to that of a sparrow, with all of the features necessary for flight.

2004 Peter Dodson recognizes approximately 40 valid species of ceratopsians. Dodson also noted another 50 invalid names, half of which were assigned by Edward Drinker Cope and Othniel Marsh.

2004 Philip Currie and D. J. Varricchio describe *Atrociraptor* ("cruel thief") from the Late Cretaceous three miles from RTM in Alberta, Canada. The type specimen, a 7-inch-long skull, is closely related to *Deinonychus*.

2004 Qingjin Meng and colleagues describe an adult *Psittacosaurus* associated with 34 juveniles, apparent evidence of parental care.

2004 Researchers suggest that there was a major increase the diversification of dinosaurs during the final 18 million years of the Cretaceous.

2004 *Saturday Night Live* includes a skit noting, "This week, Georgia's Board of Education approved a plan that allows teachers to keep using the word *evolution* when teaching biology. Though, as a compromise, dinosaurs are now called 'Jesus Horses.'"

2004 Scientists use a core from the Chicxulub Crater to suggest that the asteroid impact dates back more than 300,000 years before the K-T boundary and "thus did not cause the end-Cretaceous mass extinction as commonly believed."

2004 The Makoshika Dinosaur Museum opens in Glendive, Montana; today, that museum is part of the newly created, 14-stop Montana Dinosaur Trail (Figure 54).

Figure 54 The rich heritage of dinosaur discoveries in Montana is showcased by the Montana Dinosaur Trail, a tour that links the state's 14 dinosaur-related museums. This trail, organized in 2004, includes museums ranging from the historic Carter County Museum in Ekalaka (population 345) to the world-class Museum of the Rockies in Bozeman, which houses one of the largest collections of dinosaur fossils in the world. The sign shown in this photo advertises Jordan's Garfield County Museum, which is not far from where Barnum Brown (Figure 17) found *T. rex* in 1902. To visit all of the museums on the Montana Dinosaur Trail, you will have to drive 1,348 miles. Interestingly, the second-largest museum in Montana that features dinosaurs is not part of the Montana Dinosaur Trail; that museum, the Glendive Dinosaur and Fossil Museum in Glendive, Montana, advocates young-Earth creationism (e.g., that humans lived with dinosaurs). (Courtesy Randy Moore)

Five years later, the Glendive Dinosaur and Fossil Museum—with the head of *T. rex* crashing through its front wall—opened nearby and began promoting young-Earth creationism.

2004 The Pratt Museum of Natural History closes, and two years later reopens as the Amherst College Museum of Natural History. The museum includes the mounted hadrosaur, *Kritosaurus* (from Alberta, Canada), skulls of *T. rex* and *Triceratops*, dinosaur tracks, and parts of the sauropod *Diplodocus*.

2004 Various state organizations (e.g., Department of Commerce, Montana Fish & Wildlife, and several tourism-based agencies) create the 14-stop Montana Dinosaur Trail to promote the state's extensive collections and displays of dinosaur fossils.

2004 Xu Xing and colleagues describe *Dilong*[45] *paradoxus* ("emperor dragon"), a basal tyrannosaur and the first tyrannosaur having a preserved covering of filaments. *D. paradoxus* (IVPP 14243) was found in Lower Cretaceous sediments of China's Liaoning Province. As of 2013, dinosaur eggs unearthed in China accounted for one-third of the total dinosaur eggs found.

2005 A study of bone growth in *Plateosaurus* reports that different individuals grew at different rates. Some individuals grew fast, but others grew slower at rates similar to those of modern crocodiles.

2005 Alberta's RTM and CKUA Radio ask John Acorn to prepare a series of 80 broadcasts about prehistoric life in Alberta, Canada. Acorn's *Deep Alberta* described dinosaurs and dinosaur-hunters in Alberta.

2005 Burkhard Pohl arranges a long-term loan of a specimen of *Archaeopteryx* for the Wyoming Dinosaur Center.

2005 The Akron Fossils and Science Center opens in Akron, Ohio, to promote "creation science." The museum sponsors science fairs and an exhibit of "man tracks" from Glen Rose, Texas, claiming that humans lived with dinosaurs.

2005 Caroline Strömberg of Stockholm's Swedish Museum of Natural History uses analyses of 65-million-year-old coprolites to argue that dinosaurs ate at least five different species of grasses, as well as conifers and palms. All five species of the grasses grow throughout the world today.

2005 Famed dinosaur-hunter John Ostrom dies from complications of Alzheimer's disease in Litchfield, Connecticut, at age 77. In 1964, Ostrom had discovered *Deinonychus* in central Montana; a model of that dinosaur now decorates YPM. After seeing the movie *Jurassic Park*, Ostrom noted that the movie's *Velociraptor* was "really our own *Deinonychus* parading around under an assumed name."

2005 In Douglas Preston's "Tyrannosaur Canyon," a pristine specimen of *Tyrannosaurus* (with preserved soft-tissue) discovered in New Mexico is contaminated with a foreign virus carried to Earth by the asteroid that hit at the K-T extinction. Preston suggests that the asteroid was "pushed toward Earth . . . to make way for the evolution of human beings."

2005 In *Nature,* Oliver Rauhut of the University of Munich and his colleagues report the discovery of a massive, short-necked dinosaur (related to *Diplodocus*) from Patagonia. The discovery indicated that not all herbivorous dinosaurs had long necks.

2005 In the BBC documentary *The Truth about Killer Dinosaurs,* producers show that the sickle claw of an artificial *Velociraptor* leg could penetrate—but not tear—a prey (simulated by a pork belly). The producers of the documentary concluded that the claw of *Velociraptor* was not used to disembowel prey.

2005 Ken Ham of AiG announces, "We're putting evolutionists on notice: we're taking the dinosaurs back. They're used to teach people that there's no God, and they're used to brainwash people." Ham's Creation Museum continues to sell "We're taking the Dinosaurs Back" bumper stickers.

2005 Mary Higby Schweitzer discovers soft tissue in a marrow cavity in *T. rex.* McGill University paleontologist Hans Larsson described the finding as one that will help paleontologists "enter the field of molecular biology and really slingshot paleontology into the modern world." However, paleontologist Thomas Kaye claimed that the soft tissue was actually permineralized biofilm produced by bacteria while degrading the original specimen.

2005 Otto Gleich and his colleagues' studies of inner-ear anatomy suggest that the hearing of *Archaeopteryx* would have resembled that of modern birds, while that of a 75-ton *Brachiosaurus* would have been best near 700 Hz. *Allosaurus* heard best at 1.1 kHz, but was deaf beyond about 3 kHz, the pitch of a human scream.

2005 Peter Jackson's (b. 1961) long-awaited remake of *King Kong* includes stampeding sauropods and battling tyrannosaurs. "Vastatosaurus rex," a modern-day descendant of *T. rex*, gets most of the attention in the movie. Among other changes, the scene from the original movie in which the rescuers first encounter *Stegosaurus* was omitted from the remake. Although the original *King Kong* (1933) received no Academy Award nominations, the remake won three such awards, including one for Best Special Visual Effects.

2005 Researchers at AMNH discover a 130-million-year-old fossil of a mammal whose stomach contains a small dinosaur.

2005 Tamaki Sato and her colleagues report in *Science* a pregnant oviraptor from the Upper Cretaceous containing a pair of preserved eggs. This finding showed that dinosaurs had paired oviducts (a trait shared by crocodilians) and probably developed only one egg in each oviduct.

2005 The "Thermopolis specimen" of *Archaeopteryx* (WDC CSG 100) is described in *Science* as "a well-preserved *Archaeopteryx* specimen with theropod features" and having well-preserved bones, as well as impressions of feathers from the wings and tail. This specimen, which had been owned since the 1970s by a former employee of a Solnhofen limestone quarry, showed that, unlike modern perching-birds, *Archaeopteryx* lacked a reversed first toe, thereby limiting its ability to perch on branches and suggesting a terrestrial lifestyle. The specimen was bought by an anonymous donor for Frankfurt's Senckenberg Museum, and was then loaned to the Wyoming Dinosaur Center in Thermopolis, Wyoming. Although fossils are protected in most of Germany, Bavaria (where this specimen was found) does not have such a law.

2005 The award-winning documentary *Caveman: V.T. Hamlin & Alley Oop* chronicles the life of V. T. Hamlin, the creator of the comic strip *Alley Oop*.

2005 "Claude Bell's Dinosaurs" (Figure 36) in Cabazon, California, are purchased by Orange County (California) developer Gary Kanter, who renames the attraction "World's Biggest Dinosaurs" and transforms the park "from a tourist stop to a place of worship." The creationism-based park rejects evolution while promoting young-Earth creationists' claims about dinosaurs living with humans. Ms. Dinny's belly houses a museum and gift shop that promotes a version of creationism called "intelligent design."

2005 The Invicta Plastics Corporation of England ends production its long-lived line of plastic, 1:45-scale dinosaur models for BMNH. Invicta dinosaur toys were not painted, but were detailed and scientifically up to date. The final dinosaur of the series—*Lambeosuarus*—was produced in 1993.

2005 The magazine *Film Review* ranks *Jurassic Park* as one of the five most important films of the past 55 years. That same year, the movie *King of the Lost World* was advertised as a "modern retelling of Sir Arthur Conan Doyle's fantasy-adventure classic." Doyle's book was also acknowledged as "the epic story that inspired *King Kong* and *Jurassic Park*."

2005 While working in the dinosaur-rich flood-plains of the Yunnan province in southern China, paleontologists discover a near-complete skeleton of the

30-foot-long *Yizhousaurus sunae*, an ancestor of sauropods. The skeleton was more than 200 million years old.

2005 Xu Xing and Zhang Fucheng describe *Pedopenna,* a feathered maniraptoran discovered in China. The 3-foot-long dinosaur had bird-like traits, including bird-like legs, as well as long feathers on its hind feet, thereby suggesting that the feathers were ornamental, or even vestigial.

2005 Canadian paleontologist Robert Reisz (b. 1947) and his colleagues describe embryos extracted from *Massospondylus* eggs discovered 28 years earlier in South Africa. The near-hatchlings had no teeth, indicating that they probably required postnatal care.

2005 Zygote Games introduces *Bone Wars: The Game of Ruthless Paleontology,* a card game created by biologist Diane Kelly and science-fiction writer James Cambias.

2006 American amateur paleontologist Nathan Murphy announces his discovery of a raptor that he named "Sid Vicious." The fossil had actually been discovered four years earlier by a customer on a private ranch in Montana, where Murphy ran a fossil-hunting business. To avoid having to share profits from the discovery, Murphy said that he found the fossil elsewhere. Murphy later went to jail for his illegal fossil-collecting work.

2006 Commercial collectors find near-complete fossils of two dinosaurs (*Nanotyrannus lancensis* and *Triceratops*) locked in combat in Montana. After no museum would meet their price (for example, NMNH was offered the fossils for $15 million), the collectors announced in 2013 that the "Montana Dueling Dinosaurs" would be auctioned. Bidding stopped at $5.5 million, failing to meet the undisclosed minimum bid. Although the fossils were projected to sell for $7 million to $9 million, paleontologist Jack Horner noted that the fossils had not been excavated appropriately by scientists and were therefore "worthless."

2006 German paleontologists working near Hannover, Germany, report *Europasaurus holgeri,* a small sauropod that grew to 20 feet long.

2006 In *Night at the Museum*, a *T. rex* skeleton comes to life and wants to play fetch with one of its bones.

2006 In *Proceedings of the National Academy of Sciences,* Steve Wang and Peter Dodson estimate that there were 1,850–2,200 genera of non-avian dinosaurs, of which approximately 527 have been described. Wang and Dodson, who argued that dinosaurs' populations did not decline in the 10 million years before the K-T extinction, predicted that 75 percent of the discoverable genera of dinosaurs will be known within 60–100 years, and 90 percent within 100–140 years.

2006 In *The New Answers Book*, young-Earth creationist Ken Ham claims that the "behemoth" in Job 40 was "something like a sauropod dinosaur" that "God made *T. rex* on Day 6" of creation week, that "before sin, all animals, including the dinosaurs, were vegetarians," and that "cats and dogs cannot breed to make a new type of creature. This is God's design, and it is one fact that makes evolution impossible."

2006 Matthew Mossbrucker discovers footprints of a juvenile *Apatosaurus* near Morrison, Colorado. These prints suggested that these juveniles could run on their hind legs.

2006 Montana State University unveils the largest skull of *T. rex*. That skull is 59 inches long, which is 4.6 inches longer than the skull of the more famous "Sue" of the Field Museum in Chicago.

2006 Oil drillers strike a dinosaur fossil 1.4 miles below the North Sea off the coast of Norway. This was the deepest known dinosaur.

2006 Paleontologists George Poinar and Arthur Boucot find parasites (e.g., similar to *Entamoeba*) and eggs of trematodes and nematodes in coprolites from a dinosaur-rich area in Bernissart, Belgium.

2006 Paleontologists report a *Stegosaurus* ("roof lizard") from Portugal.

2006 Tampa's Museum of Science and Industry unveils "Recyclosaurus," which is made of recycled cans and milk cartons.

2006 AiG tells its followers that when dinosaurs are used to spread the Christian gospel, they become "Missionary Lizards."

2006 The discovery of a giant sauropod in Teruel, Spain, documents that giant dinosaurs roamed Europe in the late Jurassic. Prior to this discovery, giant dinosaurs had been found primarily in Africa and the New World.

2006 The first of the "T. Rex Café" restaurants opens in Kansas City, Kansas, and features "Chocolate Extinction," "Layers of the Earth Lasagna," "Stegosaurus Steak," "Footprints Flatbread," and "Gigantosaurus Burgers." In coming years, franchises opened at Walt Disney World and elsewhere.

2006 The giant 80-million-year-old dinosaur *Maxakalisaurus* is unearthed by Brazilian paleontologists. *Maxakalisaurus* was a small-headed dinosaur that was related to dinosaurs such as *Saltasaurus* ("Salta lizard," named for the Salta province of Argentina where it was discovered), a late Cretaceous sauropod.

2006 There are now 137 valid genera of sauropods, half of which have been named in the previous 13 years.

2006 Xu Xing names the bipedal theropod *Guanlong* ("crown dragon") from the Late Jurassic of modern China. *Guanlong* was one of the earliest known tyrannosauroids, and has been featured in the National Geographic documentary *Dino Death Trap*, the *Discovery* channel miniseries *Dinosaur Revolution*, and the movie *Ice Age: Dawn of the Dinosaurs* (2009). In that movie, *Guanlong* was included as "a twist on a *Velociraptor*." *Dinosaur Revolution* also included flocks of colorful, timid *Eoraptor*.

2006 Young-Earth creationist Kent "Dr. Dino" Hovind (owner of Dinosaur Adventure Land) is convicted of 58 federal charges, including tax evasion. Hovind's wife Jo was convicted of 44 of the charges. Hovind had ridiculed the government's ability to sentence him to prison, claiming he could "make their lives miserable." However, at his sentencing Hovind cried and begged for leniency, proposing that his followers would pay his fines. Hovind was fined $611,954 ($604,876 in restitution to the Internal Revenue Service, and $7,078 to cover the prosecution costs for Hovind's trial) and sentenced to 10 years in federal prison. Hovind's appeal to the U.S. Court of Appeals for the 11th Circuit was denied.

2007 A dinosaur egg is sold at Christie's in Paris for $97,500. That same year, a 30-foot-long, 70-million-year-old mosasaur from North Africa was auctioned in Los Angeles for more than $350,000.

2007 Afghanistan issues a series of postage stamps featuring prehistoric mammals, Barnum Brown, Edward Drinker Cope, and Othniel Marsh.

2007 Clifford Wilson, Barbara Wilson, and John Heffner claim in *The Paluxy River Dinosaur Controversy* that "the evidence is overwhelming: Men and dinosaurs walked together." In that same book, Duane Gish of ICR rejected modern biology, physics, and geology by claiming "humans and dinosaurs did indeed walk together thousands, not millions, of years ago. . . . This throws out the legitimacy of the Geological Column as a dating index."

2007 David Varricchio of Montana State University finds a fossilized family of two adults and a juvenile *Oryctodromeus* in a burrow, indicating that dinosaurs could dig. The small, bipedal dinosaurs were in spiral-shaped tubes resembling those of living mammals such as prairie dogs. This discovery of the first burrowing dinosaurs is among the best evidence yet discovered for long-term parental care in dinosaurs.

2007 Fossil collector Nathan Murphy, who had in 2000 discovered the duckbilled dinosaur "Leonardo," resigns as curator of the Judith River Dinosaur Foundation in Malta, Montana, as officials begin investigating his fossil-collecting activities.

2007 ICR concludes its Radioisotopes and the Age of the Earth (RATE) study by claiming that accelerated radioactive decay "such as during the Genesis flood, the Fall of Adam, or early Creation week" explains the apparent conflict between the Bible-based conclusion that Earth is young and the science-based conclusion that Earth is ancient. ICR added that "details regarding the Flood are beginning to unfold," "scientific discoveries continue to erode Darwinism," and "there is now abundant evidence that man and dinosaurs lived at the same time."

2007 In a revised edition of *Texas Tracks and Artifacts*, Robert Helfinstine and Jerry Roth argue that humans lived with dinosaurs, and claim that Roland Bird had excavated "some non-dinosaurian tracks that landed in the basement of [AMNH], never to see the light of day. Why? From the evolutionary viewpoint of the museum, these tracks looked too much like human footprints to be shown to the public."

2007 In *Cruisin' the Fossil Freeway*, Kirk Johnson notes that Jack Horner "discovered so many *T. rex* skeletons [along the south shores of Ft. Peck Reservoir] that he rendered the king of dinos blasé in the process."

2007 Jack Chick's booklet *There Go the Dinosaurs* claims that explanations for dinosaurs' extinction involving an asteroid impact are stories "told by people who don't trust God. And God should know. After all, He was there." Chick claimed that dinosaurs "were created, not evolved" by God, that dinosaurs were aboard Noah's Ark, that the air after Noah's flood was deficient in oxygen and made the dinosaurs "slower and easier to catch," and that "the real issue isn't where the dinosaurs went, but where YOU will go when you die."

2007 Mark Norell of AMNH notes, "The more that we learn about [dinosaurs] the more we find that there is basically no difference between birds and their closely related dinosaur ancestors like *Velociraptor*. Both have wishbones, brooded their nests, possess hollow bones, and were covered in feathers. If animals like *Velociraptor* were alive today our first impression would be that they were just very unusual looking birds."

2007 Nathan Smith and Diego Pol name *Glacialisaurus*, a sauropodomorph from the Early Jurassic of Antarctica. The discovery of this dinosaur suggested that primitive and advanced sauropods lived with each other in the Early Jurassic.

2007 Near Petersburg, Kentucky, AiG opens a 60,000-square-foot, $27 million Creation Museum filled with life-size dinosaur models, live exhibits, and other collections that "proclaim the authority of the Bible from its very first verse" (Figure 55). The museum claims that a belief in young-Earth creationism is the only true path to salvation, and that all else undermines the word of God. In the museum, Methuselah—Noah's grandfather who died just before the Flood at age 969—warns visitors of God's upcoming judgment. Exhibits compare "human

Figure 55 In 2007, Christian fundamentalist Kenneth Alfred Ham (b. 1951) opened his $27,000,000 Creation Museum in Petersburg, Kentucky, to promote young-Earth creationism. All of the museum's permanent employees must sign a statement affirming their belief in young-Earth creationism. Ham and his museum claim that humans lived with dinosaurs, and a poll in 2000 by the National Science Foundation reported that almost half of U.S. citizens agreed. More than 250,000 people visit Ham's Creation Museum each year. (AP Photo/Ed Reinke)

reason" with "God's word" and show children cavorting with dinosaurs, which were aboard Noah's Ark after being created on the sixth day of the biblical creation week. This museum, which endorses James Ussher's claim that creation occurred in 4004 BC, portrays war, death, and the pains of childbirth as the wages of primal sin. Visitors learn that most fossils were created by the Flood (God's worldwide judgment), that dinosaurs and all other animals were vegetarian before Adam's sin, and that young-Earth creationism is the only view of creationism that does not destroy churches. Critics claimed that the museum equates religious faith with ignorance. The museum attracted 404,000 visitors during its first year of operation.

2007 Oil field worker Harry Nibourg opens Big Valley Creation Science Museum in Alberta, Canada, just 35 miles from RTM. The museum, which was Canada's first permanent creation-museum, promotes young-Earth creationism and includes exhibits claiming that humans lived with dinosaurs. Elsewhere in Canada, Creation Truth Ministries (established in 2002) sponsors a traveling exhibit titled "Secrets of Creation" that promotes the same claims.

2007 Paleontologist and curator David Evans of Toronto's ROM finds a partial skeleton of *Barosaurus* (ROM 3670) that had been excavated decades earlier by Earl Douglass at the Carnegie Quarry and obtained in a trade in 1962 with CMNH. Today, *Barosaurus* is a centerpiece in ROM.

2007 Paleontologists in Eromanga, Australia, announce the discovery of two 82-foot-long titanosaurs, the largest dinosaurs ever found in Australia. That same year, scientists in southern Australia found 115-million-year-old footprints of a giant carnivore.

2007 Paleontologists working in Mongolia discover quill knobs (i.e., bumps on wing bones where feathers anchor) on the forearm a well-preserved *Velociraptor,* thus confirming the presence of feathers. Paleontologist Alan Turner concluded, "A lack of quill knobs does not necessarily mean that a dinosaur did not have feathers. Finding quill knobs on *Velociraptor*, though, means that it definitely had feathers. This is something we'd long suspected, but no one had been able to prove."

2007 Sauropod tracks near Glen Rose, Texas, are reevaluated and assigned to a new genus, *Paluxysaurus*.

2007 Seven traces of collagen proteins from a bone of *T. rex* are shown to resemble those of birds.

2007 The 42-foot-long sauropod *Spinophorosaurus* is discovered in Niger.

2007 The Dinosaur Hall of Berlin's Museum für Naturkunde reopens after two years of renovation, returning *Brachiosaurus brancai* to public view. This reconstruction is the largest mounted dinosaur skeleton in the world that includes original fossils. The remounted skeleton had originally been mounted in 1937 by Werner Janensch.

2007 The dubious genus of sauropod *Xenoposeidon* is announced by the British Palaeontological Association. This genus, from the Early Cretaceous of England, is known from one partial vertebra having unusual features that distinguish it from other sauropods. The fossil, which was hyped as "the world's most amazing sauropod," had been discovered by Phillip Rufford in the early 1890s, but received little attention until it was named by Mike Taylor and Darren Naish.

2007 The sauropod *Turiasaurus riodevensis* is reported from the Late Jurassic in Spain. *Turiasaurus,* which weighed 40–48 tons, was one of the largest animals to ever live in Europe.

2007 To remedy the nomenclatorial problems that resulted from a feather being the original type-specimen of *Archaeopteryx*, two committees of scientists petition ICZN to make the "London specimen" of *Archaeopteryx* the new type specimen of *Archaeopteryx*. The "London specimen" of *Archaeopteryx*, which is housed in BMNH, is arguably the world's most valuable fossil.

2007 U.S. and Czech researchers suggest that a giant asteroid named Baptistina was shattered by a smaller asteroid. One of the resulting fragments 3–9 miles wide hit Earth 65 million years ago at the Yucatán peninsula; this impact was correlated with the K-T mass extinction that eliminated all non-avian dinosaurs.

2007 When Alan Feduccia and his colleagues claim that primitive feathers on the Early Cretaceous theropod *Sinosauropteryx* are merely degraded parts of the animal (i.e., that *Sinosauropteryx* lacked feathers), paleontologist Kevin Padian dismisses Feduccia's claim as "nonsense."

2007 Xiao-chun Wu and colleagues describe *Eotriceratops xerinsularis* ("dawn *Triceratops*"), a transitional form in the evolution of horned dinosaurs from the Late Cretaceous in southern Alberta, Canada.

2007 Xu Xing and collaborators describe *Gigantoraptor,* a huge oviraptorosaur from the Late Cretaceous in Inner Mongolia. The bird-like *Gigantoraptor,* which was 25-foot-long and weighed 3,000 pounds, was 35-times heavier than *Oviraptor*. This discovery contradicted assumptions that dinosaurs became smaller as they evolved additional features resembling those of birds. Xu found the thighbone of *Gigantoraptor* in 2007 while reenacting the discovery of *Sonidosaurus* for a Japanese documentary.

2007 CMNH's Dinosaur Hall reopens after a two-year renovation to feature "Dinosaurs in Their Time," an acclaimed permanent exhibit featuring the museum's famous *Diplodocus* and a confrontation between two *T. rex*. As at the Field Museum, visitors to the exhibit are greeted by *Herrerasaurus*.

2008 A group of University of Utah geologists describes more than 1,000 footprints on a "dinosaur dance floor" in Early Jurassic rocks along the Utah–Arizona border. However, when dinosaur-footprint experts examined the site a few weeks later, they could find no dinosaur footprints; the alleged prints were erosional potholes.

2008 A new species of dinosaur is being named every 2–3 weeks.

2008 A study finds that 52 percent (i.e., 726) of the dinosaur species named at the end of 2004 are invalid (e.g., the species are not supported by distinctive fossils). Of the 1,401 named species, 230 (i.e., 16.4%) were regarded as synonyms, 340 (i.e., 24.3%) were *nomen dubium,* 47 (i.e., 3.4%) were *nomen nudum,* and 58 (i.e., 4.1%) were not dinosaurs. Almost 40 percent (i.e., 388) of the 1,036 genera were invalid.

2008 AiG urges people not to be swayed by scientific evidence, adding that "God's word" is undermined by pornography, abortion, homosexuality, lawlessness, "Man's opinion," and evolution. AiG later blamed school violence on the teaching of evolution.

2008 *Albertonykus* is named and described from bones discovered 98 years earlier in Late Cretaceous sediments in Alberta, Canada. The 75-centimeter-long *Albertonykus,* which has short, single-clawed forelimbs, is the earliest-known alvarezsaurid from North American. (Alvarezsaurs were bird-like theropods.)

2008 As creationists in Texas again push to have their views incorporated into the state's public schools, Alvis Delk re-examines his "Alvis Delk Cretaceous Footprint" and finds a human footprint beside the dinosaur track. Delk told reporters that "I know [the track] is going to change history" and "will disprove Darwin's theory." That same year, when he was injured and needing money to pay his bills, Delk sold the track to Carl Baugh, who featured it prominently in his Creation Evidence Museum just outside of Glen Rose, Texas, to support his claims that humans and dinosaurs lived at the same time.

2008 Canadian paleontologist Darla Zelenitsky and her colleagues examine dinosaurs' brains (generated by CT scans of skulls) and announce that tyrannosaurs and dromaeosaurs had the largest olfactory bulbs, and were therefore the best smellers among carnivorous dinosaurs.

2008 Chris Organ of Harvard University claims that "with more data, [we may] be able to place *T. rex* on the evolutionary tree between alligators and chickens and ostriches."

2008 David Varricchio and his colleagues deduce that nesting theropods produced unusually large clutches of eggs for their body mass, a pattern common in birds in which males alone care for the eggs. The researchers concluded that male theropods incubated eggs. Five years later Charles Deeming and his colleagues reexamined the data and found that theropods were not among male-only brooders, and in 2013 Marcello Ruta and Geof Birchard argued that dinosaur parents shared brooding equally.

2008 Dinosaur World, a chain of outdoor amusement parks, opens just outside of DVSP and the Creation Evidence Museum near Glen Rose, Texas. The attraction features a mile-long trail through more than 20 acres that display 100–150 life-size dinosaurs, and an indoor museum that includes fossils. Other franchises of Dinosaur World are in Cave City, Kentucky, and Plant City, Florida. The success of these museums prompted similar businesses to open elsewhere.

2008 Fucheng Zhang and his colleagues announce the discovery of the 17-inch-long maniraptoran *Epidexipteryx hui* ("Hu's display feather") from Upper Jurassic sediments in China. *Epidexipteryx* is the earliest known example of ornamental feathers in the fossil record.

2008 Fossil Cabin (Figure 26) near Como Bluff is listed on the National Register of Historic Places.

2008 Fucheng Zhang and his colleagues from Beijing's Chinese Academy of Science describe *Epidexipteryx*, a bizarre theropod covered by down and having a tail sporting four long plumes.

2008 Graeme T. Lloyd and his colleagues analyze dinosaur coprolites and conclude that most herbivorous dinosaurs ate gymnosperms. The researchers suggested that these dinosaurs' extinction was influenced by their failure to diversify as ecosystems were changing. Lloyd also suggested that the observed peak of dinosaur diversity during the mid- and Late Cretaceous (i.e., the 50 million years before their extinction) may be due largely to sampling bias. According to these researchers, dinosaurs' diversity was not declining at the end of the Cretaceous. That same year, Lloyd used an evolutionary tree to conclude that dinosaurian evolution was stagnant (i.e., there was little change in their diversity) during most of the Cretaceous. The Early Jurassic included the most diverse dinosaurs and the evolution of the major groups of dinosaurs.

2008 ICR's application to award graduate degrees in Texas is unanimously rejected by the Texas Higher Education Coordinating Board. ICR president John Morris responded to the rejection by claiming that ICR has "a higher mandate (Mark 12:30)" and asked his supporters to pray for their success as they considered appealing the decision. (In 2009, ICR filed two lawsuits to overturn the decision.) Later in the year, ICR launched its National Creation Science Foundation that provided money "to advance the study of origins science." All research funded by the foundation "must be conducted from a young-Earth, global Flood perspective, and investigators must abide by the biblical and creation science tenets" of ICR, including that humans lived with dinosaurs. The foundation's first grant was given to young-Earth creationist Steve Austin.

2008 In "Did Dinosaurs Turn into Birds?", David Menton claims that birds did not evolve from dinosaurs because God created birds on Day 5, and dinosaurs on Day 6, of creation week.

2008 In *Dinosaurs Revealed*, Larry Dye "The Creation Guy" claims that teeth of *T. rex* "were designed to rip apart branches . . . Animals like a brachiosaur would keep leaf foliage under control, and [*T. rex*] would keep branch growth under control."

2008 In *Paleobiology,* Michael J. Benton reports that 726 (i.e., 51.8%) of the 1401 species of dinosaurs named through 2004 are now regarded as invalid. In the same article, Benton suggested that humans have identified one-fourth to one-third of known dinosaur species, and reported that 340 (24.3%) of the 1401 species of dinosaurs are considered dubious (*nomen dubium*, and have no status), 47 (3.4%) are considered "naked" (*nomen nudum*, and having no status because the organism has not been formally described), 230 (16.4%) are considered synonyms, and 58 (4.1%) are not dinosaurs.

2008 In *What Bugged the Dinosaurs*, George and Roberta Poinar speculate that "the cumulative, cascading effects of many diseases" transmitted by insects killed the dinosaurs.

2008 Jack Horner, Holly Woodward, and Mark Goodwin propose that the pachycephalosaurs *Stygimoloch* and *Dracorex* (described two years earlier by Robert Bakker and his colleagues) are juveniles of *Pachycephalosaurus*. Horner and his colleagues suggested that many other species are juveniles or sub-adults of known species, and therefore that dinosaurs may have been less diverse than previously believed. In 2010, Horner and his colleagues made similar claims when they announced that *Triceratops* was a juvenile of *Torosaurus*.

2008 Mike Benton and his colleagues suggest that dinosaurs opportunistically benefitted from the mass extinctions 228 and 200 million years ago that decimated their competitors. These results suggested that the rise of dinosaurs was due to their competitors' disappearance rather than the "superiority" of dinosaurs.

2008 Nan Arens and Ian West argue that extinction was most frequent when there was volcanism and extraterrestrial impact, but neither volcanism nor impact alone increased extinction. Even with this caveat, volcanism and the impact could not explain all of the K-T extinctions. Arens and West suggested that climate and sea-level changes might have been involved by producing longer-term stresses on the environment and that "single causes for mass extinction—although appealing in their simplicity—may be inadequate to explain . . . the many episodes of mass extinction."

2008 Paleontologist Lisa Cooper and her colleagues announce that hadrosaurs (e.g., the iconic *Hypacrosaurus*) grew to adulthood in 10–12 years; this was half the time it took for tyrannosaurs (their primary predators) to reach full-size. Hadrosaurs had few obvious protections against predators (e.g., no tail-clubs as on ankylosaurs, no horns as on ceratopsians, and no spikes as on stegosaurs), and may have been able to reproduce as soon as 2–3 years after birth. That same year, Andrew Lee and Sarah Werning reported that dinosaurs could reproduce before reaching their full size.

2008 Paleontologists in Mongolia unearth *Protoceratops* fossils with two serrated teeth of a scavenging *Velociraptor*.

2008 Paul Sereno describes the 85-million-year-old *Aerosteon*, a large theropod from Argentina having avian-style lungs. Unlike the lungs of mammals, those of birds are attached to the vertebral column and have a one-way flow of air (mammals have a bidirectional, in-and-out flow of air). The lungs of birds were thought to be an adaptation for flight, but they—just like wishbones and feathers—evolved before flight.

2008 Polish paleontologists digging in a brickyard in southern Poland unearth the remains of a 200-million-year-old dinosaur that is an ancestor of *T. rex*.

2008 Protein from a 68-million-year-old *Tyrannosaurus rex* is shown to resemble that of chickens and ostriches, further supporting the claim that birds are the living descendants of dinosaurs.

2008 Scott Sampson describes *Gryposaurus monumentensis*, a duckbill having more than 800 teeth. The 30-foot-long, 10-foot-tall dinosaur had been found six years earlier in Grand Staircase-Escalante National Monument. Sampson described *Gryposaurus monumentensis* as "the Cretaceous version of a weed-whacker."

2008 Shanan Peters argues that "it is the ocean, and in particular the epic ebbs and flows of sea level and the sediment of the course of geologic time, that is the primary cause of the world's periodic mass extinctions over the past 500 million years."

2008 Steve Brusatte and Paul Sereno describe *Eocarcharia* (a theropod related to *Allosaurus*) and *Kryptops* (a theropod related to *Ceratosaurus*) from the Middle Cretaceous of what is now the Sahara Desert.

2008 The four-winged, peacock-sized *Anchiornis huxleyi*, which is 30 million years older than *Archaeopteryx*, is declared the closest dinosaurian relative. However, the subsequent discovery of a better-preserved specimen prompted a reevaluation, which concluded that *Anchiornis huxleyi* was a troodontid dinosaur. *A. huxleyi*, which lived 155–160 million years ago, was named in honor of Thomas Huxley, an early proponent of Darwinian evolution who argued that birds arose from dinosaurs.

2008 The *NOVA* documentary *The Four-Winged Dinosaur* features *Microraptor* and includes analyses of its flight based on models built by AMNH artist Mick Ellison.

2008 The type-species of *Austroraptor cabazai* is described from Late Cretaceous sediments in Argentina. *A. cabazai* is the largest dromaeosaur ever found in the Southern Hemisphere. The short arms of *A. cabazai* led to many comparisons with *T. rex*.

2008 Thomas Kaye and his coworkers argue that Mary Higby Schweitzer may have confused dinosaurs' proteins with a biofilm produced by bacteria that lined the voids occupied the dinosaur's cells and blood vessels. Schweitzer rejected Kaye's claims.

2008 *Triceratops* "Cliff" is bought for $942,797 and given to the Boston Museum of Science. The 23-foot-long specimen was discovered in 2004 and is named for the grandfather of its anonymous donor. Cliff is one of only four nearly complete *Triceratops* displayed publicly in the world.

2008 Upper Deck launches its "Dinosaur King Trading Card Game."

2008 Yale University's Jakob Vinther and his colleagues discover fossilized melanosomes in fossils of dinosaurs and prehistoric birds (melanosomes are the cellular structures that make and store the pigment melanin). Two years later, Vinther, Li Quango, and their colleagues reported that the newly discovered troodontid *Anchiornis huxleyi* had black-tipped forelimb feathers, a red-brown crest of feathers atop its head, and white stripes on feathers of its legs. (As Vinther noted, "There was hardly any part of the creature that wasn't feathered.") This was the first colored description of an entire dinosaur, and indicated that when feathers appeared in the fossil record, they were already spotted and striped.

2008 Young-Earth creationist Carl Baugh announces that his Creation Evidence Museum has a "pristine" 11-inch-long footprint of a human that overlaps that of a theropod (*Acrocanthosaurus*) track in Cretaceous rock near the Paluxy River at Glen Rose, Texas. Baugh claimed that both prints were made in mud or soft clay "within minutes, or at most hours of each other" 4,500 years ago near the time of Noah's Flood. Baugh did not explain how the dinosaur track maker, which Baugh claimed weighed "a few tons," would have left a track almost as deep as the human track, nor does he explain how the dinosaur and human were walking around during a worldwide flood. This specimen, known as "The Alvis Delk Print," is displayed prominently at Baugh's museum.

2008 Ken Ham, who claims that "it is not hard to imagine [that humans lived with dinosaurs] if we are Christians who believe God's Word," argues that humans "tamed [dinosaurs] to help with transportation, maybe even farming, hauling heavy loads (the strong ones!), and other things." In *The Answers Book for Kids*, Ham again claimed that "a total of 100 dinosaurs" were aboard Noah's Ark, that "a number of animals may have become vicious because of the effects of sin," that Noah was able to manage all of the ark's animals because God "caused the animals [on the ark] to hibernate much of the time," and that dinosaurs became extinct "because we sinned against God."

2009 AMNH's Stephen Brusatte (b. 1984) and his colleagues describe *Alioramus altai*, a 70-million-year-old, small-bodied tyrannosaur having eight horns. The long-snouted fossil, which was found in Mongolia in 2001, was 13 feet long and weighed 800 pounds.

2009 An analysis of the skulls of the dinosaurs *Dracorex hogwartsia, Stygimoloch spinifer*, and *Pachycephalosaurus wyomingensis* reveals they are different stages of *P. wyomingensis*, whose horns disappear and domed head grows over time. This finding prompted speculation that up to one-third of recognized species of dinosaurs are juvenile forms of other species.

2009 Another study concludes that holes in the skull of *T. rex* "Sue" were caused by a *Trichomonas*-like parasite.

2009 *Beishanlong*, which resembled a giant ostrich, is discovered in Early Cretaceous sediments in China's Gansu Province. *Beishanlong* is one of the largest ornithomimosaurs (bird-mimic dinosaurs) ever discovered.

2009 Commercial fossil collector Nathan Murphy is sentenced to three years of probation and ordered to pay $17,325 in restitution for stealing 13 dinosaur fossils in 2006 from Montana's Hell Creek badlands. Murphy's conviction came a month after he was sentenced to 60 days in jail for stealing a raptor fossil that he called "Sid Vicious" from private land in Malta, Montana. Murphy originally claimed that his theft was an honest mistake, but later admitted that he lied.

2009 In a paper titled "Evidence of combat in *Triceratops*," researchers report that lower bones on the outside of *Triceratops*' skulls were injured most often, an observation consistent with these animals locking horns.

2009 In a remake of *Land of the Lost*, *T. rex* "Grumpy" transforms from threatening to comedic. When the paleontologist (Will Ferrell) notes that tyrannosaurs had walnut-sized brains, Grumpy left a giant walnut for Ferrell to find.

2009 In *Censored Science: The Suppressed Evidence*, Bruce Malone claims that dinosaurs lived with humans, adding "it is only the denial of a worldwide flood which prevents modern scientists accepting the reality of the coexistence of man and dinosaurs."

2009 In *Dinosaurs for Kids*, Ken Ham claims that dinosaurs were "created on the same day that God made the first two people, Adam and Eve! Do you realize what that means? It means dinosaurs and people lived together! . . . [This was the time] a T-Rex could put its head beside Adam, open his mouth, and show all those teeth . . . and Adam wouldn't have been afraid. Adam knew that T-Rex (as we now call him) only ate plants."

2009 In *Evolution and Human Fossil Footprints*, Aaron "Maverick" Judkins describes fossilized human-footprints in ancient rocks from the Paluxy River and elsewhere. (The Paluxy "man tracks" have often been used to claim that humans lived with dinosaurs.) Judkins also promotes UFOs and claims to have "archaeological evidence" that "fallen angels" have interacted with people.

2009 In *Science* magazine, Sterling Nesbitt (b. 1982) and colleagues describe the dog-sized theropod *Tawa hallae*, a primitive precursor of *Tyrannosaurus rex* found in 215-million-year-old sediments near Ghost Ranch, New Mexico, five years earlier. The dinosaur was named in honor of Tawa, the Hopi word for the Puebloan sun god, and Ruth Hall, whose work helped found the Ruth Hall Museum at Ghost Ranch. The beds in which *Tawa* was found included other theropods more closely related to different groups from South America and Europe than to each other, suggesting that the earliest theropods originated and diverged in what is now South America more than 220 million years ago when the continents were part of Pangaea.

2009 Jack Horner's *How to Build a Dinosaur: Extinction Doesn't Have to Be Forever* suggests how to recreate a dinosaur by genetically manipulating the DNA of a chicken. In 2013, Oxford biochemist Alison Woollard claimed that it would be theoretically possible to re-create dinosaurs by genetically manipulating and "de-evolving" the genome of birds.

2009 While at Grand Staircase-Escalante National Monument in southern Utah, high school senior Kevin Terris finds a 75-million-year-old fossil of the duck-billed *Parasaurolophus*. "Joe," as the fossil came to be known, is the most complete specimen of its type ever found.

2009 Josh Mathers and his colleagues report the first bone-bed of *Triceratops* in southeastern Montana. There had been several earlier reports of bone beds of other ceratopsians, but none for *Triceratops*.

2009 Paleontologist Raimund Albersdörfer buys the "Daiting specimen" of *Archaeopteryx,* which had been known since 1996 only from a cast. The specimen was displayed publicly for the first time in 2009 at the Munich Mineral Show.

2009 Paleontologist Steven Salisbury, veterinarian Ewan Wolff, and their colleagues announce that the death of *T. rex* "Sue" was associated with infections of her mouth (that are common in hawks today).

2009 Paleontologists discover four new tyrannosaurs. In 2010, paleontologists reported two more, as well as 16 new horned-dinosaurs. Among the ceratopsians was China's Late Cretaceous *Sinoceratops*, the first example of a ceratopsian outside North America.

2009 *Paluxysaurus* is made the Lone Star State Dinosaur of Texas.

2009 Paul Barrett, Alistair McGowan, and Victoria Page conclude that "many of the peaks and troughs apparent in dinosaur diversity-curves may not represent genuine biological events, but should probably regarded as artifacts caused by variations in the amount of fossil-bearing rock preserved through time."

2009 The Great North Museum opens at Hancock in Newcastle in the United Kingdom. Its first exhibit is a replica of *T. rex*, one of only four such skeletons on display in the country (the others are in BMNH, Manchester Museum, and the Oxford University Museum of Natural History). The specimen at Great North Museum is a replica of "Big Mike" that is displayed at the Museum of the Rockies in Bozeman, Montana. (In 2014, the Museum of the Rockies housed 12 skeletons of *T. rex*.)

2009 Paul Sereno and his international team of colleagues report the 9-foot-long, 125-million-year-old *Raptorex kriegsteini* ("king of thieves"), a small, scaled-down theropod that was 100-times smaller than *T. rex*. *Raptorex* had been smuggled out of China and bought by collector Henry Kriegstein, who donated the specimen to science. *Raptorex* lived 60 million years before *T. rex* and had a large head, strong legs, feet specialized for running, and small forelimbs tipped with claws. Although scientists had thought that some traits of *T. rex* (e.g., its large back-legs, small forearms, and large head) were adaptations for its large size, Sereno's discovery suggested that the *T. rex* predation blueprint evolved in smaller dinosaurs 60 million years before the giant tyrannosaurs emerged. As Sereno noted, "It was a running set of jaws. . . . No one had any idea there was something like *Raptorex* lurking around." The following year, BHI's Peter Larson claimed that *Raptorex* was a baby *Tarbosaurus*.

2009 The opening of Australia's Age of Dinosuars Museum of Natural History coincides with the announcement of three new species of Australian dinosaurs. Two of these, the theropod "Banjo" (*Australovenator wintonensis*) and the sauropod "Matilda" (*Diamantinasaurus matildae*) were featured on postage stamps in 2013. The dinosaurs were named for characters created by poet Banjo Paterson (1864–1941), who in 1985 wrote "Waltzing Matilda," the best-known Australian song of all time.

2009 *Proceedings of the National Academy of Sciences* reports *Alioramus altai*, a smaller relative of *T. rex* having a horn and long snout.

2009 Scientists at IVPP describe *Limusaurus inextricabilis* ("mud lizard who could not escape"), a toothless, 2-meter-long theropod from the Jurassic of western China. This was the first definite ceratosaur from eastern Asia.

2009 Scientists in China find the remains of feathered theropods in Middle-Late Jurassic rocks. Other feathered theropods had been found in Early Cretaceous rocks.

2009 South Africa issues 10 3-D postage-stamps featuring dinosaurs. This was the first time that a pair of viewing glasses was supplied with stamps. The stamps featured *Suchomimus, Afrovenator, Heterodontosaurus, Jobaria*, and *Ouranosaurus*.

2009 The 150-year anniversary of the publication of Charles Darwin's *On the Origin of Species* (and the 200-year anniversary of Darwin's birth) prompts BMNH to replace the statue of Richard Owen atop the double staircase overseeing BMNH with a statue of Darwin.

2009 The cover of the April issue of *Smithsonian* features the *T. rex* collected by Barnum Brown for AMNH in 1908 to promote its story "Dinosaur Wars: Who Owns America's Fossils?" The article discussed how "prospectors—and poachers—are excavating fossils in a cash-fueled free-for-all that often pits them against scientists and the law."

2009 Dinolandia opens near Cracow, Poland and features more than 50 life-size dinosaurs, including a 65-foot-long *Diplodocus* ("Dippy") and a re-creation of the scene similar to that from *Jurassic Park* in which *T. rex* and raptors attack a Land Rover.

2009 The Paleontological Resources Protection Act (Public Law 111-11) codifies requirements that the collection of vertebrate fossils (and other rare and scientifically significant fossils) on public lands be done only by researchers with permits. The researchers must agree to deposit the fossils in public institutions.

2009 U.S. president Barack Obama (b. 1961) signs into law the Paleontological Resources Protection Act, the first law to specifically protect fossils. The law set penalties of up to five years for stealing bones or other fossils from public lands.

2009 University of Kansas paleontologist David Burnham and his coworkers suggest that *Sinornithosaurus*, a turkey-sized feathered dinosaur unearthed in China, was venomous. Today, there are more than 20 genera (mostly theropods) of feathered dinosaurs.

2009 Xiao-Ting Zheng and his colleagues at China's Shandong Tianyu Museum of Nature discover *Tianyulong confuciusi*, a 130-million-year-old bipedal ornithischian dinosaur having primitive feather-like structures on its tail, back, and neck.

2009 Xu Xing and his colleagues describe *Anchiornis* ("near bird"), a small, feathered, troodontid dinosaur discovered in western Liaoning (about 320 km north of Beijing). The type-species is the 155-million-year-old *Anchiornis huxleyi*, named in honor of Thomas Huxley, an early advocate of the close evolutionary relationship between dinosaurs and birds. This fossil, which is older than *Archaeopteryx*, is the earliest known feathered dinosaur.

2009 Nick Longrich and Phil Currie describe *Hesperonychus*, a small (3 feet long) theropod discovered by Elizabeth Nichols in DPP in 1982. *Hesperonychus* is the smallest known carnivorous dinosaur of North America.

2010 A paper in *Science* signed by 41 scientists reaffirms an extraterrestrial impact as a cause of the end-Cretaceous mass extinction. Within two days, more than 25 paleontologists and others argued in a rebuttal that the pattern and timing of extinctions at the K-T boundary were best explained by multiple causes.

2010 A study of tyrannosaur bones concludes that Late Cretaceous tyrannosaurs (including *T. rex*) were cannibals.

2010 A team of Chinese, British, and Canadian paleontologists examines a dozen species of theropods and concludes that these dinosaurs had wrists like birds. The evolution of this trait occurred alongside increases in arm-length and feathers on the arms of some of these theropods. The researchers, who concluded that feathered arm-folding preceded flight, noted that "some characteristics that biologists used to think of as distinctively avian—feathers and air sacs are other examples—are actually quite deeply rooted in theropod evolution."

2010 A University of Texas/*Texas Tribune* survey reports that nearly one-third of Texans believe that dinosaurs lived with humans, and another 30 percent are unsure of whether they lived together.

2010 AMNH's Stephen Brusatte (b. 1984) and his coworkers report 249-million-year-old footprints left by a cat-sized dinosaur. The tracks, found in modern-day Poland, move the age of the first dinosaurs to just after the Permian–Triassic boundary, the greatest mass-extinction in Earth's history.

2010 Displays of the giant sauropods *Ruyangosaurus giganteus* and *Huanghetitan ruyangensis* debut at the Cincinnati Museum Center. The fossils had been found four years earlier in China.

2010 During the past decade, there have been more than 200 law enforcement incidents about fossils involving the U.S. Bureau of Land Management.

2010 Fifty-six new genera of non-avian dinosaurs are named in 2001, up from 44 in 2009.

2010 Geologist Thomas Algeo argues that the meteor impact at the end of the Cretaceous vaporized sulfur-rich sediments that produced extremely acidic precipitation. This environmental change, combined with volcanic eruptions, helped drive non-avian dinosaurs to extinction.

2010 Grzegorz Niedzwiedzki, Richard Butler, and Steve Brusatte report the oldest fossilized footprints of the dinosaur lineage in Early Triassic (250-million-year-old) rocks in central Poland. These tracks, which were 1–4 centimeters long and from a quadruped, were referred to the ichnogenus *Prorotodactylus,* which was not a true dinosaur, but one of dinosaurs' closest cousins.

2010 In *Journal of Vertebrate Paleontology*, Jack Horner and John Scannella of MOR argue that the iconic *Triceratops* was a juvenile version of *Torosaurus*,[46] and that *Nedoceratops* was an intermediate between the young adult *Triceratops* and the old adult *Torosaurus* stage (i.e., that *Nedoceratops* should be called *Triceratops*). Amidst the ensuing publicity associated with the discovery (e.g., there was a "Save the *Triceratops*" Facebook page), Horner was quick to add, "*Triceratops* was named before *Torosaurus* so it has precedence . . . *Triceratops* stays." A few months later, ceratopsian expert Andy Farke of the Raymond M. Alf Museum of Paleontology argued that *Nedoceratops* was different from *Triceratops* (e.g., *Nedoceratops* lacked a nasal horn, has vertical brow-horns, and slot-like openings in its frill).

2010 In *Nature*, American paleontologist Sterling Nesbitt and his colleagues report the discovery of *Asilisaurus kongwe* ("ancestor lizard"), a small (9 feet long and 6 feet high), 240-million-year-old, quadrupedal archosaur found in southern Tanzania. Unlike known archosaurs, *A. kongwe* had peg-like teeth, suggesting that it had a herbivorous or omnivorous diet. *A. kongwe* was the oldest known skeleton in the dinosaur lineage.

2010 In *The Fossil Record*, John Morris (b. 1946) and Frank Sherwin argue that *T. rex* "Sue" died "in the great Flood of Noah's day," that the fossil record of dinosaurs and other organisms reveals the "truth of the Bible's message to us," that faith in their god "allows us to do better science," and that fossilized dinosaur tracks "fit the biblical model better than an evolutionary one." That same year, *Acts & Facts* argued that "dinosaurs on the Ark probably would have been young adults, no bigger than a cow perhaps. Thus, there was plenty of room on board the Ark. But the world after the Flood was much different than before, with much less vegetation and a colder, harsher climate. Evidently the dinosaurs gradually died out. Perhaps they were even hunted to extinction by humans."

2010 Jurassic Forest, an educational and entertainment park featuring 40 animatronic, life-size dinosaurs, opens north of Edmonton, Canada. The area is a famous site of dinosaur discoveries.

2010 *Jurassic Park III* director Joe Johnston announces that *Jurassic Park IV* (set to be released in 2015) will begin a second *Jurassic Park* trilogy.

2010 MIT graduate student Benjamin Black and his colleagues propose that the end-Permian mass extinction—the "Great Dying" that eliminated 70 percent of terrestrial species and 90 percent of marine species—was partially caused by ozone-depleting gases (e.g., chlorine, fluorine) spewed by massive volcanic eruptions.

2010 On October 13, the U.S. National Parks Service and American Geological Institute celebrate the first annual National Fossil Day.

2010 Paleontologists from the United States and Canada discover *T. rex* bones scarred by *T. rex* tooth-marks.

2010 Paleontologists report that *Anchiornis huxleyi* had feathers resembling those of modern-day birds that adorned the creature's feet, lower legs, and forelimbs. This arrangement may have made the dinosaur clumsy on the ground, and bolstered the claim that flight originated from trees down.

2010 Paleontologists studying the eggshells of *Troodon formosus*, a small North American theropod, report that the eggs were incubated in a similar way to brooding birds. The eggs were oriented vertically, with only the egg bottoms in mud. These findings suggested that some bird-like nesting behaviors evolved in theropods prior to the origin of birds.

2010 Paleontologists use a fragmentary skeleton to report *Chromogisaurus*, a primitive sauropodomorph from the Late Triassic.

2010 Readers of *Entertainment Weekly* rank *Jurassic Park* the greatest summer-movie of the previous two decades.

2010 Researchers at the St. George (Utah) Dinosaur Discovery Site find the first handprint of a crouching carnivorous dinosaur. This 198-million-year-old fossil confirmed that the dinosaur's front limb hit the ground on its side (like a karate chop) rather than hanging like a monkey's (with palms near-parallel to the ground), and suggested that these dinosaurs were bipedal from relatively early in their history.

2010 Robert Reisz and his colleagues find the oldest known embryos of dinosaurs (almost 200 million years old), as well as the oldest known embryos of any land-dwelling vertebrate. The 8-inch embryos (from 16-foot-long adults), which were found in the southwestern China Yunnan Province, lacked teeth, suggesting that they were fed by adults. The embryos were tentatively linked with *Lufengosaurus*, an early relative of *Apatosaurus*. In 2013, Reisz and his colleagues reported that the embryos grew fast and moved in their eggs in ways similar to that of modern birds.

2010 Scott Sampson and his coworkers report the herbivore *Kosmoceratops*, a 76-million-year-old horned dinosaur discovered in Utah's Giant Staircase-Escalante National Monument (1.9 million acres in southern Utah protected in 1996 by President Bill Clinton). The two-ton, 15-foot-long *Kosmoceratops* was discovered in 2007 and had 15 full-sized horns on its 6-foot-long skull: one horn over its nose, one protruding from each cheek bone, one over each eye, and ten across the frill at the back of its skull. When *Kosmoceratops* lived in what is now modern-day Utah, the area was part of an island continent separated from the rest of the North America by a sea. *Time* magazine dubbed *Kosmoceratops*, a relative of the more familiar *Triceratops*, "the Latest and Strangest Fossil from America's Own Backyard," but several other newspapers and magazines described the 15-foot-long, three-ton *Kosmoceratops richardsoni* as "the horniest dinosaur ever discovered." The larger *Utahceratops gettyi*, which was found nearby, had a 7-foot-long head, weighed up to 4.5 tons, and had a more typical arrangement of horns, with one large one above its nose and two outward-pointing horns above its eyes. Many paleontologists believe that horns were used to attract mates and for fighting other members of the same species.

2010 Six countries account for 75 percent of all recent discoveries of dinosaurs: China: 140 species; the United States: 137 species; Argentina: 76 species; Mongolia: 68 species; Canada: 46 species; and the United Kingdom: 38 species. The abundance of dinosaur discoveries in the United States was attributed to (1) large areas of exposed land from the Late Triassic to the Late Cretaceous (i.e., when dinosaurs lived), (2) much of this area being arid and semiarid land where erosion exposes new fossils (in many other countries, much of the land is covered with vegetation), (3) paleontologists having searched for dinosaurs in the United States for more than a century, and (4) the availability of resources and money.

2010 *Stegosaurus* is featured in the first Series 4 Prequel Webisode of *Primeval*. The dinosaur inaccurately had a horn on its head because the producers used the skull of *Embolotherium* (an Eocene mammal) from a previous episode.

2010 Stephen Brusatte and his colleagues study 300 traits in 19 tyrannosaur fossils (e.g., *Tyrannosaurus, Albertosaurus, Tarbosaurus*) and conclude that *T. rex* "is quite abnormal when compared with other" tyrannosaurs. For most of their evolutionary history, tyrannosaurs were small.

2010 The Cretaceous, 105-million-year-old sauropod *Abydosaurus mcintoshi*, excavated from DNM, is named by David Chure and his colleagues. Unlike most sauropods, *Abydosaurus* ("Abydos lizard") is known from its skull and vertebrae; its name honors John McIntosh, a sauropod expert, and the Egyptian city Abydos (now El Araba el Madfuna) that in mythology held the head of the god Osiris. *Abydosaurus*, a relative of *Brachiosaurus*, was the first complete skull to be described for a Cretaceous sauropod from the Americas. Complete skulls exist for only eight of the approximately 120 species of sauropods, probably because the skulls are made of thin, fragile bones bound by soft tissue.

2010 The discovery of *Sarahsaurus* and *Seitaad* suggests that sauropods migrated into North America several times during the Early Jurassic rather than just being part of one move northward across the continent.

2010 The River of Death and Discovery Dinosaur Museum Society is established to develop bonebeds and build a museum in Wembley, Alberta, Canada. The $35-million Philip J. Currie Dinosaur Museum is scheduled to open in late 2014.

2010 The Royal Society lists famed fossil-hunter Mary Anning as the third-most influential woman scientist in British history.

2010 The Shangdong Tianyu Museum of Nature in eastern China's Shandong Province is declared by Guinness World Records to be the largest dinosaur museum in the world. The museum includes more than 300,000 square feet of exhibit space and displays more than 1,100 dinosaur specimens.

2010 Thomas Carr and Thomas Williamson report *Bistahieversor sealeyi*, one of the first tyrannosaurs from the American Southwest. *Bistahieversor* lived 75 million years ago during the Late Cretaceous. In 2011, Carr, Williamson and their colleagues described a second kind of tyrannosaur from the Southwest (*Teratophoneus curriei*), which was found in Utah's Grand Staircase-Escalante National Monument. In 2000, Carr shared the Lanzendorf Prize for Scientific Illustration.

2010 University of Cambridge paleontologist Roger B. J. Benson and Tom Rich of Australia's Museum Victoria report the first tyrannosaur fossil found in the Southern Hemisphere.

2010 Workers in Cuenca, Spain, uncover *Concavenator* (Latin for "the hunchback hunter from Cuenca"), a 130-million-year theropod. The 20-foot-long dinosaur had quilled elbows, a belly with alligator-like scales, a hunchback, and dimples on its forearms, suggesting that the dinosaur was feathered. This discovery pushed back the first known appearance of the feathers in the fossil record. Feathers appeared long before flight-capable birds or dinosaurs, suggesting that feathers were involved in thermoregulation and/or behavior (e.g., attracting a mate).

2010 Xu Xing of IVPP argues that feathered dinosaurs, like modern birds, may have molted as they grew, and that young dinosaurs may have looked unlike their elders (e.g., that *Similicaudipteryx* may have exhibited an age-related shift in its plumage).

2010 Yale University postdoctoral student Nicholas Longrich unveils the dinosaur *Mojoceratops*, a 75-million-year-old plant-eater whose name reflects its "flamboyant" appearance, including its heart-shaped headgear. Decades ago, the fossil—which was rediscovered in 2008 in the basement of AMNH—had been manipulated with plaster to resemble *Chasmosaurus*. Although press reports of Longrich's naming of *Mojoceratops* focused on the unusual name (most dinosaurs' genera are named for a distinctive feature, people, or for where the dinosaurs were found), Longrich noted that "mojo" refers to a magic charm for getting mates, and suggested that the dinosaur's frills were probably for courtship.

2010 Fucheng Zhang and his colleagues report preserved melanosomes (pigment-containing organelles) in a fossilized *Confuciusornis*, a primitive bird that lived with dinosaurs. This discovery suggested that *Confuciusornis* was gray, black, and reddish brown.

2010 Fucheng Zhang, Mike Benton, and their coworkers suggest in *Nature* that *Sinosauropteryx* had a red-and-white-striped tail that may have been used to attract mates. This work, which used samples from several different specimens, was the first study to reveal coloration in dinosaurs. The next week, in *Science*, Jakob Vinther and his colleagues reported the colors of the entire plumage of *Anchiornis*.

2010 Zoltan Csik, Matyas Vremir, and their coworkers in Romania unearth a turkey-sized relative of *Velociraptor*. The stocky, Late Cretaceous carnivore, which was named *Balaur bondoc* ("stocky dragon") after the dragon-like balaur of Romanian folklore, had large, hyperextensible claws on both the first and second digits of its feet (other dromaeosaurids have sickle claws only on their second toes).

2011 After being closed for five years for remodeling, the new 10,500-square-foot Quarry Exhibit Hall reopens at DNM, 96 years after the creation of the original 80-acre DNM. The exhibit, which is built over the Carnegie Dinosaur Quarry, features a 200-foot-long cliff face from the Jurassic. Today, DNM covers 210,000 acres.

2011 After reconstructing a phylogeny to include the 155-million-year-old, chicken-sized *Xiaotingia zhengi*, Xu Xing and his colleagues cautiously suggest that the crow-sized *Archaeopteryx* be classified not as a bird, but instead as a birdlike dinosaur. Much of what has been known about the early evolution of birds was influenced by *Archaeopteryx*; if *Archaeopteryx* is considered a birdlike dinosaur, feathered fossils such as *Epidexipteryx* could be the oldest birds. That same year (2011), paleontologists reported the discovery of the 11th specimen of *Archaeopteryx*. Like other *Archaeopteryx*, this specimen was found in southern Germany.

2011 After screening thousands of pieces of amber discarded by a coal mine near Grassy Lake in southern Alberta, Canada, Ryan McKellar and his colleagues find pigments in 11 amber-encased feathers of dinosaurs and early birds. The specimens included unbranched filaments (proposed as the first stage of feather evolution) and those of modern diving birds.

2011 *American Experience* premiers *Dinosaur Wars*, a documentary showcasing the hostilities between Othniel Marsh and Edward Drinker Cope during their famous "Bone War." Although the Cope–Marsh feud stimulated growth of paleontology and entertained readers of newspapers, it also produced many sloppy, incomplete, confusing, and inaccurate publications. It took paleontologists decades to resolve many of the taxonomic problems produced by Cope and Marsh.

2011 AMNH's "The World's Largest Dinosaurs" exhibit features a variety of super-sized sauropods, including *Mamenchisaurus*. That same year, Honolulu's Bishop Museum featured "Dinosaurs Unearthed," an exhibit including moving replicas of dinosaurs. Outside, a 22-foot-tall *T. rex* attacked a 10-foot-tall *Triceratops*.

2011 Analyses of protein from bones of *T. rex* and a duckbill hadrosaur support the claim that the proteins are 80-million-year collagen; amino-acid sequences in the ancient tissues matched those in protected regions of collagen of humans and rats, where they might degrade slowly.

2011 At the annual meeting of the SVP, Ryan Carney announces that feathers of *Archaeopteryx* were black.

2011 Colorado College's Henry Fricke and his colleagues suggest in *Nature* that sauropods in western North America migrated to uplands to avoid summer droughts. Such migrations might have helped explain how sauropods reached their giant sizes. Fricke's study was based on oxygen isotopes in enamel of 150-million-year-old *Camarasaurus* differing from those of minerals in nearby sediments.

2011 Dinosaurs Alive!, the world's largest animatronic dinosaur-themed park, opens at Kings Island in Mason, Ohio. The park's 4,000-foot-long trail features more than 60 life-size dinosaurs. There are now several Dinosaurs Alive! parks in the United States and Canada.

2011 Glen Kuban concludes that the carving of "Stegosaurus" at Cambodia's Ta Prohm temple (built in the late 1100s) depicts a chameleon or rhinoceros, not a dinosaur. This carving has been discredited repeatedly, but continues to be used by some creationists to claim that humans lived with dinosaurs.

2011 In *Answers* magazine, young-Earth creationist Marcus Ross (b. 1976) claims that *T. rex* "ate only plants," that Adam's Fall transformed *T. rex* into a carnivore, and that fossils of *T. rex* date to "Noah's time, nearly 1,700 years after Adam's Fall." Ross also claimed that the short arms of *T. rex* were one of two options available to God (the other being a longer tail). According to Ross, the "marvelous designs" of extinct dinosaurs "praise their Creator."

2011 In *Extinction and Radiation: How the Fall of Dinosaurs Led to the Rise of Mammals*, J. David Archibald concludes that there is no single pattern of extinction and survival at the K-T boundary, and that multiple causes—especially volcanism, marine regression, and extraterrestrial impact—provide the best explanation for the phylogenetic and ecological differences between those who survived the K-T extinction and those who did not. Although the impact at the end of the Cretaceous was the tipping point for the deteriorating environment, none of the other four mass extinctions was associated with an impact, and all of them were associated with major global marine regressions.

2011 In *Science*, Paul Sereno and his coworkers announce their discovery of a bipedal, dog-sized, 230-million-year-old (i.e., Late Triassic) theropod named *Eodromaeus* ("dawn runner"). *Eodromaeus*, which was found in the Andes foothills of what is now Argentina, lived near the dawn of dinosaurs, was 4 feet long, weighed 10–15 pounds, and ate plants and animals. The fossils of *Eodromaeus* had been found in 1991 in Ischigualasto, an arid valley that had yielded *Eoraptor* two decades earlier (which Sereno and others had named in 1993). Sereno noted that some of the traits of *Eodromaeus*, such as its grasping hands, air pockets in the skull, and a balancing tail, "you find 150 million years later in *T. rex*, right before the asteroid hit. . . . It was a little dinosaur, but it carried a big evolutionary stick." *Eoraptor*, a similar-sized dinosaur, lived at the same time and in the same area, but *Eoraptor* ate plants. Later in 2011, Sereno announced the discovery of a crocodile-dominated ecosystem from the Cretaceous in North Africa.

2011 Kentucky governor Steve Beshear (b. 1944) announces that AiG will build "Ark Encounter," a 160-acre (expandable to 800 acres), $150-million theme-park based on Noah's Ark and other icons of young-Earth creationism. The park, which will include a 500-foot-long, 75-foot-wide, $25-million full-scale replica of Noah's Ark and will portray humans living with dinosaurs, anticipates opening in 2015 Officials project that Ark Encounter will attract more than 1.5 million customers per year.

2011 Neck bones of "Ashdown maniraptoran" are found in Cretaceous sediments (145–100 million years old) in the southern United Kingdom. This feathered, birdlike theropod, which was less than 40 centimeters long, was one of the world's smallest dinosaurs. At the time of its discovery, the tiniest dinosaur was *Anchiornis*, a birdlike creature that lived in China 160–155 million years ago. In 2013, John Hutchinson and his colleagues argued that some dinosaurs began to stand and move like birds as they evolved bigger arms.

2011 Octávio Mateus (b. 1975) and his colleagues describe the sauropod *Angolatitan adamastor*, one of the largest creatures in history and the first dinosaur discovered in Angola.

2011 On average, a new dinosaur is discovered every eight weeks.

2011 Paleontologist Heinrich Mallison reports that *Kentrosaurus* could swing its tail in a 75° arc and, in doing so, puncture tissue and break bones of attackers.

2011 Paleontologists working in Utah find four intact skulls of a previously unknown species that appears to be a transitional form between early and late sauropods.

2011 *Pentaceratops*, which was excavated 70 years ago from 74-million-year-old rocks in northwestern New Mexico, is suggested by Yale University's Nick Longrich to be *Titanoceratops ouranos*, an older relative of *Triceratops*. The dinosaur's skull, among the largest of any known land-animal, was 2.6 meters long.

2011 Petroglyphs left several hundred years ago by Native Americans at Utah's Kachina Bridge—which young-Earth creationists claim to show humans living with dinosaurs—are shown to be smeared pictures. Some of the lines are and etchings by humans, whereas others are natural stains. Officials at young-Earth organizations such as the Creation Evidence Museum dismissed the study and continued to claim that dinosaurs lived with humans a few thousand years ago.

2011 Tony Fiorillo of Dallas's Perot Museum of Nature and Science names *Pachyrhinosaurus perotorum*, which had been found five years earlier along Alaska's North Slope. The dinosaur had distinctive frills, a dome not on other pachyrhinosaurs, and "a horn going the wrong way." Elsewhere in Dallas, the Museum of Earth History claims that humans lived with dinosaurs.

2011 The area near Laiyang in Shandong Province where some of China's first dinosaur fossils were found in 1923 (e.g., *Tanius sinensis*) is made a national park.

2011 Researchers at Johns Hopkins University describe *Propanoplosaurus marylandicus*, the youngest nodosaur ever discovered. The 13-centimeter-long specimen had been discovered four years earlier by Ray Stanford.

2011 Staff of the Morrison Natural History Museum discover tracks of an infant *Apatosaurus* in foothills west of Denver, Colorado. These tracks appear to show baby dinosaurs walking on their back feet.

2011 Surgeon Benjamin Carson (b. 1951) uses "The World's Biggest Dinosaurs Museum" (Figure 36) in Cabazon, California, to claim that there is no evidence for evolution. As Carson noted, "It's just not there."

2011 The bird-like, 3-foot-tall *Talos sampsoni*, discovered three years earlier, is announced from the Late Cretaceous of the Grand Staircase-Escalante National Monument.

2011 The expensive, prehistoric *Terra Nova*, produced by Steven Spielberg, premiers on Fox television. In the show, the Shannons travel back in time 85 million years and mingle with dinosaurs. The show was canceled after one season (11 episodes).

2011 The large hipbone of the newly described *Brontomerus* ("thunder thighs") is suggested to have supported massive muscles used for kicking and attacking predators.

2011 The lump in the chest of *Thescelosaurus* suggested in 2000 to be a fossilized heart is shown to be cemented sand.

2011 The Natural History Museum of Los Angeles County opens its 14,000-square-foot Dinosaur Hall, which houses more than 300 fossils and 20 complete mounts of dinosaurs and other creatures. The museum includes the world's only display of differently aged specimens of *T. rex* (an 11-foot-long, two-year-old baby, a 20-foot-long, 14-year-old juvenile; and a 34-foot-long, 17-year-old young adult). More than 50 specimens of *T. rex* have been found covering a span of two million years.

2011 The Texas Natural Science Center (outside the Texas Memorial Museum) in Austin begins restoring some of the dinosaur tracks collected by Roland Bird near Glen Rose, Texas. The tracks, which will be housed in a climate-controlled gallery inside the Texas Memorial Museum, had been displayed since 1941 in a small building that was not climate-controlled. Absorption of water from the underlying soil cracked and eroded the slab of tracks.

2011 Using new techniques for three-dimensional imaging, researchers at London's Royal Veterinary College and the Field Museum estimate that *T. rex* gained as much as 3,950 pounds per year as a teenager. When alive, the Field Museum's *T. rex* "Sue" weighed almost nine tons (30% more than earlier estimates).

2011 When other imaging techniques cannot reveal a dinosaur bone's internal structure, Kristi Curry Rogers drills into the bone and discovers that it is hollow. The bones (osteoderms), from a 50-foot-long *Rapetosaurus* from Madagascar, were apparently used as a reservoir for minerals such as calcium. In 2001, Curry Rogers had named *Rapetosaurus* for the mischievous Malagasy folklore giant, Rapeto.

2011 Yale University paleontologists find a horn of *Triceratops* just 5 inches below the K-T boundary, indicating that some dinosaurs lived up to the K-T event. The previous "fossil gap" further below the K-T boundary (3 meters in North America as of 1994) had prompted some researchers to suggest that dinosaurs gradually declined and were extinct before the asteroid hit.

2012 A new computer-simulation suggests that *T. rex* bit down with a force of almost 12,800 pounds, which was about the weight of an adult *T. rex*. This was more than six times greater than that of *Allosaurus*, and more than three times more force that previous estimates, and made *T. rex* the hardest-biting terrestrial animal ever known. Other record-holders include 41,000 pounds by an extinct shark (*Carcharodon megalodon*, or *Carcharocles megalodon*) and 23,000 pounds by an ancient crocodilian (*Deinosuchus*).

2012 *All Yesterdays* depicts dinosaurs in unfamiliar, speculative scenes to give readers new perspectives about the Mesozoic.

2012 AMNH paleontologist Stephen Brusatte and three colleagues' studies of seven major non-avian dinosaur groups (comprising almost 150 species that included hadrosaurs and ceratopsids) conclude that some groups of dinosaurs were declining toward the end of the Cretaceous. Other groups, including many theropods and midsized herbivores, were not declining. One group, the sauropods, was growing.

2012 American chemist Ronald Breslow suggests that "elsewhere in the universe there could be life forms based on D-amino acids and L-sugars. Such life forms could well be advanced versions of dinosaurs, if mammals did not have the good fortune to have the dinosaurs wiped out by an asteroidal collision, as on Earth."

2012 An international team of paleontologists, employing techniques used to measure stress on objects such as cars, analyzes a skull of *Diplodocus* and concludes that the 25-ton animals maximized their intake of food by stripping leaves and swallowing them whole.

2012 Analyses of 11 teeth from two species of sauropods indicate that the dinosaurs had body temperatures similar to that of modern humans (i.e., 100.8°F and 96.3°F). Caltech's Rob Eagle noted, "If you [had] walked up to a dinosaur, it would feel like a cow."

2012 Carl Baugh's Creation Evidence Museum of Texas announces plans to build a 25-foot-long, 1:20 scale replica of Noah's Ark. That same year, Baugh announced that he would soon "bring online" the "world's first hyperbaric biosphere," a giant container that Baugh claims simulates atmospheric conditions before Noah's flood. This biotron is claimed to have tripled the lifespan of fruit flies and detoxified poisonous snakes. The ark, complete with several rooms housing dinosaurs, was completed in 2013.

2012 Columbia University paleontologists Paul Olsen (b. 1953) and Dennis Kent argue that the reign of dinosaurs began with a meteor strike at the end of the Triassic. Although dinosaurs had evolved 25 million years earlier, the meteor strike removed much of their competition, after which the dinosaurs radiated into new niches.

2012 DinoPark opens atop the Harfa Mall in Prague, Czech Republic. The 7,000-square-foot rooftop-exhibit features more than 30 dinosaurs. Outside the mall, life-size models of *Hadrosaurus* and *Brachiosaurus* lure passersby to the attraction.

2012 During 2012, paleontologists name 32 species of non-avian dinosaurs and five new species of hadrosaurs.

2012 Eric Prokopi sells *Tarbosaurus* at a Manhattan action for $1,050,000 to real-estate developer Coleman Burke, but soon thereafter Prokopi—a self-described "commercial paleontologist"—is charged by Homeland Security officials with smuggling fossils (including the duckbill *Saurolophus*) from Mongolia. (The Gobi Desert is the only place that *Tarbosaurus* has ever been found.) Although Prokopi claimed to be just "a guy from Gainesville, Florida . . . trying to support my family, not some international bone smuggler," authorities described him as "a one-man black market in prehistoric fossils." Prokopi billed the fossil as "75 percent complete," but it was made from bones of several dinosaurs; Prokopi, who tried to deceive authorities by describing the fossils as assorted reptile fossils, later pled guilty to three felonies, and agreed to return his stolen fossils to Mongolia, where fossils are state property. (Among Prokopi's fossils that were sold illegally was a 67-million-year-old skull of *T. bataar* bought by American actor Nicolas Cage in 2007 for $276,000.) When the stolen fossils were returned to Mongolia, they were seen by more than 500,000 visitors in just four months at the newly created national Mongolian dinosaur museum. Paleontologists have found only about 20 intact specimens of *Tarbosaurus*.

2012 Florida State University paleobiologist Gregory Erickson and his colleagues report that teeth of 70-million-year *Edmontosaurus* had six types of tissue that wore down at differing rates, thereby causing the teeth to resemble those of modern horses and other grazers (that have four main types of tissues).

2012 Fossils of the feathered, flightless, wolf-sized *Sinocalliopteryx gigas*, a carnivorous dinosaur from Liaoning, China, are found with bones of the cat-sized, feathered *Sinornithosaurus* and the crow-sized *Confuciusornis*. This suggested that *Sinocalliopteryx* was an active hunter.

2012 Geochronologists show that massive volcanic eruptions at the opening of the Atlantic Ocean 201 million years ago triggered a mass extinction that cleared the way for the rise of dinosaurs.

2012 ICR's *Acts & Facts* claims that, because birds and dinosaurs were created by God on different days of creation week, (1) structures on feathered dinosaurs are probably "decayed skin," not feathers, and (2) "dinosaurs could not have evolved into birds." In the same publication, Frank Sherwin and Brian Thomas claimed that sauropods were created by God, and did not evolve.

2012 Oliver Rauhut (b. 1969) and colleagues describe *Sciurumimus,* a small, feathered theropod from Late Jurassic sediments in Germany. The fuzzy juvenile of *Sciurumimus,* a close relative of *Megalosaurus*, hinted that feathers were older and more widespread in dinosaurs than previously believed.

2012 Paleontologist Xu Xing and colleagues describe the theropod *Yutyrannus huali* ("beautiful feathered tyrant"), the largest known dinosaur covered with feather-like structures (proto-feathers) discovered to date. *Yutyrannus* was 40-times larger than the largest previously known feathered dinosaur. *Y. huali*, which was discovered in 125-million-year sediments in northeastern China, was 9 meters long, weighed 1.5 tons, and was about one-fourth the mass of *T. rex*. Corwin Sullivan of IVPP described *Yutyrannus* as "rather shaggy-looking, a killer fuzzball with deadly teeth and claws." The proto-feathers, which were too spindly for flight, probably insulated the animals during a cold part of the Early Cretaceous.

2012 Paleontologists suggest that the hind wings of four-winged dinosaurs such as *Microraptor gui* probably stayed tucked under their body until they were used to chase prey or dodge branches. Raising one leg and tucking the other could have sped the dinosaur's aerial movements. In 2013, the discovery of *Microraptor* with a stomach full of fish prompted researchers to suggest that the dinosaur hunted like a housecat.

2012 Researchers led by David Wilkinson argue that sauropods' farts released 520 million tons of methane into the atmosphere per year. This amount is similar to the total amount of methane produced today by natural and human activities.

2012 Robert Telleria's *The Visual Guide to Scale Model Dinosaurs* is the first catalogue of collectible sculptures and model kits of dinosaurs and other prehistoric animals.

2012 Sarah Brown and her colleagues use charcoal deposits in the fossil record to conclude that fires were widespread during the Cretaceous. These fires were probably a typical part of dinosaurs' lifestyles and may have forced migrations.

2012 Science historian Charles Milner's *Charles R. Knight: The Artist Who Saw through Time* celebrates the artist who Stephen Jay Gould claimed had more influence on our ideas of extinct creatures than any paleontologist.

2012 Sterling Nesbitt and an international team of paleontologists formally describe what may be the world's earliest dinosaur: *Nyasasaurus parringtoni*, a Labrador dog-sized animal that lived 245 million years ago (10–15 million years earlier than previous candidates for earliest dinosaur). The fossils (NHMUK R6856) had been collected near Tanzania's Lake Nyasa in the 1930s by Cambridge's Rex Parrington, and were first described in 1956 in Alan Charig's doctoral dissertation, after which they were kept in BMNH.

2012 The 130-million-year-old dinosaur trackway near James Price Point in Western Australia is threatened by construction of a natural-gas plant. The site is significant because it includes thousands of tracks of several major groups of dinosaurs.

2012 The Houston Museum of Natural Science opens a 30,000-square-foot, $85-million Hall of Paleontology. The museum includes "Wyrex," a *T. rex* with well-preserved hands, feet, and small patches of skin impressions.

2012 The movie *Jurassic Park* (made for $63 million in 1993) has grossed $357,067,947 domestically, making it the fifth-highest grossing film of all time in the United States. For comparison, the U.S. domestic gross for *The Lost World; Jurassic Park* (made for $73 million in 1997) was $229,086,679, and that for *Jurassic Park III* (made for $93 million in 2001) was $181,171,875. *Jurassic Park* spawned several new dinosaur-themed books, comics, toys, video games, cookies, and other products.

2012 The movie *The Dinosaur Project* describes an expedition to Congo by the British Cryptozoological Society to find Mokèlé-mbèmbé. After a flock of flying reptiles causes the group's helicopter to crash, the explorers find a small, two-legged dinosaur (as in *Gorgo,* the dinosaur is only a baby). *T. rex* dominates the DVD's cover, but was not in the movie.

2012 The Natural History Museum of Utah reopens in a new building overlooking Salt Lake City.

2012 The Royal Canadian Mint issues 25,000 collectable coins featuring *Pachyrhinosaurus lakustai,* a four-ton, 26-foot-long dinosaur discovered in Alberta in 1972. The skeleton of the dinosaur on the $29.95 coin glowed in the dark.

2012 Tokyo-stomping *Gojira* (1954), who inspired 27 sequels (including *Godzilla*) and numerous spinoffs, is restored on DVD. *Gojira,* an amalgam of words for gorilla and whale, was a dark and bitter allegory about nuclear war.

2012 University of Barcelona paleontologist Meike Köhler and her colleagues report that annual growth-rings of 41 species of warm-blooded mammals (e.g., reindeer and gazelles) resemble those in dinosaurs' bones. Köhler's results undermined claims that differences in growth lines support dinosaurs being cold-blooded.

2012 When television evangelist Pat Robertson is told of a follower's fear that her children and husband will not go to heaven "because they question why the Bible could not explain the existence of dinosaurs," Robertson responds, "You've got all the carcasses of dinosaurs frozen in time out in the Dakotas. They're out there. So, there was a time when these giant reptiles were on the Earth, and it was before the time of the Bible. So, don't try and cover it up and make like everything was 6,000 years. That's not the Bible. If you fight science, you are going to lose your children." Young-Earth creationists such as AiG's Ken Ham, who claim that dinosaurs lived with humans, denounced Robertson's comments as "destructive."

2012 Workers at AMNH describe the 2.5-pound *Albinykus baatar,* one of the smallest theropods ever discovered. The fossil came from the eastern Gobi Desert.

2012 Yale University's Nicholas Longrich and Daniel Field's analyses of immature and adult *Triceratops* and *Torosaurus* conclude that the two dinosaurs are separate species. Jack Horner, who in 2010 claimed that *Triceratops* was a juvenile version of *Torosaurus* and attributed *Triceratops*' frill to youth, suggested that the Yale researchers had used a criterion (i.e., bone structure) that may not be a reliable way to gauge age.

2013 AiG reaffirms its claim that, in the beginning, dinosaurs and all other animals were vegetarians and that dinosaurs lived with humans.

2013 Although *T. rex* is arguably the most recognizable dinosaur, 63 percent of precollege students (and 72% of college students) draw the tyrannosaur upright like the cartoon character *Barney* rather than pitched forward as it was in real life. Cornell paleontologist Robert Ross concluded that students' perceptions of *T. rex* are stuck in the early 1900s.

2013 Australian mining magnate Clive Palmer (b. 1954) is granted permission to build the "world's largest" park of robotic dinosaurs. The park, at Palmer's Coolum Resort, includes more than 150 moving and roaring dinosaurs.

2013 The San Diego Natural History Museum cancels an auction of a dozen dinosaur fossils (valued at $15 million) after enraging dinosaur fossil experts. The auction, which included some discovered by Charles M. Sternberg, would likely have made the fossils inaccessible to scientists.

2013 German researchers use a CT scanner and a 3D printer to duplicate a dinosaur fossil wrapped in plaster. The new nondestructive technique is less time-consuming than conventional methods, and is promising for objects too rare or fragile to be moved.

2013 Australian paleontologist Roger Seymour and his colleagues conclude that dinosaurs needed warm blood to fuel the muscles required for their lifestyles.

2013 Canadian paleontologists describe the turkey-sized *Albertadromeus syntarsus*, the smallest herbivorous dinosaur to be found in Canada.

2013 Carl Baugh's Creation Evidence Museum of Texas resumes excavations along the Paluxy River to document that humans lived with dinosaurs. The excavation uncovered three "man tracks" near the tracks of dinosaurs.

2013 Dartmouth researchers report that the comet associated with the K-T mass extinction was much smaller than originally thought.

2013 David Burnham and his colleagues unearth a hadrosaur with a *T. rex* tooth in its tail, suggesting that *T. rex* hunted its prey.

2013 David Evans and his colleagues report *Acrotholus* ("high dome"), the oldest known pachycephalosaur ever found in North America. *Acrotholus* roamed what is now Alberta, Canada, 85 million years ago.

2013 A skeleton of "Misty," a 55-foot-long *Diplodocus longus* found in 2009 in Wyoming by the children of German paleontologist Raimond Albersdörfer, sells for $651,000. The iconic dinosaur will be displayed at Denmark's Natural History Museum in Copenhagen.

2013 In Alberta's Dinosaur Provincial Park, paleontologists find the smallest intact skeleton ever found of a ceratopsid. The specimen of *Chasmosaurus belli*, now named "Kevin," was three years old and 1.5 meters long.

2013 Decades after geologists rejected Carl Baugh's discoveries of dinosaur tracks with alleged human ("man") tracks, ICR's John Morris cautiously concedes that "man tracks" near Glen Rose, Texas, "are too ambiguous to make a clear determination." Nevertheless, Morris again proclaimed, "ICR holds rigorously to the view that dinosaurs lived at the same time as man."

2013 Episode 33 of ICR's *That's a Fact* claims the Bible says—and science confirms—that dinosaurs did not live before humans, but instead were created on the same day as (and lived with) humans. At Ken Ham's Creation Museum ("Prepare to Believe!), a leading seller is a shirt showing *Velociraptor* and the text "God Created Me on Day 6."

2013 Geologists at the University of Alaska-Fairbanks discover thousands of tracks from a variety of dinosaurs in Late Cretaceous rock along Alaska's Yukon River. AiG claimed that "the global flood actually makes sense of discoveries like these," and AiG geologists noted that the rock layer containing the prints was deposited "not long before Day 150" of the Biblical flood.

2013 Haley Coomes ranks *Micropachycephalosaurus* (MY-co-PACK-ee-SEFF-a-low-SORE-us) as the most difficult dinosaur-name to pronounce.

2013 In *Chronicles of Dinosauria,* David Wetzel and Richard Dobbs claim that dinosaurs in the classic "death pose" (i.e., with their head back and recurved tail) resulted from asphyxiation because the dinosaurs "drowned in the great Flood."

2013 In Ulan Bator (the capital of Mongolia), the Lenin Museum (including a giant statue of Soviet dictator Vladimir Lenin [1870–1924]) is replaced with an exhibit of fossils, including a 70-million-year-old *Tyrannosaurus bataar*. (The museum had been open since 1980, when Mongolia was a Soviet satellite.) The *T. bataar* (which is also known as *Tarbosaurus bataar* because of a dispute over taxonomy) was returned to Mongolia when Erik Prokopi pled guilty in 2012 to stealing the fossils.

2013 Jan Werner and Eva Maria Griebeler use models of extant birds and reptiles to conclude that theropods probably produced one clutch of eggs per year, and that hadrosaurs and sauropods produced several. Most dinosaurs probably produced fewer than 200 eggs per year, and giant sauropods probably produced less than 400 eggs per year.

2013 Jordan Mallon of the Canadian Museum of Nature examines 100 dinosaur skulls and concludes that different species of large plant-eating dinosaurs coexisted by specializing in eating different types of vegetation.

2013 Researchers find the first mass-burial site of *Triceratops* near Newcastle, Wyoming. Near the site is a mass grave containing hundreds of hadrosaurs.

2013 Kickstarter offers 3D-printed sculptures of *T. rex* skulls. A 3-inch model cost $38, and a 10-inch model cost $300.

2013 Maureen O'Leary and her colleagues construct a new family tree based on 27 genes and 4,541 anatomical traits showing that lineages of the 5,100 species of modern placental mammals arose after dinosaurs went extinct 65 million years ago. (Placental mammals such as bats, horses, whales, and humans produce well-developed young.) Although some scientists had concluded from molecular data that some lineages of modern placental mammals originated as early as 100 million years ago, paleontologists were skeptical because there were no reports of such fossils older than 65 million years.

2013 Paleontologists working near the town of Lufeng in southern China find a bed of *Lufengosaurus* (an early cousin of sauropods) bones, eggshells, and embryos. The Early Jurassic (190–197 million years old) site yielded 24 femurs of embryos, but few skulls. The bones were porous, suggesting increased blood-flow and rapid growth. *Lufengosaurus* was the first dinosaur mounted in China, and the first dinosaur to appear on a postage stamp anywhere in the world.

2013 Pascal Godefroit and colleagues describe *Eosinopteryx*, a feathered, flightless theropod from Late Jurassic sediments of China's Liaoning Province. The small (30-centimeter-long) non-avian dinosaur showed that dinosaurs and birdlike dinosaurs occupied different niches. *Eosinopteryx* had a small wingspan, a bone-structure that would have restricted its ability to flap its wings, and toes suited to walking on the ground. Unlike *Anchiornis, Eosinopteryx* lacked flight feathers, and instead had flexible, downy filaments that probably functioned in sexual displays or insulation. The discovery of *Eosinopteryx*, which predated other dinosaurs that evolved into flying birds in the Early Cretaceous, questioned the claim that *Archaeopteryx* was pivotal to the evolution of modern birds.

2013 Paul Renne (b. 1957) and his colleagues at the University of California at Berkeley's Geochronology Center use a high-precision dating analysis of debris from the Chicxulub impact to conclude that the K-T extinction and the K-T impact both occurred approximately 66,038,000 years ago, and no more than 33,000 years apart. These two dates are as close "as one can possibly demonstrate with existing dating techniques." As Renne said, "The impact was clearly the final straw, the tipping point. We've shown that [the impact and extinction] are synchronous to within a gnat's eyebrow, and therefore than an impact clearly played a major role in the extinction. But it probably wasn't just the impact. The impact was the *coup de grace*." Earth's atmospheric carbon cycle returned to normal within 5,000 years after the impact, and the oceans were back to normal 1–2 million years later. Later in 2013, researchers used multibeam sonars to create the first detailed map of the impact crater, which is now covered by almost 1,000 meters of marine sediments.

2013 Scientists at the University of Manchester report that DNA cannot be extracted from amber, thereby dashing hopes (or fears) that dinosaurs could be resurrected as they were in Michael Crichton's *Jurassic Park* (1990).

2013 Phil Manning and his colleagues use powerful X-ray techniques to determine that feathers of *Archaeopteryx* were lightly colored with black tips (rather than all black as previously thought).

2013 Princeton University's Evan Saitta uses CT scans of stegosaur bone plates to provide the first evidence for sexual selection in dinosaurs.

2013 University of Alberta becomes the first Canadian institution to offer a massive open online course (MOOC) for credit in Canada with its *Dino 101: Dinosaur Paleobiology*, which is taught by Phil Currie.

2013 Hundreds of fossilized eggs of at least four species of sauropods are found in the Lleida Province of northern Spain. This area was a marsh during the Late Cretaceous.

2013 Researchers interpret claw marks on a 100-million-year-old riverbed in China as indicating that some theropods swam by dog-paddling.

2013 Royal Mail issues 10 stamps commemorating the United Kingdom's more than 200 years of dinosaur discoveries. The stamps feature dinosaurs such as *Iguanodon, Hypsilophodon, Megalosaurus*, and *Cetiosaurus*.

2013 Scott Persons, Philip Currie, and Mark Norell's analysis of a feathered, two-legged *Oviraptor* from Mongolia suggests that the dinosaurs used their muscular tails for courtship displays.

2013 Scott Sampson and his colleagues at the Natural History Museum of Utah announce *Nasutoceratops titusi*, a plant-eating horned dinosaur found in Utah's Grand Staircase-Escalante National Monument. The 5-meter-long *Nasutoceratops* ("big-nose horned face"), which had been discovered in 2006, lived on the swampy landmass of Laramidia (about one-fifth the size of Africa) 76 million years ago. The new dinosaur, which had a stout snout, simple frill, and unusually long horns, represented a previously unknown branch of the horned dinosaur family. Horned dinosaurs evolved in only 5 to 8 million years, and became the most diverse family of dinosaurs on Laramidia.

2013 IVPP's Hai-Lu You finds *Yunganglong datongensis*, the seventh known basal hadrosaur ever found. The other six include three from Asia and three from North America.

2013 *Lythronax argestes*—a two-ton tyrannosaur covered with scales and feathers—found in 2009 in southern Utah is unveiled at the Natural History Museum of Utah. The discovery showed that giant tyrannosaurs were around 10 million years earlier than previously thought.

2013 The Australian PGA Championship leaves its traditional Queensland location following a dispute about the placement of ads and Jeff, a 20-meter-long dinosaur that had been built between the ninth and tenth holes.

2013 The bird-like, 1.6-meter-long *Aurornis xui* is reported after being found the previous year in northeastern China. Researchers claimed that the Middle Jurassic (160-million-year-old) *A. xui* was the most primitive bird and that it affirmed *Archaeopteryx* as a bird. Few reports, however, noted that the chicken-size *Aurornis* ("dawn bird") had been acquired from a fossil dealer, and the authors of the *Nature* paper in which it was announced acknowledged that it could be 35 million years younger than reported. There is a vast black market for fossils and forgeries in China (e.g., "Archaeoraptor" proclaimed by National Geographic in 1999 as a "missing link" between birds and dinosaurs), and many of the forgeries are masterful.

2013 There are now 1,415 named dinosaurs, up from 1,344 two years earlier. Of the 1,415 named dinosaurs, 897 are established names, 129 are doubtful names, 85 are junior synonyms, 38 are preoccupied names, 6 are rejected or forgotten names, 129 are vernacular names, and 111 are non-dinosaurian (valid or invalid) names.

2013 Two decades after its original release, Steven Spielberg's box-office hit and cultural phenomenon *Jurassic Park* returns to movie theaters in 3-D. The three movies of the *Jurassic Park* franchise have taken in more than $1.9 billion worldwide. *Jurassic Park 4: Jurassic World* is scheduled for release in summer 2015.

2013 University of Pennsylvania researchers suggest that three species of *Psittacosaurus* are actually one. Fossils of *Psittacosaurus* are abundant; as of 2011, more than 300 skeletons of *Psittacosaurus* had been collected (it is also an index fossil for early Cretaceous sediments of central Asia).

2013 BMNH's Paul Barrett and ROM's David Evans argue that feathers on dinosaurs were the exception rather than the rule. Most ornithischians and sauropods had scales or armor.

2013 University of Toronto researchers announce the 85-million-year-old, dog-sized, bipedal *Acrotholus audeti*, the oldest pachycephalosaur in North America.

2013 College students Christie Sims and Alara Branwen (both pen names) publish several erotic books starring dinosaurs, with titles that include *Ravished by the Triceratops, Taken by the T-Rex*, and *In the Velociraptor's Nest*. All of the books' descriptions at Amazon.com warned readers that "this is a tale of monster sex. This story was written to unlock your darkest fantasies and innermost desires. It is not for the faint of heart and is not your mother's erotica. All of the sexual descriptions found in this book are very explicit in nature. It's not suitable for someone under 18 years of age. Read at your own risk." Sims described herself as "a plain old, everyday Midwestern girl" with "lusty thoughts of big, strong, powerful monsters having their way with beautiful maidens." Branwen said her writing was based "on her own desires, or from personal experiences."

2013 University of Utah paleontologist Matthew Cobley uses analyses of ostriches to argue that long-necked dinosaurs such as *Diplodocus* had less-flexible necks than previously thought.

2013 Using a new technique to estimate tooth formation, researchers conclude that sauropods replaced their teeth faster than other types of dinosaurs. A 100-foot-long sauropod could have produced a new tooth in every position every 1–2 months.

2013 Wind-tunnel tests suggest that the small, four-winged *Microraptor* would have been an efficient glider even without feathers. This suggested that plumage might not have evolved for flight but may have instead been an important social key for sexual selection.

2013 Yale University's Nicholas Longrich and his colleagues report *Leptorhynchos gaddisi*, a turkey-sized feathered dinosaur from 75-million-year-old rocks in Big Bend, Texas. That same year, Longrich also reported the 78-million-year-old *Judiceratops*, the earliest known cousin of *Triceratops*.

2013 AiG claims that a fossil of *Allosaurus* recently donated to its Creation Museum is evidence of the worldwide flood that happened "a few thousand years ago" as described in the Bible. The fossil, which was described as "one of the four best-preserved *Allosaurus* skulls ever discovered," will be used by the Creation Museum as "a testimony to the creative power of God and . . . evidence to the truth of a worldwide catastrophic flood of the earth about 4,500 years ago as described in the Bible."

2013 In Utah, Field Museum paleontologists find the four-ton, 30-foot-long *Siats meekerorum*, a large precursor of *T. rex*.

2013 The most completely preserved ankylosaur ever found is Europe is discovered in a coal mine in Spain. The dinosaur was named *Europetta carbonensis*, meaning "Europe's shield from the coal."

2013 A mummified skeleton of *Edmontosaurus regalis* found in Alberta, Canada, includes a never-before-seen crest atop its skull. Researchers likened the structure to a rooster's red comb, and speculated that it may have been used for sexual or social signaling. As David Evans of ROM noted, "*Edmontosaurus* will never be that plain, unadorned duck-billed dinosaur ever again."

2013 When amateur paleontologist Nathan Myhrvoid cannot replicate the finding of several studies of dinosaur growth, he discovers that the questionable studies involved lost data, statistical errors, data from undocumented sources, and other errors (e.g., peak growth-rates of *T. rex* had been overestimated by a factor of two).

2013 Paleontologists from Burpee Museum find a fossilized oviraptor near Ekalaka, Montana. At the time of this discovery, CMNH had the only completely mounted North American oviraptor in the world. The 6-foot-tall (at the hips) discovery in Ekalaka was named "Pearl," and was described as a "parrot on steroids."

2013 Singapore Science Centre hosts "Titans of the Past," which features the largest *T. rex* skull to ever leave the United States.

2013 Japan's Yoshitsugu Kobayashi describes a nest of therizinosaurs found two years earlier in the Gobi Desert that includes 17 clutches of eggs. This site, which is the largest known colony of nonavian theropods ever discovered, showed that therizinosaurs were social animals.

2013 After comparing bones of dinosaurs, reptiles, and mammals, scientists from Richard Stockton College conclude that dinosaurs had more layers of cartilage in their joints. This made their joints softer and thicker than those of mammals, and meant that dinosaurs could grow larger because their joints could withstand more pressure and support more weight.

2013 Nicholas Chan of Australia's Macquarie University argues that as descendants of theropods evolved flight, their tails became a liability and shrank, and that this loss limited the size of flightless birds.

2013 When Jack Horner cuts open fossilized bones of several dinosaurs, he discovers that most of the animals—including all six *T. rex* in his sample—were still growing when they died. These adults were getting bulkier rather than taller or longer. Although robust and gracile forms of otherwise similar skeletons had often been attributed to different species or different genders (i.e., robust forms were considered male, and gracile forms female), Horner's work suggested that they represented different ages.

2013 University of Alberta paleontologist Scott Persons argues that feathered dinosaurs such as oviraptors used tail plumage to attract mates much like modern-day turkeys and peacocks.

2013 More than 80 tracks left by small theropods are found at the base of a cliff southeast of Seoul, South Korea, near several Bangudae Petroglyphs. Nine sites featuring dinosaur tracks in South Korea are national monuments.

2014 Canadian researchers led by ROM's David Evans announce the discovery of *Acheroraptor temertyorum,* a carnivorous cousin of *Velociraptor*. The fossils, which were found in Montana's Hell Creek Formation four years earlier, represent one of the last non-avian dinosaurs. The nine-foot-long *A. temertyorum* lived with *T. rex* and *Triceratops* 66 million years ago.

2014 Paleontologists unearth the first remains of dinosaurs in Saudi Arabia. Tailbones of a 73-million-year-old titanosaur, along with teeth of a 20-foot-long theropod, were found in the northwest part of the Gulf kingdom.

2014 AiG and ICR renew their claims that dinosaurs were aboard Noah's Ark, that there was no death (and that dinosaurs were vegetarians) before Adam's sin, and that dinosaurs lived with humans because Adam, Eve, and dinosaurs were all made on Day 6 of Creation Week. More than two million people have visited AiG's Creation Museum.

2014 Anthony Fiorillo and Ronald Tykoski of the Perot Museum of Nature and Science in Dallas, Texas, announce the discovery of the 22-foot-long, 1,000-pound *Nanuqsaurus hoglundi* from what is now northern Alaska. This pygmy tyrannosaur, which is the first known tyrannosaur from either pole, lived 70 million years ago. *Nanuqsaurus* had a 25-inch-long skull (for comparison, the skull of *T. rex* was more than twice as long). *Nanuqsaurus,* which was collected in 2006, means "polar bear lizard" in the language of the Iñupiat people of Alaska's North Slope. The name *hoglundi* honors Forrest Hoglund, who raised $185 million to build the Perot Museum.

2014 A new model suggests that the asteroid impact at the end of the Cretaceous produced lethal precipitation that killed the dinosaurs.

2014 The Austrian Mint releases its third "Back from the Dead" coin. The obverse of the popular 20-euro coin featured ribs, bones, and a reconstruction of *Struthiosaurus austriacus*, and the reverse featured *Zalmoxes* defending its young from *Struthiosaurus austriacus*.

2014 Emma Schachner and Matt Lamanna announce the discovery of three skeletons of the beaked, sharp-clawed *Anzu wyliei*, a birdlike dinosaur that roamed the Dakotas 66 million years ago with *T. rex*. The 11-foot-long, 500-pound raptor, which was excavated from the Hell Creek Formation, was labeled the "chicken from hell." *Anzu* refers to a birdlike demon in Mesopotamian mythology, and *wyliei* honors the grandson (named Wylie) of a CMNH trustee. *Anzu wyliei* is the largest oviraptosaur found in North America. As Lamanna noted, *Anzu wyliei* is "as close as you can get to a bird without being a bird."

2014 A giant billboard in Melbourne, Australia, showing Jesus petting a dinosaur enrages Christians amid a controversy over religious instruction in schools. Dinosaurs are an icon of young-Earth creationists, who believe that humans lived with dinosaurs a few thousand years ago.

2014 Scientists from Edinburgh University announce "Pinocchio rex" (*Qianzhousaurus sinensis*), a 9-meter-long, 66-million-year-old, long-nosed tyrannosaur found at a construction site in the city of Ganzhou in southern China. The dinosaur's scientific name was inspired by Qianzhou, the ancient name for Ganzhou.

2014 AiG's Creation Museum opens its $500,000 exhibit of an *Allosaurus* "to tell the account of history according to the Bible."

2014 Oxford University paleontologist Roger Benson and his colleagues unveil the body weights for 426 different species of dinosaurs. The biggest of them all was the plant-eating *Argentinosaurus,* which weighed as much as 90 tons. The smallest dinosaur was the sparrow-sized *Qiliania,* which weighed 15 grams.

2014 A $160-million remake returns *Godzilla* to movie theaters throughout the world. Unlike in previous versions of *Godzilla,* the scaly, 355-foot-tall monster was created digitally.

2014 *Leinkupal laticauda,* a long-necked relative of *Diplodocus* and *Apatosaurus,* is found in early Cretaceous sediments in Argentina. The word *laticauda* means *wide tail*, and the word *Leinkupal* means *vanishing family*, so named because previously diplodocid relatives came from the Jurassic (i.e., the period before the Cretaceous).

2014 Paleontologists in Argentina unearth fossilized bones of an herbivorous dinosaur believed to be the largest creature to ever walk the Earth. The 95-million-year-old dinosaur was 130 feet (40 meters) long, 65 feet (20 meters) tall, and weighed 85 tons.

2014 Rutgers University paleontologist Dennis Kent and his colleagues argue that dinosaurs in Argentina and in what is now the western United States moved toward the equator because of increases in the concentration of atmospheric carbon dioxide. Similar movements of dinosaurs in response to climate had been reported earlier in Canada, Iceland, and Europe.

2014 Montana State University paleontologists L. J. Krumenacker and David Varricchio report the most productive and diverse dinosaur-fossil site known in Idaho. The site, located in Bonneville County, is in the 98-million-year-old Wayan Formation.

2014 Using a "camera obscura" and a powerful X-ray, scientists produce the first three-dimensional images of the entire skeleton and feathers of *Archaeopteryx*.

NOTES

1. *Tertiary* is a discontinued term (replaced by *Paleogene,* or *Pg*), but its symbol (*T*) is still commonly used in discussions of the extinctions at the end of the Cretaceous (*K*).

The letter *K* was used for Cretaceous because *C* was already used to designate the Carboniferous.

2. Mantell lacked the tooth's fossilized jaw, a piece of evidence that would have proved whether the tooth came from a reptile. Mammals' jaws, even if teeth are missing, have differently shaped spaces for the different kinds of teeth (e.g., incisors, molars, canines). A reptile does not have several types of teeth, so all of the sockets for teeth have the same shape.
3. For comparison, £100 in 1900 equaled approximately £7,200 in 2014.
4. Buckland's claims assumed that the creature was proportioned like modern lizards.
5. That same year, Owen named the sauropod *Cardiodon,* but the type specimen was lost. The first sauropods from other continents whose genus remains valid include *Astrodon* (North America, 1859), *Titanosaurus* (Asia, 1877), *Argyrosaurus* (South America, 1893), *Algoasaurus* (Africa, 1904), and *Rhoetosaurus* (Australia, 1926).
6. Mantel's spine was also the subject of a paper about the pathology of spinal deformities.
7. Although Buckland was buried in Oxfordshire, England, Westminster Abbey includes a plaque commemorating Buckland's life.
8. The head of *Hylaeosaurus* is a replica made from Hawkins's original model, which is displayed outside the park's visitors' center.
9. In 1802, Hawkins's father had attended a dinner hosted by Charles Willson Peale inside a skeleton of *Mastodon*.
10. Famed American paleontologist Othniel Marsh (and others, such as Henry Hutchinson and Charles Schuchert) was later critical of Hawkins's Crystal Palace sculptures. Marsh's criticisms were ironic, for he later adorned *Apatosaurus* with the head of *Camarasaurus*.
11. Hayden, Colorado, and Yellowstone's Hayden Valley are named for Hayden.
12. To Owen, evolution symbolized progress, and if the largest reptiles lived in the past, then there had been no progress and, therefore, no evolution.
13. The first illustrated edition of Verne's book (i.e., the seventh edition) was not published until 1867. Verne's novel included no female participants. *Journey* was Verne's second novel.
14. Although *Camarasaurus* had spoon-shaped teeth, *Apatosaurus* had pencil-shaped teeth. Other inaccuracies included a shorter tail (as in *Camarasaurus*), substituting a forelimb and foot from *Camarasaurus*, and constructing the neck of 12 vertebrae (as in *Camarasaurus*) instead of the 15 known for *Apatosaurus*. Marsh never noted his use of secondary specimens in this restoration.
15. The Lance Formation in Wyoming is equivalent to the Hell Creek Formation in Montana.
16. The most complete *Torosaurus,* unveiled in 1992 and consisting of 85 percent real bones, can be seen at the Milwaukee Public Museum. The fossils were found in the Hell Creek Badlands of northeastern Montana in 1981. *Torosaurus* was Wisconsin's second mounted dinosaur.
17. These displays include 28 specimens in the Hall of Ornithischian Dinosaurs, six specimens in the Hall of Saurischian Dinosaurs, and the *Barosaurus* in the Theodore Roosevelt Rotunda.
18. When it was discovered that Cope had incipient syphilis, Cope's body was not made the type specimen.
19. Knight produced some of his greatest murals when he could barely see, and was totally blind the final two years of his life.

20. Despite the fame of the Marsh–Cope "bone war," these two paleontologists did not dominate the discipline as much as other scientists did with other groups of animals. For example, just four scientists named almost half of the more than 6,100 species of thrips named between 1901 and 1993, and one of these scientists named more than 1,000 of these tiny, winged insects during his life.
21. Riggs used no formal names for sediments during his work in Grand Valley. For rocks now known as the Morrison Formation, Riggs used the names (in ascending order) green clay, cross-bedded sandstone, and variegated clay.
22. For example, versions of Knight's *Tyrannosaurus* vs. *Triceratops* (Figure 19) later appeared in Sinclair's booklet of dinosaur stamps, a booklet that accompanied the exhibit titled "The World a Million Years Ago," and the dinosaur park in Rapid City, South Dakota.
23. For more than five years, workers hand-carried more than 4,300 bamboo cases more than 50 kilometers from the quarry to the coast for shipping. Many of the cases remain unopened at the Berlin Museum for Nature.
24. The map of Doyle's "Lost World" resembled that of the Weald, a region of southwestern England. The "central lake," which was once believed to be a Cretaceous swamp, included rocks containing *Iguanodon,* and the Weald's Lavant Caves were the holding-places of the Lost World.
25. *Protoceratops* was the first dinosaur discovered in Mongolia; its white bones are relatively easy to see in the area's bright red sand.
26. Similar standoffs appeared in several movies and documentaries, including *Planet of Dinosaurs* (1978), *Fantasia* (1940), *Dino-Riders* (1988), and *Land of the Lost* (1991).
27. For example, the cover of the booklet showed *Stegosaurus* confronting *Allosaurus,* the Jurassic equivalent of Knight's famous *T. rex* vs. *Triceratops* (Figure 19). Allen's version of *T. rex* vs. *Triceratops* appeared on pp. 6–7 of the Sinclair booklet.
28. Earnest Adams was the first recorded discoverer of sauropod tracks, and his brother George was the first recorded discoverer of theropod tracks.
29. The half-ton track given to SMU is now displayed at the Dallas Museum of Natural History.
30. The tracks originally found in the area in 1909 were in a tributary of the Paluxy River called the Wheeler Branch, which is northeast of Glen Rose and not part of Dinosaur Valley State Park. Although many of the dinosaur tracks have eroded away (including those shown in Figure 39), many remain visible in the bed and banks of a seven-mile stretch of the Paluxy River upstream from Glen Rose.
31. Only one scene from the story (i.e., "Rhedosaurus" destroying the lighthouse) is in the film.
32. The standoff between *Stegosaurus* and *Ceratosaurus,* which had appeared in *Journey to the Beginning of Time* (1954), was repeated in *When Dinosaurs Roamed America* (2001), *Walking with Dinosaurs* (1999), *Dinosaur Valley Girls* (1996), and *Planet of Dinosaurs* (2001).
33. This illustration was part of a series that began in 1956 in *Illustrated London News*. Parker's illustrations of Jurassic and Cretaceous animals were later featured in Edwin Colbert's *Dinosaurs: Their Discovery and Their World* (1961).
34. Gorgo does not die in *Gorgo* because Lourie's daughter cried when, after seeing *Beast from 20,000 Fathoms,* "the big, nice beast" was killed.
35. The United States is the biggest exporter of dinosaur bones. Many other countries—including Canada, China, Argentina, and Mongolia—have laws banning the export of their fossils.

36. In 2011, Vetri pleaded no contest to attempted voluntary manslaughter of her husband and was sentenced to nine years in state prison.
37. Other actresses who reprised similar roles have included Dana Gillespie, Darah Marshall, Carol Landis, and Jennifer O'Dell.
38. *A Cold Look at the Warm-Blooded Dinosaurs* was edited by Roger D. K. Thomas and Everett C. Olson. Before joining the University of Chicago, Olson (1910–1993)—a student of Alfred S. Romer—was a barker at "The World a Million Years Ago" at the Chicago World's Fair.
39. In fact, the first recorded discoverer of the sauropod tracks was Earnest "Bull" Adams, who helped Bird find the tracks in Glen Rose, Texas.
40. There were numerous claims that some stamps had misspelled the word *Pteranodon*. These claims were false; as the American Philatelic Society noted: "All the (Pteranodon stamps) have the correct spelling of Pteranodon. None exist for an incorrect spelling."
41. Gurche's works include the cover of Robert Bakker's *The Dinosaur Heresies* and the painting of *Barosaurus* at American Museum of Natural History (AMNH).
42. A "valid" name is one based on diagnostic materials described according to rules of International Code of Zoological Nomenclature (ICZN). The invalidation rate of species exceeds that of genera because the minor anatomical differences often used to justify new species are easier to reject than the more substantial differences used to validate genera.
43. The dinosaur was intended to be named "Utahraptor spielbergi" in exchange for funding for paleontological research. However, when the parties could not agree on a sum, the dinosaur was named *U. ostrommaysi*. In 2000, the name was changed from *U. ostrommaysi* (singular) to *U. ostrommaysorum*.
44. The movie *Jurassic Park* was less accurate than Crichton's famous novel. Speculations and inaccuracies in the movie include *Brachiosaurus* chewing side to side (it could chew only up and down), *T. rex* seeing only motion, the giant, featherless *Velociraptor*, and the dwarf, frilled, poison-spitting *Dilophosaurus*. Also, the ailing *Stegosaurus* in the novel *Jurassic Park* was replaced by *Triceratops* in the movie. *Stegosaurus* appears in the movie only as a misspelled name on a vial ("Stegasaurus").
45. In China, *-long* is used to name dinosaurs in much the same way that the Latin *-saur(us)* is used in the West.
46. *Torosaurus* had a larger frill with two holes (*fenestrae*, which is Latin for *windows*).

Appendix: The Geologic Timescale

Most of the major geological timespans were named in the 1800s by geologists who were either indifferent or hostile to evolution. Some of these names commemorate areas where rocks of a particular age—and containing distinctive fossils—were first collected and studied. For example, the Cambrian Period was named in 1835 by Adam Sedgwick while he was studying rocks from Northern Wales (*Cambria* is a Roman-Latin term for *Welsh*). Similarly, the Permian was named after the town of Perm on the western edge of the Ural Mountains, and in 1799, Alexander von Humboldt applied the name *Jurassic* to the predominantly limestone rocks exposed in the Jura Mountains of France and Switzerland.

The names of some other geological timespans commemorate the main types of substances deposited at a particular time. For example, the Carboniferous was named in 1822 for the coal beds and associated strata of north-central England, and the Cretaceous was named for the chalk deposited as the Cretaceous progressed (*Kreta* is German for *Creta*, the Latin word for *chalk*). As first noted by William Smith in 1796, different geological ages can be distinguished and recognized by their distinctive fossils.

Finally, other periods are named in a numerical scheme. For example, north of the Jura Mountains, the strata below the Jurassic consist of limestone sandwiched between sandstone and clay. In 1834, Friedrich von Alberti named this three-layered zone the Triassic.

The International Stratigraphic Chart

Era	*Period and Symbol* (Year Named)*	*Epoch (Year Named)*	*Millions of Years (Ma) from Start to Present***	*Major events*
Cenozoic (Cz) "Recent Life"	Quaternary (Q)*** (1829)	Holocene (1833)	0.0117	Human activities reduce biological diversity to the lowest levels Flowering plants dominate land Rise of agriculture and civilizations
		Pleistocene (1839)	1.806	Modern humans appear Continents in modern positions Repeated glaciations and lowering of sea level Apelike ancestors of humans appear in Africa
	Neogene (N) (1856)	Pliocene (1833)	5.332	Increasingly dry, cool climate
		Miocene (1833)	23.03 ± 0.5	First hominins appear
	Paleogene (Pg)**** (1856)	Oligocene (1854)	33.9 ± 0.1	Radiation of mammals, birds, snakes, angiosperms, teleost fishes, pollinating insects
		Eocene (1833)	55.8 ± 0.2	
		Paleocene (1874)	66.5 ± 0.3	Continents near modern positions; climate increasingly cool
*The K-T (or, more accurately, the K-Pg ****) mass extinction eliminates more than 60 percent of all species. This is the most famous mass extinction, not because of its magnitude (the Permian mass extinction wiped out many more species), but because of its most famous victims: dinosaurs. Crocodiles, lizards, birds, and mammals are relatively unaffected.*				
Mesozoic (Mz)	Cretaceous (K) (1822)		145.5 ± 4.0	Increasing diversity of mammals, birds, social insects, and angiosperms Continued radiation of dinosaurs Primitive birds replace pterosaur Breakup of Gondwana

	Jurassic (J) (1799)	199.6 ± 0.6	First birds, lizards, and angiosperms Gymnosperms dominate land Diversification of dinosaurs Mammals common but small Breakup of Pangaea into Gondwana and Laurasia
The Triassic mass extinction eradicates about 20 percent of all species, thereby opening niches that allow the diversification of dinosaurs.			
	Triassic (Tr) (1834)	251.0 ± 0.4	Increasing diversity of marine life Continents begin to separate Gymnosperms rise to dominance Diversification of reptiles First dinosaurs First mammals and Crocodilia
Paleozoic (Pz) "Ancient Life"	Permian (P) (1841)	299.0 ± 0.8	Glaciations and low sea level Diversification of insects, including beetles and flies Decline of amphibians Continents aggregate into Pangaea, creating the Appalachians Appearance of gymnosperms
The Permian mass extinction kills more than 90 percent of all species. The Permian mass extinction is the most devastating mass extinction.			
	Carboniferous (C) (1822)	359.2 ± 2.5	Extensive forests of early vascular plants dominate land Early orders of winged insects Appearance of reptiles Diversification of amphibians Gondwanaland and small northern continents form

(Continued)

The International Stratigraphic Chart

Era	*Period and Symbol* (Year Named)*	*Epoch (Year Named)*	*Millions of Years (Ma) from Start to Present***	*Major events*
	Devonian (D) (1839)		416.0 ± 2.8	First amphibians, seed plants, insects, ferns, trees, and ammonoids Diversification of bony fishes Early tetrapods First wingless insects
The Devonian mass extinction eliminates as many as two-thirds of all species. Most of these extinctions are marine species; there are far fewer losses on land.				
	Silurian (S) (1835)		443.7 ± 1.5	Origin of jawed fishes Earliest vascular plants First millipedes and arthropleurids on land
	Ordovician (O) (1879)		488.3 ± 1.7	Diversification of invertebrates; arthropods and mollusks dominate the sea Early fish, corals, and echinoderms First green plants and fungi on land Ice Age at end of period
The Ordovician-Cambrian mass extinction eliminates most marine species; many groups lose more than half of their species. Biologists know relatively little about the cause and impact of this mass extinction because most animals were soft-bodied and therefore were unlikely to become fossils.				
	Cambrian (C) (1835)		542.0 ± 1.0	First appearance of many animal phyla in the Cambrian Explosion Diversification of marine animals and algae Abundant marine invertebrates First shelled organisms

Proterozoic "First Life"	2500	Earliest eukaryotes Trace fossils of animals Multicellular animals First invertebrates Atmosphere becomes oxygenic Possible "Snowball Earth" period
Archaean "Old"	4000	Origin of life First fossils Diversification of bacteria Photosynthetic bacteria produce O_2 Evolution of aerobic respiration
Hadean (informal) "Hell"	Lower limit not defined	Oldest known rocks and minerals Formation of Earth (c. 4560 Ma)

* Paleontologists often refer to *faunal stages* rather than *geological periods*.

** These dates are accurate ± 1%, but the boundary dates continue to change as geologists learn more about rocks and refine their methods of radiometric dating. The dates listed here are from the *International Stratigraphic Chart of the International Commission on Stratigraphy 2004*.

*** The definition of the Quaternary and the Pleistocene are under revision. The historic Tertiary includes the Paleogene and Neogene, but now has no official rank.

**** The Paleogene is a new name for what was previously called the *Tertiary* (T). Although the Tertiary is a discontinued term, its abbreviation (T) is still commonly used when discussing the mass extinction at the end of the Cretaceous.

Glossary

abelisaurs a group of Cretaceous theropods having short arms and short, deep skulls; *Carnotaurus* and *Genusaurus* were abelisaurs.

adaptation an inherited trait that confers a fitness benefit by improving an organism's chances of survival and reproduction.

Age of Dinosaurs an informal term for the Mesozoic Era (251–65.5 million years ago). During this era, non-avian dinosaurs were Earth's dominant form of life.

allosaurs "different lizard family," a subgroup of theropods (formally known as "allosauroids") that included dinosaurs such as *Allosaurus* and *Sinraptor*. Allosaurs lived during the Jurassic and Cretaceous.

Alvarez event See *K-T event.*

Alvarezsaurus "Alvarez's lizard," a genus of small, long-legged running dinosaurs having heavily muscled forelimbs and a large claw on the thumb of each hand. *Alvarezsaurus* was named for Don Gregorio Alvarez.

amber a translucent or opaque yellow, fossil resin from coniferous trees.

angiosperms flowering plants. Angiosperms evolved during the Cretaceous.

ankylosaurs "fused lizard group," a group of plant-eating, armored ornithischian dinosaurs that were protected by bony plates, shields, or lumps in their skin.

anticline a ridge-shaped fold of stratified rock in which the sides slope down and away from the crest.

Archaeopteryx one of the oldest and most primitive fossil birds yet discovered. *Archaeopteryx,* one of the most famous fossils in the world, lived 150 million years ago; it was about 6 inches high, 12–18 inches long, and weighed 3–7 pounds. The first *Archaeopteryx* fossil was discovered in 1861 in Upper Jurassic limestone in Germany. *Archaeopteryx* ("ancient feather"), which was named in 1863 by Richard Owen, had feathers, wings, and hollow bones like a bird, and teeth, legs, and a bony tail like a small coelurosaur.

archetype an abstract concept of a primitive form or body plan from which a group of organisms presumably developed. Anatomist Richard Owen described an archetype as a "divine idea" and "primal pattern."

archosaurs "ruling reptiles," a group of reptiles having pointed snouts, narrow skulls, and teeth set in sockets. Archosaurs, which appeared in the Late Permian, include crocodiles, birds, and their extinct relatives.

articulated skeleton a skeleton in which the bones are in their original positions. Articulated skeletons are the *crème de la crème* of fossils.

avian describing, resembling, or relating to birds (Aves).

badlands land cut by erosion, lacking plants, and often exposing fossils.

biogeography the study of the geographic distribution of organisms and the changes in those distributions over time. Biogeography comprises a major source of evidence for evolution.

biostratigraphy a branch of stratigraphy that correlates and assigns relative ages to rock strata based on fossils in the strata. See also *faunal succession*.

bipedal walking or running on hind limbs, especially in an upright, human manner.

bird a member of the class Aves. Birds are the only remaining member of the dinosaur clade. The origin of modern birds from animals resembling *Tyrannosaurus rex* is one of the most remarkable examples of an evolutionary transition.

bonebed a sediment having many fossil bones, bone fragments, and similar organic remains

brachiosaurs a subgroup of sauropod dinosaurs having front legs longer than rear legs.

camarasaurs a subgroup of sauropod dinosaurs having weight-saving scoops and hollow areas in their backbones.

carnosaur a generalized name for large theropods that lived during the Jurassic and Cretaceous periods. *Tyrannosaurus* and *Allosaurus* were carnosaurs.

catastrophism the geological theory espoused in the 18th and 19th centuries by Georges Cuvier, Louis Agassiz, and others that Earth's geologic features have resulted primarily from sudden, widespread (worldwide), violent, or unusual catastrophes (presumably caused by capricious natural forces) that are beyond our current experiences (e.g., a worldwide flood). These catastrophes were followed by the creation of new forms of life. Compare with *uniformitarianism*.

caudal vertebrae vertebrae at the rear end of the spinal column, usually forming the animal's tail.

Cenozoic "recent life," the most recent period of Earth's history lasting form 65.5 million years ago to the present.

ceratopsians horned ornithischian dinosaurs; a group of plant-eating dinosaurs having facial horns, beak-like snouts, and large shoulder shields or neck-frills. *Triceratops* was a ceratopsian.

ceratosaurs a group of theropods, most of which have horns or rounded crests above their eyes and four digits on their front limbs.

Carcharodontosaurus "shark-toothed family," a group of theropods that included some of the largest predators ever known.

Chicxulub crater the remnant of a crater near the Yucatán Peninsula formed when an asteroid struck earth 65 million years ago.

clade a taxonomic group consisting of all (and only) those organisms that descended from a most recent common ancestor. A clade can be likened to a branch on the tree of life. Also called a monophyletic group. Dinosaurs are a clade that includes the common ancestor of *Triceratops* and birds and all their descendants. This means that if you start with *Triceratops* and finches (the last members of each branch of dinosaurs' evolutionary tree) and then back up to their last common ancestor, all of the organisms in between are dinosaurs. Birds are the descendants of dinosaurs.

cladistics phylogenetic systematics; a method for classifying life based on common measurable traits showing the evolutionary history of different organisms.

clavicle collarbone; the bone joining the shoulder blade to the sternum.

coelurosaur "hollow tail lizard," a generalized name for small meat-eating dinosaurs that lived near the end of the Cretaceous and which had bird-like bones. *Coelophysis* was a coelurosaur.

continental drift the movement of continents in geologic time over Earth's surface due to *plate tectonics*.

coprolite fossilized feces. Coprolites reveal the diet of extinct animals.

Cretaceous the final period of the Mesozoic era. The Cretaceous, which lasted from 145 to 66 million years ago, ended with the K-T event, which was also associated with the extinction of non-avian dinosaurs. *Velociraptor, Triceratops, Tyrannosaurus, Microraptor,* and *Deinonychus* are dinosaurs that lived during the Cretaceous.

Cretaceous system the group of stratified rocks normally below the oldest Paleogene (i.e., Tertiary) deposits and above the Jurassic system. The Cretaceous system closed the Mesozoic era. The system, which is famous for its abundant chalk, is usually represented by a K (for *Kreide,* which is German for *chalk*).

Deccan traps large areas of flood basalt in the Indian subcontinent that were deposited in a series of massive volcanic eruptions at the end of the Cretaceous. The Deccan traps, which are one of the largest volcanic features on Earth, are believed by many to be involved in the K-T mass extinction.

deep time the extreme antiquity of Earth, as reflected by the geologic record, measured in millions or billions of years.

dinosaur any of the many extinct, Mesozoic, terrestrial reptiles of the Saurischia or Ornithischia. Dinosaurs were one of the most successful terrestrial groups in Earth's history.

diplodocids subgroup of sauropods having long tails that had twin, ski-like chevrons on the underside of each bone. *Diplodocus* and *Supersaurus* were diplodocids.

dromaeosaurs "running lizard family," any of the fast, agile, medium-sized, bird-like theropods that included *Deinonychus* and *Velociraptor*. Dromaeosaurs, which are sometimes referred to as "raptors," also had a large brain, large eyes, and a sickle-shaped talon on the second toes of each foot.

ectotherm a "cold blooded" organism that derives most of its body heat from external sources. Crocodiles and lizards are ectotherms. Compare with *endotherm*.

endotherm a "warm blooded" organism that derives most of its body heat from its own metabolism. Birds and mammals are endotherms. Compare with *ectotherm*.

eon a long period of geological time. For example, the Phanerozoic Eon includes the Paleozoic, Mesozoic, and Cenozoic eras.

evolution change in the genetic makeup of a population over time. Evolution occurs over successive generations, not in the lifetimes of individual organisms.

extant in existence; not extinct.

extinct having no living members; not extant. Non-avian dinosaurs are extinct.

extinction the permanent disappearance of a species or a group of species.

faunal succession a scientific principle originated in 1816 by William Smith based on the observation that fossils in sedimentary rock strata are arranged in a specific, reliable order that can be identified over wide areas. For example, fossilized humans are never found in the same strata as dinosaurs because the two species lived millions of years apart. The principle of faunal succession, which allows geologists to identify and date strata by the fossils within the strata, is explained by the theory of evolution. Faunal succession became the basis for *biostratigraphy*.

femur thighbone.

fenestra a gap or opening into a hard substance such as a bone.

fossil a preserved remnant, impression, or trace of an organism that lived in the past. Today, there are more than 250,000 known species of fossil animals, most of which were shelled creatures that lived in shallow seas. Few organisms become fossils.

fossil record the cumulative evidence about organisms and conditions of the past provided by fossils.

fossilization complex processes that transform the remains of an organism into a solid artifact.

frill in horned dinosaurs, the ornamented sheet of bone extending back from the skull.

furcula the fused clavicles (collarbones) that form the "wishbone" of birds and some non-avian theropods.

gap creationism the religious belief that there was a large time period between the creation stories described in Genesis 1:1 and Genesis 1:2. Gap creationism posits that the creation described in Genesis 1:1 was destroyed before Genesis 1:2, when God created Adam, Eve and the rest of the world in six days.

gastroliths stones swallowed deliberately by an animal (e.g., birds), usually to help grind food; gastroliths are often found with the gut of herbivorous dinosaurs.

gene a unit of heredity. Most genes are specific sequences of DNA (or RNA in some viruses) that contain information needed to make a protein.

geologic time the 4.6-billion-year timescale used by scientists to describe events in Earth's history.

Gondwana the southern half of Pangaea.

Great Chain of Being a hierarchical system of classification in which divinely created organisms have unchangeable positions that reflect their degrees of perfection. This popular idea, which originated with the ancient Greeks, was used in the Middle Ages and Renaissance to place nature in a religious context, as well as to justify social inequities and the divine rights of kings and nobility. The influence of the Great Chain of Being persists with claims that some organisms are "higher" or "lower" than other organisms. Today, we know that evolution is not about links between steps in a hierarchal ladder or chain, and that no species is "higher" or "lower" than others. That is, cockroaches, jellyfish, and ferns are not "lower" or "older" than humans, nor are they more "primitive." *Anything* alive today has been subject to evolution by natural selection just as long as *everything* alive today has been subject to evolution by natural selection. The Great Chain of Being is also called *scala naturae*.

ground-up theory a theory arguing that avian flight originated when animals ran and leaped after prey.

hadrosaurs plant-eating ornithischian dinosaurs commonly called "duckbills," having wide mouths, and sometimes having crests of bone on their heads. Hadrosaurs were abundant and diverse near the end of the Cretaceous; their eponymous feature was their bill, a broad area at the front of their mouth that supported a horny bill superficially like that of ducks. Hadrosaurs also had many teeth compressed together to form a grinding surface.

Haversian bone bone tissue that is highly vascularized and associated with active, endothermic lifestyles.

Hell Creek Formation a Late Cretaceous rock formation in Montana, North Dakota, South Dakota, and Wyoming. The Hell Creek Formation is named for exposures along Hell Creek near Jordan, Montana. Many dinosaurs and other fossils have been found in this formation. The largest collection of Hell Creek fossils is in the Museum of the Rockies in Bozeman, Montana.

heterodontosaurs a group of primitive, herbivorous, small ornithopods.

humerus upper arm bone.

hypsilophodontids a group of small, fast, herbivorous ornithopods that lived during the Jurassic and Cretaceous periods.

ichnology the study of footprints and other trace fossils.

ichnotaxonomy the naming and classification of trace fossils.

ichthyosaur a large, marine reptile from the Mesozoic period. These reptiles, which had dolphin-like heads and fish-like bodies, spent their entire lives in water, but they breathed air. Ichthyosaurs had four limbs modified as paddles.

iguanodontids a group of large, strong, thumb-spiked, herbivorous ornithopods that lived in the Cretaceous period. *Iguanodon* and *Camptosaurus* were iguanodontids.

in situ Latin for *in place*; in its original place.

inheritance of acquired traits the claim that traits acquired during a parent's lifetime can be passed to offspring. This was the basis for the first (albeit inaccurate) recorded scientific theory of evolution. The inheritance of acquired traits, which is associated with Jean-Baptiste Lamarck, was popular until the establishment of modern genetics. Also known as *Lamarckism*.

iridium a rare platinum-group metal that occurs in low levels in nature, but in higher concentrations in meteorites and volcanic rocks. Large amounts of iridium at the K-T boundary are interpreted as evidence of a meteoric impact.

junior synonym a name that describes the same taxon as a previously published name. "Brontosaurus" is a junior synonym of *Apatosaurus.*

Jurassic the second of the three geological periods of the Mesozoic. The Jurassic lasted from 200 to 145 million years ago. *Stegosaurus, Diplodocus, Allosaurus, Apatosaurus, and Camarasaurus* are dinosaurs that lived during the Jurassic.

K-T boundary the abrupt change in rock types at the boundary of the Cretaceous and Tertiary (now Paleogene) periods. The K-T boundary is sometimes marked by a thin layer of clay.

K-T event an ancient cataclysm involving at least one large meteorite that hit Earth near the present-day Yucatán Peninsula approximately 65 million years ago at the end of the Cretaceous (the third and final period of the Mesozoic era). This event triggered a mass extinction, the most famous victims of which were the non-avian dinosaurs. Also called the *Alvarez event* for its early proponents, Luis and Walter Alvarez.

Lamarckism a discredited, but historically influential, proposal of Jean-Baptiste Lamarck claiming that changes in traits during an organism's life can be passed to the organism's offspring. See also *inheritance of acquired traits.*

lambeosaurs crested duck-billed dinosaurs.

Laurasia the northern half of Pangaea.

maniraptors a large group of theropods having wrists that could swivel, long arms, and long arm- and tail-feathers. Maniraptors ("grasping hands") included *Deinonychus.*

mass extinction the collective extinction of large numbers of species in a relatively short geologic period of time. There have been several mass extinctions during Earth's history, the largest of which occurred 250 million years ago, at the end of the Permian; this mass extinction ("The Great Dying") eliminated more than 80 percent of the species on the earth. However, the most famous mass extinction occurred 65 million years ago at the end of the Cretaceous. That extinction, which killed two-thirds of existing species, is best known for its most famous victims: the non-avian dinosaurs. See also *Appendix.*

Mesozoic the geologic era that lasted from 251 to 66 million years ago and which is divided into the Triassic, Jurassic, and Cretaceous periods. Non-avian dinosaurs evolved in the Triassic and became extinct at the end of the Cretaceous. The first mammals, birds, and flowering plants also evolved during the Mesozoic. The Mesozoic, which means "middle life," was preceded by the Paleozoic ("first life") and followed by the Cenozoic ("new life"). See *Appendix.*

missing link a popular term used during Charles Darwin's era to denote hypothetical organisms that linked different groups of organisms, and especially humans with anthropoid apes. Most biologists no longer use the term *missing link* because it implies that organisms are linked by a hierarchal chain or ladder, when, in fact, organisms share common ancestors. The use of *missing link* also inaccurately implies that if a certain fossil has not yet been found, then evolution cannot be valid. See *transitional form.*

monophyletic group a group that that has descended from a single ancestor. See *clade.*

mosasaur a group of extinct, predatory, aquatic reptiles that lived during the Cretaceous. Mosasaurs had short necks, streamlined bodies, long tails, and four paddle-like limbs.

natural selection an evolutionary mechanism that produces differences in survival and reproduction among organisms with different heritable traits. Natural selection is the mechanism for adaptive evolutionary change proposed by Charles Darwin in *On the Origin of Species* (1859).

nomen dubium "dubious name"; a name that is of unknown or doubtful application.

nomen nudum "naked name"; a name that has appeared in print but has not formally been published to meet criteria for official naming.

nomen oblitum "forgotten name"; a name that has not appeared in the scientific literature for at least 50 years since it was published. *Manospondylus gigas* is a *nomen oblitum* for *Tyrannosaurus rex*.

non-avian dinosaurs all dinosaurs except birds.

old-earth creationism the claim that the earth is billions of years old, during which time the changes in the earth and the earth's organisms have been directed by God. Unlike *young-earth creationists,* old-earth creationists accept the scientific evidence documenting an ancient earth and universe.

ornithischian dinosaurs bird-hipped dinosaurs; one of the two major groups of dinosaurs in which the lower part of the pubis slants down and backward, as in modern birds. (Despite this similarity, birds evolved from saurischian dinosaurs.) Ornithischian dinosaurs include stegosaurs, ankylosaurs, hadrosaurs, and ceratopsians. Compare with *saurischian dinosaurs*.

ornithopods the main group of ornithischian "bird foot" herbivorous dinosaurs that includes hadrosaurs, *Hypsilophodon* and its cousins, and *Iguanodon*. Ornithopods had three-toed feet and walked upright on its hind limbs.

orthogenesis a disputed claim that variations in evolution follow a unilinear, innate direction and are not merely sporadic and fortuitous. Advocates of classic orthogenesis (e.g., Lamarck) argue that life has an inherent tendency to evolve in a unilinear way because of some innate "driving force." Also known as *progressive evolution*.

osteoderm bones shaped like spikes, plates, and scales that form within the skin of many dinosaurs and other reptiles.

pachycephalosaurs a group of relatively small, herbivorous dinosaurs having thickened bones atop their skull.

paleontologist a scientist who studies fossil plants and animals.

paleontology the scientific study of fossils to reconstruct the history of life.

Pangaea the supercontinent formed at the end of the Paleozoic period, consisting of all Earth's landmasses. Pangaea existed during the Permian and Triassic when all the modern continents were conjoined. During the Cretaceous, Pangaea split to form Gondwana and Laurasia. This breakup of Pangaea during the Late Jurassic caused the isolation (and therefore the separate evolution) of different groups of dinosaurs and other organisms from each other.

pelvis the hipbones of most vertebrates, including the pubis, ilium, and ischium.

phylogenetics the study of evolutionary relationships between groups of organisms.

phylogeny the evolutionary history of a species or higher taxonomic group of organisms.

Piltdown Man a fraudulent fossil consisting of a human skull and an ape's jaw, allegedly discovered in England and presented in 1912 as a genuine hominid from the early Pleistocene. Although Piltdown Man was shown in 1953 by scientists to be a hoax, many creationists continue to cite Piltdown Man as evidence that evolutionary biology is fraudulent.

plaster of Paris powdered calcium sulfate. When mixed with water, plaster of Paris crystalizes into a hard shell that can protect fossils.

plate tectonics the theory that Earth's crust consists of movable plates that can join or separate over geologic time. The movements of plates explain continental drift, earthquakes, volcanoes, mountain building, and some aspects of biogeography.

preparator a person who prepares dinosaur models or reconstructions for display.

principle of superposition the claim by Nicolas Steno in 1669 that in undisturbed sedimentary rock, older strata are lower in the geologic column than younger strata.

prosauropod saurischian herbivorous dinosaurs that lived before sauropods in the Triassic and Early Jurassic periods; these dinosaurs usually had small heads, a bulky body, a small head, a long tail, and pillar-like legs. The term *prosauropod* is being increasingly abandoned.

psittacosaurs "parrot lizard family," a primitive group of bipedal ceratopsians known from Cretaceous rocks of Asia.

pterosaurs a group of flying reptiles from the Mesozoic having skin-covered wings.

pubis one of the three bones in the pelvis. In most saurischians, the pubis points down and forward, whereas in ornithischians it points toward the rear.

quadruped an animal that uses all four limbs to walk or run, rather than its hind limbs alone.

racial senility a discredited theory popular in the early 1900s that claimed that some lineages, after developing all of their evolutionary potential, evolved useless, extravagant structures.

radiometric dating a common way of estimating the age of a fossil or rock by analyzing the elemental isotopes and the products of their decay within the accompanying rock.

raptor a term correctly applied to several birds of prey, including eagles, owls, and hawks. The movie *Jurassic Park* (1993) popularized the use of *raptor* to describe small, predatory dinosaurs such as *Velociraptor* and *Utahraptor*.

saurischian dinosaurs lizard-hipped dinosaurs. Saurischian dinosaurs were one of the two major groups of dinosaurs, in which the lower part of the pubis slants down and forward. Saurischians include sauropods and theropods. Birds evolved from Saurischian dinosaurs. Compare with *ornithischian dinosaurs*.

sauropod a large, four-legged herbivorous dinosaur having a small head, bulky body, long neck, long tail, and four pillar-like legs. Sauropods thrived during the Late Jurassic and Cretaceous periods, and were the largest land animals of all time. Sauropods included *Apatosaurus, Brachiosaurus,* and *Diplodocus*.

sedimentary rocks rocks formed when tiny particles of sand, mud, or other sediments settle and are cemented together. In general, sedimentary rocks are the only rocks to contain fossils.

sexual selection differences in fitness as a result of differences in the ability to obtain mates.

shocked quartz quartz crystals deformed by intense pressure. Some kinds of shocked quartz are associated with meteoric impacts.

Spinosaurus "spine lizard," a genus of large, bipedal theropods having conical teeth and crocodile-like skulls.

spontaneous generation the supposed production of life from nonliving matter and without biological parentage.

stegosaur an armor-plated, plant-eating ornithischian dinosaur having slabs or points of bone along its back. Stegosaurs had rows of tall plates along their neck, back and tail, and long spikes at the tips of their tails.

strata layers of sedimentary rock.

stratigraphy the study of layered rocks and sediments. See also *biostratigraphy.*

supernova an explosion that occurs when a large star runs out of hydrogen and explodes. Some scientists have suggested that a supernova caused or helped to cause the K-T mass extinction.

taphonomy a branch of paleontology that studies the processes that occur from the death of an organism to its discovery as fossil remains.

taxon a named group of organisms of any rank within a classification system.

taxonomy the theory and practice of naming and classifying groups of organisms, according to their similarities.

tetrapod a vertebrate animal having four limbs.

therizinosaurs a specialized group of dinosaurs having a broad hips, beak-like mouths, stubby legs, and large claws on the forelimbs. Therizinosaurs, which lived mainly during the Cretaceous period, appear to have undergone an evolutionary transition from carnivory or omnivory to herbivory. Therizinosaurs are also called segnosaurs.

theropod a predatory, carnivorous (i.e., meat-eating) saurischian dinosaur that walked upright on its hind limbs and had grasping hands with clawed fingers. Theropods ranged from the huge *Tyrannosaurus* to the tiny *Compsognathus*. Like birds, theropods had three-toed feet, a wishbone, and hollow bones.

titanosaurs a diverse subgroup of sauropods such as *Argentinosaurus* that thrived throughout the world in the Cretaceous period.

trackway row of footprints or tracks.

transitional form an organism having anatomical features intermediate between those of two major groups of organisms in an evolutionary sequence. Transitional forms show evolutionary sequences between lineages by having characteristics of ancestral and newer lineages. Since all populations are in evolutionary transition, a transitional form represents a particular evolutionary stage that is recognized in hindsight. *Archaeopteryx* is a transitional form between birds and dinosaurs, and *Tiktaalik* is a transitional form between land-dwelling tetrapods evolved from fish ancestors. Compare with *missing link*.

trees-down theory a theory arguing that avian flight originated with gliding animals that evolved the capacity for powered flight.

Triassic the first period of the Mesozoic era. Dinosaurs evolved during the Triassic period, which lasted from 251 to 200 million years ago. *Coelophysis, Herrerasaurus, Eoraptor,* and *Plateosaurus* are dinosaurs that lived during the Triassic.

Troodons "wounding tooth family," a group of birdlike, sickle-clawed maniraptors.

type specimen the original specimen upon which the description of a new species is based. Examples of type specimens include Taung Child (*Australopithecus africanus*) and Java Man (*Homo erectus*).

tyrannosaurs a group of coelurosaurs that includes *Tyrannosaurus* and other large carnivorous dinosaurs. *Gorgosaurus, Daspletosaurus, Albertosaurus,* and *Tarbosaurus* were tyrannosaurs.

uniformitarianism a theory suggested by James Hutton and developed by Charles Lyell summarized by the phrase "the present is the key to the past"—that is, the earth's geologic features have developed over long periods through a variety of slow geologic processes involving common events such as rain, volcanic activity, and wind. Uniformitarianism does not imply that change occurred at a uniform rate or deny the occurrence of localized catastrophes. Indeed, much evidence attests to occasional catastrophic events in the earth's history, most famously the impact of a meteor or comet off the Yucatán Peninsula approximately 65 million years ago that led to the extinction of much Mesozoic life (including non-avian dinosaurs). Such events that depart from gradual geologic change do not necessarily support young-earth creationism or biblical catastrophes such as a worldwide flood. Compare with *catastrophism*.

young-earth creationism the claim that earth is 6,000–10,000 years old, that the six days of creation described in Genesis each lasted 24 hours, and that catastrophic events such as Noah's Flood produced the Grand Canyon and other geological features. Young-earth creationists, who reject many aspects of modern biology, geology, physics, and other sciences, are biblical literalists who believe that modern organisms were created by God, and that life did not evolve. Young-earth creationism, which involves divine intervention that suspends the laws of nature, also contradicts theistic evolution, which claims that God works through natural laws. Young-Earth creationists claim that humans lived with dinosaurs because humans and all land animals were created on Day 6 of Creation Week.

Bibliography

Acorn, John. 2007. *Deep Alberta: Fossil Facts and Dinosaur Digs*. Edmonton, Alberta, Canada: University of Alberta Press.

Allmon, Warren D. 2007. The pre-modern history of the post-modern dinosaur: Phases and causes in post-Darwinian dinosaur art. *History of the Earth Science Society* 25 (1), 5–35.

Alvarez, L. W., W. Alvarez, F. Asaro, and H. V. Michel. 1980. Extraterrestrial cause for the Cretaceous-Tertiary Extinction. *Science* 208, 1095–1108.

Alvarez, Walter. 1997. *T. rex and the Crater of Doom*. Princeton, NJ: Princeton University.

Andrews, Roy Chapman. 1923. *All about Dinosaurs*. New York: Random House.

Andrews, Roy Chapman. 1943. *Under a Lucky Star: A Lifetime of Adventure*. New York: Viking Press.

Archibald, J. David. 2011. *Extinction and Radiation: How the Fall of Dinosaurs Led to the Rise of Mammals*. Baltimore, MD: The Johns Hopkins University Press.

Averett, Walter R. 1987. *Paleontology and Geology of the Dinosaur Triangle*. Grand Junction, CO: Museum of Western Colorado.

Averett, Walter R. 1991. *Guidebook for Dinosaur Quarries and Tracksites Tour*. Grand Junction, CO: Museum of Western Colorado.

Bakker, Robert T. 1968. The superiority of dinosaurs. *Discovery* 3 (2), 11–22.

Bakker, Robert T. 1975. Dinosaur renaissance. *Scientific American* 232 (4), 58–78.

Bakker, Robert T. 1986. *The Dinosaur Heresies: New Theories Unlocking the Mystery of Dinosaurs and Their Extinction*. New York: William Morrow and Co.

Barrett, Paul M., Alistair J. McGowan, and Victoria Page. 2009. Dinosaur diversity and the rock record. *Proceedings of the Royal Society B* 276, 2667–2674.

Barth, Amy. 2010. In the footsteps of giants. *Discover* (July/August), 67–71.

Baugh, Carl E. 1987. *Dinosaur: Scientific Evidence That Dinosaurs and Men Walked Together*. Orange, CA: Promise Publishing.

Baugh, Carl E. 1999. *Why Do Men Believe Evolution against All Odds?* Bethany, OK: Bible Belt Publishing.

Baugh, Carl E. and Clifford Wilson. 1994. *Footprints and the Stones of Time*. Oklahoma City, OK: Hearthstone Publishing.

Baur, G. 1890. A review of the charges against the Paleontological Department of the U.S. Geological Survey, and of the defense made by Prof. O. C. Marsh. *American Naturalist* 24, 298–304.

Bausum, Ann. 2000. *Dragon Bones and Dinosaur Eggs: A Photobiography of Explorer Roy Chapman Andrews*. Washington, DC: National Geographic Society.

Beierle, Frederick P. 1977. *Man, Dinosaur, and History*. Prosser, WA: Perfect Printing.

Benton, M. J. 2008a. Fossil quality and naming dinosaurs. *Biology Letters* 4, 729–732. doi:10.1098/rsbl.2008.0402

Benton, M. J. 2008b. How to find a dinosaur, and the role of synonymy in biodiversity studies. *Paleobiology* 34, 516–533.

Benton, M. J. 2010. Naming dinosaur species: The performance of prolific authors. *Journal of Vertebrate Paleontology* 30, 1478–1485. doi:10.1080/02724634.2010.501462

Berry, Mark F. 2005. *The Dinosaur Filmography*. Jefferson, NC: McFarland & Company.

Bible-Science Association. 1973. *Dinosaurs and Sin*. Caldwell, ID: Bible-Science Association.

Bible-Science Society. 1974. *Human Footprints and Dinosaur Tracks*. Caldwell, ID: Bible-Science Association.

Bird, Roland T. 1941. A dinosaur walks into the Museum. *Natural History* 47, 74–81.

Bird, Roland T. 1953. We captured a live brontosaur. *National Geographic* 105 (5), 707–722.

Bird, Roland T. 1985. *Bones for Barnum Brown: Adventures of a Dinosaur Hunter*. Ft. Worth, TX: Texas Christian University Press.

Bohor, B. F., E. E. Foord, P. J. Modreski, and D. M. Triplehorn. 1984. Mineralogic evidence for an impact event at the Cretaceous-Tertiary boundary. *Science* 224, 867–869.

Bramwell, Valerie and Robert M. Peck. 2008. *All in the Bones: A Biography of Benjamin Waterhouse Hawkins*. Philadelphia, PA: The Academy of Natural Sciences of Philadelphia.

Brett-Surman, M. K., Thomas R. Holtz, Jr., James O. Farlow, and Bob Walters (Eds.). 2012. *The Complete Dinosaur*, Second Edition. Bloomington: Indiana University Press.

Brinkman, Paul D. 2010. *The Second Jurassic Dinosaur Rush: Museums and Paleontology in America at the Turn of the Twentieth Century*. Chicago: University of Chicago Press.

Brochu, Christopher A. 2001. Progress and future directions in archosaur phylogenetics. *Journal of Paleontology* 75 (6), 1185–1201.

Brown, Barnum. 1938. The mystery dinosaur. *Natural History* 41, 190–202, 235.

Brown, Don. 2003. *Rare Treasure: Mary Anning and Her Remarkable Discoveries*. New York: Houghton Mifflin.

Brown, Lilian. 1950. *I Married a Dinosaur*. New York: Dodd, Mead & Company.

Budd, Graham E. 2001. Royal fossils: The Royal Society and progress in palaeontology. *Notes and Records of the Royal Society* 55, 51–67. doi:10.1098/rsnr.2001.0125

Burdick, Clifford C. 1950. When giants roamed the earth. *Signs of the Times*, July 25.

Burdick, Clifford C. 1976. Reaction to Neufeld. *Origins Research* 3 (1), 7–8.

Burness, G. P., J. Diamond, and T. Flannery. 2001. Dinosaurs, dragons, and dwarfs: The evolution of maximal body size. *Proceedings of the National Academy of Sciences of the United States of America* 98, 14518–14523.

Burroughs, Edgar Rice. 1921. *Tarzan the Terrible*. Chicago: A.C. McClurg and Company.

Cadbury, Deborah. 2000. *Terrible Lizard: The First Dinosaur Hunters and the Birth of a New Science*. New York: Henry Holt and Co.

Cadbury, Deborah. 2010. *The Dinosaur Hunters*. London: Fourth Estate.

Cain, Dana and Mike Fredericks. 1999. *Dinosaur Collectibles*. Norfolk, VA: Antique Trader Books.

Caleb Curriculum. 1985. *Dinosaurs: And the People Who Knew Them*. Herrin, IL: Caleb Publishers.

Carlisle, David Brez. 1995. *Dinosaurs, Diamonds, and Things from Outer Space: The Great Extinction*. Stanford, CA: Stanford University Press.

Carozzi, Albert V. 1970. New historical data on the origin of the theory of continental drift. *Geological Society of American Bulletin* 81, 283–286.

Carpenter, Kenneth and Philip J. Currie. 1986. *Dinosaur Systematics: Approaches and Perspectives*. Cambridge, UK: Cambridge University Press.

Carpenter, Kenneth, Karl F. Hirsch, and John R. Horner. 1994. *Dinosaur Eggs and Babies*. Cambridge, UK: Cambridge University Press.

Chambers, Paul. 2002. *Bones of Contention: The Fossil That Shook Science*. London: John Murray.

Charig, Alan. 1979. *A New Look at the Dinosaurs*. New York: Mayflower Books.

Charig, Alan, A. N. Milner, C. A. Walker, and P. J. Whybrow. 1986. *Archaeopteryx* is not a forgery. *Science* 232, 622–626.

Chatterjee, S. 1991. Cranial anatomy and relationships of a new Triassic bird from Texas. *Philosophical Transactions of the Royal Society of London B* 332, 277–346.

Chiappe, Luis M. 2007. *Glorified Dinosaurs: The Origin and Early Evolution of Birds*. Sydney, Australia: University of New South Wales Press.

Chure, Daniel, Brooks B. Britt, John A. Whitlock, and Jeffrey A. Wilson. 2010. First complete sauropod dinosaur skull from the Cretaceous of the Americas and the evolution of sauropod dentition. *Naturwissenschaften*. doi:10.1007/s00114–010–0650–6

Clark, James M. and Xu Xing. 2009. Evolutionary transitions among dinosaurs: Examples from the Jurassic of China. *Evolution Education and Outreach* 2, 236–247.

Cohen, Daniel. 1987. *Hollywood Dinosaur*. New York: Pocket Books.

Colbert, Edwin H. 1961. *Dinosaurs: Their Discovery and Their World*. New York: Dutton.

Colbert, Edwin H. 1968. *Men and Dinosaurs: The Search in the Field and Laboratory*. New York: Dutton.

Colbert, Edwin H. 1983. *Dinosaurs: An Illustrated History*. Maplewood, NJ: Hammond.

Cope, E. D. 1875. *The Vertebrata of the Cretaceous Formations of the West*. Report of the U.S. Geological Survey of the Territories, Vol. 2.

Cope, E. D. 1877. On a gigantic saurian from the Dakota epoch of Colorado. *Paleontological Bulletin No. 25,* 5–10.

Cope, E. D. 1884. *The Vertebrata of the Tertiary Formations of the West*. Report of the U.S. Geological Survey of the Territories, Vol. 3.

Cuppy, Will. 1941. *How to Become Extinct*. Chicago: University of Chicago Press.

Currie, Philip J. (Foreword). 2000. *Dinosaur Imagery: The Science of Lost Worlds and Jurassic Art—The Lanzendorf Collection*. San Diego, CA: Academic Press.

Currie, Philip J. and Eva B. Koppelhus (Editors). 2005. *Dinosaur Provincial Park: A Spectacular Ancient Ecosystem Revealed*. Bloomington: Indiana University Press.

Currie, Philip J. and Eva B. Koppelhus. 2010. Introduction to *Albertosaurus* special issue. *Canadian Journal of Earth Science* 47, 1111–1114.

Currie, Philip J., Eva B. Koppelhus, Martin A. Shugar, and Joanna L. Wright (Editors). 2004. *Feathered Dragons: Studies on the Transition from Dinosaurs to Birds*. Bloomington: Indiana University Press.

Czerkas, Stephen. 2006. *Cine-Saurus: The History of Dinosaurs in the Movies*. Blanding, UT: The Dinosaur Museum.

Czerkas, Sylvia J. (Editor). 2002. *Feathered Dinosaurs and the Origin of Flight*. Blanding, UT: The Dinosaur Museum.

Czerkas, Sylvia J. and Stephen A. Czerkas. 1991. *Dinosaurs: A Global View*. New York: Mallard Press.

Czerkas, Sylvia M. and Donald F. Glut. 1982. *Dinosaurs, Mammoths, and Cavemen: The Art of Charles R. Knight*. New York: E. P. Dutton.

Darwin, Charles. 1859. *On the Origin of Species by Means of Natural Selection; Or the Preservation of Favoured Races in the Struggle for Life*. London: John Murray.

Dean, Dennis R. 1999a. *Gideon Mantell and the Discovery of Dinosaurs*. Cambridge, MA: Cambridge University Press.

Dean, Dennis R. (Editor). 1999b. *The First "Dinosaur" Book: Richard Owen on British Fossil Reptiles (1842)*. Delmar, NY: Scholars' Facsimiles & Reprints.

Debus, Allen A. 2006. *Dinosaurs in Fantastic Fiction: A Thematic Survey*. Jefferson, NC: McFarland & Company.

Debus, Allen A. and Diane E. Debus. 2002a. *Dinosaur Memories: Dino-trekking for Beasts of thunder, Fantastic Saurians, "Paleo-people," "Dinosaurabilia," and Other "Prehistoria."* San Jose, CA: Authors Choice Press.

Debus, Allen A. and Diane E. Debus. 2002b. *Paleoimagery: The Evolution of Dinosaurs in Art*. Jefferson, NC: McFarland & Company.

Desmond, Adrian J. 1975. *The Hot-Blooded Dinosaurs*. New York: Warner Books.

Desmond, Adrian J. 1982. *Archetypes and Ancestors*. Chicago: University of Chicago Press.

Dingus, Lowell. 2004. *Hell Creek, Montana: America's Key to the Prehistoric Past*. New York: St. Martin's Press.

Dingus, Lowell. 2008. *The Dinosaur Hunters: The Extraordinary Story of the Man and Women Who Discovered Prehistoric Life*. New York: Metro Books.

Dingus, Lowell and Mark A. Norell. 2010. *Barnum Brown: The Man Who Discovered Tyrannosaurus rex*. Berkeley, CA: University of California Press.

Dingus, Lowell and Timothy Rowe. 1998. *The Mistaken Extinction: Dinosaur Evolution and the Origin of Birds*. New York: W.H. Freeman and Company.

Dinosaur Committee, Atlantic Union Conference of Seventh-Day Adventists Office of Education. 1983. *A Creationist View of Dinosaurs*. South Lancaster, MA: Atlantic Union Conference.

Dodson, Peter. 1990. Counting dinosaurs: How many kinds were there? *Proceedings of the National Academy of Sciences of the United States of America* 87, 7608–7612. doi:10.1073/pnas.87.19.7608

Dodson, Peter. 1996. *The Horned Dinosaurs: A Natural History*. Princeton, NJ: Princeton University Press.

Dougherty, Cecil N. 1979. *Valley of the Giants*. Cleburne, TX: Bennett Printing.

Doyle, Arthur Conan. 1914. *The Lost World*. London: Smith, Elder and Co.

Edgar, Blake. 1999. *Dinosaur Digs*. Singapore: APA Publications.

Efremov, I. A. 1940. Taphonomy: New branch of paleontology. *Pan-American Geologist* 74, 81–93.

Fairbank, Jeremy C. T. 2004. William Adams and the spine of Gideon Algernon Mantell. *Annals of the Royal College of Surgeons of England* 86, 349–352.

Farke, A. 2011. Anatomy and taxonomic status of the Chasmosaurine ceratopsid *Nedoceratops hatcheri* from the Upper Cretaceous Lance Formation of Wyoming, U.S.A. *PLoS ONE* 6 (1). doi:10.1371/journal.pone.0016196

Farlow, James O. 1987. *A Guide to Lower Cretaceous Dinosaur Footprints and Tracksites of the Paluxy River Valley, Somervell County, Texas*. Waco, TX: Baylor University Press.

Farlow, James O. 1993. *The Dinosaurs of Dinosaur Valley State Park*. Austin, TX: Texas Parks and Wildlife Department.

Farlow, James O. and M. K. Brett-Surman (Editors). 1997. *The Complete Dinosaur*. Bloomington: Indiana University Press.

Fastovsky, David E. 1996. *The Evolution and Extinction of the Dinosaurs*. Cambridge; New York: Cambridge University.

Fields, Wilbur. 1980. *Paluxy Field Explorations*. Joplin, MO: Wilbur Fields.

Fiffer, Steve. 2000. *Tyrannosaurus Sue: The Extraordinary Saga of the Largest, Most Fought over T. rex Ever Found*. New York: W.H. Freeman.

Finsley, Charles. 1999. *Discovering Texas Dinosaurs*. Houston, TX: Gulf Publishing.

Fitzgibbon, Monty. 2008. *Tyrannosaur*. Cambridge, MA: Candlewick Press.

Fortey, Richard. 2008. *Dry Storeroom No. 1: The Secret Life of the Natural History Museum*. New York: Alfred A. Knopf.

Foster, John. 2007. *Jurassic West: The Dinosaurs of the Morrison Formation and Their World*. Bloomington: Indiana University Press.

Frankel, Charles. 1999. *The End of the Dinosaurs: Chicxulub Crater and Mass Extinctions*. Cambridge, UK: Cambridge University Press.

Frazer, Persifor. 1900. Life and letters of Edward Drinker Cope. *American Geologist* 26, 67–128.

Freeman, Michael. 2004. *Victorians and the Prehistoric World*. New Haven, CT: Yale University Press.

Gentet, Robert E. 1974. *Dinosaurs before Adam?* Pasadena, CA: Ambassador College.

Gillette, David D. 1994. *Seismosaurus: The Earth Shaker*. New York: Columbia University Press.

Gillette, David D. and M. G. Lockley (Editors). 1989. *Dinosaur Tracks and Traces*. Cambridge, UK: Cambridge University Press.

Gish, Duane T. 1972. *Evolution: The Fossils Say No!* San Diego, CA: Creation-Life.

Gish, Duane T. 1992. *Dinosaurs by Design*. Colorado Springs, CO: Master Books.

Gish, Duane T. 1995. *Evolution: The Fossils Still Say No!* El Cajon, CA: Institute for Creation Research.

Glut, Donald F. 1980. *The Dinosaur Scrapbook*. Secaucus, NJ: The Citadel Press.

Glut, Donald F. 1999. *Carbon Dates*. Jefferson, NC: McFarland & Company.

Glut, Donald F. 2001. *Jurassic Classics: A Collection of Saurian Essays and Mesozoic Musings*. Jefferson, North Carolina: McFarland and Company.

Godefroit, P., H. Demuynck, G. Dyke, D. Hu, F. O. Escuillié, and P. Claeys. 2013. Reduced plumage and flight ability of a new Jurassic paravian theropod from China. *Nature Communications* 4, 1394. doi:10.1038/ncomms2389

Gore, Rick. 1989. What caused the earth's great dyings? *National Geographic* 175, 662–699.

Gould, Stephen Jay. 1991. *Bully for Brontosaurus: Reflections in Natural History*. New York: W.W. Norton and Company.

Green, Fitzhugh. 1930. *Roy Chapman Andrews: Dragon Hunter*. New York: G. P. Putnam's Sons.

Haines, Tim. 1999. *Walking with Dinosaurs*. London: BBC Worldwide.

Halls, Kelly Milner. 2006. *The Random House Dinosaur Travel Guide*. New York: Random House.

Ham, Ken. 1998. *The Great Dinosaur Mystery Solved! A Biblical View of These Amazing Creatures*. Green Forest, AR: Master Books.

Ham, Ken. 2001. *Dinosaurs of Eden: A Biblical Journey through Time*. Green Forest, AR: Master Books.

Ham, Ken. 2009. *Dinosaurs for Kids*. Green Forest, AR: Master Books.

Ham, Ken and Cindy Malott. 2008. *The Answers Book for Kids, Volume 2: 22 Questions from Kids on Dinosaurs and the Flood of Noah*. Green Forest, AR: Master Books.

Ham, Kenneth. 2006. *The New Answers Book: 25 Top Questions on Creation/Evolution and the Bible.* Green Forest, AR: Master Books.

Hastings, Ronnie J. 1988. Rise and fall of the Paluxy man tracks. *Perspectives on Science and Christian Faith* 40 (3), 144–155.

Hatcher, J. B. 1893. The Ceratops beds of Converse County. *American Journal of Science* 145, 135–144.

Hatcher, J. B. 1903. Osteology of *Haplocanthus,* with description of a new species, and remarks on the probably habits of the Sauropoda and the age and origin of the *Alantosaurus* beds. *Memoirs of the Carnegie Museum* 2, 1–72.

Hatcher, J. B. 1907. *The Ceratopsia: Based on Preliminary Studies by Othniel Charles Marsh. Edited and Completed by Richard S. Lull.* Monograph of the U.S. Geological Survey, Vol. 49.

Heilmann, Gerhard. 1926. *The Origin of Birds*. London: Witherby.

Helfinstine, Robert F. and Jerry D. Roth. 2007. *Texas Tracks and Artifacts: Do Texas Fossils Indicate Coexistence of Men and Dinosaurs?* Anoka, MN: R&J Publishing.

Hise, James Van. 1993. *Hot-Blooded Dinosaur Movies*. Las Vegas, NV: Pioneer Books.

Hitchcock, Edward. 1848. An attempt to discriminate and describe the animals that made the fossil footmarks of the United States, and especially of New England. *Transactions of American Academy of Arts and Sciences* 3, 129–259.

Hitchcock, Edward. 1858. *Ichnology of New England: A Report on the Sandstone of the Connecticut Valley, Especially Its Fossil Footmarks*. Boston, MA: W. White.

Holland, W. J. 1913. *To the River Plate and Back: The Narrative of a Scientific Mission to South America, with Observations upon Things Seen and Suggested.* New York: G. P. Putnam's Sons.

Horner, Jack. 2001. *Dinosaurs under the Big Sky*. Missoula, MT: Mountain Press Publishing Co.

Horner, Jack and James Gorman. 2009. *How to Build a Dinosaur: Extinction Doesn't Have to Be Forever.* New York: Dutton.

Horner, John R. and Don Lessem. 1993. *The Complete T. rex*. New York: Simon & Schuster.

Horner, John R. and Edwin Dobb. 1997. *Dinosaurs Lives: Unearthing an Evolutionary Saga*. San Diego, CA: Harcourt Brace & Company.

Horner, John R. and James Gorman. 1988. *Digging Dinosaurs*. New York: Workman Publishing Group.

Hoyle, F. and C. Wickramasinghe. 1986. *Archaeopteryx, the Primordial Bird: A Case of Fossil Forgery*. London: Christopher Davies.

Huxley, Thomas H. 1868. On the animals which are most nearly intermediate between birds and reptiles. *Annals and Magazine of Natural History* 2 (Ser. 4), 66–75.

Jacobs, Louis L. 1993. *Quest for the African Dinosaurs*. New York: Random House.

Jacobs, Louis L. 1995. *Lone Star Dinosaurs*. College Station: Texas A&M University Press.

Jaffe, Mark. 2000. *The Gilded Dinosaur: The Fossil War between E.D. Cope and O.C. Marsh and the Rise of American Science*. New York: Crown Publishers.

Jasinski, Laurie E. 2008. *Dinosaur Highway: A History of Dinosaur Valley State Park*. Ft. Worth, TX: TCU Press.

Ji, Q. and S.-A. Ji. 1996. On the discovery of the earliest bird fossil in China and the origin of birds. *Chinese Geology* 233, 30–33. (in Chinese)

Johnson, Kirk R. and Ray Troll. 2007. *Cruisin' the Fossil Freeway*. Golden, CO: Fulcrum Publishing.

Judkins, Aaron. 2009. *Evolution and Human Fossil Footprints*. Oklahoma City, OK: Bible Belt Publishing.

Keener, James. 1988. *Dinosaur Triangle: Land of the "Terrible Lizards."* Grand Junction, CO: Grand River Publishing.

Klein, Nicole, et al. (Editors). 2011. *Biology of the Sauropod Dinosaurs: Understanding the Life of Giants*. Bloomington: Indiana University Press.

Knight, Charles R. 1907. *Diplodocus* restored: The largest creature that ever roamed the earth. *Scientific American* 96 (24), 485.

Knight, Charles R. 1942. Parade of life through the ages. *National Geographic* 81 (2), 141–184.

Knight, Charles R. 2005. *Charles R. Knight: Autobiography of an Artist*. Ann Arbor, MI: G.T. Labs.

Kofahl, Robert E. 1980. *Handy Dandy Evolution Refuter*. San Diego, CA: Beta Books.

Kohl, Michael F. and John S. McIntosh. 1997. *Discovering Dinosaurs in the Old West: The Field Journals of Arthur Lakes*. Washington, DC: Smithsonian Institution Press.

Kuban, Glen J. 1986a. A summary of the Taylor Site evidence. *Creation/Evolution* 6 (1), 10–18.

Kuban, Glen J. 1986b. The Taylor Site "man tracks." *Origins Research* 9 (1), 1–10.

Lanham, Url. 1973. *The Bone Hunters*. New York: Columbia University Press.

Lankester, E. Ray. 1905. *Extinct Animals*. London: A. Constable & Co.

Larson, Peter and Kristin Donnan. 2002. *Rex Appeal: The Amazing Story of Sue, the Dinosaur That Changed Science, the Law, and My Life*. Montpelier, VT: Invisible Cities Press.

Leidy, J. 1856. Notices of Remains of Extinct Reptiles and Fishes, Discovered by Dr. F.V. Hayden in the Badlands of the Judith Rover, Nebraska Territories. *Proceedings of the Academy of Natural Sciences of Philadelphia* 8, 72–73.

Leidy, J. 1858. *Hadrosaurus foulkii*, a new saurian from the Cretaceous of New Jersey. *Proceedings of the Academy of Natural Sciences of Philadelphia* 10, 212–218.

Lightman, Bernard. 2007. *Victorian Popularizers of Science*. Chicago: University of Chicago Press.

Lloyd, Graeme T., et al. 2008. Dinosaurs and the Cretaceous terrestrial revolution. *Proceedings of the Royal Society B* 275, 2483–2490.

Lockley, M. G. 2000. *Dinosaur Tracks and Other Fossil Footprints of Europe*. New York: Columbia University Press.

Lockley, M. G., B. J. Fillmore, and L. Marquardt. 1997. *Dinosaur Lake: The Story of the Purgatoire Valley Dinosaur Tracksite Area*. Denver, CO: Colorado Geological Society.

Lockley, Martin and Adrian Hunt. 1995. *Dinosaur Tracks and Other Fossil Footprints of the Western United States*. New York: Columbia University Press.

Lockley, Martin G. 1991. *Tracking Dinosaurs: A New Look at an Ancient World*. Cambridge, UK: Cambridge University Press.

Lockley, Martin G. 2001. *A Field Guide to Dinosaur Ridge*. Morrison, CO: Friends of Dinosaur Ridge.

Lubenow, Marvin. 1992. *Bones of Contention*. Grand Rapids: Baker.

Lucas, Spencer G. 1994. *Dinosaurs: The Textbook*. Dubuque, IA: William C. Brown Publishers.

Lull, Richard S. 1918. The development of vertebrate paleontology. *American Journal of Science* XLVI, 193–221.

Mackal, Roy P. 1987. *A Living Dinosaur: In Search of the Mokele-Mbembe*. Leiden, the Netherlands: E.J. Brill.

MacLary, John. 1938. Dinosaur trails of Purgatory. *Scientific American* 156, 72.

Maidment, S. 2010. Stegosauria: A historical review of the body fossil record and phylogenetic relationships. *Swiss Journal of Geosciences*. doi:10.1007/s00015–010–0023–3

Malik, Peter. 1981. *The Ape Men*. Uniontown, PA: Penn Highlands Publications.

Malone, Bruce A. 2009. *Censored Science: The Suppressed Evidence*. Midland, MI: Search for the Truth Publications.

Manning, Phillip. 2008. *Grave Secrets of Dinosaurs: Soft Tissues and Hard Science*. Washington, DC: National Geographic.

Mantell, Gideon. 1825. Notice on the *Iguanodon,* a newly discovered fossil reptile, from the sandstone of Tilgate Forest, in Sussex. *Philosophical Transactions of the Royal Society of London* 115, 179–186. doi:10.1098/retl.1825.0010

Marsh, O. C. 1877. Notice of a new and gigantic dinosaur. *American Journal of Science* 14, 87–88.

Marsh, O. C. 1880. *Odontornithes: A Monograph on the Extinct Toothed Birds of North America*. Washington, DC: Government Printing Office.

Marsh, O. C. 1883. Principal characters of American Jurassic dinosaurs, Part IV: Restoration of *Brontosaurus*." *American Journal of Science* 26 (Ser. 3), 81–85.

Marsh, O. C. 1891a. Principal characters of American Jurassic dinosaurs, Part IV: Restoration of *Triceratops*." *American Journal of Science* 41 (Ser. 3), 339–342.

Marsh, O. C. 1891b. Principal characters of American Jurassic dinosaurs, Part IV: Restoration of *Stegosaurus*." *American Journal of Science* 42 (Ser. 3), 179–181.

Marsh, O. C. 1896a. *Dinocerata: A Monograph of an Extinct Order of Gigantic Mammals*. Monograph of the U.S. Geological Survey, Vol. 10.

Marsh, O. C. 1896b. *The Dinosaurs of North America*. 16th Annual Report of the U.S. Geological Survey, Part I, 133–244.

Matthew, William Diller. 1915. *Dinosaurs—With Special Reference to the American Museum Collections*. New York: American Museum of Natural History.

Mayr, G., B. Pohl, and D. S. Peters. 2005. A well-preserved *Archaeopteryx* specimen with theropod features. *Science* 310, 1483–1486.

McGinnis, Helen J. 1982. *Carnegie's Dinosaurs: A Comprehensive Guide to Dinosaur Hall at Carnegie Museum of Natural History, Carnegie Institute*. Pittsburgh, PA: Geyer Printing Company.

McGowen, Charles H. 1976. *In Six Days*. New Wilmington, PA: Son-Rise Publications and Distributing, Inc.

McGowan, Christopher. 2001. *The Dragon Seekers: How an Extraordinary Circle of Fossilists Discovered the Dinosaurs and Paved the Way for Darwin*. Cambridge, MA: Perseus Publishing.

McIntosh, John S. 1981. *Bulletin of Carnegie Museum of Natural History: Annotated Catalogue of the Dinosaurs (Reptilia, Archosauria) in the Collections of Carnegie Museum of Natural History*. Pittsburgh, PA: Carnegie Museum of Natural History.

McIntyre, Melody J. 1981. *Dinosaur ABC's*. San Diego, CA: Creation-Life.

Mendyk, S. 1985. Robert Plot: Britain's "Genial Father of Country Natural Histories." *Notes and Records of the Royal Society* 39, 159–177.

Milner, Richard. 2012. *Charles R. Knight: The Artist Who Saw through Time*. New York: Abrams.

Mitchell, W. J. T. 1998. *The Last Dinosaur Book*. Chicago: University of Chicago Press.

Moody, R. T. J., E. Buffetaut, D. Naish, and D. M. Martill (Editors). 2010. *Dinosaurs and Other Extinct Saurians: A Historical Perspective. Geological Society Special Publication No. 343*. London: The Geological Society.

Moore, Randy. 2014. Did humans live with dinosaurs? Excavating "man tracks" along the Paluxy River. *American Biology Teacher* 76 (4), 243–246.

Moore, Randy and Mark D. Decker. 2008. *More Than Darwin: An Encyclopedia of the People and Places of the Evolution-Creationism Controversy*. Westport, CT: Greenwood Press.

Moore, Randy, Mark Decker, and Sehoya Cotner. 2010. *Chronology of the Evolution-Creationism Controversy*. Santa Barbara, CA: ABC-CLIO.

Moore, Randy and Janice Moore. 2006. *Evolution 101*. Westport, CT: Greenwood.

Morris, John and Ken Ham. 1988. *What Really Happened to the Dinosaurs?* Green Forest, AR: Master Books.

Morris, John D. 1980. *Tracking Those Incredible Dinosaurs . . . and the People Who Knew Them*. San Diego, CA: Christian Life Publishers.

Morris, John D. 1988. Continued research on the Paluxy Tracks. *Acts and Facts* 17 (12). Special edition.

Morris, John D. and Frank J. Sherwin. 2010. *The Fossil Record: Unearthing Nature's History of Life*. Dallas, TX: Institute of Creation Research.

Naish, Darren. 2009. *The Great Dinosaur Discoveries*. Berkeley, CA: University of California Press.

Nesbit, Sterling, J., et al. 2012. The oldest dinosaur? A Middle Triassic dinosauriform from Tanzania. *Biological Letter* 9 (1). doi:10.1098/rsbl.2012.0949

Neufeld, Berney. 1975. Dinosaur tracks and giant men. *Origins Research* 2 (2), 64–76.

Norell, Mark. 2005. *Unearthing the Dragon: The Great Feathered Dinosaur Discovery*. New York: Pearson Education.

Norell, Mark A., Lowell Dingus, and Eugene Gaffney. 1995. *Discovering Dinosaurs: Evolution, Extinction, and the Lesson of Prehistory*. Berkeley, CA: University of California Press.

Norell, Mark A., E. S. Gaffney, and Lowell Dingus. 1995. *Discovering Dinosaurs at the American Museum of Natural History*. New York: Nevraumont Publishing Company.

Norman, David. 1985. *The Illustrated Encyclopedia of Dinosaurs: An Original and Compelling Insight into Life in the Dinosaur Kingdom*. New York: Crescent Books.

Novacek, Michael. 2002. *Time Traveler: In Search of Dinosaurs and Ancient Mammals from Montana to Mongolia*. New York: Farrar, Straus and Giroux.

O'Connor, Ralph. 2007. *The Earth on Show: Fossils and the Poetics of Popular Science, 1802–1856*. Chicago: University of Chicago Press.

Officer, Charles B. and Jake Page. 1996. *The Great Dinosaur Extinction Controversy*. Reading, MA: Addison-Wesley Publishing.

Olsen, E. 1962. Late Permian terrestrial vertebrates, U.S.A. and U.S.S.R. *Transactions of the American Philosophical Society* 52 (2), 1–224.

Osborn, Henry Fairfield. 1898. Biographical memoir of Joseph Leidy, 1823–1871. *Biographical Memoirs of the National Academy of Sciences* 7, 335–396.

Osborn, Henry Fairfield. 1905. *Tyrannosaurus* and other Cretaceous carnivorous dinosaurs. *Bulletin of the American Museum of Natural History, Author's Edition* 21 (Article 14), 259–265.

Osborn, Henry Fairfield. 1906. Tyrannosaurus, Upper Cretaceous carnivorous dinosaur (second communication). *Bulletin of the American Museum of Natural History* 22 (16), 281–296.

Osborn, Henry Fairfield. 1931. *Cope: Master Naturalist: The Life and Letters of Edward Drinker Cope with a Bibliography of His Writings Classified by Subject.* Princeton, NJ: Princeton University Press.

Osborn, Henry Fairfield. 1935. Thomas Jefferson as a paleontologist. *Science* 82, 533–538.

Ostrom, John H. 1971. *Report to the National Park Service on Mesozoic Vertebrate Paleontological Sites for Possible Inclusion in the Registry of Natural Landmarks.* New Haven, CT: Peabody Museum of Natural History, Yale University Press.

Ostrom, John H. 1974. Archaeopteryx and the origin of flight. *Quarterly Review of Biology* 49, 27–47.

Ostrum, John H. and John S. McIntosh. 1966. *Marsh's Dinosaurs: The Collections from Como Bluff.* New Haven, CT: Yale University Press.

Ottaviani, Jim and Big Time Attic. 2005. *Bone Sharps, Cowboys, and Thunder Lizards: Edward Drinker Cope, Othniel Charles Marsh, and the Gilded Age of Paleontology.* New York: G.T. Labs.

Owen, Richard. 1860. *Paleontology, or a Systematic Summary of Extinct Animals and Their Geological Relations*. Edinburgh, Scotland: Adam and Charles Black.

Parker, Steve. 2003. *Dinosaurus: The Complete Guide to Dinosaurs*. Buffalo, NY: Firefly Books.

Parker, Steve and Leslie Mertz. 2008. *Extreme Dinosaurs*. Irvington, NY: HarperCollins.

Parsons, Keith M. 2001. *Drawing Out Leviathan. Dinosaurs and the Science Wars*. Bloomington: Indiana University Press.

Parsons, Keith M. 2004. *The Great Dinosaur Controversy: A Guide to the Debates*. Santa Barbara, CA: ABC-CLIO.

Paul, G. S. 1988. *Predatory Dinosaurs of the World.* New York: Simon & Schuster.

Paul, G. S. 2002. *Dinosaurs of the Air: The Evolution and Loss of Flight in Dinosaurs and Birds*. Baltimore, MD: Johns Hopkins University Press.

Paul, G. S. 2010. *The Princeton Field Guide to Dinosaurs*. Princeton, NJ: Princeton University Press.

Penfield, G. T. and Z. A. Camargo. 1981. Definition of a major igneous zone in the central Yucatan platform with aeromagnetics and gravity. *Abstract of the Annual Meeting of the Society for Exploration Geophysicists* 51, 37.

Phillips, Phil. 1994. *Dinosaurs, the Bible, Barney and Beyond.* Lancaster, PA: Starburst Publishers.

Piggott, J. R. 2004. *Palace of the People: The Crystal Palace at Sydenham 1854–1936.* London: C. Hurst and Company.

Plate, Robert. 1964. *The Dinosaur Hunters: Othniel C. Marsh and Edward D. Cope.* New York: McKay.

Poinar, George O. and Roberta Poinar. 2008. *What Bugged the Dinosaurs? Insects, Disease, and Death in the Cretaceous*. Princeton, NJ: Princeton University Press.

Powell, James L. 1998. *Night Comes to the Cretaceous: Dinosaur Extinction and the Transformation of Modern Geology*. New York: W.H. Freeman & Co.

Preston, Douglas J. 1986. *Dinosaurs in the Attic: An Excursion into the American Museum of Natural History*. New York: St. Martin's Press.

Psihoyos, Louie and John Knoebber. 1994. *Hunting Dinosaurs*. New York: Random House.

Raup, David M. 1986. *The Nemesis Affair: A Story of the Death of Dinosaurs and the Ways of Science*. New York: W.W. Norton & Company.

Raup, David M. 1992. *Extinction: Bad Genes or Bad Luck?* New York: W.W. Norton & Company.

Raup, David M. and J. J. Sepkosi, Jr. 1984. Periodicity of extinctions in the geologic past. *Proceedings of the National Academy of Sciences* 81, 801–805.

Rea, Tom. 2001. *Bone Wars: The Excavation and Celebrity of Andrew Carnegie's Dinosaur.* Pittsburgh, PA: University of Pittsburgh Press.

Rogers, Katherine. 1999. *The Sternberg Fossil Hunters: A Dinosaur Dynasty*. Missoula, MT: Mountain Press Publishing Co.

Rudwick, Martin J. S. 2001. Jean-André de Luc and nature's chronology. *Geological Society, London*, Special Publications, 190, 51–60.

Rupke, Nicolaas A. 1994. *Richard Owen: Victorian Naturalist*. New Haven, CT: Yale University Press.

Rupke, Nicolaas A. 2009. *Richard Owen: Biology without Darwin, A Revised Edition*. Chicago: University of Chicago Press.

Ryan, M. J., B. J. Chinnery-Allgeier, and D. A. Eberth. (Editors). 2010. *New Perspectives on Horned Dinosaurs: A Royal Tyrrell Museum Ceratopsian Symposium*. Bloomington: Indiana University Press.

Sampson, Scott D. 2009. *Dinosaur Odyssey: Fossil Threads in the Web of Life*. Berkeley, CA: University of California Press.

Sanz, José Luis. 2002. *Starring T. rex! Dinosaur Mythology and Popular Culture*. Bloomington: Indiana University Press.

Scannella, J. and J. Horner. 2010. *Torosaurus* Marsh, 1891, is *Triceratops* Marsh, 1889 (Ceratopsidae: Chasmosaurinae): Synonymy through ontogeny. *Journal of Vertebrate Paleontology* 30 (4), 1157–1168. doi:10.1080/02724634.2010.483632

Schuchert, Charles. 1938. *Biographical Memoir of Othniel Charles Marsh 1831 1899*. Washington, DC: National Academy of Sciences.

Schuchert, Charles and Clara Mae LeVene. 1940. *O.C. Marsh: Pioneer in Paleontology.* New Haven, CT: Yale University Press.

Scully, Vincent, Rudolph F. Zallinger, Leo J. Hickey, and John H. Ostrom. 1990. *The Great Dinosaur Mural at Yale: The Age of Reptiles*. New York: Harry N. Abrams, Inc. Publishers.

Seeley, H. G. 1888. On the classification of the fossil animals commonly named Dinosauria. *Proceedings of the Royal Society of London* 43, 165–171.

Segraves, Kelly L. 1975. *The Great Dinosaur Mistake*. San Diego, CA: Beta Books.

Segraves, Kelly L. 1980. *Everything You Always Wanted to Know about Dinosaurs, Flying Saucers, the Beginning and the End*. San Diego, CA: Beta Books.

Semonin, Paul. 2000. *American Monster: How the Nation's First Prehistoric Creature Became a Symbol of National Identity*. New York: New York University Press.

Shuker, Karl P. N. 2008. *Dinosaurs and Other Prehistoric Animals on Stamps: A Worldwide Catalogue*. North Devon, UK: CFZ Press.

Simpson, George G. 1942. The beginnings of vertebrate paleontology in North America. *Proceedings of the American Philosophy Society* 86 (1), 130–188.

Sinclair Oil Corporation. 1966. *A Great Name in Oil: Sinclair through Fifty Years*. Hightstown, NJ: F.W. Dodge Company.

Spalding, David A. E. 1993. *Dinosaur Hunters: Eccentric Amateurs and Obsessed Professionals*. Toronto, Ontario, Canada: Key Porter Books.

Spamer, Earle E. 1999. Know thyself: Responsible science and the lectotype of *Homo sapiens* Linnaeus, 1758. *Proceedings of the Academy of Natural Sciences of Philadelphia* 149, 109–114.

Sternberg, Charles H. 1909. *The Life of a Fossil Hunter*. New York: Henry Holt.

Sternberg, Charles H. 1917. *Hunting Dinosaurs in the Bad Lands of the Red Deer River, Alberta, Canada*. Lawrence, KS: World Company Press.

Stokes, William Lee. 1985. *The Cleveland-Lloyd Dinosaur Quarry: Window to the Past*. Washington, DC: U.S. Government Printing Office.

Swinton, W. E. 1951. Gideon Mantell and the Maidstone Iguanodon. *Notes and Records of the Royal Society* 8, 261–176. doi:10.1098/rsnr.1951.0020

Switek, Brian. 2013. *My Beloved Brontosaurus*. New York: Scientific America.

Tanke, Darren and Bruce Rothschild. 2002. *Dinosores: An Annotated Bibliography of Dinosaur Pathology and Related Topics—1838–2001*. Albuquerque, NM: New Mexico Museum of Natural History.

Taquet, Philippe. 1994. *Dinosaur Impressions: Postcards from a Paleontologist*. (translated into English by Kevin Padian in 1998). Cambridge, UK: Cambridge University Press.

Taylor, Carol. 1973. *Fossil Footprints Found*. Caldwell, ID: Bible-Science Association.

Taylor, Paul S. 1981. *The Great Dinosaur Mystery and the Bible*. Elmwood, IL: Films for Christ Association.

Taylor, Stanley E. 1973. *Footprints in Stone*. Mesa, AZ: Films for Christ Association.

Thackray, John and Bob Press. 2001. *The Natural History Museum: Nature's Treasurehouse*. London: Natural History Museum.

Thulborn, Tony. 1990. *Dinosaur Tracks*. New York: Chapman and Hall.

Torrens, Hugh. 1992. Where did the dinosaur get its name? *New Scientist* 134, 40–44.

Van Hise, James. 1993. *Hot Blooded Dinosaur Movies*. Las Vegas, NV: Pioneer Books.

Wallace, David Rains. 1999. *The Bonehunters' Revenge: Dinosaurs, Greed, and the Greatest Scientific Feud of the Gilded Age*. Boston, MA: Houghton Mifflin.

Wallace, Joseph. 1987. *The Rise and Fall of the Dinosaur*. New York: W.H. Smith Publishers.

Wang, Steve C. and Peter Dodson. 2006. Estimating the diversity of dinosaurs. *Proceedings of the National Academy of Sciences of the United States of America* 103 (37), 13601–13605.

Ward, Peter Douglas. 2000. *Rivers in Time: The Search for Clues to Earth's Mass Extinctions*. New York: Columbia University Press.

Watson, John Allen. 2001. *Man, Dinosaurs, and Mammals Together*. Crosbyton, TX: Mt. Blanco Publishing.

Webster, Donovan. 2009. The dino wars. *Smithsonian* 40 (1), 48–57.

Weishampel, David B. and Coralia-Maria Jianu. 2011. *Transylvanian Dinosaurs*. Baltimore, MD: John Hopkins University Press.

Weishampel, David B. and Nadine M. White. 2003. *The Dinosaur Papers, 1676–1906*. Washington, DC: Smithsonian Press.

Weishampel, David B., P. Dodson, and H. Osmólska (Editors). 2004. *The Dinosauria*, Second Edition. Berkeley, CA: University of California Press.

Whitcomb, John C. and Henry M. Morris. 1961. *The Genesis Flood: The Biblical Record and Its Scientific Implications*. Grand Rapids, MI: Baker Book House.

Wieland, G. 1925. Dinosaur extinction. *American Naturalist* 59 (665), 557–565.

Wilford, John Noble. 1985. *The Riddle of the Dinosaurs*. New York: Alfred A. Knopf.

Williston, Samuel W. 1878. American Jurassic dinosaurs. *Transactions of the Kansas Academy of Science* 6, 42–46.

Wilson, Clifford, Barbara Wilson, and John Heffner. 2007. *The Paluxy River Dinosaur Controversy: Did Dinosaurs and Men Walk Together? Final Nails in the Coffin of Evolution*. Victoria, Australia: Pacific Christian Ministries.

Winchester, Simon. 2001. *The Map That Changed the World: William Smith and the Birth of Modern Geology*. New York: HarperCollins.

Woetzel, David and Richard Dobbs, Jr. 2013. *Chronicles of Dinosauria: The History and Mystery of Dinosaurs and Man*. Green Forest, AR: Master Books.

Woodmorappe, John. 1996. *Noah's Ark: A Feasibility Study*. Santee, CA: Institute for Creation Research.

Wortman, Jacob L. 1899. Othinel Charles Marsh. *Science* 9, 562, 565.

Wyman, J. 1855. Notice of fossil bone from the red sandstone from the Connecticut River Valley. *American Journal of Science* 20, 394–395.

Yates, Adam M. and James W. Kitching. 2003. The earliest known sauropod dinosaur and the first steps towards sauropod locomotion. *Proceedings of the Royal Society B* 270, 1753–1758.

Zhang, F., Z. Zhou, X. Xu, X. Wang, and C. Sullivan. 2008. A bizarre Jurassic maniraptoran from China with elongate ribbon-like feathers. *Nature* 455, 1105–1108. doi:10.1038/nature07447

Zimmerman, Virginia. 2008. *Excavating Victorians*. Albany, NY: State University of New York.

Index

Note: Some of the index entries are quotations; these are italicized and appear within quotation marks.

AAAS. *See* American Association for the Advancement of Science
Aachenosaurus, xxii
Aberhart, William "Bible Bill," 195
Abydosaurus, 394
Academy of Natural Sciences (ANS), 9, 19, 38, 43, 47, 51–52, 53, 56, 106
Ackerman, Forrest, 166, 223
Acrocanthosaurus, 143, 252, 323, 385
Adams, Earnest Tolbert "Bull," 202, 415 n.28
Adams, George, 143, 202, 259
Adams, William Davenport, 73
Aegyptosaurus, 145
Aepisaurus, 46
Afrovenator, 308, 333
Agathaumas, 112
Age of Dinosaurs. *See* Mesozoic Era
Age of Mammals, xviii
The Age of Reptiles, 157, 212, 215, 229, 236, 260, 281, 339
Age of the Earth, xviii–xx, xxviii n.1, 4, 8, 10, 118, 134, 146, 151, 230
AiG. *See* Answers in Genesis
Alamosaurus, 121, 172, 323, 357
Albertonykus, 381
Albertosaurus, 94, 97, 102, 126, 133, 143, 144, 154, 162, 187, 211, 215, 234, 259, 310
Alen, P. G., 178, 187
Algoasaurus, 130
All about Dinosaurs, 224
"*All dinosaurs not on Noah's Ark were drowned*," 322
Allen, James E., 192
Alley Oop, 186, 188, 234, 248, 340, 373
Allosaurus, 28, 61, 67, 80, 81, 87, 88, 93, 110, 111, 118, 128, 135, 140, 146, 148, 149, 158, 162, 169, 171, 179, 181–82, 194, 196, 197, 200, 201, 205, 212, 223, 237, 243, 245, 257, 268, 284, 285, 320, 364, 372, 415 n.27
"*Alluringly scary, but sufficiently safe*," 338
Along the Moonbeam Trail, 164
Alvarez, Luis and Walter, 3, 16, 33, 231, 264, 278–80, 282–84, 295, 296, 314, 326, 327, 342, 346, 365
Ameghino, Florentino, 92
American Anthropometric Society, 101, 117
American Antiquities Act, 135, 197, 265
American Association for the Advancement of Science (AAAS), 34, 79, 109, 273
American degeneracy, 9
American Museum of Natural History (AMNH), xii, 51, 68–69, 76, 81, 82, 89, 92, 100, 103, 106, 108, 111, 112, 123, 126, 128, 154, 165, 230, 239, 340, 364. *See also* Andrews, Roy Chapman; Osborn, Henry Fairfield
American Philosophical Society (APS), 7, 12, 43, 52, 83
"*America's first great evolutionary theorist*," 271
Amherst College Museum of Natural History, 43, 49
Ammosaurus, 94
AMNH. *See* American Museum of Natural History
Amphicoelias, 85, 102, 113, 115, 210
Anatosaurus, xxii, 138, 172, 211, 276
Anatotitan, xxii, 90, 91, 125, 197
Anchiceratops, 153
Anchiornis, 384, 390, 392, 407
Anchisaurus, 21, 108, 150

Andersson, Johan Gunnar, 163, 169
Andrews, Roy Chapman, 72, 100, 108, 114, 135, 137, 163, 165, 167, 168, 169, 171, 175, 184, 187, 191, 193, 207, 209, 218, 224, 239, 275, 363. *See also* American Museum of Natural History
An Investigation into the Principles of Knowledge, 10
Ankylosaurus, 144, 212, 246
Annals of the Old Testament, 2. *See also* Age of the Earth
Anning, Mary, 6, 18, 19, 24, 28, 41, 42, 205, 394
ANS. *See* Academy of Natural Sciences
Answers in Genesis (AiG), 237–38, 277–78, 285, 301, 306, 333–34, 343, 356, 357, 372, 375, 377, 381, 397, 404, 410, 412. *See also* Ham, Ken; Young-Earth Creationism
Antarctica, 256–57, 296, 314, 320, 335
Antarctosaurus, 76, 169
Anti-Preparedness Committee, 158
Antrodemus, 69
Apatornis, 90
Apatosaurus, xi, xx, xxi, xxii, 62, 63, 69, 79, 80, 84, 86, 87, 90, 106, 107, 110, 113, 115–18, 121, 122, 128, 131, 135, 141, 142, 146, 150, 151, 153–56, 159, 170, 171, 173, 183, 187, 188, 196, 200, 207, 212, 223, 224, 231, 235, 240, 250, 255, 267, 273, 277, 301, 303, 321, 325, 331, 335, 393, 399, 414 n.10, 414 n.14. See also *Brontosaurus*
Appleton, Samuel, 49
Appleton Cabinet, 49
APS. *See* American Philosophical Society
Archaeopteryx, 49, 56–60, 64, 71, 118, 151, 177, 228, 252, 260, 266, 371, 395, 407, 409
  Berlin specimen, 78, 81, 351
  Bürgermeister-Müller (chicken wing) specimen, 356
  Daiting specimen, 387
  Eichstätt specimen, 78, 222
  Haarlem specimen, 49–50, 76
  London specimen, 56, 58, 228, 229, 240, 297, 369, 380
  Maxberg specimen, 229–30, 235
  Munich specimen, 330
  Solnhofen specimen, 308
  Thermopolis specimen, 328, 373. *See also* Solnhofen
Archaeopteryx, the Primitive Bird
Archaeoraptor, 354, 359, 366, 367, 409
Archeological Resources Protection Act, 278
Archetype and Homologies of the Vertebrate Skeleton, 43
Archosaurs, xvii
Argentinosaurus, xix, 273, 300, 301, 344
Argyrosaurus, 104, 414 n.5
The Artist in His Museum, 16
Ashmolean Museum, 2, 3
Asilisaurus, xvii
Asimov, Isaac, 208, 219, 235, 274, 305
"*The asteroidal theory has changed all that*," 280
Astor, John Jacob, IV, 105
Astrodon, 51, 54, 115, 305, 414 n.5
Atherstone, W. G., 41
Atlantosaurus, 79, 90, 105, 115, 300
At the Earth's Core, 162, 168
Aurornis, 409
Avian dinosaurs, xx
Avimimus, 284

Baby—Secret of the Lost Legend, 293, 301
Bacon, Sir Francis, 2
Bahariasaurus, 145
Bain, A. G., 41
Bakewell, Robert, 27
Bakker, Robert, 126, 139, 164, 210, 212, 229, 235, 249, 254, 256–58, 262–64, 267, 273, 287, 295, 297, 302, 306, 314, 316, 323, 325, 327, 330, 334, 335, 343, 347, 358, 363, 364, 368, 416 n.41
Baldwin, David, 90
Ballou, William Hosea, 99, 108, 113
Bambi Meets Godzilla, 257
BAND. *See* Birds Are Not Dinosaurs
Bannister, H. M., 112

Barlow, Caleb, 61
Barney & Friends, 122, 301–2, 326
Barnum, P. T., 14, 16
Barosaurus, 99, 100, 111, 118, 141, 164, 171, 172, 185, 187, 194, 195, 223, 243, 256, 318, 364, 365, 379, 414 n.17, 416 n.41
Barringer, Daniel, 134
Barsbold, Rinchen, 261
Baryonyx, 86, 288, 322, 362
Baugh, Carl, 204, 238, 249, 267, 284, 300, 301, 305, 306, 314, 322, 324, 335, 337, 342, 356, 361, 367, 381, 385, 401, 405. *See also* Creation Evidence Museum; Young-Earth Creationism
Bayerische Staatssammlung für Paläontologie und Geologie, 146
B.C., 234
The Beast from 20,000 Fathoms, 202, 220, 221, 222, 223, 224–25, 232, 236, 243, 415 n.34
The Beast of Hollow Mountain, 185, 231, 251
Beckles, Samuel, 47
Becquerel, Antoine Henri, 106
Beecher, Charles, 127
Beierle, Fred, 279
Beipiaosaurus, 354
Bell, Claude. *See* Cabazon dinosaurs
Benton, Michael J., 383, 395
Berman, David S., 142, 174, 267, 273
Bernissart, 85–86, 91, 92, 153, 375
"*The best white man I ever saw*," 75
BHI. *See* Black Hills Institute of Geologic Research
Biblical flood. *See* Noah's flood
Bierce, Ambrose, 144
Big Bone Lick, 5, 6, 8, 13
Bird, Roland, 137, 143, 161, 188, 200, 202–4, 206, 209, 219, 222, 224, 245, 249, 258, 259, 262, 270, 275, 303, 307, 377, 416 n.39. *See also* Glen Rose, Texas
Birds: as dinosaurs, xx
  evolution of, xviii
  toothed, 73, 77, 90, 118, 185
"*Birds are evolved from dinosaurs*," 64
Birds Are Not Dinosaurs (BAND), 59, 263
"*Birds have come down to us through the Dinosaurs*," 79
The Birth of a Flivver, 161
Black Hills Institute of Geologic Research (BHI), 275, 316, 326, 327
Bleak House, 46
BMNH. See Natural History Museum
Boltwood, Bertram, 135. *See also* Age of Earth
Bonaparte, José, 214, 257, 278, 280, 295, 354
Bone Cabin Quarry, 110, 113, 123, 124
Bone wars, 52, 66, 73, 84, 100, 142, 374, 396. *See also* Cope, Edward Drinker; Marsh, Othniel
Bones for Barnum Brown, 202
The Book of Healing, 1
Bothriospondylus, 53
Boylan, Grace and Thomas, 191, 215. *See also* Fossil Cabin
Brachiosaurus, 63, 86, 114, 119–20, 122, 128, 136, 142, 155, 159, 172, 177, 200–201, 212, 235, 300, 333, 372, 394, 401, 416 n.44
Brachyceratops, 121, 154
Bradbury, Ray, 220, 222, 224, 325
Bradbury, Stanton M., 121
Branca, Wilhelm, 138
Bray, John, 154
The Break-Up of the Large and Small Worlds, 2
Brees, Christopher, 289–90, 336
Brett-Surman, Michael, 346
Brigham Young University Museum of Paleontology (BYU), 241
Bringing Up Baby, 202
The British Cuvier, 37
British Museum (Natural History). *See* Natural History Museum
Brontosaurus, xii, xxii, xxiii, 29, 30, 62, 63, 80, 84, 88, 92, 95, 101, 104, 117, 127, 128, 131, 133, 141, 150, 151, 153, 159, 173, 179, 184, 189, 190, 193, 195, 197, 198, 202, 210, 212, 224, 229, 234, 235, 244, 312, 321. See also *Apatosaurus*

Brookes, Richard, 7
Broom, Robert, 41, 130–31, 151
Brown, Barnum, 15, 74, 77, 84, 89, 92, 94, 100, 102, 103, 106, 108–12, 116, 124–27, 130, 132, 134, 135, 136, 137, 143, 144, 146, 149, 153, 158, 159, 161, 166, 171, 178, 181, 183, 184, 188, 189, 192, 194–96, 199, 200, 205, 207, 211, 239, 242, 244–47, 249, 273, 350, 364, 370, 376, 389
Brusatte, Steve, 384, 386, 390, 391, 393, 400
Brute Force, 152
Bryan, William Jennings, 130, 167–68, 357
BSP. *See* Bayerische Staatssammlung für Paläontologie und Geologie
Buckland, William, xii, 17, 19, 20, 21–22, 24–25, 27, 28, 33, 35, 37, 40, 41, 45, 47, 51, 104, 256, 333, 414 n.4, 414 n.7
Buffon, Georges-Louis, 6, 9
Bugg, George, 27
Burdick, Clifford, 218–19, 249, 266–67, 343
Burian, Zdenêk, 122, 195, 210, 219, 229, 232, 238, 313
Burpee Museum of Natural History, 362
Burroughs, Edgar Rice, 162, 168, 183, 267, 270, 271
Busk, George, 51
"*Buy this for Pittsburgh*," 116
"*By, Joe, it was a pretty good job*," 235
BYU. *See* Brigham Young University Museum of Paleontology

Cabazon dinosaurs, 246–47, 259, 340, 373, 399
Camarasaurus, 81, 83, 87, 92–93, 113–15, 121, 141, 154, 156, 164, 165, 171, 173, 174, 179, 182, 185, 187, 197, 229, 234, 267, 284, 295, 307, 308, 414 n.10, 414 n.14
Camptosaurus, xxiii, 61, 89, 95, 113, 130, 182, 185, 194, 195, 205, 223
Canada-China Dinosaur Project, 298–99
Cannon, George L., 97, 98
Carcharodontosaurus, 145, 179, 213, 333
Cardiff Giant, 69
Carlin, William, 87
Carnegie, Andrew, xii, 7, 55, 63, 106, 107, 116, 123–24, 127, 128–29, 131, 137, 156, 161, 164, 240. *See also* Carnegie Museum of Natural History
Carnegie, Louise, 141
Carnegie Museum of Natural History (CMNH), xii, 63, 80, 89, 106, 107, 116, 123, 126, 127, 129, 136, 139, 144, 150, 157, 158, 164, 172, 235, 355, 380
Carnosaur, 332, 335, 341–42
Carnosauria, 154
Carnotaurus, 361
Carolini, Rubén Dario, 330, 339
Carpenter, Kenneth, 54, 286, 296, 299, 314, 334
A Catalogue of the Extraneous English Fossils, 4
Catastrophism, 19, 21, 55
Caudipteryx, 349
Centennial Exposition, 77
Central Asia Expeditions, 167, 184, 191, 209
Central Park, 67–68
Centrosaurus, 85, 130, 153, 163, 182, 205, 272
Cerasinops, 287
Ceratops, 61, 97
The Ceratopsians, 272, 388
Ceratosaurus, 93, 94, 101, 108, 145, 147, 152, 164, 172, 223, 231, 268, 342, 415 n.32
Cetiosaurus, 40, 44, 64, 72, 132, 283, 408
Charig, Alan, 243, 297
Chasmosaurus, 143, 154, 159, 181, 256, 260, 395
Chiappe, Luis, 351
Chick, Jack, 358, 377
Chicxulub crater, 134, 274, 282, 316, 318, 333, 337, 370. *See also* Alvarez, Luis and Walter; Extinction
Chromosaurs, 305
Chure, Dan, 337
Cigarette cards. *See* Trading cards

Clade, xvii, 300
Claosaurus, 103, 108, 123, 172
Clapp, George H., 208
Clavicle, 177
Cleveland-Lloyd Quarry, 140, 179, 181, 182, 205, 214, 236–37, 240, 241, 243, 248, 255, 268, 294, 301, 344
"*Clever girl*," 332
CMNH. *See* Carnegie Museum of Natural History
Coelophysis, 69, 90, 98, 216–17, 272, 281, 283, 285, 292, 316, 350
Coelosaurus, 61
Coelurus, 88
Coggeshall, Arthur, 129, 131, 136, 150, 234
Colbert, Edwin Harris "Ned," 98, 101, 177, 183, 196, 207, 214, 215, 222, 224, 228, 241, 244, 245, 256–57, 271, 335, 338, 363, 415 n.33
Colbert, Margaret, 207
Collini, Cosimo Alessandro, 8
Collinson, Peter, 6
Colorado National Monument, 119, 121, 146
Colorado state dinosaur, 286
Colorado state fossil, 82
Columbian Museum of Chicago. *See* Field Museum
Como Bluff, 66, 76, 82, 83–4, 86–7, 88, 90, 95, 98, 110, 114, 116, 191, 251, 276, 323
Compsognathus, 53, 56, 60, 61, 64, 68, 70, 71, 73, 80, 84, 90, 108, 222, 259, 342
Concavenator, 394
Concerning the System of the Earth, Its Duration, and Stability, 9
Connecticut state fossil, 318
Connecticut Valley, 21, 40, 53, 56, 67, 70, 81
Continental drift, 1, 2, 165, 200, 237, 257
Conybeare, William, 17, 22, 23, 25
Coombs, Walter, 139
Coombs, Walter P., Jr., 274, 281, 292
Cope, Edward Drinker, xii, xxii, 15, 24, 30–36, 38–43, 61–63, 65, 66, 68, 73, 74, 76–79, 87–89, 97–101, 104, 107–9, 112, 122, 124, 127, 137, 151, 167, 175, 184, 201, 217, 235, 244, 245, 252, 268, 269, 271, 282, 296, 303, 316, 330, 334, 366, 367, 369, 376, 396, 414 n.18
Cope: Master Naturalist, 185
Copeia, 151
Cope's Law, 38
Coprolites, 25, 28, 350, 371
Corythosaurus, 143, 144, 150, 242, 246, 248, 250, 348
The Country of the Iguanodon, 36
Creation and Its Mysteries Unveiled, 53
Creation Evidence Museum, 204, 249, 267, 284–85, 324, 337, 361, 367, 381, 401, 405. *See also* Baugh, Carl; Young-Earth Creationism
Creationism: Gap, 55, 268
  Young-Earth, xii, 2, 3, 59, 162, 204, 213, 237, 238, 241, 245, 247, 249, 254, 265, 266, 271, 279, 281, 282, 298, 319, 324, 342, 357, 358, 370, 374, 377–78, 385, 391, 396, 397, 404
A Creationist View of Dinosaurs, 288
Creation Museum, 237, 238, 278, 324, 334, 357, 377–78, 406
Creation museums, xii, 237, 238, 324, 350
Creation Studies Institute, 308
Creston General Store, 189, 190
Cretaceous Period, xix, 22, 413 n.1, 417, 418, 421
Crichton, Michael, 292, 315, 338, 346, 407, 416 n.44
Crompton, Alfred, 243
Cross, Whitman, 105
Cryptovolans, 364
Crystal Palace, 46–47, 48, 49, 60, 63, 67, 86, 91, 104, 108, 197, 302, 414 n.10
Culbertson, T. A., 43
Culver, James Erwin, 102
Currie, Phil, 126, 171, 183, 272, 285, 288, 292, 298, 299, 304, 313, 314, 327, 341, 350, 358, 359, 369, 390, 394, 408
Curtis, Wardon Allan, 117
Cutler, William Edmund, 172, 180

Cuvier, Georges, xii, 8, 11, 14, 15, 18, 19, 20–21, 23, 25, 26, 27, 47, 51, 80, 90
Czar Nicholas II, 146
Czerkas, Stephen, 299, 302, 360
Czerkas, Sylvia, 286, 302

Dacentrurus, 76
Dahlgren, Carl, 102
Dames, Wilhelm, 78
Dana, James Dwight, 63
Danforth, Jim, 260
Darwin, Charles, 5, 6, 8, 10, 14, 17, 30, 32–36, 38, 40, 42, 49, 51, 56, 58, 61, 67, 69, 71, 73, 74, 77, 89, 115, 118, 129, 131, 353, 389
Darwin, Erasmus, 8, 24
Darwin, Robert, 4
Daspletosaurus, 126, 165
Daubenton, Louis Jean Marie, 7
Davenport Ranch, 203, 206
Davies, William, 76
Da Vinci, Leonardo, 1
Dawley, Herbert M., 161–62
Dawson, George Mercer, 75
Deane, James, 33, 34, 40–41, 53
de Beer, Gavin, 228
de Blainville, Henri Marie Ducrotay, 20–21
de Camp, Lyon Sprague, 230
de Caumont, Arcisse, 27
de Chardin, Pierre Teilhard, 182, 184, 203
de Chomon, Segundo, 139
"*Decidedly ungodly message*," 306
"*Degenerated into an exercise of name-calling*," 315
Deinocheirus, 260
Deinodon, 47, 50
Deinonychus, 151, 168, 173, 178, 188, 240, 247–48, 254, 257, 261, 267, 269, 304, 320, 332, 342, 359, 371
"*Deinonychus must have been anything but reptilian*," 247–48
de la Beche, Henry Thomas, 28, 34, 46
de Lapparent, Albert, 237
de Laubenfels, Max W., 231
Delfs, Ed, 225
Delgado, Marcel, 113, 163, 166, 174–75, 185, 190, 231
A Delineation of the Strata of England, 20
Delk, Alvis, 361, 381, 385
de Luc, Jean-André, 8
de Maupertuis, Pierre, 5
Denver Museum of Natural History, 139, 202, 321
Depéret, Charles, 108
de Ricqlès, Armand, 254, 264, 279
Desmond, Adrian, 267, 285
De solido, 3
Diaz-Romeral, Armando, 289
Diceratops, 101, 133
Dickens, Charles, 46
"*The Dick Tracy of the Mesozoic Age*," 215
Diclonius, 172
Dilong, 371
Dilophosaurus, 288, 315, 331, 342, 416 n.44
Dimetrodon, xvii
Dinny, 186. See also *Alley Oop*
Dino (Sinclair Oil), xii, 184, 187, 192, 199
Dino-Riders, 122, 306
Dinornis. See Moa
Dinosaur Adventure Land, 321, 362, 376. *See also* Hovind, Kent
The Dinosaur and the Missing Link, 158, 161
The Dinosaur Book, 214
Dinosaur Diamond National Scenic Byway, 364
The Dinosaur Dictionary, 262
Dinosaur Eggs and Babies, 334
Dinosaur Gardens Prehistoric Zoo, 193
The Dinosaur Heresies, 297, 416 n.41
Dinosaur National Monument (DNM), 71, 87, 135, 141, 156, 157, 164, 168, 171, 173, 174–75, 181, 185, 201, 223, 232, 234, 242, 246, 267, 337, 364, 395
Dinosauroid, 186
Dinosaur Park (Rapid City, South Dakota), xii, 198–99, 253, 415 n.22
Dinosaur Planet, 26
Dinosaur Provincial Park (DPP), 94, 117, 119, 127, 149, 156, 228, 265, 272, 276, 284, 292, 313

Dinosaur Revolution, 376
Dinosaur Ridge, 75, 78, 79, 80, 183, 325–26
Dinosaurs, 159
Dinosaurs: best-known, 108
confusion with other animals, xvii
defining traits, xvii
diseases, 383
diversity, xx–xxi
fate of, xxviii
heart, 151, 356, 399
last alive, xix
naming, xxi–xxiii
when they lived, xviii
where they lived, xviii–xix
Dinosaurs: Those Terrible Lizards, 271
Dinosaurs Alive!, 90
Dinosaurs and Other Prehistoric Reptiles, 250
Dinosaurs and Sin, 245, 263
"*The dinosaurs are not extinct*," 267
"*The dinosaurs are still alive*," 292
Dinosaurs before Adam?, 265
Dinosaurs by Design, 322
Dinosaurs from China, 304
"*Dinosaurs must have been on the Ark*," 353
Dinosaur Society, 329, 347
Dinosaurs of Eden, 319
The Dinosaurs of North America, 93, 109
Dinosaurs on stamps, frequency, 250
Dinosaurs Past and Present, 297
"*The dinosaurs seem to have suffered*," 105
Dinosaur State Park, 21, 251, 256
Dinosaur towns, xii, 367. *See also* Glen Rose, Texas; Rapid City, South Dakota; Vernal, Utah
Dinosaur Train, 73, 122, 339
The Dinosaur Triangle, 294
Dinosaur Valley Girls, 344, 415 n.32
Dinosaur Valley Museum, 308, 348
Dinosaur Valley State Park (DVSP), 203, 206, 242, 246, 248–49, 251–53, 254–56, 259, 284, 303, 361. *See also* Glen Rose, Texas
Dinotopia, 322, 349, 366
Diplodocus (airplane), 192
Diplodocus (dinosaur), xii, 7, 61, 65, 66, 69, 79, 84, 89, 94, 101, 106, 107, 110, 113, 114, 116, 118, 120, 121, 123, 128, 130, 131, 136–39, 141, 143, 147, 152, 153, 155–56, 158, 162, 164, 169, 170, 171, 180, 183, 185, 194, 210, 217, 223, 233, 234, 240, 244, 279, 284, 296, 309, 355, 364, 400, 405. *See also* Dippy
Diplodocus Marsh, 124
Dippy (*Diplodocus*), 107, 116–17, 123, 129, 146, 180, 192, 355. *See also* Carnegie Museum of Natural History; *Diplodocus*
Disney, 126, 186, 192, 194, 205, 207, 208, 242, 248, 251, 292, 312, 318, 345, 361, 375
"*Divinely plotted path of adaptive changes*," 40
Dodson, Peter, 261, 268, 300, 304, 315, 329, 356, 369, 374
Dollo, Louis, xxiii, 52, 85–86, 91, 96, 103, 121
Dollodon, 86
Dollo's Law, 103, 176D'Omalius, Jean Baptiste Julien, 22
"*Don't let anything happen to my drawings*," 223
"*Doomed to help the cause he hated*," 29
Dörr, Johann, 78
Douglass, Earl, 80, 87, 127, 141, 156, 170, 173, 201, 205, 232, 379
Doyle, Arthur Conan, 26, 32, 72, 112, 115, 123, 132, 140, 163, 168, 239, 341, 373. See also *The Lost World*
DPP. *See* Dinosaur Provincial Park
Dracorex, 368, 386
Drake, Charles L., 293
Dream of the Rarebit Fiend, 133
Drinker nisti, 316
Dromaeosaurus, 153
Dry Mesa Quarry, 235, 261, 278
Dryosaurus, 141, 172, 205, 223
Dryptosaurus, 62, 79, 94, 102, 108, 112. See also *Laelaps*
Dryptosauruses Fighting, 106
Dunkle, David, 210

Duplicate species, 64, 213
DVSP. *See* Dinosaur Valley State Park
Dwight, Elihu, 15
Dynamosaurus, 122, 239. See also *Tyrannosaurus*
Dystrophaeus, 54

"*Each stratum contained organized fossils peculiar to itself*," 12
Edmontia, 143
Edmonton Formation (Horseshoe Formation), 161
Edmontonia, 173
Edmontosaurus, 103, 124, 139, 147, 149, 154, 161, 172, 211, 227, 325, 366, 369, 401, 410
Egg Mountain, 277, 288
Eggs, xx, 41, 54, 82, 167, 169–70, 171, 175, 272, 275, 276, 277, 301, 304, 329, 375
Einiosaurus, 294, 339
"*Either birds or reptiles, or more probably both*," 53
Elaphrosaurus, 142, 177
Elasmosaurus, 117, 190, 296
Elementary Geology, 55
Engelhardt, Johann Friedrich, 32
Eodromaeus, 397
Eoraptor, 308, 319, 397
Eosinopteryx, 407
Eotriceratops, 380
Epidendrosaurus, 366
Epidexipteryx, 382, 396
Essay on the Theory of the Earth, 21
Essay towards a Natural History of the Earth, 3
Eubrontes (?) titanopelopatidus, 160
Euhelopus, 181, 230
Euparkeria, 151
Euskelosaurus, 47, 50
Evans, John, 43, 60
"*Everybody in the free world wanted a dinosaur*," 259
Extinct Animals, 112, 115, 132
Extinct Monsters, 104
Extinction, 3, 5, 8, 11, 12, 13, 15, 33, 175, 181, 208, 213, 231, 244, 246, 255, 256, 260, 262–64, 267, 268, 275, 276, 278–80, 283, 286, 291, 292, 295, 299, 310, 313, 314, 316, 319, 325, 336, 338, 342, 345, 391, 392, 396, 405, 407

Fantasia, 89, 98, 126, 154, 186, 194, 205, 208, 415 n.26
Farlow, James, 267, 272, 283, 312, 314, 346
Farwell's Dinosaur Park, 253
"*Father of the dinosaurs*," 126
Faunal succession, 12, 20
"*Fearfully great, a lizard*," 39
"*Feathered animal differing essentially from our birds*," 57
Feduccia, Alan, 263, 342, 367, 380
Felch, M. P., 93
Field, Marshall, 55, 104, 113, 166
Field Columbian Museum. *See* Field Museum
Field Museum, 68, 89, 111, 114, 120, 122, 131, 134, 166, 177, 198, 200, 210, 223, 231, 235, 317, 326, 333, 345
Figuier, Guillaume Louis, 58, 59, 60
Final Report on the Geology of Massachusetts, 38
Fiorillo, Tony, 156, 336, 398, 412
First accurate image of a sauropod, 72
First accurate restoration of horned dinosaurs, 133
First American textbook of anatomy, 9
First American to apply Linnaean names to fossil vertebrates, 24
First AMNH publication to feature work by Charles Knight, 108
First anatomical description of a dinosaur, 51–52
First anatomical description of an ichthyosaur, 22
First animated cartoon featuring a dinosaur, 133
First application of a Linnaean name to an American fossil, 10
First application of stratigraphic principles to geologic record, 3
First attempt to reconstruct a dinosaur, 30
First baby dinosaurs ever found, 275
First book about dinosaurs for nonspecialists, 159

First book devoted primarily to saurians, 27
First book having drawings of prehistoric animals, 44
First book to establish "lost world" theme, 132
First Canadian postage stamp to feature a dinosaur, 310
First cladistics study to include birds and dinosaurs, 218
First claim that *Archaeopteryx* is fake, 58
First college to offer courses about dinosaurs, 179
First colored description of an entire dinosaur, 385
First color-movie featuring stop-action dinosaurs, 228–29
First comic to feature anatomically correct dinosaurs, 221
First complete dinosaurs, 57, 86
First complete skull of a sauropod discovered, 79
First complete theropod discovered, 56
First corporate-sponsored geological educational materials, 187
First description of an ankylosaur, 31
First description of an articulated fossil reptile, 4
First description of an herbivorous dinosaur, 26
First description of a pterosaur, 8
First description of a Triassic dinosaur, 33
First description of *Brontosaurus*, 88
First description of *Tyrannosaurus*, 132
First diagram of dinosaurian evolution, 189
First digital reconstruction of an entire dinosaur, 357
First dinosaur bone given scientific name, 3
First dinosaur bones collected in Canada, 94
First dinosaur bones from Dinosaur Provincial Park, 112
First dinosaur bones from Hungary, 105
First dinosaur bone that can still be seen in a museum, 4
First dinosaur described from Australia, 101
First dinosaur discovered in Argentina, 104
First dinosaur discovered in Southern Hemisphere, 47
First dinosaur discovered in the United States, 9
First dinosaur eggs discovered, 167
First dinosaur embryos discovered, 16
First dinosaur exhibited in Japan, 243
First dinosaur formally recognized in North America, 47, 50
First dinosaur found in Antarctica, 296
First dinosaur found in China, 127
First dinosaur found in Germany, 32
First dinosaur found in Japan, 195
First dinosaur found in North America, 16
First dinosaur found in the Morrison Formation, 54
First dinosaur from Germany, 32
First dinosaur in a published work of fiction, 46
First dinosaur mounted in China, 407
First dinosaur named from China, 169, 181
First dinosaur named from good skeletal material in North America, 51
First dinosaur named from more than a few bones or teeth, 31
First dinosaur nesting sites found in Canada, 304
First dinosaur recognized in New Zealand, 266
First dinosaur remains discovered in North America, 43
First dinosaurs discovered in Spain, 53
First dinosaurs found in central India, 27
First dinosaurs found in France, 27
First dinosaurs from Laos, 197
First dinosaur skeleton mounted in Europe, 87
First dinosaur specimens still preserved in a North American museum, 21
First dinosaur theme-park in the United States to feature life-sized animals, 198–99
First dinosaur threatening humans, 152
First dinosaur to be mounted in a Canadian museum, 149

First dinosaur to go into outer space, 294
First dinosaur toy sold to promote a movie, 174
First dinosaur tracks discovered in the Morrison formation, 117
First dinosaur tracks found in Africa, 90
First dinosaur tracks found in Australia, 192
First dinosaur tracks found in Brazil, 150
First dinosaur tracks found in China, 182
First dinosaur tracks found in Israel, 244
First dinosaur tracks found in Mexico, 274
First dinosaur tracks found in South America, 197
First discoveries of dinosaurs in South America, 92
First discovery of a sauropod bone, 17
First discovery of burrowing dinosaurs, 376
First display of a fully articulated dinosaur, 19, 64–65
First drawing of a dinosaur, 26
First essentially complete dinosaur skeleton mounted in the United States, 103
First evidence of birds in the Jurassic, 57
First evidence of colonial nesting by dinosaurs, 277
First expedition to American West to collect fossils, 70
First extensive treatment of dinosaurs in popular fiction, 132
First feathered dinosaur to be reported, 340
First federally funded geological survey and explorations, 25, 63
First fossil recognized in North America, 50
First fossil skeleton mounted in America, 8
First fossils of a ceratopsian, 118
First freestanding mount of a theropod, 88
First geologic map, 14
First hadrosaur named, 51
First horned dinosaur known from an entire skull, 98
First identification of a pterodactyl, 18–19
First illustrated book to mention fossils in England, 3
First illustration of an American vertebrate fossil, 6
First Indian to name a dinosaur, 193
First journal devoted entirely to vertebrate paleontology, 232
First known discovery of dinosaur remains in North America, 43
First known specimen of a pterosaur, 8
First known vertebrate fossil in an American college, 9
First large-scale discovery of dinosaurs, 82
First large Triassic dinosaur to be discovered, 32
First life-sized models of dinosaurs, 48
First major work of paleontology, 19
First microstructural study of dinosaur eggs, 82
First models of dinosaurs marketed to the public, 47
First mounted dinosaur in Europe, 68
First mounted sauropod, 131
First mounted skeleton in the world, 10, 16
First movie to include prehistoric animals, 133
First museum in America, 7
First museum in Asia specializing in dinosaurs, 304
First museum to display a fossil vertebrate skeleton in America, 5, 14
First naming of Dinosauria, 39
First national, public, and secular museum, 7
First natural history museum, 2
First natural science institution in the Western Hemisphere, 19
First near-complete brachiosaur discovered, 155
First nonprofit society devoted to dinosaurs, 329
First organized vertebrate paleontology expedition, 5
First ornithischian dinosaur to be discovered, 27
First outdoor re-creation of prehistoric Earth, 189
First paleontologist featured on a postage stamp, 253

First paper devoted entirely to vertebrate paleontology, 4
First permanently mounted sauropod skeleton, 123
First person to collect a dinosaur from the Morrison formation, 87, 160
First person to describe dinosaur remains in North America, 52–53
First phylogenetic analysis of sauropods, 327
First popular article about dinosaurs to include restorations, 102–3
First popular book about dinosaurs, 214
First postage stamps in United States to feature dinosaurs, 259
First postage stamp to feature a dinosaur, 209
First prime-time animated series on television, 238
First professional vertebrate paleontologist, 24
First professional vertebrate paleontologist in China, 179
First professor of paleontology in the New World, 61
First public exhibition of *Tyrannosaurus*, 158
First publically displayed articulated dinosaur, 52
First publication dealing primarily with giant fossil reptiles, 23
First published reconstruction of a North American carnivorous dinosaur, 101
First reconstruction of a paleo-ecosystem, 28
First reconstruction of a sauropod, 81
First reconstruction of a titanosaur, 182
First recorded fossil-vertebrates, 5
First remains of dinosaurs in Western Hemisphere, 16
First reported dinosaur eggshells, 54
First report of a feathered dinosaur, 340
First restoration of *Brontosaurus*, 92–93
First roadside dinosaur, 189, 190
First robotic dinosaurs, 176
First sauropod discovered outside England, 46
First sauropod named from North America, 51, 83
First sauropod to be discovered, 40
First scientific book about dinosaurs for general public, 194
First scientific comparisons of the masses of a sauropod, 112
First scientific description of a dinosaur, xii, 22, 24
First scientific estimate of the mass of dinosaurs, 244
First scientific journal published in the United States, 21
First scientific organization in the United States, 12
First scientific reconstruction of a dinosaur, 30
First scientific treatment of American fossils, 7
First scientist to apply taphonomy to dinosaur fossils, 261
First skeletal mount, 10
First small ornithopod discovered, 69
First stamps to feature *T. rex*, 250
First stamp to feature a dinosaur, 234
First stamp to feature a prehistoric animal, 222
First stamp to feature prehistoric life in the United States, 229
First statement of the principle of faunal succession, 12
First *Stegosaurus* found in North America, 81
First study to reveal color in dinosaurs, 395
First successful application of radiometric dating, 135
First technical description of professional quality about American vertebrate fossil, 13
First trading cards to feature dinosaurs, 241
First unequivocal claim about Pangaea, 53
First validly published name applied to a dinosaur, 7
First vertebrate fossils seen by Europeans, 1
First woman harassed by a dinosaur, 116, 175

First woman recognized for her work in American paleontology, 39
First woman to describe several new dinosaurs, 268
First woman to find and describe a dinosaur, 145
First woman to get governmental funds for research, 41
Flaming Cliffs, 167, 168, 169, 171, 175, 217, 304, 313, 334
Flammarion, Camille, 95–96
Flash Gordon, 192
The Flintstones, 234, 237–38, 319, 334, 336
"The Fog Horn," 220, 224
Footprints and the Stones of Time, 335
Footprints in Stone, 254, 262, 264, 283, 297, 306
"*Formed into an Animal Mould*," 4
Fossil Cabin, 191, 215, 318, 382
The Fossils of the South Downs, 23
Foulke, William Parker, 51, 289, 367. *See also* Haddonfield, New Jersey; *Hadrosaurus foulkii*
Fox, William, 61, 91
Fraas, Eberhard, 123, 138, 330
Franczak, Brian, 314
Franklin, Benjamin, 12
Frauds, 69, 73, 99, 279, 297, 354, 359
Frohawk, Frederick, W., 132
"*From a tourist stop to a place of worship*," 247

Galton, Peter M., 177, 258, 264, 272, 278
Garbani, Harley, 252–53
Garden of Eden, 2, 5
Gastroliths, 135
Gauthier, Jacques, A., 300
Gegenbaur, Karl, 60
The Genesis Flood, 219, 241, 357
Geological Society of London, 15, 16, 24, 33
Geological Survey of Canada, 97
Geology for Beginners, 40
Gertie on Tour, 161
Gertie the Dinosaur, 150, 154, 155, 161
Gervais, Paul, 46, 82
The Ghost of Slumber Mountain, 158, 161–62, 164, 185
Ghost Ranch, 98, 217, 228, 270, 272, 283, 285, 338, 387
Giannotti, John, 52
The Giant Behemoth, 225, 236, 243
Giebel, Christoph, 58
"*Gigantic forms of terrestrial Saurians*," 38
"*Gigantic restorations of extinct inhabitants*," 49
Gigantis, the Fire Monster, 227
Gigantoraptor, 380
Gigantosaurus, 68, 172, 273, 353
Gil, Emilio Portex, 146, 180
Gilbert, David, 23, 26
Gilmore, Charles Whitney, xxiii, 121, 127, 128, 143, 145, 151, 154, 157, 160, 164, 169, 171–72, 174–75, 181, 187, 207, 214, 215, 299, 303
Giraffatitan, 136, 155
Glendive Dinosaur and Fossil Museum, 370–71. *See also* Young-Earth Creationism
Glen Rose, Texas, xii, 137, 143, 145, 160, 162, 178, 189, 202, 206, 218–19, 222, 224, 226, 246, 250, 251, 253, 254, 258, 264–66, 270, 272, 279, 286, 293, 296, 297, 299, 301, 305, 308, 310, 312, 323, 335, 337, 350, 364, 367, 371, 379, 399. *See also* Paluxy River
Glut, Don, 262, 345
Gobi Desert, 165, 215, 217
Godzilla, King of the Monsters!, 220–21, 226–27
Godzilla pose, 122, 134
Gojira, 220–21, 226–27, 404. See also *Godzilla, King of the Monsters!*
Gondwana, 200, 257, 358
Goodrich, George, 141
Gordon, Elizabeth, 104
Gorgo, 225, 243, 415 n.34
Gorgosaurus, 133, 143, 150, 154, 161, 162, 165, 210, 214, 231
Gould, Charles, 95
Gould, Stephen Jay, 111, 271, 280, 282, 310, 312, 321, 338

Granger, Walter, 100, 110, 117, 130, 131, 137, 170, 208–9
Grant, Ulysses S., 75, 76
Gray, Jim, 315
Great Chain of Being, 7
The Great Dinosaur Mystery and the Bible, 278, 283, 293, 312, 338, 351
The Great Dinosaur Mystery Solved, 351
"*The greatest bone bed in existence*," 113
"*The greatest collector of fossil vertebrates that ever lived*," 209
"*The greatest embarrassment of paleontology*," 342
"*The greatest naturalist America has produced*," 76
"*The greatest tool for the creationist explanation*," 262
"*Greatest vertebrate paleontologist*," 343
Greenough, George Bellas, 14
Gresslyosaurus, 50
Griffith, D. W., 116, 152
Griphosaurus, 57, 58
Gryposaurus, 159, 384
GSC. *See* Geological Survey of Canada
Guettard, Jean Êtienne, 6
"A Gun for Dinosaur," 230
Gurche, John, 312, 416 n.41
Gurney, James, 322, 349, 366

Häberlein, Ernst Otto, 78
Häberlein, Karl, 56, 78
Haddonfield, New Jersey, 51–52, 62, 65, 72, 77, 260, 289–90, 336
Hadrosaurus foulkii, 51–52, 289–90, 307, 319, 336, 367
Hall, E. T., Jr., 208
Hallett, Mark, 289, 297, 303
Halley, Edmond, 4
Ham, Ken, 59, 238, 277, 290, 306, 319, 324, 351, 353, 356, 372, 375, 378, 386, 404, 406. *See also* Answers in Genesis; Young-Earth Creationism
Hamlin, Vincent Trout, 186, 248
Hammer, Michael, 356
Hammond, Joel, 300
Haplocanthosaurus, 128, 225
Harder, Heinrich, 137
Harlan, Richard, 13, 24
Harry and His Bucket Full of Dinosaurs, 57
Harryhausen, Ray, 192, 212, 221, 223, 225, 231–32, 233, 245, 251, 257, 252, 295, 322, 369
Hart, Johnny, 234
Hatcher, John Bell, 55, 94, 97, 98, 99, 100, 101, 103, 104, 109, 113, 115, 118, 121, 122, 123, 127, 128, 130, 131, 133, 136–37, 323, 339
Haversian system in dinosaur bones, 233
Hawkins, Benjamin Waterhouse, 46–47, 48, 55, 57, 58, 63, 64, 67, 69, 71, 75–77, 87, 91, 104, 108, 189, 197, 266, 273, 414 n.9
Hay, Oliver P., 127, 137, 139, 175
Hayden, Ferdinand Vandeveer, 47, 50, 69, 70, 414 n.11
Hayes, Rutherford B., 82
"*He died for us. Dinosaurs didn't.*," 312
Heilmann, Gerhard, 151, 152, 176–77, 178
"*The heirs of Bacon and the acquitters of Galileo*," 55
Hell Creek Formation, 16, 124, 414 n.15
Helopus, 169, 181, 230
Hendrickson, Sue, 316–17, 326, 360
Hennig, Edwin, 136
Hennig, Willi, 218
Herbert, Hilary A., 103
Herrera, Victorino, 236
Herrerasaurus, xx, 236, 307, 380
Hesperonychus, 390
Hesperornis, 72–73, 77, 80, 90, 136
Hesperosaurus, 296
Heterodontosaurus, 243
"*He was using me . . . to furnish him brains*," 58
Hildebrand, Alan, 282, 313, 318
Hill, R. S., 92
Hirsch, Karl, 334
Historical Disquisition, 15
Historic Sites Act, 197, 244
Hitchcock, Charles Henry, 61
Hitchcock, Edward, 15, 21, 33–35, 38–44, 49, 52–53, 55, 56, 67, 81, 224, 256, 258

Hitchcock, Edward, Jr., 60
Hoffet, Josué, 197
Holland, William, 106, 116, 119, 120, 129, 137, 141, 145, 149, 153, 155, 159, 164, 174, 187
Holmes, Arthur, 146, 181. *See also* Age of the Earth
Holmes, Oliver Wendell, 45
Holtz, Thomas, 352, 358
Homo sapiens, type specimen, 110, 235, 334
Hooke, Robert, 2
Hop Brook Bridge, 94, 258
Hopson, James, 188
Hornaday, William Temple, 124
Horner, John R. "Jack," xxiii, 159, 160, 171, 214, 234, 253, 262, 267, 275, 277, 286, 287, 288, 291, 292, 294, 298, 304, 310, 315, 325, 327, 329, 331, 334, 344, 349, 358, 359, 374, 377, 383, 387, 391, 411
Horton, Juanita (Bessie Love), 116, 174–75
The Hot-Blooded Dinosaur, 257, 267
Houdini, Harry (Erik Weisz), 168
Hovind, Kent, 317, 321, 362, 368, 376. *See also* Young-Earth Creationism
Howe, Barker, 185
Howe Quarry, 185, 194, 195, 320, 364
Hoyle, Fred, 297
Hubbell, H. F., 88, 128
Hulke, John Whitaker, 72, 91
Human Footprints and Dinosaur Tracks, 263
Human history, xiii, xxviii n.1
  living with dinosaurs, 278, 282, 283, 293, 305, 308, 310, 317, 319, 320, 321, 324, 335, 353, 366, 386, 390. *See also* Young-Earth Creationism
Hunter, Oscar, 92
Hunting Dinosaurs, 334
Hutton, James, 1, 9, 10, 14, 20, 28, 264, 270. *See also* Uniformitarianism
Huxley, Thomas Henry, 29, 40, 43, 45, 47, 49, 50, 52, 53, 55, 59, 63–64, 64, 65, 67–70, 77, 78, 80, 84, 86, 89, 109, 115, 117, 132, 228, 247, 285, 303
Hylaeosaurus, 30, 31, 33, 36, 38, 39, 41, 46, 49, 53, 57, 61, 73, 276, 292, 414 n.8
Hypacrosaurus, 144, 157, 304
Hypselosaurus, 68
Hypsilophodon, 43, 69, 70, 91, 195, 408

"*I am prepared to go to the stake if requisite*," 55
"*I am quite convinced of my mistake*," 23
"*I am skeptical, actually*," 361
Ice Age, 133
Ichnology of New England, 52
Ichthyornis, 73, 80, 90
Ichthyosaurs, xvii, 4, 18, 19, 22, 46
ICR. *See* Institute for Creation Research
ICZN. *See* International Code of Zoological Nomenclature
Ideal Scene of the Lower Cretaceous Period, 59
"*I didn't sell the fossil to you*," 316
"*I do not know how to say this nicely*," 362
"*I don't mind being controversial*," 285
Iguanodon, 17, 19, 22, 25, 26–28, 30–32, 49, 52, 58, 59, 72, 73, 85–86, 87, 91, 95, 96, 105, 108, 120, 121, 123, 140, 148, 153, 177, 217, 226, 286, 295, 318, 408
Iguanosaurus, 23
"*I have done all I could in laying the foundation*," 52
"*I have done my humble part*," 186
"*I have made a grand discovery*," 31
"*I have never seen anything like it from the Cretaceous*," 125
"*I have seen nothing like it before*," 138
"*I knew it would be laughed at, or worse*," 283
Illustrations of the Geology of Sussex, 27
Illustrations of the Huttonian Theory of the Earth, 14
I Married a Dinosaur, 218
"*In a place I never expected it*," 275
Inquiry into Fossil Remains, 19
"*Insatiable appetite with practically no brains*," 214
Institute for Creation Research (ICR), 277, 279, 293, 296–98, 321, 324, 343, 351, 353, 356, 358, 376, 377, 382, 402, 412

Institute for Vertebrate Paleontology and Paleoanthropology (IVPP), 182, 233, 298, 363
International Code of Zoological Nomenclature (ICZN), xxi, xxii, 81, 102, 105, 152, 416 n.42
"*In the chaos of Noah's Flood*," 288
Irritator, 341
Isaac, J. C., 77
Isle of Wight, 22, 31, 39, 43, 325, 362
"*It carried a big evolutionary stick*," 397
"*It is a grand case for me*," 56
"*It is nearer bird than reptile, but is neither*," 59
"*It's more than a curiosity*," 2
"*It would feel like a cow*," 400
IVPP. *See* Institute for Vertebrate Paleontology and Paleoanthropology
"*I was examining merely the track of a bird*," 34
"*I was the first to put feathers on dinosaurs*," 363
"*I would propose the name* Dinosauria," 39

Jacobs, Louis, 284, 323, 330, 338
Jameson, Robert, 21
Janensch, Ernst Martin Werner, 136, 142–43, 155, 178, 183, 200
Jardin des Plantes, 8
Jefferson, Thomas, xii, 8, 9, 12, 13, 15, 16, 52
Jensen, James A., 233, 235, 241, 256, 261, 277, 288, 294, 307
Jesup, Morris, 68, 100, 137
Jobaria, 349
Johnston, Christopher, 54
Joly, John, 4, 118
Jonas, Louis Paul, 220, 242, 250
Jones, Vivian and Ed, 235, 261
Journal of Paleontology, 179
Journal of Vertebrate Paleontology, 280
Journeys and Adventures of Doctor Festus, 31
Journey to the Beginning of Time, 228
Journey to the Center of the Earth (book), 58, 60, 414 n.13
Journey to the Center of the Earth (movie), 60, 139, 207, 236
Judith River, 16, 29, 47, 50, 76, 78, 85, 165
Jurassic Fight Club, 122
The Jurassic Foundation, 352
Jurassic Park (book), 32, 149, 292, 315, 338
Jurassic Park (movies), xiii, 133, 168–69, 173, 241, 267, 298, 303, 306, 319–20, 327, 329, 331, 341, 349, 364, 373, 392, 403, 409, 416 n.44
Jurassic Period, xix, 10, 417, 419
"*Just about everything you believe is wrong*," 327

K-T boundary and extinction, xx, xxviii n.2, 16, 291, 293, 299, 303, 304, 307, 309, 310, 313, 318, 320, 326–28, 333, 349, 396, 400
Kaisen, Peter, 94, 95, 110, 127, 144, 150, 155, 162, 168, 197, 200
Kanter, Gary, 373
Keaton, Buster, 169
Kentrosaurus, 142, 143, 172, 181, 212, 283
Kermack, Kenneth A., 220
Kerr, Robert, 10
Kielan-Jaworowska, Zofia, 250
King Alfonso XIII, 146
King Dinosaur, 207, 225, 229
King Edward VII, 7, 128, 131
King Kong, 111, 112, 113, 123, 132, 133, 141, 150, 161, 166, 185, 189–91, 219, 222–24, 245, 268, 334, 372
King William IV, 14
Kircher, Athanasius, 2
Kirkland, James, 176, 329, 350, 363
Kish, Eleanor, 268, 283, 310
Kitching, James, 268, 272
Knight, Charles, 62, 69, 74, 79, 88, 100, 101, 104, 106–8, 110–12, 115, 120, 123, 132, 134, 136, 137, 146, 159, 160, 163, 169, 174, 176, 177, 185, 187, 189, 192, 195, 198, 210, 212, 213, 214, 217, 219, 221, 222, 227, 229, 241, 254, 257, 266, 283, 286, 303, 305, 321, 325, 335, 350, 351, 357, 367, 369, 403, 414 n.19, 415 n.22
Knight, Wilbur, 98, 119

Knipe, Henry R., 133
Kosmoceratops, 393
Kovalevsky, Vladimir, 99–100
Kuban, Glen, 279, 396
Kukufeldia, 42
Kuwasseg, Josef, 45

"*Lacking in the most basic scholarly standards*," 362
Laelaps, 29, 61, 65, 69, 79, 94, 102, 112, 185
Lakes, Arthur, 75, 78–79, 80, 81, 83, 87, 88, 101, 105, 160, 183, 202, 325, 368
Lamarck, Jean-Baptiste, 17, 24, 39, 55
Lambe, Lawrence, 50, 94, 102, 112, 117, 127, 130, 149, 151, 152, 154, 161, 163
Lambeosaurus, 161, 231, 292
Lambert, Marshall, 197
The Land before Time, 133, 301, 306
Land of the Lost, 90, 265
Landry, Sarah, 267
The Land That Time Forgot, 162, 267
The Land Unknown, 132, 224, 233
Lang, Walter, 245, 266
Lankester, E. Ray, 112, 115, 131, 132, 140
Lanzendorf, John J., 292, 357
Laosaurus, 108
"*The largest of them all*," 120
Larson, Gary, 28
Larson, Peter, 300, 316, 327, 341, 352, 360, 388
The Last Dinosaur Book, 351
The Last Dinosaur, 271, 362
Le Mesle, George, 90
Le Moyne, Charles, 5
Leach, Sheryl, 301–2. See also *Barney & Friends*
Leaellynasaura, 309–10
Leakey, Louis S. B., 172
Lehman, Thomas M., 309
Leicester Codex, 1
Leidy, Joseph, xxi, 42, 47–48, 49, 50, 51, 53, 54, 56, 61, 65, 69, 91, 98, 101, 117
Leptoceratops, 153, 163
Lessem, Don, 273, 329, 354
Lessemsaurus, 354
Lewis and Clark, 5, 12, 15, 16, 24, 126
L'Histoire Naturelle, 6
Lhwyd, Edward, 4
Life in the Primeval World, 73
The Life of a Fossil Hunter, 142
"*Life on Earth has been disturbed by terrible events*," 12
Limnosaurus, 121. See also *Telmatosaurus*
Limusaurus, 389
Linnaeus, Carl, 4, 5, 334
Linnean Society, 5
Lister, Martin, 3
The Little Dinosaurs of Ghost Ranch, 338
Lloyd, Graeme T., 382
Lloyd, Malcolm, Jr., 205, 237
Loch Ness Monster, 18
Lockley, Martin, 286, 287, 288
Lomax, Edward L., 119
Long, Charles J., 184
Long, Robert, 291
Longfellow, Henry Wadsworth, 33
Longrich, Nick, 390, 395, 404, 410
The Lost World (book), 26, 32, 72, 112, 115, 123, 132, 140, 147–49, 163, 171, 213, 338, 341, 346, 415 n.24. *See also* Doyle, Arthur Conan
The Lost World (movies), 111, 112, 116, 123, 133, 148–50, 158, 163, 166, 168, 171, 174, 177, 182, 211, 220, 223, 236, 240, 248, 334, 344
Louisiana Purchase, 15
Love, Bessie. *See* Horton, Juanita
Lubbock, John, 131
Lucas, Frederic A., 123
Lucas, Oramel, 83, 84
Lufengosaurus, 209, 234, 393, 407
Lull, Richard Swann, 21, 99, 101, 124, 127, 133, 136, 146, 150, 158, 165, 173, 182, 189, 195, 211, 224, 259
Lydekker, Richard, 27, 72, 85, 93, 104
Lyell, Charles, 1, 11, 25, 27, 28, 44, 103, 264, 270. *See also* Uniformitarianism
Lyson, Tyler, 352
Lystrosaurus, 216, 256

Mackal, Roy, 303
Mackie, S. J., 57

MACN. *See* Museo Argentino de Ciencias Naturales
Madsen, James H., Jr., 240, 270, 294
Magyarosaurus, 157
Maiasaura, 276, 285, 288, 294, 303, 331
Maidstone, 24, 32, 217
Major, Charles, 175
Majungasaurus, 108
Makela, Robert "Bob," 275
Maleev, Evgeny, 126, 217, 226
Mamenchisaurus, 225
Mammoth, xvii, 9
Man, Dinosaur, and History, 279
Mandschurosaurus, 126
"*Man lived with* Allosaurus," 263
Man's Origin, Man's Destiny, 249
Manospondylus, xxii, 102
Mantell, Gideon Algernon, 6, 17, 19, 22–23, 26, 27, 29, 30–33, 37–39, 43, 45, 46, 47, 52, 53, 86, 217, 303, 360, 414 n.2; Mantell's spine, 24, 45, 414 n.6
Mantell, Mary Ann, 23, 44, 45, 217, 360
Mantell, Walter, 45, 369
"*The man who made fossil hunting a profession*," 77
Map that changed the world, 14
Markgraf, Richard, 145
Marsh, Othniel, xii, 9, 15, 24, 28, 29, 38, 52, 55, 58, 60, 61–63, 66, 69, 72, 74, 77–79, 81, 87, 88, 92, 94, 97 100, 102, 103, 122, 124, 127, 136, 141, 154, 173, 244, 245, 269, 272, 316, 366, 368, 369, 376, 396, 414 n.10, 414 n.14
Marshall, Harry, 181
Marsh pick, 186, 207
Marsh's Dinosaurs, 252
Martin, John, 36, 37, 40, 46
Marx toys, 163, 229, 231, 234, 260, 274, 281, 298
Maryanska, Teresa, 268
Maryland state dinosaur, 54, 305, 369
Masiakasaurus, 363
Mason, Ruth, 165
Massospondylus, 47, 50, 268, 270, 272, 374
Mastodons, 1, 6, 9, 12, 19
Matheron, Pierre Émile Philippe, 41, 68
Matlack, Timothy, 9
Matley, Charles A., 169
Matthew, William Diller, 104, 107, 111, 126, 131, 146, 153, 159, 162, 165, 179, 207
Mays, Chris, 329
McCay, Winsor, 133, 146, 150, 154, 161
McDonald, Andrew, 42
McFall, Emmitt, 284
McGill, Caroline, 233
McIntosh, John, 141, 174, 180, 225, 267, 273, 347
McKenna, Malcolm, 214, 313
McLearn, F. H., 170
Medals of Creation, 40, 44
Meek, Fielding, 112
Megalonyx, 13
Megalosaurus, 3, 6, 7, 8, 11, 12, 14, 20, 21, 22, 25, 27, 30, 31, 36, 38, 39, 46, 53, 58, 59, 61, 72, 73, 105, 148, 172, 179, 333, 402, 408
Megatherium, 10, 11
Melville, Herman, 33, 45
Menke, H. W., 119–20
Merchandising, 131, 176, 195, 219. *See also* Toys
Mesozoic Era, xviii–xix, 38, 418–419
Messmore and Damon, 176, 178, 188, 193, 220
Meteoric impact, xix, 3, 9, 16, 21, 56, 134, 144, 181, 230, 231, 256, 289, 292, 401. *See also* Chicxulub crater; Extinction
Microraptor, 354, 360, 364, 385, 402, 410
Microvenator, 178, 188
Mill Canyon Dinosaur Trail, 299
Milner, Angela, 304
Minchin, Chow, 232
Missionary lizards, 375
Misspellings, xxii, 200, 201, 416 n.40
Moa (*Dinornis*), 30, 33, 45, 55, 60, 158
Mohabey, Dhananjay, 301
Mojoceratops, 395
Mokèle-mbémbè, 282, 294, 301, 303, 403
Monoclonius, 143, 144, 163
Monolophosaurus, 288
Mononychus, 303, 327

The Monster of Lake LaMetrie, 117
Monsters of the Past, 169, 171
Montana Dinosaur Trail, 197, 370, 371
Montanoceratops, 163
Moody, Pliny, 14, 39
Mook, Charles, 165
MOR. *See* Museum of the Rockies
A More Ancient Dorset, 28
Morgan, J. Pierpont, 51, 55, 68, 109, 111, 113, 115, 132
Morosaurus, 115, 121, 195. See also *Camarasaurus*
Morris, John, 204, 266, 279, 293, 296, 298, 306, 353, 382, 391, 405. *See also* Institute for Creation Research; Young-Earth Creationism
Morrison Formation, 54, 72, 81–83, 85, 87, 88, 102, 104, 105, 110, 112, 115, 117, 119, 121, 141, 160, 183, 195, 223, 235, 261
Mosasaurs, xvii, 8, 46, 118
Moss, Charlie, 145, 162, 218
Moss, Grady, 145
Mossbrucker, Matthew, 142
"*The most amazing collector I have ever known*," 108
"*The most gigantic of all quadrupeds*," 25
Mount Temple, 159
Möwe, 159
Mudge, Benjamin Franklin, 44, 72, 73, 76, 79, 81, 83
Mundus Subterraneus, 2
Murchison, Roderick, 39
Murphy, Nathan, 357, 374, 376, 386
Museo Argentino de Ciencias Naturales, 257
Museum of Creation and Earth History, 324
Museum of the Rockies (MOR), 233–34, 286, 315, 358, 370
Mygatt-Moore Quarry, 284, 346

Nakajima, Haruo, 226. See also *Gojira*
Nanotyrannus, 126, 197, 211, 374
Nash Dinosaur Track Site, 188
National Museum of Natural History (NMNH), 28, 101, 111, 128, 141, 145, 141, 171, 172, 181, 315
National Science Foundation (NSF), 219, 238, 360, 366
Natural History Museum (BMNH), xxviii n.5, 5, 6, 7, 14, 23, 27, 28, 30–32, 36, 42, 43, 50–51, 56, 61, 90–91, 101, 102, 111, 115, 129, 131, 141, 145, 151, 159, 171–72, 180, 183, 239, 325, 389
The Natural History of Oxfordshire, 3, 4
The Natural History of Waters, Earths, Stones, Fossils, and Minerals, 7
"*Nature does nothing in vain*," 2
Nazis, 179. *See also* World War II Bombings
Neanderthals, 51, 139
Nedoceratops, xxiii
Nesbitt, Sterling, 387, 391, 403
Neufeld, Berney, 258, 266
Neville, Henry, 104
Newberry, John Strong, 54
New Jersey state dinosaur, 307, 319
New Mexico state fossil, 281
Newspaper coverage of dinosaurs, 124
Newton, Isaac, 3, 28
New Zealand Te Papa Tongarewa, 37, 45
Nigersaurus, 355
Night at the Museum, 133
Nintendo, 57
Nipponosaurus, 195
NMNH. *See* National Museum of Natural History
Noah's flood, xii, 1, 2, 3, 5, 12, 14, 16, 21, 22, 24, 25, 35, 53, 219, 283, 307. *See also* Young-Earth Creationism
Noah's Raven, 14
Noble, John W., 97, 99
Nodosaurus, 165
Nomen dubium, xxii, 64, 68
Nomen ex dissertation, xxii
Nomen nudum, xxii
Nomen oblitum, xxii, 7, 102
Nomen rejectum, xxii
Non-avian dinosaurs, xx
Nopcsa, Franz, 41, 61, 105, 121, 127, 138, 157, 180
Nopcsa, Illona, 105
Norell, Mark, 313, 327, 338, 369, 377, 408
Norman, David, 293

Notes on the State of Virginia, 9
Notice on the Iguanodon, 26
"*No true modern physiological counterparts*," 347
"*Not with pistols or fists, but in print*," 74
Novacek, Michael, 171, 313, 328
Novum Organum, 2
NSF. *See* National Science Foundation
Nyasasaurus, 403
A Nymphoid Barbarian in Dinosaur Hell, 314

O'Brien, Willis "O'Bie," 111, 149, 158, 161, 163, 166, 174, 190, 211–12, 231, 232, 238, 245, 369
Obruchev, Vladimir, 159
O. C. Marsh: Pioneer in Paleontology, 185
Officer, Charles B., 293, 342
O'Hare (Chicago) International Airport, 120, 201, 333
Oklahoma state fossil, 336
Olsen, George, 169–70, 171
Omosaurus, 53, 7
One Million B.C., 154, 206–7, 219, 220, 225, 233, 236, 305
One Million Years B.C., 206, 231, 232, 251, 260
"*On the eve of some important discoveries*," 82
On the Origin of Species, 17, 29, 38, 40, 54, 56, 73, 86, 129, 389. *See also* Darwin, Charles
Oppel, Albert, 57
Organic Remains of a Former World, 15
The Origin of Birds, 151, 152, 176
Ornithichnology, 34
Ornithischians, xviii, 40, 86, 96
Ornitholestes, 113, 124, 130, 196, 245, 246
Ornithomimus, 61, 98
Ortelius, Abraham, 1
Orthogenetic evolution, xii, 71, 182, 224
Oryctology, 34
Osborn, Henry Fairfield, xii, 51, 56, 68, 70, 71, 79, 81, 87, 89, 94, 98, 100, 102, 104, 106, 108, 110, 114, 117, 121, 123, 124, 127, 129, 130, 132, 137, 139, 146, 153, 159, 160, 165–67, 182, 184, 187, 196, 215, 245, 247–48, 256, 259, 351
Osmólska, Halszka, 268
Ostrom, John, 49, 79, 118, 173, 178, 184, 188, 214, 217, 240, 242, 254, 256, 258, 266, 270, 277, 282, 329, 331, 352, 371
Othnielia (*Othnielosaurus*), 119, 272, 316
Ouranosaurus, 247, 269
Out of Time's Abyss, 162
Oviraptor, 69, 170, 171, 173, 206, 313, 328, 334, 338, 345, 373, 380, 408, 411
Owen, David Dale, 43
Owen, Richard, xii, 6, 19, 23, 25, 29, 44–46, 49, 50, 53, 56, 60, 61, 64, 75, 76, 82, 90, 91, 93, 94, 102, 109, 115, 129, 189, 197, 228, 230, 239, 285, 297, 368, 389, 414 n.5, 414 n.12
  Names Dinosauria, 31, 39

Pacheco, Joviano, 150
Pachycephalosaurus, 202, 230, 383, 386
Pachyrhinosaurus, 163, 218, 398, 403
Pack, Arthur, 228
Padian, Kevin, 334, 349, 380
Palaeontographical Society, 42
Palaeoscincus, 47
Paleogene, xxviii n.2
Paleoneurology, 72
Paleontological Resources Protection Act, 389
Paleontologist Barbie, 309–310
Palissy, Bernard, 1
Paluxy River, 143, 145, 162, 189, 202–4, 206, 209, 219, 226, 254–55, 259, 263–65, 269, 261, 279, 281, 282, 284, 286, 293, 299, 301, 306, 361, 376, 385, 387, 415 n.30. *See also* Bird, Roland; Glen Rose, Texas
Paluxysaurus, 252, 347, 379
Pangaea, xix, 53, 147, 200, 257, 328, 387
Panoplosaurus, 164
Parasaurolophus, 169, 188, 197, 210, 335, 387
Parker, Neave, 122, 219, 239, 259, 415 n.33

Parkinson, James, 15, 22
Parks, William A., 152, 163, 169, 176, 178, 199
Parksosaurus, 199
Parque Cretácio, 336–37
Patterson, Claire, 134, 230. *See also* Age of the Earth
Patterson, George, 173
Patton, Don, 305, 306, 342
Peabody, George, 55, 63, 93, 119
Peale, Charles Willson, 5, 8, 14, 15, 16, 19, 52, 414 n.9
Peale, Rembrandt, 15, 16, 19
Pelecanimimus, 334
Pelorosaurus, 43, 44, 46
Peña, Roque Saenz, 146
Penfield, Glen, 274, 318
Pentaceratops, 170, 285, 309, 398
Pentland, Joseph, 19
The People That Time Forgot, 162, 267, 272
Permian-Triassic mass extinction, xviii
Peron, Pierre, 90
Peterson, Olaf A., 72
Petrified Forest National Park, 130, 135, 215, 291, 296
Phillips, John, 2, 7, 38
Phillips, Phil, 306, 334
Philosophie Zoologique, 17–18
Placet, François, 2
Planet of the Dinosaurs, 149, 274, 415 n.26, 415 n.32
Plaster of Paris cast, 34, 80, 110, 112, 195, 201–2
Plateosaurus, 32, 33, 35, 41, 50, 139, 164, 165, 170, 212
Plate tectonics. *See* Continental drift
Platt, Joshua, 6
Playfair, John, 14, 28
Plesiosaur, xvii, 65
Plesiosaurus, 4, 18, 25, 27
Pleurocoelus, 143, 347
Plot, Robert, 3, 7
Poekilopleuron, 36, 40
Poicilopleuron, 69
Polacanthus, 61, 91, 292
Polk, James K., 41
"*The pope of 18th-century zoologists*," 6
Pouech, Abbé Jean-Jacques, 54
Powell, John Wesley, 72, 88, 97, 99, 100, 101, 103
Prehistoric Animals, 232
A Prehistoric Monster in a Modern Town, 96
Prehistoric Monster Trail, 48–49
Prehistoric Peeps, 133
Prehistoric Times, 330
"*A previous world to ours, destroyed by some kind of catastrophe*," 12
The Primitive World in Its Different Period of Formation, 45
Princeton University, 57, 64, 68, 69, 70, 76, 104, 205, 267
Principle of superposition, 1, 3
Principles of Geology, 28–29, 44
Prison, 1, 14, 77, 180, 321, 327, 341, 362, 368, 376, 416 n.36
"*Prize fighter of antiquity*," 134
Proceedings of the American Philosophical Society, 7
Proceratosaurus, 144, 148
Prokopi, Eric, 401, 406
Protoceratops, 74, 167, 170, 189, 210, 217, 261, 313, 318, 334, 384, 415 n.25
Psihoyos, Louie, 330, 334
Psittacosaurus, xxiii, 170–71, 197, 281, 365, 369, 409
Pteranodons, xx
Pterodactyls, xx, 18, 46, 118
Pterodactylus, 49
Pterosaurs, xvii, 8, 65, 70, 87
Purgatoire River tracks, 196, 286, 287
"*Purity of race*," 130
Pycraft, William Plane, 132

Rabbit Valley, 284, 286, 294–95
Racial senility, 140, 173, 175, 181, 224, 259. *See also* Extinction
Rapid City, South Dakota, xii, 99, 100, 190, 198, 199, 253, 289. *See also* Dinosaur Park
Raptor Red, xvii, 343
"*The rarest and most historic dinosaur book in English*," 27

Raup, David, 264, 283, 291
Ravished by the Triceratops, 409
Raymond M. Alf Museum of Paleontology, 205, 391
Readasaurus, 306
Reck, Hans, 143
Red Cloud, 75
Red Deer River Valley, 94, 97, 133, 143, 144, 145, 149, 150, 151, 152, 163, 178, 186, 228
Reed, William Harlow "Bill," 87, 89, 92, 93, 98, 111, 116, 122, 127, 128, 159, 294
Reig, Osvaldo, 236
Reisz, Robert, 393
Reliquiae Diluvianae, 24
Renne, Paul, 407
Report on British Fossil Reptiles, 36–37, 39–40
Reptilicus, 244
Researches on Fossil Bones, 18
Rhedosaurus, 221, 224
Richardson, George F., 31
Riggs, Elmer, 89, 106, 113, 119–20, 121, 122, 128, 131, 136, 138–39, 152, 235, 299, 308, 333, 415 n.21
Riggs Hill, 120, 200, 201, 294, 295, 299
Riou, Édouard, 58, 59, 75
Rite of Spring, 207–8. See also *Fantasia*
Roberts, Charles, 147
Robinet, Jean-Baptiste, 7
Rogers, Kristi Curry, 400
ROM. *See* Royal Ontario Museum
Romer, Alfred S., 169, 188, 195, 207, 214, 229, 230, 257, 261, 416 n.38
Roosevelt, Theodore, 51, 68, 135, 137, 171, 318
Rowley, John, 104
Royal Belgium Institute of Natural Sciences, 85–86
Royal Ontario Museum (ROM), 149, 163, 215, 243
Royal Tyrrell Museum of Paleontology (RTM), 133, 171, 234, 283, 292, 294, 310
RTM. *See* Royal Tyrrell Museum of Paleontology
Russell, Dale, 126, 154, 165, 214, 215, 259, 260, 268, 271, 285, 294, 298, 310, 327
Russell, Loris, 214, 231, 249
Rütimeyer, Ludwig, 50
Ryals, Jim, 202
Ryder, John, 81, 165

Sabath, Karol, 266
Sacrison, Stan, 275, 300
Sampson, Scott, 294, 339, 363, 384, 393, 408
The Sandstone Bird, 35
Sattler, W. B., 136
Saull, William D., 31
Saurischians, xviii, 131, 167, 211
Saurolophus, 143, 144, 151, 217
Sauroposeidon, 358
Scannella, John, xxiii, 391
Scansoriopteryx, 366
Scelidosaurus, 53, 54–55, 57, 108
Schaaffhausen, Hermann, 51. *See also* Neanderthals
Scharf, George, 31
Schulz, Wilhelm, 53
Schweitzer, Mary Higby, 354, 358, 359, 372, 385
"*Science not difficult to learn*," 20
"*Scientific truth is religious truth*," 55
Scilla, Agostino, 3
Scriptural Geology, 27
Scrotum humanum, 7
Sedgwick Museum, 4, 8, 17, 96
Seeley, Harry Govier, 40, 68, 75, 86, 96, 98, 101, 170
Segisaurus, 177
Seismosaurus, 279, 296, 328
Sell Rite Giftware (SRG) toys, 213
Sensiba, Alfred, 92
Sepkoski, Jack, 291
Sereno, Paul, 307, 319, 333, 340, 346, 347,349, 352, 353, 355, 358, 384, 388, 397
"*The sex of angels*," 279
Sheehan, Peter, 319
"*She sells seashells by the seashore*," 42
Shoemaker, Eugene, 134

Shuler, Ellis, 1960
Siccar Point, 9
Sichuan Province, 95
Sieber, Max, 125
Silliman, Benjamin, 21, 63, 118
Simpson, George Gaylord, 6, 13, 207, 209, 218, 236, 254, 343
Sina, Iba, 1
Sinclair, Harry Ford, xii, 77, 126, 159, 180, 230. *See also* Sinclair Oil and Refining Company
Sinclair Oil and Refining Company, 77, 89, 123, 159, 178, 179, 184, 187, 189, 192, 193, 196, 199, 200, 202, 206, 229, 236, 242, 244, 245–49, 259, 270
Sinornithosaurus, 353, 389, 402
Sinosauropteryx, 339, 340, 341, 354, 380, 395
Sketches of Creation, 75
Sleeman, William H., 27
"*The slow agency of existing causes*," 28
Smets, Gerhard, xxii–xxiii
Smit, Joseph, 104, 133
Smith, James Edward, 5
Smith, Nathan, 21
Smith, William, 12, 13, 14, 17, 31, 33, 38
Smithson, James Louis Macie, 27–28
Smithsonian Institution, 41, 68, 215
Snider-Pellegrini, Antonio, 53
Society of Vertebrate Paleontology, 130, 202, 207
Solnhofen, 56–57, 58, 59, 78, 356
"*A sort of dinosaur with wings*," 266
"A Sound of Thunder," 222
South Dakota state fossil, 309
Species Plantarum, 4
Spielberg, Steven, 227, 272, 303, 315, 319–20, 331, 346, 348–49, 399, 409
Spinosaurus, 158, 213, 364
Spotila, J. R., 215
Springfield Armory National History Site, 60
Stamposaurus, 312
Stamps, 14, 15, 68, 169, 186, 192, 195, 209, 215, 222, 229, 240
"*The standard against which other dinosaurs are measured*," 358
Star-spangled dinosaur, 118
Staurikosaurus, 180, 216
Stearn, W. T., 235
Stegoceras, 127, 143
Stegosaurus, 29, 61, 67, 72, 75, 76, 81, 87, 89, 93, 97, 101, 102, 104, 108, 110, 111, 112, 113, 117, 118, 121, 123, 126, 130, 132, 133, 141, 146, 148, 156, 157, 158, 179, 182, 183–84, 190, 195, 197, 198, 200, 207, 210, 234, 245, 250, 272, 278, 299, 315, 318, 331, 353, 375, 393, 415 n.27, 415 n.32, 416 n.44
  two brains, 81–82, 147
Steno, Nicolas, 1, 2, 3, 20, 307
Stenonychosaurus, 253
Stenonychosaurus, xxi, 186
Stephanosaurus, 127
Stereocephalus, 127. See also *Euoplocephalus*
Sternberg, Charles H., xii, 44, 50, 72, 76, 77, 78, 94, 105, 110, 136, 138, 140, 142, 143, 144, 145, 146, 147, 150, 154, 157, 160, 162, 178, 180, 186, 201, 212–13, 325
Sternberg, Charles M., xxi, 138, 139, 143, 149, 151, 163, 164, 165, 173, 183, 186, 196, 199, 215, 218, 227, 228, 258, 276, 283, 284
Sternberg, George, 127, 142, 143, 144, 153, 188, 284
Sternberg, Levi, 142, 143, 149, 150, 163, 196, 284
Sternberg, Raymond, 206
Sternberg Museum of Natural History, 153
St. George Dinosaur Discovery Site, 359, 369, 392
Stokes, L. Grant, 240
Stokes, William Lee, 182, 214, 236–37, 268
Stokesosaurus, 268
A Stone Age Adventure, 154
Stout, William, 269
Stovall, J. W., 185, 195
Strata Identified by Organized Fossils, 12, 20
Stromer, Ernst, 145, 158, 179, 213

Struthiomimus, 153, 155, 160
Stukeley, William, 4
"*The stupidest member of a very moronic family*," 214
Stutchbury, Samuel, 23, 33
Styracosaurus, 143, 144, 151, 154, 210
Successive catastrophes theory, 22
"*Such is the life of a fossil man*," 112
Suchomimus, 346, 352
Sue (*T. rex*), 120, 315–16, 326, 327, 332, 341, 345, 348, 360, 391
Sullivan, Emmet A., 198, 253
Supersaurus, 235, 300, 301
Sussex Scientific Institution and Mantellian Museum, 27
SVP. *See* Society of Vertebrate Paleontology
Swinton, William Elgin, 194, 240, 259
Sydenham, 46–47, 197, 266, 273. *See also* Crystal Palace
Sykes, Sir Christopher, 8
Synapsids, xvii
Systema Naturae, 4–5, 7, 235

Taft, William, 146
Tagart, Edward, 140
Talbot, Mignon, 145
Tan, H. C., 169
Taphonomy, 206, 219, 258
  taphonomic bias, 54
Taquet, Philippe, 247, 269, 278, 334
Tarbosaurus, 126, 217
Tarzan at the Earth's Core, 183
Tarzan the Terrible, 165, 168
Taylor, Bert L., 147
Taylor, Paul, 283, 293, 312, 338, 351
Taylor, Stanley, 254, 262, 264
Teapot Dome scandal, 180
Teenage Caveman, 152, 219
Tendaguru Expeditions, 138, 142, 163, 172, 177, 180, 181
"*Tendency to progression*," 17
Tennyson, Lord Alfred, 43
Tenontosaurus, 200
Texas Memorial Museum, 210, 307, 399
Texas state dinosaur, 347, 388
Texas Tracks and Artifacts, 335
Teylers Museum, 49–50, 260
Thecodontosaurus, 33, 40, 164, 205
Theory of the Earth, 9, 10
There Go the Dinosaurs, 377
"*There's something very wrong with our dinosaurs*," 297
Therizinosaurus, 217, 226
Thesaurus Geographicus, 1
Thescelosaurus, 101, 121, 151, 199, 356, 399
"*These must be dinosaur eggs. They can't be anything else*," 170
"*This nonsense is amazingly resilient*," 330
Thoreau, David, 45
The Three Ages, 169
Tippett, Phil, 291, 295, 319
Titanosaurus, 72, 79, 108, 157, 169, 300, 414 n.5
"*To doubt evolution today is to doubt science*," 79
Toepffer, Rodolphe, 31
Tornier, Gustav, 137
Torosaurus, xxiii, xxviii n.4, 97, 98, 103, 130, 321, 339, 383, 391, 414 n.16, 416 n.46
Torvosaurus, 115, 223, 261, 268
Toys, 174, 246, 253–54, 265, 301, 306, 308–10, 325, 327, 334, 359, 373. *See also* Marx toys; Sell Rite Giftware
Trachodon, 47, 50, 94, 127, 149, 172, 198
Trading cards, 103, 241, 259, 309, 325, 338, 366, 385
Trans-Continental Railroad, 68
"*The transformation from the Saurian to the bird*," 57
T. rex and the Crater of Doom, 346
"*T. rex originally ate vegetables*," 342
Triassic Dinosaurs of Europe, 139
Triassic Dinosaurs of the World, 139
Triassic Life of the Connecticut Valley, 158
Triassic Period, xix, 32, 417, 419
Triceratops, xxi, xxiii, 29, 61, 92, 93, 97, 98, 104, 108, 109, 118, 122, 124, 125, 139, 147, 152, 161, 163, 176, 184, 185, 189, 193, 197, 198, 207, 212, 231, 244, 250, 314, 318, 321, 328,

331, 351, 357, 371, 374, 383, 385–87, 391, 393, 396, 400, 404, 406, 415 n.27, 416 n.44
Troodon (*Troödon*), xxi, 47, 50, 133, 197, 289, 392
Turner, George, 221
Turok, Son of Stone, 227
Tweed, William "Boss," 67
Tyrannosaurus, xx, xxi, xxii, 23, 75, 96, 114, 122, 125, 126, 132, 134, 138, 143, 144, 153, 161, 171, 177, 184, 189, 190, 193, 196, 198, 205, 207, 208, 210–12, 224, 226, 230, 233, 234, 240, 244, 246, 249, 253, 270, 271, 274, 287, 285, 292, 298, 301, 306, 312, 316, 318, 323, 333, 345, 358, 362, 368, 372, 384, 387, 393, 406, 415 n.22
"*Tyrant king of the dinosaurs*," 132
Tyrrell, Joseph B., 94, 97, 133, 234

UCMP. *See* University of California, Museum of Paleontology
UFHNH. *See* Utah Field House of Natural History State Park Museum
Ultrasaurus (*Ultrasauros*), 277, 294
Uncle Beazley, 242, 275
Unger, Franz, 44
Uniformitarianism, 1, 8, 11, 14, 28, 264, 270
Union Pacific Railroad, 66, 68, 82, 90, 119
United States Geological Survey (USGS), 50, 63, 67, 68, 70, 88, 90, 91, 100, 102, 103, 121
University of California, Museum of Paleontology (UCMP), 179, 253
Unknown Island, 132, 217
Untamed Women, 207, 219, 222
Untermann, Billie, 117
Untermann, Gerhard Ernest, 117, 217
Urvögel, 56
USGS. *See* United States Geological Survey
Ussher, James, 2
Utah Field House of Natural History State Park Museum (UFHNH), 82, 210, 217, 233, 234, 271, 285, 294, 309, 352
Utah state fossil, 88, 305
Utahraptor, 285, 316, 320, 329, 343, 416 n.43
Utahsaurus, 140
Utterback, William H., 101

Vain Speculation Undeceived by Sense, 3
The Valley of Gwangi, 98, 212, 231, 257
Valley of the Dinosaurs, 265
van Bendeden, Pierre-Joseph, 85
Varricchio, David, 376, 381, 413
Velociraptor, 69, 167, 168, 170, 173, 261, 315, 329, 372, 379, 395, 406, 416 n.44
"*Verily a stranger to science and himself blind*," 5
Vernal, Utah, xii, 72, 116, 117, 141, 156, 181, 217, 233, 234, 257, 271, 294, 309, 310. *See also* Utah Field House Natural History
Verne, Jules, 58, 60, 139, 236, 414 n.13
Vetri, Victoria, 260, 416 n.36
Vickers-Rich, Patricia, 309
Vindiciae Geologicae, 22
Vinther, Jakob, 385, 395
Vogt, Peter, 262
von Alberti, Friedrich August, 32
von Haller, Albrecht, 5
von Huene, Friedrich, 76, 139, 144, 154, 157, 164, 165, 170, 178, 180, 182, 186, 209
von Humboldt, Alexander, 10, 14, 15, 51
von Meyer, Christian Erich Hermann, 30, 32, 33, 35, 49, 56–57, 76, 164, 260, 303
von Quenstedt, Friedrich August, 46
von Siemens, Werner, 78

Wagner, Johann Andreas, 56, 57, 58
Wagner, John L., 143
Walcott, Charles Doolittle, 156, 170
Walker, Larry, 347
Walking with Dinosaurs, 26, 90, 122, 153, 320, 329, 356
Wallace, Alfred Russel, 6, 77, 115
Wall Drug Store, xi, 198, 255
Wang, Steve, 374

Wankel, Kathy, 309
Ward's Natural Science Establishment, 47, 63
Washington, George, 12, 14, 15
Waterhouse, George Robert, 56
Waterhouse, John William, 6
"*We can now say that birds are theropods*," 354
"*We find no vestige of a beginning,—no prospect of an end*," 11
Wegener, Alfred, 53, 147, 166, 200, 237. *See also* Continental drift
Weishampel, David, 188, 329
Welch, Raquel, 235, 252
Wellnhofer, Peter, 308, 314, 330
"*We need endothermic dinosaurs*," 267
"*We're out of a job*," 344
"*We're taking the dinosaurs back*," 372
Wesley, John, 7
Weston, Thomas Chesmer, 97
What Really Happened to the Dinosaurs, 306
When Dinosaurs Roamed America, 122, 320, 415 n.32
When Dinosaurs Ruled the Earth, 206, 260
Where Dinosaurs and the Bible Meet, 362
Whitaker, George, 215–16
White, J. B., 181
White, Ora L., 39
White House, xii, 12, 16
"*Who's going to walk off with a tyrannosaurus?*," 327
Wickramasinghe, Chandra, 297
Wieland, George, 99, 175
Wilder-Smith, A. E., 249
Willett, O. W., 226
Williams, Maurice, 316–17, 326
Williams, Mike, 126, 225
Willis, John, 137
Williston, Samuel, 73, 79, 80, 82, 83–84, 85, 86, 89, 95, 99, 103, 106, 114, 119, 124, 127, 128, 162, 202
Wilson, Woodrow, 156–57
Wiman, Carl, 181
Winchell, Alexander, 75
Winkley, William, 202
Winston, Stan, 169
Wistar, Caspar, 9, 13
Wistar Institute of Anatomy and Biology, 117
Wolbach, Wendy, 293, 309
Wollaston Medal, 14, 31, 33
Wonders of Geology, 36, 37
Woodward, Alice, 133
Woodward, Arthur Smith, 132, 140, 144
Woodward, Henry, 74, 104
Woodward, John, 3, 4
Woodwardian Museum. *See* Sedgwick Museum
The World a Million Years Ago, 176, 178–79, 189, 415 n.22, 416 n.38
The World before the Creation of Man, 95
The World before the Deluge, 58, 59, 75
World's Fair: Chicago, 123, 178–79, 189, 192, 194, 248, 416 n.38
  New York, 242, 245, 246, 248, 275
  Philadelphia, 77, 87
  St. Louis, 123, 130. *See also* Crystal Palace
"*The world's greatest discoverer of T. rex*," 253
World War II bombing, 36, 126, 145, 158, 159, 179, 205, 208, 212, 213
The World We Live In, 223, 224
Wortman, Jacob L., 92, 94, 104, 107, 108, 111–13, 116, 120–21, 123
WPA Works Progress Administration, Washington, DC, USA (closed in 1943), xxviii
Wyman, Jeffries, 21
Wyoming Dinosaur Center, 328, 340, 371, 373
Wyoming Fossil Fields Expedition, 119
Wyoming state dinosaur, 336

Xenoceratops, 225
Xing, Xu, 368, 369, 371, 374, 376, 380, 390, 395, 402

Yale Peabody Museum of Natural History (YPM), 21, 75, 78, 87, 92, 93, 99, 103, 117, 127, 133, 161, 173, 184, 215, 224, 344

Yanornis, 354, 366
Yefremov, Ivan Antonovich, 206, 215, 217, 219, 265
"*You have made our museum famous*," 260
Young, C. C., 179, 182, 203, 209, 225, 232, 263
Young-Earth Creationism, xii, 2, 162, 237, 247, 281, 311, 324, 333, 357, 370, 371, 377, 397
YPM. *See* Yale Peabody Museum of Natural History
Zallinger, Rudolph, 111, 122, 157, 212, 215, 218, 223, 224, 227, 229, 231, 233, 236, 250, 260, 263, 281, 339. See also *The Age of Reptiles*
Zalmoxes, 127
Zangerl, Rainer, 225
ZDM. *See* Zigong Dinosaur Museum
Zhang, Fecheng, 365, 374, 382, 395
Zhiming, Dong, 274, 304
Zigong Dinosaur Museum (ZDM), 304

About the Author

Randy Moore earned a PhD in biology from UCLA, after which he worked as a biology professor at several large universities. He edited *The American Biology Teacher* for 20 years, teaches courses about evolution and creationism, and has authored and coauthored several books about evolution and the evolution-creation controversy, including *Evolution 101* (2006), *More than Darwin: The People and Places of the Evolution-Creationism Controversy* (2008), *No Prospect of an End: A Chronology of the Evolution-Creationism Controversy* (2009), *Arguing for Evolution* (2011), and *Understanding Galápagos: What You'll See and What It Means* (2013). Randy is H.T. Morse–Alumni Distinguished Teaching Professor of Biology at the University of Minnesota.